Leslie,

I apologize for finally

pagos.

to you that I brought you .

While on the islands I had a moment to reflect on
the amazing journey I had gone through and
really think about those who had influenced
me the most. You were my Evolution instructor
at Berkeley and you tirelessly put up with
my many questions through the semester and
helped me find my academic path. Understanding
the evolution (and development) of herps has
been my goal since and dedicate this
book to you and the other influential
scientists that I continue to admire.

Raf

Fascinating Life Sciences

This interdisciplinary series brings together the most essential and captivating topics in the life sciences. They range from the plant sciences to zoology, from the microbiome to macrobiome, and from basic biology to biotechnology. The series not only highlights fascinating research; it also discusses major challenges associated with the life sciences and related disciplines and outlines future research directions. Individual volumes provide in-depth information, are richly illustrated with photographs, illustrations, and maps, and feature suggestions for further reading or glossaries where appropriate.
Interested researchers in all areas of the life sciences, as well as biology enthusiasts, will find the series' interdisciplinary focus and highly readable volumes especially appealing.

More information about this series at http://www.springer.com/series/15408

Janine M. Ziermann
Raul E. Diaz Jr • Rui Diogo
Editors

Heads, Jaws, and Muscles

Anatomical, Functional, and Developmental Diversity in Chordate Evolution

Editors
Janine M. Ziermann
Department of Anatomy
Howard University
Washington, DC
District of Columbia
USA

Raul E. Diaz Jr
Department of Biology
La Sierra University
Riverside
California
USA

Rui Diogo
Department of Anatomy
Howard University
Washington, DC
District of Columbia
USA

ISSN 2509-6745 ISSN 2509-6753 (electronic)
Fascinating Life Sciences
ISBN 978-3-319-93559-1 ISBN 978-3-319-93560-7 (eBook)
https://doi.org/10.1007/978-3-319-93560-7

Library of Congress Control Number: 2018965739

This Springer imprint is published by the registered company Springer Nature Switzerland AG
The registered company address is: Gewerbestrasse 11, 6330 Cham, Switzerland

Preface

The vertebrate head is the most complex part of the animal body, and its diversity in nature reflects a variety of lifestyles, feeding modes, and ecological adaptations. The past decades brought many new insights into the development and evolution of the head, jaw, and associated muscles. Our head is crucial for our communication with other living beings and mediating our interaction with the environment through the use of our jaws, eyes, ears, nose, and brain that allow us to eat, speak, breathe, and express ourselves. Furthermore, the evolution of the vertebrate head is a fascinating story that has been captivating generations of scientists and the broader public interested in knowing more about the appearance of the jaws, the differentiation of sensory organs (e.g., the eyes and nose), the varying complexity of brains or skulls, and the variations in nerves, muscles, and/or blood supply.

In this book, experts joined forces to integrate, for the first time, state-of-the-art knowledge on the anatomy, development, function, diversity, and evolution of the head and jaws and their muscles within all major groups of extant vertebrates. Considerations about and comparisons with fossil taxa, including emblematic groups such as dinosaurs (Chap. 10), are also a landmark of this book, which will be a leading reference for many years to come. This book will take you on a journey to discover the origin and diversification of the head, which evolved from a seemingly headless chordate ancestor (Chap. 1). Building on the recent discovery of the cardiopharyngeal field in urochordates and on the comparative anatomy of chordate and vertebrate muscles, Chapter 1 focusses on the broader comparative and developmental anatomy of chordate muscles and the origin of vertebrate cephalic muscles.

Despite their structural diversity, the heads develop in a highly conserved fashion in vertebrate embryos. Major sensory organs like the eyes, ears, nose, and brain develop in close association with surrounding tissues such as bones, cartilages, muscles, nerves, and blood vessels. Ultimately, this integrated unit of tissues gives rise to the complex functionality of the musculoskeletal system as a result of sensory and neural feedback, most notably in the use of the vertebrate jaws, a major vertebrate innovation lacking in extant hagfishes and lampreys. In particular the origin of the vertebrate jaw is still controversial; several hypotheses need further experimental testing in order to be contradicted or further supported (Chaps. 2 and 3). Interestingly, recent discoveries showed that some members of the earliest jawed vertebrates, the placoderms, have had teeth and hypobranchial muscles similar to those in extant chondrichthyans (Chap. 2). These features, together with the evolution of jaws and, later in evolution, of a neck, allowed jawed vertebrates to be more efficient in food and water intake (Chap. 4).

The cranium diversified in cartilaginous fishes (sharks, rays, skates; Chap. 4) and bony fishes (Osteichthyes, which include ray-finned fish—Actinopterygii—that are covered in Chapter 5, and lobe-finned fish—the Sarcopterygii). The Sarcopterygii include lungfishes (Chap. 6), which are essential for our understanding of the major transition from fishes living in an aquatic environment to tetrapods living mostly on land. The new land environment is related to new requirements for the skull, jaw, and their muscles. While larval amphibians are somewhat more similar to lungfishes (Chap. 7), adult amphibians clearly have several adaptations for living on land (Chap. 7). Reptiles are so diverse in their lifestyles, feeding modes, and ecological adaptations that they are covered in several chapters in this book (Chelonia: turtles in Chap. 8; Lepidosauria: tuatara, snakes, lizards, worm lizards, in Chap. 9; and Archosauria: birds and crocodiles, in Chap. 10). Mammals with a special focus on facial muscles and on primates, which include our own species, are discussed in Chapter 11.

All chapters cover unique aspects about the evolution and diversification of the vertebrate head and head muscles. Several chapters integrate paleontological findings that help to understand changes that lead to the huge variety in the heads of extant taxa, seen today. Furthermore, some gene regulatory networks are surprisingly conserved (Chap. 1) and present from the earliest chordates to mammals. Many chapters include developmental data from modern experimental methods to present hypothesis about homology of structures, therefore shedding light on their evolutionary development.

Washington, DC — Janine M. Ziermann
Riverside, CA — Raul E. Diaz Jr
Washington, DC — Rui Diogo

Acknowledgments

A book like this is not possible without the support and contribution of many colleagues and friends. We, therefore, thank all the chapter authors and peer reviewers for their outstanding contributions to this book. The following reviewers reviewed one or multiple chapters (in alphabetical order): Virginia Abdala, Carole Burrow, Gerardo Antonio Cordero, David Cundall, Anthony Herrel, Tatsuya Hirasawa, Casey Holliday, Philippe Janvier, Peter Johnston, Shigeru Kuratani, John A. Long, Philip Motta, Sebastien Olive, Alan Pradel, Lauren Sallan, Jayc Sedlmayr, Vivian Shaw, Christopher A. Sheil, and Sam van Wassenbergh. We also thank Vignesh Iyyadurai Suresh from Springer Nature for shepherding this project forward with such great care and efficiency. Finally, we thank our families and friends for all their support throughout our careers.

Contents

Contributors

Bhart-Anjan S. Bhullar Department of Geology and Geophysics, Yale University, New Haven, CT, USA

Catherine A. Boisvert School of Molecular and Life Sciences, Curtin University, Bentley, WA, Australia

Alice M. Clement School of Biological Sciences, College of Science and Engineering, Flinders University, Adelaide, Australia

Raul E. Diaz Jr Department of Biological Sciences, Southeastern Louisiana University, Hammond, LA, USA

Natural History Museum of Los Angeles County, Los Angeles, CA, USA

Rui Diogo Department of Anatomy, Howard University, Washington, DC, USA

Gabriel S. Ferreira Senckenberg Center for Human Evolution and Palaeoenvironment, Eberhard Karls Universität, Tübingen, Germany

Fachbereich Geowissenschaften, Eberhard-Karls-Universität, Tübingen, Germany

Laboratório de Paleontologia de Ribeirão Preto, FFCLRP, Universidade de São Paulo, Ribeirão Preto, SP, Brazil

Alessia Huby Laboratory of Functional and Evolutionary Morphology, University of Liège, Quartier Agora, Institut de Chimie, Liège, Belgium

Zerina Johanson Department of Earth Sciences, Natural History Museum, London, UK

Peter Johnston Department of Anatomy and Medical Imaging, University of Auckland, Auckland, New Zealand

Eric Parmentier Laboratory of Functional and Evolutionary Morphology, University of Liège, Quartier Agora, Institut de Chimie, Liège, Belgium

Vance Powell CASHP, Department of Anthropology, George Washington University, Washington, DC, USA

Daniel Smith-Paredes Department of Geology and Geophysics, Yale University, New Haven, CT, USA

Paul A. Trainor Stowers Institute for Medical Research, Kansas City, MO, USA

Department of Anatomy and Cell Biology, University of Kansas Medical Center, Kansas City, KS, USA

Kate Trinajstic School of Molecular and Life Sciences, Curtin University, Bentley, WA, Australia

Ingmar Werneburg Senckenberg Center for Human Evolution and Palaeoenvironment, Eberhard Karls Universität, Tübingen, Germany

Fachbereich Geowissenschaften, Eberhard-Karls-Universität, Tübingen, Germany

Museum für Naturkunde, Leibniz-Institut für Evolutions- und Biodiversitätsforschung, Humboldt-Universität zu Berlin, Berlin, Germany

Janine M. Ziermann Department of Anatomy, Howard University College of Medicine, Washington, DC, USA

Department of Anatomy, Howard University, Washington, DC, USA

About the Editors

Janine Ziermann is an Assistant Professor at Howard University College of Medicine in Washington, DC. She received her PhD in Germany studying the evolution and development of head muscles in larval amphibians. This was followed by a postdoc in the Netherlands and one in the USA to further study vertebrates. Her research focuses on the evolution and development of the cardiopharyngeal field, a field that gives rise to the head, neck, and heart musculature. Additionally, she aims to use the knowledge from her research to understand the pathology of congenital defects, which often affect both head and heart structures. She won several awards, including the "American Association of Anatomists (AAA) and the Keith and Marion Moore Young Anatomist's Publication Award" (YAPA). Furthermore, she single authored or coauthored more than 30 papers in top journals such as *Nature*, *Biological Reviews*, books, book chapters, and commentaries.

Raul E. Diaz Jr is an Assistant Professor at Southeastern Louisiana University and a Research Associate at the Natural History Museum of Los Angeles County. He received his BS from the University of California at Berkeley and his MA from the University of Kansas where he studied frog skeletal development through metamorphosis. His PhD was completed through the University of Kansas Medical Center though his research was conducted at the Stowers Institute for Medical Research where he developed the veiled chameleon as a useful model for studying squamate reptile embryonic development. His research has meandered through his academic career from tropical field biology and systematics to anatomy and cell biology and development and genomics of reptiles and amphibians and now combines all these disciplines (as well as modern techniques in microscopy and imaging) to bridge the disciplines of comparative morphology, development, and biomedical research.

Rui Diogo Associate Professor at Howard University was one of the youngest researchers to be nominated as Fellow of the American Association of Anatomists, and has won several prestigious awards, being the only researcher selected for first/second places for Best Article of the Year in the top anatomical journal, two times in just 3 years (2013/2015). He is the author or coauthor of more than 100 papers in top journals such as *Nature* and of numerous book chapters; he is the coeditor of 5 books and the sole or first author of 13 books covering subjects as diverse as fish evolution, chordate development, human medicine and pathology, and the links between evolution and behavioral ecology. One of these books was adopted by medical schools worldwide, *Learning and Understanding Human Anatomy and Pathology: An Evolutionary and Developmental Guide for Medical Students*, and another one has been often listed as one of the best ten books on evolutionary biology in 2017, *Evolution Driven by Organismal Behavior: A Unifying View of Life, Function, Form, Mismatches, and Trends.*

Evolution of Chordate Cardiopharyngeal Muscles and the Origin of Vertebrate Head Muscles

1

Janine M. Ziermann and Rui Diogo

1.1 Introduction

The origin and evolution of chordates and vertebrates, and in particular the origin of the vertebrate head, have fascinated researchers for centuries (e.g., Gill 1895; Minot 1897; Gregory 1935; Holland and Holland 1998, 2001; Holland et al. 2008; Koop et al. 2014; Ziermann et al. 2014 and citations within). To investigate and understand the origin, evolution, and diversity of chordates, one has to research the origin, development, and comparative anatomy of hard (e.g., skeleton) and soft tissues (e.g., muscles, nervous system, cardiovascular system). Moreover, the finding that urochordates (e.g., tunicates, *Ciona*) and not cephalochordates (also known as amphioxus or lancelets, e.g., *Branchiostoma*) are the closest sister group of vertebrates (Fig. 1.1; Delsuc et al. 2006) has dramatically changed our understanding on the origin and evolution of both chordates and vertebrates. Cephalochordates are the sister taxon of Olfactores (= urochordates + vertebrates; Fig. 1.1), and amphioxus (lancelet) is therefore one of the best models to analyze chordate and vertebrate evolution (Koop and Holland 2008). Amphioxus adults have morphological features that are more easily compared with features found in vertebrates, and their genome sequence has more archetypal characters of ancestral chordates preserved as compared to either tunicates or vertebrates (e.g., Garcia-Fernàndez and Holland 1994; Shimeld and Holland 2000; Putnam et al. 2008; Candiani 2012). For instance, amphioxus has segmented muscles and pharyngeal gill slits, a dorsal notochord, and a hollow nerve cord (Shimeld and Holland 2000). However, other vertebrate characters such as the presence of a cartilaginous or bony skeleton are missing (Shimeld and Holland 2000).

The recently described **cardiopharyngeal field** gives rise to both **branchiomeric muscles** and myocardium (Diogo et al. 2015). A contribution of myogenic progenitors to cardiac and branchiomeric derivatives was shown to be present in *Ciona* (tunicates, urochordates; e.g., Stolfi et al. 2010; Razy-Krajka et al. 2014; Kaplan et al. 2015), chick (e.g., Tirosh-Finkel et al. 2006), and mouse (e.g., Tzahor 2009; Lescroart et al. 2010, 2015). With urochordates being the closest sister taxon of vertebrates (e.g., Delsuc et al. 2006), the cardiopharyngeal field was therefore present in at least the last common ancestor (LCA) of urochordates + vertebrates (Diogo et al. 2015), i.e., in LCA of Olfactores. In the amphioxus larvae, a structure that is sometimes called a "heart" (a contractile vessel; Willey 1894) lies posterior to the first three gill slits (Holland et al. 2003; Simões-Costa et al. 2005). However there are doubts about whether this "heart" is related to the

J. M. Ziermann · R. Diogo (✉)
Department of Anatomy, Howard University, Washington, DC, USA
e-mail: janine.ziermann@howard.edu; rui.diogo@howard.edu

J. M. Ziermann et al. (eds.), *Heads, Jaws, and Muscles*, Fascinating Life Sciences,
https://doi.org/10.1007/978-3-319-93560-7_1

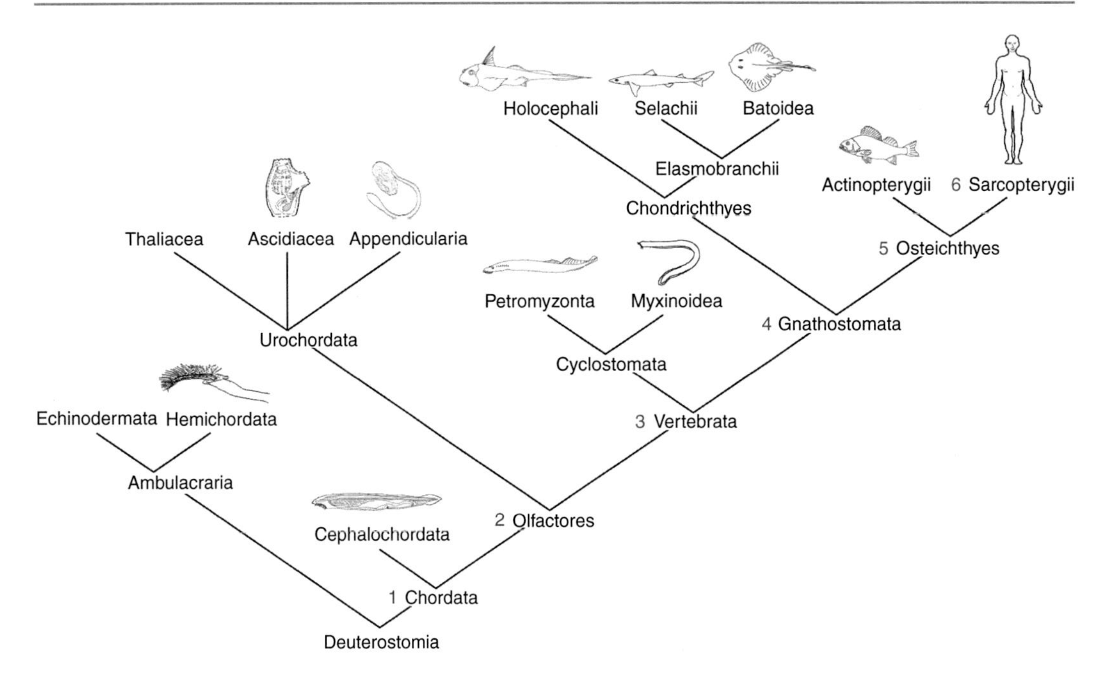

Fig. 1.1 Some of the synapomorphies of the Chordata and its subgroups, according to our own data and review of the literature (modified from Diogo et al. 2015, who own its copyright and gave permission to be used here): (1) Somites and branchiomeric muscles. (2) Placodes, neural crest-like cells, and cardiopharyngeal field (CPF; NB although within non-vertebrate chordates, conclusive evidence for these features was only reported in urochordates, some of them may have been already present in the LCA of extant chordates: see text) giving rise to first heart field and second heart field and to branchiomeric muscles (possibly not all of them, i.e., inclusion of oral/velar muscles into CPF might have occurred during vertebrate evolution: see text). (3) Skull, cardiac chambers, and differentiation of epibranchial and hypobranchial somitic muscles. (4) Jaws and differentiation between hypaxial and epaxial somitic musculature; paired appendages and fin muscles; origin of the branchiomeric muscle **cucullaris**. (5) Loss of epibranchial muscles; **cucullaris** divided into levatores arcuum branchialium (extending to pharyngeal arches) and protractor pectoralis (extending to pectoral girdle), an exaptation that later allowed the emergence of the tetrapod neck. (6) Within sarcopterygians, the protractor pectoralis gave rise to the amniote neck muscles trapezius and sternocleidomastoideus

heart of other chordates and to the head muscles because it consists of a coelomic epithelium (myoepithelium) opposed to the gut (walled off from basal lamina: Holland et al. 2003). In fact, in the most recent review on the subject, Diogo et al. (2015) argued that amphioxus does not seem to have a heart that is homologue to that of urochordates + vertebrates; hence, their suggestion that although cephalochordates likely have branchiomeric muscles as vertebrates do, they do not have a true cardiopharyngeal field such as that present in the Olfactores.

The discovery of the **cardiopharyngeal field** also revealed genetic mechanisms that are conserved in vertebrates and seem to have been evolved in the LCA of Olfactores (Fig. 1.1). This complex gene network was extensively studied in *Ciona intestinalis* and mouse (Lescroart et al. 2015; reviewed in Diogo et al. 2015). In short, it shows that Olfactores had common pan-cardiopharyngeal (mesodermal) progenitors that produce the first heart field (left ventricle, atria) and the Tbx1-positive cardiopharyngeal progenitors; *Tbx1* encodes a T-box–containing transcription factor. The Tbx1-positive cardiopharyngeal progenitors differentiate into the cardiopharyngeal mesoderm, and the anterior part of the progenitor cells activate *Lhx2* (LIM homeobox 2; LIM domain is named after founding members LIN-11, Islet-1, and MEC-3), self-renew, and produce the second heart field-derived right ventricle and outflow tract and, in mouse, the first (mandibular, muscle of mastication) and second (hyoid, muscles of facial expression) arch **branchiomeric**

muscles (Diogo et al. 2015). Other crucial genes involved in the differentiation of cephalic muscles are, for example, *Islet 1*, *Nkx2-5*, and *Mesp1* (the respective genes in *Ciona*: *Islet*, *Mesp*, *Nkx4*, *Tbx1/10*) (for more details see Diogo et al. 2015; Lescroart et al. 2015). The majority of cephalic muscles in vertebrates are embryologically derived from muscle plates from the mandibular, hyoid, and branchial arches (i.e., the **branchiomeric muscles**), while some originate from anterior somatic myotomes (i.e., the epibranchial and hypobranchial muscles) and some (e.g., pharyngeal muscles) from the mesoderm surrounding the pharynx or esophagus (Edgeworth 1935). Recent studies have shown that both the pharyngeal muscles and at least part of the esophageal muscles are developmentally closely related to the muscles derived from the branchial arches (Gopalakrishnan et al. 2015).

Ziermann et al. (2014) inferred from their comparative study of the cephalic muscles in a wide range of vertebrates that the cephalic muscles present in the LCA of extant vertebrates were probably (Table 1.1): (1) mandibular muscles, an undifferentiated intermandibularis muscle sheet, labial muscles, and some other mandibular muscles, (2) at least one hyoid muscle (at least some constrictores hyoidei), (3) at least some branchial muscles (at least some constrictores branchiales), and (4) and (5) undifferentiated epibranchial and hypobranchial muscle sheets. Furthermore, it seems now clear that the adults of the LCA of extant vertebrates had an elongated motile body similar to that of the adult basal gnathostomes and of amphioxus (reviewed in Diogo and Ziermann 2015). With the new insights from the developmental processes in vertebrates and tunicates (e.g., **cardiopharyngeal field**; Diogo et al. 2015) and in amphioxus (e.g., Holland 2015), it is crucial to review in detail the muscles in the adult amphioxus (cephalochordate) and adult *Ciona* (urochordates) in a broader, more informed context to infer the ancestral states for chordates. Therefore, the first sections below will focus on our new findings, based on our recent gross anatomical dissections and comparisons, of cephalochordates, urochordates, and vertebrates. Then, in the subsequent sections, we will put these observations and comparisons together with recent developmental and comparative data to address broader evolutionary and anatomical questions and to pave the way for the next chapters on the musculature of vertebrates.

1.2 Musculature of the Sea Squirt *Ciona intestinalis* and Amphioxus *Branchiostoma floridae*

Apart from the numerous vertebrates, including cyclostomes, that we have dissected in the past (reviewed in Ziermann et al. 2014; see also next chapters), we recently dissected adult amphioxus (*Branchiostoma floridae*, cephalochordates) and adult *Ciona intestinalis* (tunicates, urochordates) specimens. The myological terminology used in the text below and in Figs. 1.2 and 1.3 follows Moreno and Rocha (2008) for *Ciona* and Willey (1894) for amphioxus, unless explicitly stated otherwise.

The adult morphology of *Ciona* was described in some detail by Moreno and Rocha (2008), so here we will just provide a summary of new findings and/or of key structures that are necessary to mention to pave the way for the below comparisons between amphioxus and/or vertebrates. The adult *Ciona intestinalis* is sessile and the individual is surrounded by a dense translucent tunica (Fig. 1.2a). *Ciona*'s body can be divided into thorax (pharynx/branchial basket) and abdomen that contains the digestive tract, the heart, and the gonads (Fig. 1.2b). The orientation of transverse and longitudinal muscle fibers is visible also on the tunica (compare Fig. 1.2a, b). The **oral siphon** (branchial siphon) is the larger opening, with a lobed margin which is the inflow opening (Fig. 1.2a–c). The **atrial siphon** is a cylindrical extension of the body as indicated by the fact that the gonoduct and the rectum end well before reaching the siphon (Fig. 1.2a). Through the atrial siphon (Fig. 1.2), gametes and feces leave the body. Both siphons have a dense area of transverse muscle fibers (Fig. 1.2a–d). In addition, *Ciona* has transverse and longitudinal

Table 1.1 Muscles inferred to be present in the last common ancestor (LCA) of extant chordates and the LCA of extant vertebrates

LCA extant chordates	*Branchiostoma floridae* (amphioxus; Cephalochordata)	LCA extant Olfactores	*Ciona intestinalis* (sea squirt; Tunicata or *Urochordata*)	LCA extant vertebrates
Anterior muscles				**Mandibular**
Muscles of oral region	External and internal oral tentacle muscles, velar sphincter muscle	Muscles of oral region	Oral siphon muscles (corresponding to oral/velar muscles in cyclostomes and/or mandibular muscles in gnathostomes[a])	Undifferentiated intermandibularis muscle sheet
				Other mandibular muscles [e.g., at least one labial muscle and probably also at least one velar and/or dorsal mandibular muscle; see Ziermann et al. 2014 for discussion]
Branchiomeric muscles (NB: mandibular muscles only became incorporated into/part of branchiomeric series in vertebrates[a])				
Muscles corresponding to atrial sphincter and/or pterygial muscle of cephalochordates/urochordates + some transverse and longitudinal muscles (genetic/morphogenetic program for differentiation into branchiomeric versus body/somitic muscles was probably not as sharply defined as in LCA of vertebrates)	Pterygial (subatrial) muscle, atrial sphincter muscle [also "trapezius muscle" *sensu* Holmes 1953, which very likely does not correspond to trapezius muscle of amniotes]	Branchiomeric muscle derivatives from cardiopharyngeal field	Atrial siphon muscles and some associated muscles (e.g., some transverse/longitudinal body muscles)	**Hyoid:** constrictores hyoidei [at least one]
				True branchial: constrictores branchiales [one or more] and also adductors branchiales and/or interbranchiales [see Ziermann et al. 2014 for discussion]
				Other branchial muscles
No clear differentiation of epibranchial and/or hypobranchial muscle groups derived from somitic musculature				Epibranchial and hypobranchial musculature

[a]The circular oral siphon muscles in urochordates, which are not part of the cardiopharyngeal field, could correspond to oral/velar muscles of amphioxus and of vertebrates such as cyclostomes and thus to the first arch (mandibular) muscles of gnathostomes, which would thus be included as part of the branchiomeric muscle series only during vertebrate evolution

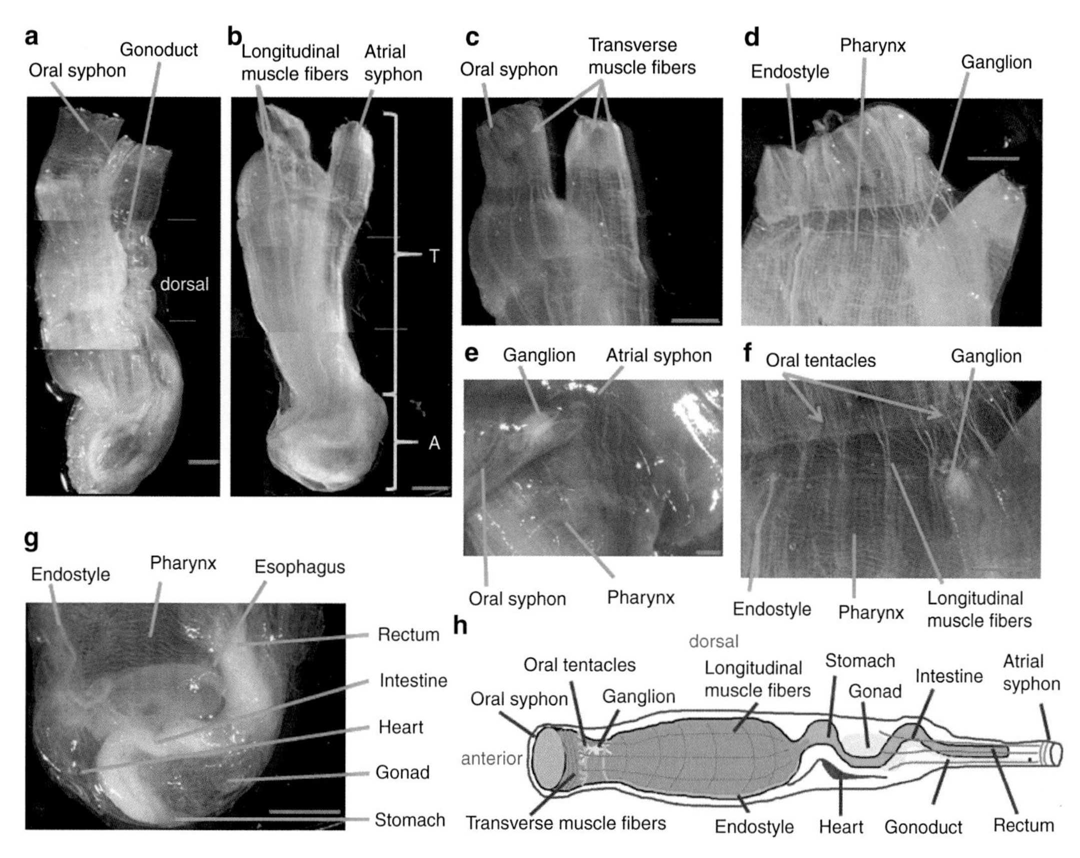

Fig. 1.2 Adult *Ciona intestinalis* (based on new dissections and data from the authors). Dorsal in all figures to the right. (**a**) *In situ* with tunica. (**b–g**) Tunica removed and specimen stained with alcian blue. (**c**) Atrial and oral siphons with dense transverse fibers. (**d**) Oral siphon laterally opened. The ventral endostyle is clearly visible beginning at the anterior end of the pharynx. (**e**) Oral siphon flexed ventrolaterally to see the dorsal ganglion (nerve center). (**f**) Red light added to increase contrast. The longitudinal fibers clearly end just anterior to the oral ring. The nerve fibers extending from the ganglion toward the oral siphon and the oral tentacles at the oral ring. (**g**) Abdominal region with the ventral heart and the rectum in the extended cylindrical tube that is bended dorsally and ends in the atrial siphon just next to the oral siphon (see a). (**h**) Theoretical scheme of an adult *Ciona* as if unfolded in the abdominal region (see g). Scale bar in *e* = 1 mm, all other 5 mm

body muscles (Fig. 1.2). The transverse muscles lying in the thoracic region are less dense than those of the siphons (Fig. 1.2c). The longitudinal fibers extend from the oral siphon to the extremity of the abdomen (Fig. 1.2b) and are parallel to the endostyle (Fig. 1.2d, f), which is a mucus-producing organ formed on the ventral midline of pharyngeal endoderm in non-vertebrate chordates and in larval lampreys. However, they do not extend through the whole oral siphon but seem to start just superior to the oral ring with the oral tentacles (Fig. 1.2f). Posterior (inferior) to the oral ring, there is apparently a "second ring" where the pharynx starts (Fig. 1.2f). Remarkably, amphioxus has a buccal ring with tentacles/cirri and a velum (see below); the latter lies in a region that topologically seems to correspond to the "second ring" of *Ciona*.

Most of *Ciona*'s adult body includes the large pharynx that ends at an esophagus dorsal to the heart (Fig. 1.2g). Here, the particular morphology of tunicates becomes obvious. The **oral siphon** is the anterior end, followed by the pharynx (branchial basket) that ends at the esophagus, which itself ends at the (not well defined) stomach (Fig. 1.2g). With the ganglion (neural complex)

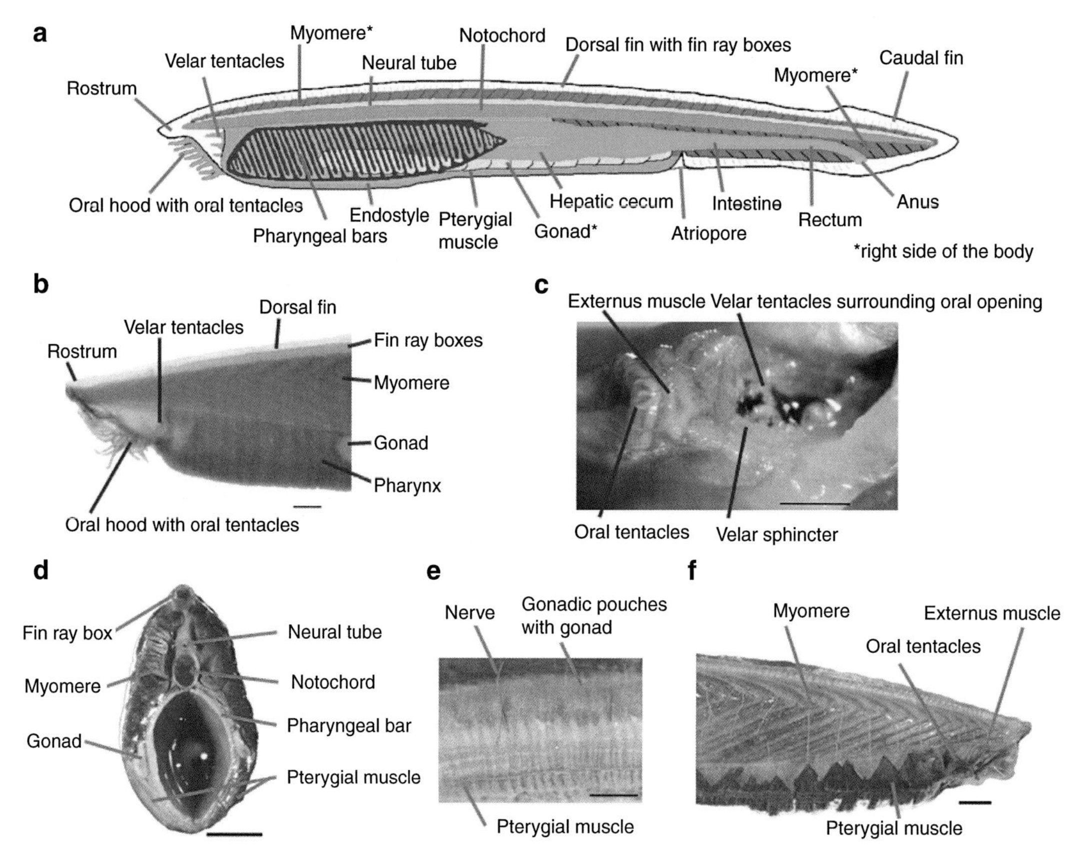

Fig. 1.3 Adult amphioxus (*Branchiostoma floridae*) (based on new dissections and data from the authors). (**a**) Scheme of an adult amphioxus. Myomeres and gonads not shown (*) on the left side of the body to show underlying structures. (**b–e**) Specimen stained with alcian blue and Lugol's solution. (**b**) View of the anterior region. (**c**) Ventral view, anterior to the left. Oral and velar region. Oral hood with tentacles reflected anteriorly. Scale bar = 0.5 mm. (**d**) Transverse section through pharyngeal region. (**e**) Ventrolateral view, anterior to the left. Nerve redrawn to increase visibility. (**f**) Right lateral view of the anterior body region. Pterygial muscle cut in midline and reflected laterally showing clear striation. b, d, e, f: Scale bar = 1 mm

located dorsally at the base of the oral siphon (Fig. 1.2e), the orientation of the animal is defined into anterior/posterior, dorsal/ventral, and left/right (Fig. 1.2a). However, the intestine curves at the bottom of the animal, and the rectum continues dorsally and anteriorly in the extended cylindrical tube leading to the **atrial siphon** (Fig. 1.2g). The gonads lie between the curled intestine, and the gonoduct has a similar course to the rectum but extends further distally (Fig. 1.2a, g). That is, when the animal is schematically "unfolded" at their abdominal area (Fig. 1.2h), it actually does not seem to have a body plan so different from that of other adult chordates (Fig. 1.2h). The ganglion (neural center) is anterodorsally; the heart is ventral to the stomach. The heart is ventral to the pharynx in gnathostome fishes, and its caudal position in *Ciona* might be due to the enlarged pharynx, which is likely a feature related to filtration feeding.

Concerning the adult features of amphioxus, these were described in some detail by Willey (1894). However, due to the new phylogenetic, morphological, genetic, and developmental insights about chordates and cephalochordates and the controversial interpretations about—and lack of detailed studies of—the cephalic musculature, it is crucial to take a fresh, comprehensive look on those muscles and related structures in amphioxus adults. Amphioxus is an elongated animal with a dorsal neural tube that extends far anterior into the cephalic region and into the tail (Fig. 1.3a). The notochord lies ventrally to the neural tube (Fig. 1.3b) and also spans the whole body but extends even further anteriorly until the anterior tip of the body. The anteroventral mouth is surrounded by oral tentacles (buccal cirri) and ends into a large pharynx (Fig. 1.3a, b). The entrance to the pharynx is surrounded by velar tentacles Fig. 1.3a–c). An endostyle spans the entire ventral length of the pharynx (Fig. 1.3a). The esophagus connects to the intestine; the hepatic caecum is situated shortly thereafter. The intestine (rectum) ends in an anus, ventrally, anteriorly to the short tail. The atrium is the most ventral organ, terminating via an atrial opening in the **atrial siphon** that contains **atrial sphincter muscles**, well anterior to the anal opening (Fig. 1.3a).

Segmented muscles (myomeres, myotomes) cover the dorsal body in its entire length and also extend into the anterior tip dorsal to the buccal cavity (Fig. 1.3b). The myotomes are longitudinal muscles used for locomotion and stretch from the notochord down to just cover the gonads (Fig. 1.3d). The cross-striated appearing **pterygial muscle** (subatrial or transverse muscle *sensu* Willey 1894) is a muscle that spans ventrally (Fig. 1.3a, d, e), covering the floor of the atrium and extending anteriorly to end in the velar sphincter (Fig. 1.3c) and caudally where it seems to form the **atrial sphincter muscle**. The pterygial muscle is divided by a median longitudinal septum into two halves that are further divided by thin transverse septa into a series of compartments that are not segmentally arranged (Fig. 1.3e). The velar sphincter seems to be the only velar muscle. The muscles of the oral hood include one externus muscle and one internus muscle (Fig. 1.3c). The outer one (externus) lies at the base of the cirri, and the fibers of one side interlace ventrally with the ones of the other side. The inner one (internus) lies between two consecutive cirri and is actually not a single muscle but is instead composed of multiple tiny muscles—each one of them is situated at the base between two oral cirri.

The central nervous system (neural tube) is a long tube above the notochord, and both structures extend far anterior in amphioxus (Fig. 1.3a). The dorsal nerve roots divide into dorsal and ventral rami that run externally to the myotomes. The dorsal rami extend from the dorsal region over the myomeres to the ventral region where they split into cutaneous branches and branches that turn medially around the metapleural fold to innervate the **pterygial muscle** (Fig. 1.3e).

1.3 Evolution and Homology of Chordate Muscles Based on Developmental and Anatomical Studies

Amphioxus and the lamprey larvae (ammocoete; cyclostomes) are filter feeders and have both a functional endostyle (Holland et al. 2008). However, in larval amphioxus the pharynx has an asymmetric development that was likely not present in the LCA of chordates (Stokes and Holland 1995; Presley et al. 1996). Almost the entire muscular system of amphioxus is composed of striated muscle fibers. Smooth muscles are found in the post-pharyngeal gut, excluding the gut diverticulum (Holmes 1953). The striated muscles can be divided into the parietal muscles, which are the myotomes, and the visceral and splanchnic muscles (Willey 1894). The visceral muscles include the pterygial (transverse or subatrial) muscle, the muscles of the oral hood and cirri, and the velar and anal sphincter muscles. Another striated muscle was described by Holmes (1953) under the confusing name **trapezius**, lying dorsal to the pharynx at so-called funnels and lateroventrally to the notochord (NB: we did not identify a muscle that follows this description). This muscle does not seem to be homologous to the **trapezius/cucullaris** of gnathostomes, because such a muscle is missing in cyclostomes and in urochordates. All striated muscles of amphioxus are built from flat lamelliform plates, which in cyclostomes are found in connection with the lateral muscles only (Willey 1894).

The fibers of amphioxus' oral and velum muscles closely resemble those fibers found in the walls of the heart of the higher vertebrates, contain striations, and are elongated fibers with nuclei, while in other striated muscles of the adult, amphioxus nuclei are rarely found (Willey 1894). This is a thought-provoking point when one takes into account the new discoveries about the strong links between the heart and **branchiomeric muscles** in urochordates and vertebrates (**cardiopharyngeal field**; see Diogo et al. 2015). In amphioxus larvae, the muscle fibers on the peritoneum on the pharyngeal floor are functionally related to the closing of the gill slits (Yasui et al. 2014). In contrast, the myoepithelial cells in hemichordate branchial muscles derive from longitudinal muscles (Cameron 2002) and give elasticity to the lacunae in the tongue bar and the blood vessels (Yasui et al. 2014). The amphioxus' anterior myotomes overlapping the region of the oral hood might correspond to the supraocularis of lampreys as this muscle is also an anterior extension of the parietal (epibranchial) somitic muscle that is only separated by the latter one through a septum (connective tissue that also separates myotomes) (Ziermann et al. 2014; Diogo and Ziermann 2015). This is a similar condition as seen with the nasalis muscle in hagfishes (anterior extension of somitic muscle parietalis) (Ziermann et al. 2014).

Both larval and adult amphioxi have **orobranchial muscles** that are developmentally and anatomically similar to the vertebrate **branchiomeric musculature** (Diogo et al. 2015) (Fig. 1.4). Yasui et al. (2014) suggested that the amphioxus larval orobranchial muscles might be anatomically more similar to the branchiomeric muscles of adult vertebrates than the adult oral, velar, and **pterygial muscle**s of amphioxus adults. These authors described five distinct larval orobranchial muscles and stated that these muscles disappear during metamorphosis and are topologically replaced by the adult oral, velar, and pterygial muscles. However, despite the observed apoptosis in the larval pharyngeal region of amphioxus (Willey 1894; Yasui et al. 2014), it is not clear if all the larval pharyngeal muscles do vanish and are missing in the adult stage. That is, the developmental origin of the adult pharyngeal muscles of amphioxus is still not resolved, and a detailed developmental study on the transformations occurring in the region of the gill slits/pharyngeal arches during metamorphosis is needed.

The adult amphioxus has two oral muscles, the externus and internus, related to the oral tentacles, and one **velar sphincter muscle**. The oral muscles seemingly develop without segmental patterning (Yasui et al. 2014). According to Willey (1894), these oral muscles of amphioxus are not like anything observed in vertebrates as they relate to the cirri. The amphioxus' **pterygial muscle** extends anteriorly and posteriorly forming the velar and **atrial sphincter**, respectively (Holmes 1953), being branchiomeric muscles *sensu* Diogo et al. (2015) (Table 1.1). This is almost the same configuration as seen in *Ciona* (urochordates) in the sense that in *Ciona* the muscles related to both the circular oral and atrial sphincters (**siphons**) express *Tbx1* and seem to correspond to vertebrate **branchiomeric muscles** (e.g., Stolfi et al. 2010; Sambasivan et al. 2011; Diogo et al. 2015). In fact, various authors have noted that the **pterygial muscle** develops ventrally in the amphioxus pharynx and is innervated by peripheral nerves that are similar to the branchiomeric nerves of vertebrates (Fritzsch and Northcutt 1993; Yasui et al. 2014). The musculature of the adult amphioxus **velar sphincter** might correspond to the **transversus oris** of adult hagfish and/or to the **annularis** of adult lamprey (Table 1.1). However, the transversus oris and annularis are not homologous to each other, although both are part of the nasal muscle group of **mandibular muscles** in cyclostomes (Ziermann et al. 2014). The amphioxus oral internus muscle consists of multiple small muscles associated with the base between two oral tentacles, what might indicate a correlation with the huge number of cephalic muscles in cyclostomes (lingual, dental, and velar muscles; see Ziermann et al. 2014 and Diogo and Ziermann 2015; NB: the nasal muscles in cyclostomes might be better explained by the splitting of myotomal structures in the head).

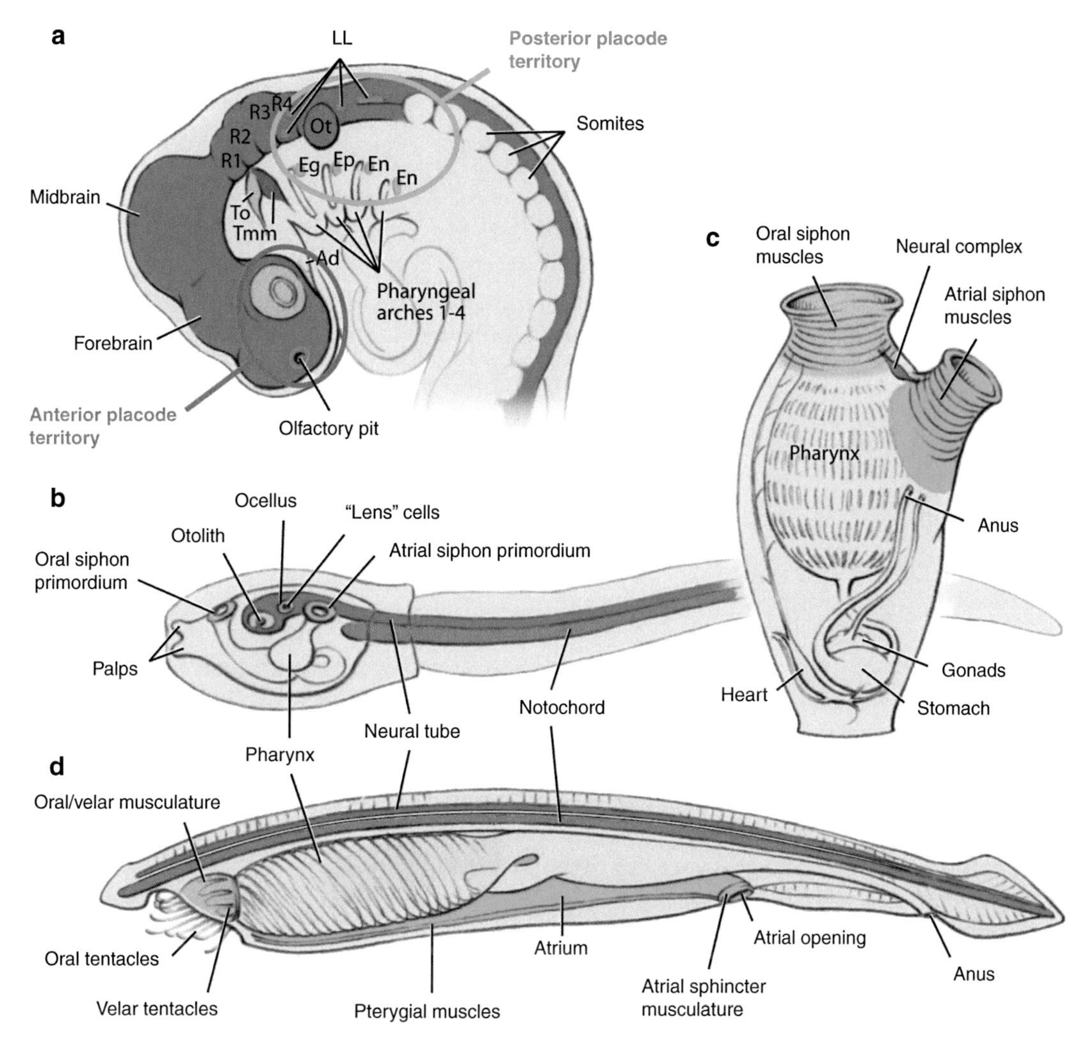

Fig. 1.4 Comparative anatomy of cephalochordates, urochordates, and vertebrates (modified from Diogo et al. 2015, who own its copyright and gave permission to be used here). (**a**) Location of ectodermal placodes in vertebrate head according to Graham and Shimeld's (2013) hypothesis (anterior to the left): olfactory placode/pit (red) at tip of forebrain; lens placode (orange) forms posteriorly as part of the eye; adenohypophyseal placode (Ad; yellow) lies ventrally to the forebrain; trigeminal placodes form alongside anterior hindbrain at the levels of rhombomeres 1 and 2 (R1, R2), the anterior one being the ophthalmic placode (To; light blue) and the posterior one the maxillomandibular placode (Tmm; purple); otic placode (Ot; brown) forms opposite the central domain of hindbrain; lateral line placodes (LL; pink) form anteriorly and posteriorly to otic placode; epibranchial placodes (green)—geniculate (Eg), petrosal (Ep), and nodose (En)—form as part of pharyngeal series. Dark blue: forebrain, midbrain, R1 (R2, R3, R4)—rhombomere 1 (2, 3, 4) and somites. (**b**) Urochordate tadpole larva (anterior to the left): notochord in red and two siphon primordia (green and orange), with putative relationships to the anterior and posterior placode territories shown in a. (**c**) Adult urochordate showing siphon primordia after metamorphosis. (**d**) Adult cephalochordate showing the hypotheses of urochordate-cephalochordate muscle homology proposed in the present review. Figures modified from Willey (1894), Sambasivan et al. (2011), and Graham and Shimeld (2013)

The **atrium** of amphioxus, which is a big space lined by ectoderm that opens via the atrial opening to the environment, shares some similarities with the atrium of urochordates (e.g., *Ciona*) (Fig. 1.4); cyclostomes do not have an atrium. The atrial opening is the outflow opening in both taxa. In amphioxus, the innervation (sensory and motor) of the atrial region, including the pterygial muscle that covers this region ventrally, is by dorsal nerve roots (Holmes 1953; Bone 1960). The motor axons also control the lateral ciliary tracts of the pharyngeal bars (Bone 1960). The atrial nervous system includes connected neurons on the visceral and parietal borders of the atrium (Holmes 1953). The atrial epithelium arises from invagination of larval ectoderm, and the majority of neurons of the atrial nervous system lie in the epithelium suggesting that this nervous system is associated with an ectodermal layer (Holmes 1953). Therefore, this atrial nervous system does not seem to be homologue to the sympathetic systems in craniate vertebrates, suggesting that the visceral nervous systems of amphioxus and vertebrates are not homologous (Holmes 1953; Bone 1960).

The **atrial cavity** in amphioxus seems to play an important role in filter feeding (Dennell 1950) and develops during the larval period (development of first gill slit on the right side until metamorphosis) (Willey 1894). The contraction of the pterygial muscle reduces the atrial cavity, and water is expelled through the atriopore (Willey 1894; Dennell 1950). The mechanism depends on closing the atriopore, which is not just a perforation of the atrium floor (Dennell 1950). Between the external aperture of the atriopore and the atrial cavity intervenes a short chamber (atrial siphon; Dennell 1950). This asymmetrical atrial siphon is ventrally cut off from the main atrium cavity. Anteriorly the false floor, disconnecting the siphon from the atrium, extends horizontally and reaches as two processes freely into the atrium (Dennell 1950). Posteriorly the siphon opens to the exterior (Dennell 1950). The floor of the atrium and the siphon walls are muscular, and a contraction of the latter results in the occlusion of the cavity and the atriopore's closure (Dennell 1950).

In order to clean the oral (buccal) cirri after feeding, these cirri are spasmodically flexed into and out of the oral hood cavity accompanied by an expulsion of water from the hood, removing the external particles (Dennell 1950). The inwards movement of the cirri results from the contraction of the oral (labial) muscles together with the powerful movements of the atrium floor via contractions of the **pterygial muscle** (Dennell 1950). The latter movement changes the volume of the atrium and with it the direction of water streaming in and out through the pharynx and the oral hood and not through the atriopore (Dennell 1950). This functional association between the oral and pterygial musculature in amphioxus is fascinating, because it might suggest that there is also a developmental link between these muscles. This would support Diogo et al.'s (2015) idea that both the oral and **pterygial muscles** of amphioxus are **branchiomeric muscles** (Figs. 1.1 and 1.4). In Table 1.1 we infer the presence of muscles in the LCA of chordates; the oral muscles of amphioxus are not included in the branchiomeric muscles as **mandibular muscles** become later included in the branchiomeric series of vertebrates (see below).

Concerning nerves, the **somatic muscles** of amphioxus are restricted only to the myotomes (Yasui et al. 2014) and have a peculiar mode of innervation, as muscle tails (extensions of the muscles) take their innervation on the ventral surface of the nerve cord (Holmes 1953; Flood 1966), while other muscles are innervated with peripheral nerves from the dorsal roots (Holmes 1953; Yasui et al. 2014). In vertebrates, branchial motor neurons are located dorsally to somatic motor neurons, although this is not as distinct as the postcranial **spinal nerves** (Yasui et al. 2014). In cephalochordates (amphioxus), the peripheral nerves from the central nervous system (CNS) show a metameric pattern, as seen in vertebrates, and do not innervate the mouth and gills directly. Instead, they extend into the metapleural folds, where they anastomose and form the oral nerve ring (Kaji et al. 2001, 2009) or metapleural longitudinal nerves (Yasui et al. 2014) before innervating the oral and branchial targets. In gnathostomes the hypobranchial muscle precursors migrate together with the hypoglossal nerve anlage along the boundary of the pharynx and the body cavity to reach the oral floor (Oisi et al. 2015). During the development of lampreys, muscle precursors arise from rostral somites and migrate caudally and ventrally along the caudal end of the pharynx at the interface to the rostral part of the body wall to turn rostrally reaching the pharyngeal wall (Marinelli and Strenger 1954). Therefore, typically both the myotomal muscle precursors and the hypoglossal nerve do not migrate into pharyngeal arches (Mackenzie et al. 1998). Branchiomeric nerves (**cranial nerves**) are associated with pharyngeal arches and are characterized by their lateral positions, while spinal nerves are associated with somites, and their dorsal roots are more medial to the dermomyotome (Oisi et al. 2015). At the head-trunk transition area in gnathostomes, the relationship between vagus nerve (cranial) and hypoglossal nerve (spinal) is reversed, and the latter nerve is more lateral (Kuratani 1997). This pattern is also found in the lamprey but greatly modified in the hagfish (Oisi et al. 2015); still, it is likely present in the LCA of gnathostomes and cyclostomes (i.e., in vertebrates).

The CNS in adult amphioxus, i.e., the neural tube, was described in detail by Willey (1894). It ends anteriorly just posterior to the anterior end of the notochord. Also, anteriorly a pair of nerves emerges from the sides of the nerve tube, followed by a pair that arises more dorsally—also called **cranial nerves**—that lies in front of the first myotomes, have no ventral roots, and seem to be only sensory: they do not innervate any muscles, are only found in the snout, and have peripheral ganglionic enlargements. All following **spinal nerve** pairs are not arranged symmetrical anymore but alternate with one another similar to the alteration of myotomes. This asymmetrical alteration becomes more pronounced posteriorly. Behind the second pair of nerves ascend dorsal and ventral nerve roots, arising dorsally and ventrally from the neural tube, respectively. The dorsal roots are compact nerves from collected nerve fibers, while the ventral fibers emerge separately in loose bundles from the neural tube. Each body segment has one pair of dorsal roots and one pair of ventral root bundles, and both types of roots are completely independent of each other in contrast to vertebrates where the dorsal and ventral roots coalesce (Willey 1894). The first dorsal spinal nerve pair (i.e., the third pair in total) passes from the neural tube to the skin through the septum that separates the first and second myotomes. All following dorsal roots show this pattern (second dorsal spinal nerve pair passes through the septum between second and third myotomes and so on). The dorsal roots divide into the ramus dorsalis and ramus ventralis shortly after leaving the neural tube, and both rami run externally to the muscles in the sub-epidermic cutis (Holmes 1953). The corresponding branches of spinal nerves in vertebrates lie medially to the muscles during the first part of their course—i.e., between the muscle and the notochord. Those **cranial nerves** of vertebrates resemble amphioxus' dorsal roots in the sense that they are externally to the somites of the head.

The ramus dorsalis divides into finer nerves innervating the skin of the back, while the ventral ramus divides in several cutaneous nerves and a visceral branch turning medially below the myotomes and passing between the myotomes and the **pterygial muscle** (Willey 1894; Holmes 1953). The dorsal **spinal nerves** of amphioxus are therefore sensory and motor nerves. The ventral spinal nerves are entirely motor nerves, and after leaving the neural tube (spinal cord), they spread fanlike and innervate the myotomes. Interestingly, in vertebrates the ventral roots are motor and the dorsal roots are sensory (Kaji et al. 2009). The visceral neurons are different in amphioxus compared to other vertebrates (craniate animals *sensu* Holmes 1953). The descending visceral branch of each segment in the atrial region of amphioxus runs over the pterygial muscle and is often described as a branching transverse nerve. This transverse nerve passes between the atrial floor epithelium and the pterygial muscle fibers and provides the motor innervation for these fibers. Experiments by Holmes (1953) showed that the motor nerves from the dorsal root induce the contractions of the atrial floor; the motor division of the descending visceral ramus comes straight from the cord to the pterygial muscle. In summary, the pterygial muscle, the velar muscles, and the oral hood muscles are innervated by the visceral branches of the dorsal nerves, as is probably also the **atrial sphincter** (i.e., the caudal fibers of the pterygial muscle surrounding the anus). With respect to their innervation, the muscles in the atrial region in amphioxus probably correspond to the **branchiomeric muscles** in vertebrates (see also Gans 1989; Diogo and Ziermann 2015).

The innervation of the larval mouth of amphioxus differs dramatically from the innervation of the adult oral region. It seems that the larval oral nerve ring plays a crucial role in patterning the nervous system in the oral region but has no homology to any structures in vertebrates (Kaji et al. 2009) and is not the precursor of the inner **oral hood nerve plexus** and the **velar nerve ring** as described by Kaji et al. (2001). The visceral branches from the dorsal **spinal nerves** that innervate the oral hood in the adult amphioxus arise from the branches of the third to sevenths dorsal nerves (Willey 1894). One set of those branches courses beneath the outer surface of the oral hood and forms frequent anastomoses which gave this network the term outer plexus (Willey 1894; Kaji et al. 2009). The other set is deep to the inner surface of the oral hood and is the inner plexus. Both plexuses are distinct from each other, besides the fact that the nerves have a common origin from the dorsal roots (Willey 1894). The outer plexus continues up into the individual buccal cirri, and the inner plexus seems to end at the base of the buccal cirri (Willey 1894). The inner plexus of both sides of the oral hood is exclusively formed by nerves arising from the left third and fourth nerves (Willey 1894; Kaji et al. 2001). The innervation of the velum is by the fourth, fifth, and sometimes sixth dorsal nerves of the left site only (Willey 1894; Kaji et al. 2009). This asymmetry seems to be related to the peculiar development of amphioxus.

The dorsal **spinal nerves** of amphioxus have some characteristics typical of the **cranial nerves** of vertebrates, but the walls of the gill slits are innervated in amphioxus by spinal nerves, while they are innervated by cranial nerves in craniate vertebrates (Willey 1894). The cerebral vesicle is a widening of the central canal in the region of the cranial nerves and is not divided into ventricles. The cerebral vesicle opens in young amphioxus by an aperture called the neuropore into the base of the olfactory pit. The neuropore closes in later stages and is only indicated by a groove at the base of a stalk connecting the olfactory pit with the roof of the brain (Willey 1894). Behind the cerebral vesicle, the central canal widens into a dorsal portion that is independent of the ventral tube. The region of the nerve tube over which the dorsal portion extends was compared to the medulla oblongata of craniate vertebrates. During the development of the CNS of vertebrates, there might be a stage that is comparable to the adult condition in amphioxus (Willey 1894). However, in vertebrates the anterior portion of the medullary tube enlarges and divides into fore-, mid-, and hindbrain.

1.4 Recent Findings in the Context of the New Head Hypothesis

Data obtained since Gans and Northcutt's (1983) paper on the new head hypothesis (NHH) can be divided into three major categories: those that support parts of the NHH, those that revive earlier ideas, and those that present new, and often surprising, scenarios. As an example of the first category, paleontological studies support the idea that a head skeleton composed of cartilage and calcified tissues derived from neural crest and sclerotomal mesoderm is an ancestral vertebrate feature (e.g., Valentine 2004). However, these studies also revealed specific evolutionary changes that differ markedly from previous assumptions; for example, the first gnathostomes (e.g., placoderms) probably possessed not only calcified endochondral bones but also dermal bones (e.g., maxillary) similar to those found only in extant bony fish (osteichthyans) (Zhu et al. 2013). Gans and Northcutt's hypothesis that the evolution of chordates and early vertebrates relates to a shift from filter feeding to suction feeding and then to a more active mode of predation has also been supported in recent decades (Northcutt 2005), but the specific phenotypic changes involved are still heatedly debated. For instance, Mallatt's (2008) neoclassical hypothesis for the origin of the vertebrate jaw is more conservative in assuming that the upper lip and its muscles in sharks are homologous with those of lampreys, while Kuratani et al.'s (2013) heterotopic hypothesis assumes that the upper lip was lost in gnathostomes and acquired *de novo* in some gnathostome groups, such as sharks.

The second category of data includes a surprising revival of ideas defended by classical authors such as Goodrich, Garstang, Gegenbaur, Edgeworth, and even Darwin. Some of these ideas were widely accepted in the late 19th and early twentieth centuries, but they were largely abandoned during the second half of the 20th century and therefore were not incorporated in the NHH. They include (a) the sister group relationship between urochordates and vertebrates (Delsuc et al. 2006), previously defended by authors such as Darwin (1871) and Garstang (1928); (b) Gegenbaur's (1878) hypothesis that the pectoral appendage (girdle + fin) originated as an integral part of the head (Gillis et al. 2009) and, therefore, that the pectoral appendage evolved independently of, and may not be serially homologous to, the pelvic appendage (Diogo et al. 2013; Diogo and Tanaka 2014; Diogo and Ziermann 2014); and (c) Edgeworth's (1935) hypothesis that at least part of the esophageal musculature and the **cucullaris** derivatives (e.g., trapezius) derive from the branchiomeric musculature and/or follow a head program [(e.g., Piotrowski and Nüsslein-Volhard 2000; Diogo and Abdala 2010; Sambasivan et al. 2011; Minchin et al. 2013) but see (e.g., Minchin et al. 2013)], which was supported by the recent clonal studies of Lescroart and colleagues (2015; see Fig. 1.6).

The third category comprises new and mostly unexpected scenarios. For instance, contrary to what was usually accepted at the time of the writing of the NHH, cranial neural crest cells, while giving rise to numerous skeletal elements of the head and serving as precursors for connective tissue and tendons, do *not* form muscles (Noden 1983, 1986; Noden and Francis-West 2006). Instead, the mesoderm-derived muscle progenitors fuse together to form myofibers within cranial neural crest-derived connective tissue in a precisely coordinated manner. Muscles of a certain arch are usually associated with connective tissue, and through this tissue also with skeletal elements, of the same arch (Köntges and Lumsden 1996).

As mentioned above, another remarkable discovery was that of the **cardiopharyngeal field** (Figs. 1.4, 1.5, and 1.6; reviewed by Diogo et al. 2015). Strikingly, the results of these analyses suggest that some **branchiomeric muscles** are more closely related to certain heart muscles than to other branchiomeric muscles, contradicting the long accepted view that the branchiomeric muscles mainly constitute a single anatomical and developmental unit (e.g., Edgeworth 1935). Furthermore, these studies suggest that the first (mandibular) arch probably was not part of the original series of branchial arches. Instead, the most rostral branchial arch of basal chordates such as cephalochordates and "prevertebrate" fossils such as *Haikouella* is thought to correspond to the second (hyoid) branchial arch of vertebrates (Mallatt 2008). According to this idea, the first arch was incorporated into the branchial arches only in more derived chordates, which explains why it is the only arch in vertebrates in which *Hox* genes are not expressed and do not pattern arch formation (Mallatt 2008; see below).

It was recently shown that *Hox1* is essential for the anterior-posterior (AP) axial identity of the endostyle in the urochordate *Ciona intestinalis* (Yoshida et al. 2017). Experiments by Yoshida et al. (2017) suggest that the identity of the anterior and posterior endostyle is determined by the expression of *Otx* (anterior) and *Hox1* (posterior). The overexpression of *Hox1* represses *Otx* expression and with that the anterior identity of the endostyle causing that end to differentiate with a posterior identity. If *Hox1* is knocked out, the posterior end of the endostyle is transformed to an anterior identity because of ectopic expression of *Otx*, and the atrial siphon and gill slits are lost (Sasakura et al. 2012; Yoshida et al. 2017). Importantly, simultaneous knockout of *Hox1* and

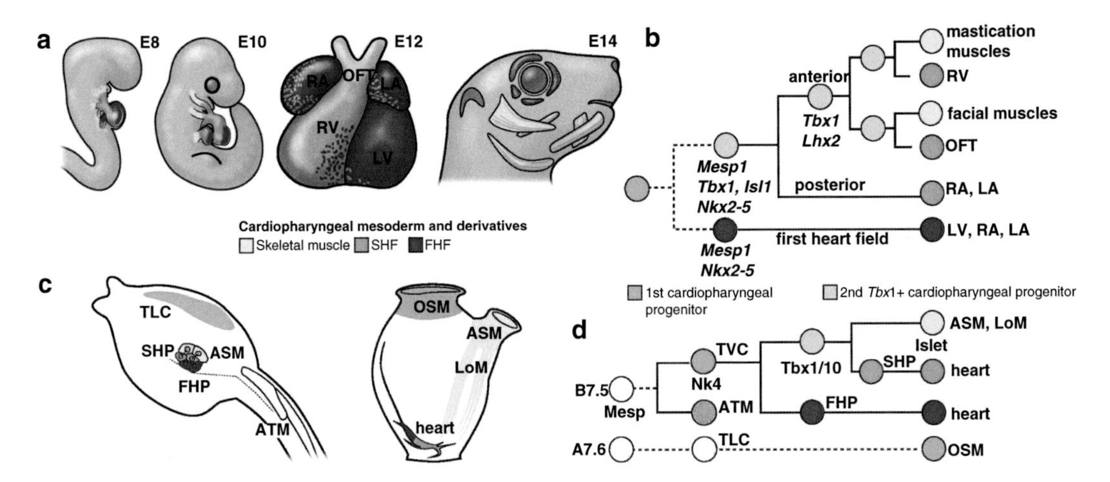

Fig. 1.5 An evolutionary conserved cardiopharyngeal ontogenetic motif (modified from Diogo et al. 2015, who own its copyright and gave permission to be used here). (**a**) Mouse embryos at embryonic days (E) E8 and E10, the four-chambered mouse heart at E12, and the mouse head at E14. Red, first heart field (FHF)-derived regions of heart [left ventricle (LV) and atria]; orange, second heart field (SHF)-derived regions of heart [right ventricle (RV), left atrium (LA), right atrium (RA), and outflow tract (OFT)]; yellow, branchiomeric skeletal muscles; and purple, extraocular muscles. (**b**) Cell lineage tree depicting the origins of cardiac compartments and branchiomeric muscles in mouse. All cells derive from common pan-cardiopharyngeal progenitors (dark green) that produce the FHF, precursors of the left ventricle (LV) and atria (RA, LA), and the second, *Tbx1*+, cardiopharyngeal progenitors (light green). Broken lines indicate that the early FHF/SHF progenitor remains to be identified in mouse. In anterior cardiopharyngeal mesoderm (CPM), progenitor cells activate *Lhx2*, self-renew, and produce the SHF-derived RV and OFT and first and second arch branchiomeric muscles (including muscles of mastication and facial expression). (**c**) Cardiopharyngeal precursors in *Ciona intestinalis* hatching larva (left) and their derivatives in the metamorphosed juvenile (right). The FHF (red) and SHF (orange) heart precursors contribute to the heart (red-orange mix), while atrial siphon muscle precursors (ASM, yellow) form atrial siphon and longitudinal muscles (LoM, yellow). Oral siphon muscles (right: OSM, blue) derive from a heterogenous larval population of trunk lateral cells (left: TLC, blue). Cardiopharyngeal mesoderm is bilaterally symmetrical around the midline (dotted line). (**d**) Cell lineage tree depicting clonal relationship and gene activities deployed in *Ciona* cardiopharyngeal precursors. All cells derive from *Mesp* + B7.5 blastomeres, which produce anterior tail muscles (ATM, gray disks, see also left panel in c) and trunk ventral cells (TVC, dark green disk). The latter pan-cardiopharyngeal progenitors express *Nk4* and divide asymmetrically to produce the first heart precursors (FHP, red disk) and second TVCs, the *Tbx1/10*+ second cardiopharyngeal progenitors (2nd TVC, light green disk). The latter divide again asymmetrically to produce second heart precursors (SHP, orange disk) and the precursors of atrial siphon and longitudinal muscles (ASM and LoM, yellow disk), which upregulate *Islet*. The OSM arise from A7.6-derived trunk lateral cells (TLC, light blue disk)

Otx leads to animals without *Hox1* expression but with *Otx* expression in the anterior and posterior endostyle, indicating that the default identity of the posterior endostyle is in fact the same as the anterior expressing *Otx* (Yoshida et al. 2017). A change in regional identity of the endoderm causes a disruption of the body wall muscle formation implying that the endostyle, a major part of the pharyngeal endoderm, is essential for coordinated pharyngeal development (Yoshida et al. 2017). Furthermore, retinoic acid receptor (RAR) and retinoic acid (RA) signaling from larval endoderm and muscle induce *Hox1* expression in the posterior endostyle and RA synthesis is required to maintain *Hox1* expression (Yoshida et al. 2017). The posterior endodermal identity and posterior RA synthesis are needed for the elongation of the body wall muscles toward the posterior end in *C. intestinalis*. In chordates *Otx* and *Hox1* transcription factors are expressed in the embryonal pharyngeal endoderm.

The mechanisms observed in *C. intestinalis* endostyle AP patterning were previously described in mouse and amphioxus pharyngeal

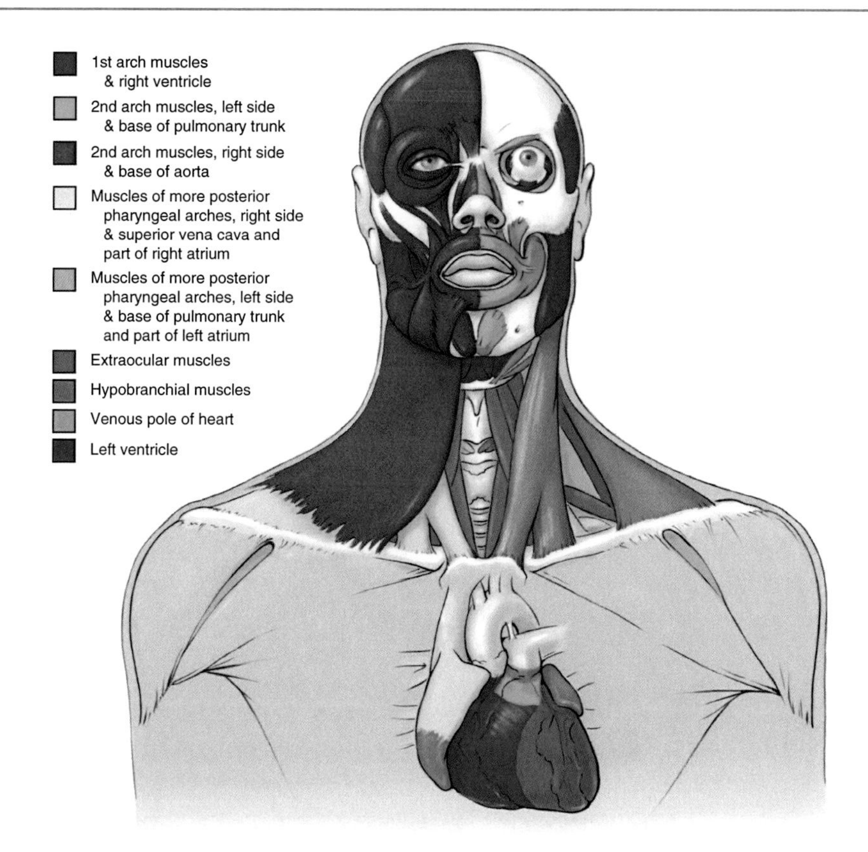

Fig. 1.6 The striking heterogeneity of the human head musculature (modified from Diogo et al. 2015, who own its copyright and gave permission to be used here to include the new data provided in Lescroart et al. 2015 paper). The head musculature includes at least seven different muscle groups, all arising from the cardiopharyngeal field and being branchiomeric, except the hypobranchial, and perhaps (but not likely) the extraocular muscles. On the left side of the body (right part of figure), the facial expression muscles were removed to show the masticatory muscles. The seven groups are: (1) *1st/mandibular arch muscles*, including cells clonally related to the right ventricle (shown in purple) and seemingly also to the extraocular muscles (see below); (2) *left 2nd/hyoid arch muscles*, with cells related to myocardium at the base of the pulmonary trunk (green); (3) *right 2nd/hyoid arch muscles*, related to myocardium at the base of the aorta (red); (4) *left muscles of the most posterior pharyngeal arches*, including muscles of the pharynx and larynx and the **cucullaris**-derived neck muscles trapezius and sternocleidomastoideus, which are related to the base of the pulmonary trunk and part of the left atrium (orange); (5) *right muscles of the most posterior pharyngeal arches*, including muscles of the pharynx and larynx and the cucullaris-derived neck muscles trapezius and sternocleidomastoideus, which are related to the superior vena cava and part of the right atrium (yellow); (6) *extraocular muscles* (pink), which are often not considered to be branchiomeric, but that according to classical embryologic studies and recent retrospective clonal analyses in mice contain cells related to those of branchiomeric mandibular muscles; and (7) *hypobranchial muscles*, including the tongue and infrahyoid muscles that derive from somites and migrate into the head and neck (dark gray, to show that they are not part of the colored cardiopharyngeal field). The venous pole of the heart is shown in blue and the left ventricle, derived from the first heart field, in brown

endoderm patterning. During mouse development, *Otx2* expression is observed in the first arch endoderm (Ang et al. 1994) and *Hox1a* and *Hox1b* expression in the caudal pharynx which are dependent on RA (Wendling et al. 2000; Niederreither et al. 2003). The cephalochordate amphioxus was shown to express *Hox1* in the endoderm repressing *Otx* expression and under RA control determining the posterior limit of the pharynx (Schubert et al. 2005). All this points toward a genetic mechanism present in the LCA of chordates needed for the proper AP axis specification of the early pharyngeal endoderm, which in turn is needed for the proper formation of pharyngeal muscles (Yoshida et al. 2017).

Within their NHH Gans and Northcutt (1983) argued that one of the main differences between vertebrates and invertebrates is that vertebrates possess complex sense organs and associated cranial ganglia, while invertebrates have poorly specialized sense organs and no neurogenic placodes (see also Chap. 2). However, studies performed in the last three decades, particularly on urochordates, strongly contradict this scenario. Apart from the discovery of a **cardiopharyngeal field** in urochordates (Figs. 1.4 and 1.5; reviewed by Diogo et al. 2015), the results of these studies have shown that urochordates also have placodes and neural crest-like cells, as summarized by Graham and Shimeld (2013) and Hall and Gillis (2013). The points made by these latter authors are briefly summarized below and in Fig. 1.4.

1.4.1 Development and Evolution of Chordate Muscles and the Origin of Jaw and Other Head Muscles in Vertebrates

In their Table 3 and Figure 1, Gans and Northcutt (1983) suggested that **branchiomeric muscles** were acquired at the origin of vertebrates. However, recent works, as well as some older and unfortunately often ignored studies, clearly show that branchiomeric muscles related to the pharynx were present in the LCA of chordates (Fig. 1.1). More than 100 years ago, it was reported that larvae of the giant cephalochordate *Epigonichthys* develop complex **orobranchial musculature**, but almost no investigation of the orobranchial musculature of this clade has been completed since (reviewed in Yasui et al. 2014). In the cephalochordate amphioxus, the larval mouth and unpaired primary gills develop five groups of **orobranchial muscles** as the larval mouth enlarges posteriorly, the oral musculature developing without segmental patterning (Yasui et al. 2014). During metamorphosis, the orobranchial musculature disappears completely, and the adult oral, velar, and pterygial (= subatrial or transverse) muscles (Fig. 1.2) develop independently. Yasui et al. (2014) suggested that the cephalochordate orobranchial muscles are probably a larval adaptation to prevent harmful intake, but they noted that the larval orobranchial muscles are perhaps more similar anatomically to the vertebrate branchiomeric muscles than are adult cephalochordate oral, velar, and **pterygial muscles**. They also noted that vestigial muscles appear transiently with secondary gill formation, suggesting an ancestral state of bilateral muscular gills and a segmental pattern of branchiomeric muscles in chordates. Six years after Gans and Northcutt's (1983) paper, Gans (1989) did recognize that the muscles of the atrial region of cephalochordates might correspond to the vertebrate branchiomeric muscles. He explained that cephalochordates and vertebrates have two patterns of motor innervation: one involves somatic motor neurons located within the basal plate of the spinal cord (somitic muscles); the other is seen in the cranial end of cephalochordates, where somatic motor axons leave the neural tube via a dorsal cranial root that proceeds ventrally to innervate the striated **pterygial muscles** of the atrial floor. Therefore, according to Gans (1989), and contrary to Gans and Northcutt's (1983) NHH, the pterygial musculature of amphioxus could be homologous with the branchiomeric muscles of vertebrates, which could well have arisen by an invasion of paraxial mesoderm to surround the pharynx laterally and ventrally, instead of by muscularization of hypomeric tissues.

At first glance, the proposed homology between vertebrate and urochordate cardiopharyngeal muscles and cephalochordate muscles might seem counterintuitive: one would expect the urochordate oral siphon and the cephalochordate oral/velar muscles, rather than the urochordate atrial siphon and the cephalochordate atrial muscles, correspond to, for example, the mandibular (first arch) muscles of vertebrates. In ascidians, water flows into the body through the oral siphon and is then expelled out of the body through the atrial siphon; therefore, it is the oral siphon that most likely corresponds to the mouth of vertebrates (Gans 1989). However, as shown in Fig. 1.4, recent studies have shown that the ascidian atrial siphon muscles derive from the **cardiopharyngeal field**, as do the **branchiomeric muscles** of vertebrates, but that the ascidian oral siphon muscles do not derive from this field (reviewed in Diogo et al. 2015). This fact seems to lend support to the idea that the mandibular arch was not part of the plesiomorphic branchial arch series of chordates (Mallatt 2008). In cephalochordates and urochordates, the oral/velar region lacks a skeleton, and the branchial bars are positioned a short distance behind the velum in a region that seems to correspond to the second branchial (= hyoid) arch of vertebrates (Figs. 1.2, 1.3, and 1.4; e.g., Mallatt and Chen 2003). That is, it makes sense that in early chordate evolution, the oral/velar muscles were not part of the **cardiopharyngeal field** (as continues to be the case in extant urochordates: Stolfi et al. 2010) and that they only became integrated into this field with the later co-option/homeotic shift of at least some oral structures and their muscles to form the first branchial (mandibular) arch. Interestingly, in basal vertebrates such as lampreys, *Tbx1/10* is expressed first in the mesodermal core of the branchial arches and pharyngeal muscles and the region of the otic vesicle, which seemingly corresponds to the atrium of non-vertebrates, and only later in development becomes expressed in the labial/oral and velar muscles (Sauka-Spengler et al. 2002). If in this case there is a parallel between ontogeny and phylogeny, these data would therefore also support the hypothesis that the inclusion of the velar/oral muscles in the **cardiopharyngeal field** and in the branchiomeric muscle group was a derived (later) event within chordate evolution.

However, there is at least one alternative scenario: that the urochordate oral siphon muscles do not correspond to any of the branchiomeric muscle groups present in extant vertebrates, i.e., the urochordate atrial siphon also includes at least some muscles that correspond to/are precursors of the vertebrate first arch muscles, as suggested by Stolfi et al. (2010). This suggestion was based on studies showing that, in derived vertebrates such as mouse and chick, the **cardiopharyngeal field** gives rise to **mandibular muscles**; it is now known that this field also gives rise to muscles of more posterior branchial arches (e.g., of the hyoid arch; e.g., Lescroart et al. 2010). Remarkably, some oral/velar muscles of adult cephalochordates are innervated by neurons from a region of the brain that is putatively homologous with the region that gives rise to the facial motor neurons innervating the muscles of the second (hyoid) arch in vertebrates (Northcutt 2005). Further studies are needed to investigate whether the cephalochordate oral/velar musculature corresponds to the oral siphon musculature (Fig. 1.1) or instead/also includes part of the atrial musculature of urochordates.

In enteropneust-type hemichordates serially arranged gill openings in the pharynx associated with musculature are found (Cameron 2002). However, this musculature is very different developmentally, anatomically, and histologically from the branchiomeric musculature of chordates (Yasui et al. 2014). In fact, the pharynx of the hemichordate *Saccoglossus* does not express *Tbx1*. According to Gillis et al. (2012), *Tbx1*-expressing pharyngeal mesoderm probably originated along the chordate stem, and the acquisition of cranial paraxial mesoderm within the pharyngeal region is probably a chordate synapomorphy. *Tbx1/10* is expressed in the pharyngeal mesoderm of cephalochordates and the atrial muscles of urochordates and *Tbx1* in the **branchiomeric muscles** of vertebrates, while *AmphiPax3/7* is expressed in the anterior and posterior somites of amphioxus (cephalochordate) and *Pax3* in all somitic muscles of vertebrates (Mahadevan et al. 2004). This distribution of gene expression indicates that the pterygial and oral/velar muscles of basal chordates and the branchiomeric muscles of vertebrates do not derive from the anterior somites and thus that the LCA of chordates already had a separation between somitic muscles (*Pax3*) and branchiomeric muscles (*Tbx1*). However, *Tbx1/10* is expressed in the atrial siphon muscles and also in so-called body wall muscles of urochordates (Stolfi et al. 2010) and in the pharyngeal mesoderm and the ventral part of some somites of amphioxus (Mahadevan et al. 2004), meaning that this separation probably was not as well defined in early chordates as it is in extant vertebrates.

However, although more defined, the separation between branchiomeric and somitic muscles, and between the head and the trunk in general, remains somewhat blurry. An illustrative example is the **cucullaris**, one of the best-studied yet most puzzling vertebrate muscles. Its amniote derivatives, the **trapezius** and sternocleidomastoideus, have played a central role in the studies of the origin and evolution of the vertebrate head and neck. These muscles share characteristics of at least five different muscle types: branchial, somitic epibranchial, somitic hypobranchial migratory, somitic limb nonmigratory ("primaxial"), and somitic limb migratory ("abaxial"). Topologically, the cucullaris resembles the epibranchial muscles of lampreys (e.g., Kusakabe et al. 2011), yet its developmental migration is similar to that of somitic hypobranchial migratory muscles (e.g., Matsuoka et al. 2005). Additionally, the trapezius receives contributions from both "primaxial" and "abaxial" cells (e.g., Shearman and Burke 2009). However, long-term fate-mapping studies have shown that muscles that are consensually accepted as branchiomeric, derived not only from posterior (e.g., laryngeal) but also from more anterior (e.g., the hyoid muscle interhyoideus) arches, receive a partial contribution from somites (Piekarski and Olsson 2007). These studies further complicate the distinction between the head/neck and trunk but in turn also show that the fact that the trapezius receives some somitic contribution does not contradict its original branchiomeric origin. Actually, the balance of available developmental, molecular, and anatomical data strongly supports the idea that the cucullaris and its derivatives are branchial, and thus branchiomeric, muscles. The cucullaris originates anatomically from the posterodorsal region of the branchial musculature and is usually innervated by the 11th (accessory) cranial nerve (e.g., Edgeworth 1935; Diogo and Abdala 2010; Ziermann and Diogo 2013, 2014). Also, the levatores arcuum branchialium of osteichthyan fishes (bony fishes), consensually considered to be branchial muscles, were clearly derived from the undivided cucullaris of plesiomorphic gnathostomes (Ziermann et al. 2014). Finally, neural crest cells from a caudal branchial arch migrate with trapezius myoblasts and form tendinous and skeletal cells within its zone of attachment (e.g., Noden and Schneider 2006). A stronger support to the branchiomeric identity of the **cucullaris** and its derivatives comes however from gene expression studies in mammals: *Tbx1* is expressed in/its lack affects the branchiomeric (e.g., laryngeal and first and second arch) muscles and the **trapezius**, while *Pax3* is expressed in/its lack affects all the somitic (i.e., limb, diaphragm, tongue, infrahyoid, and trunk) muscles, but not the trapezius (Sambasivan et al. 2011). In turn, these data also emphasize the heterogeneity of the vertebrate neck, which therefore includes branchiomeric (e.g., trapezius and sternocleidomastoideus), hypobranchial (e.g., tongue and infrahyoid), and trunk (somitic epaxial; e.g., deep neck and back) muscles.

Despite its profound implications for the NHH in particular, and for evolutionary and developmental biology and human medicine in general, heterogeneity in the vertebrate head and neck is poorly documented in textbooks, academic and medical curricula, and even many specialized research publications. In fact, one of the most crucial implications of recent studies on the **cardiopharyngeal field** is that they show that the head musculature derives from at least seven developmentally different types of primordia (Fig. 1.6). In addition, Cyclostomata, Selachii, and Holocephali (see Fig. 1.1) possess an eighth group of head muscles, designated epibranchial muscles, which derive from the anterior portion of the somites (Edgeworth 1935) (Table 1.1). Even within the same arch, muscles can follow different genetic programs; for instance, in zebrafish, *Ret* signaling is necessary for the development of only a few specific mandibular and hyoid muscles associated with the movements of the opercle (bony plates supporting the gill covers) (Knight et al. 2011). Likewise, *C-met* is crucial for the development and migration of the mammalian muscles of facial expression, derived from the second (hyoid) arch, but not for the other second-arch muscles (Prunotto et al. 2004).

1.4.2 General Remarks

An elongated motile adult stage is likely a representative condition for the adult LCA of Olfactores (Diogo and Ziermann 2015) because amphioxus is elongated and motile. Fossil evidence suggests that the hemichordate LCA had an enteropneust-(worm-)like motile adult stage as well (Caron et al. 2013), which is suggestive for an elongated motile adult as the LCA of chordates (Lowe et al. 2015). The existing scenarios regarding the origin of vertebrates were recently reviewed by Holland et al. (2015). Assuming that the LCA of chordates and the LCA of Olfactores were elongated and motile has the consequence that adult sessile ascidians (e.g., *Ciona*) would be seen as representing a derived condition due to peramorphosis (Diogo and Ziermann 2015). Furthermore, as elucidated by Diogo and Ziermann (2015), it is likely that the amphioxus larva represents a derived feature acquired during the evolutionary history of cephalochordates via addition of early developmental stages. The larval asymmetry was described as secondary or cenogenetic feature without ancestral significance (Willey 1894). One reasoning is the adaptation to the specialized feeding mode (Presley et al. 1996), where during the secondary gill formation, vestigial muscles transiently appeared (Yasui et al. 2014), supporting the addition of early developmental stages during the evolution leading to amphioxus. Furthermore, this supports the presence of bilateral muscular gills and segmented **branchiomeric muscles** in the LCA of extant chordates (Diogo et al. 2015).

Another support for the addition of embryonic stages is given by the observation that the notochord differentiates from anterior to posterior, as in other chordates, but in the embryo with eight paired myocoelomic pouches, the anterior end of the notochord is posterior to the anterior end of the body, while the anterior portion of the archenteron extends beyond the notochord. Later, the notochord extends to the tip of the head that indicates that this is a secondary phenomenon (Willey 1894). As both the larval amphioxus and the adult lamprey are highly specialized due to their feeding mode, it is likely that the LCA of vertebrates probably had muscular features resembling adult amphioxus and other features resembling those of non-adult lampreys. This is also suggested by the fact that the endostyle of ascidians and amphioxus is homologue to the hypobranchial groove of ammocoetes; those structures share a similar development, histological structure, and overall position (Willey 1894).

As noted above, Graham and Shimeld (2013) showed the homology of different placodes between urochordates and vertebrates using all available genetic and developmental studies and further evidence from fossils (Fig. 1.4). They proposed that the siphon primordia in urochordates are homologue to vertebral placodes (see also Diogo et al. 2015). The circular oral siphon muscles in urochordates, which are not part of the **cardiopharyngeal field**, could correspond to oral/velar muscles of amphioxus and of vertebrates such as cyclostomes and thus to the first arch (mandibular) muscles of gnathostomes, which were included in the branchiomeric muscle series only during vertebrate evolution (Diogo et al. 2015; Diogo and Ziermann 2015). The **pterygial muscle** of amphioxus thus probably corresponds to the atrial siphon and associated muscles of urochordates (Table 1.1), because both these muscles are in the atrial region and both end in the **atrial siphon** muscles and therefore to the muscles of the second and more posterior branchial arches of vertebrates (Diogo and Ziermann 2015).

In summary, we infer from our dissections, observations, comparisons, and literature review of developmental processes that the LCA of extant chordates had anterior muscles (muscles of the oral region) that were only in vertebrates incorporated as **mandibular muscles** in the branchiomeric series (Table 1.1; Fig. 1.1). Other **branchiomeric muscles** were already present in the LCA of extant chordates. Those muscles correspond to the **atrial sphincter** and/or **pterygial**

muscle of cephalochordates/urochordates and some transverse and longitudinal muscles. However, the genetic/morphogenetic program for differentiation into branchiomeric versus body/somitic muscles was probably not as sharply defined in the LCA of chordates as it was in the LCA of vertebrates. However, future developmental, genetic, and comparative studies are crucial to test this and the other hypotheses and suggestions proposed in the present chapter, which has as its main task to call the attention that without a detailed knowledge of the muscles of non-chordates, it is difficult to have a clear understanding of the origin and evolution of the vertebrate muscles.

Acknowledgments We would like to thank all the other members of the First International Meeting on the Evolutionary Developmental Biology of Head-Heart Muscles held in May 2014 at Howard University and in particular among them Michael Levine, Eldad Tzahor, Robert Kelly, Lionel Christiaen, Julia Molnar, and Drew Noden who collaborated with us in the *Nature* paper that was published as a result of that meeting, and that is a key basis for the present chapter. We are also thankful to numerous other colleagues that have discussed with us subjects related to the issues included in this chapter. Our gratitude goes furthermore to Peter Johnston and Virginia Abdala for reviewing the current chapter.

References

Ang S-L, Conlon RA, Jin O, Rossant J (1994) Positive and negative signals from mesoderm regulate the expression of mouse Otx2 in ectoderm explants. Development 120:2979–2989

Bone Q (1960) The central nervous system in amphioxus. J Comp Neurol 115:27–64

Cameron CB (2002) The anatomy, life habits, and later development of a new species of enteropneust, *Harrimania planktophilus* (Hemichordata: Harrimaniidae) from Barkley Sound. Biol Bull 202:182–191

Candiani S (2012) Focus on miRNAs evolution: a perspective from amphioxus. Brief Funct Genomics 11(2):107–117. https://doi.org/10.1093/bfgp/els004

Caron J-B, Morris SC, Cameron CB (2013) Tubicolous enteropneusts from the Cambrian period. Nature 495:503–506

Darwin C (1871) The descent of man, and selection in relation to sex. J Murray, London

Delsuc F, Brinkmann H, Chourrout D, Philippe H (2006) Tunicates and not cephalochordates are the closest living relatives of vertebrates. Nature 439:965–968

Dennell R (1950) Note on the Feeding of Amphioxus (Branchiostoma bermudœ). P Roy Soc Edinb B 64:229–234

Diogo R, Abdala V (2010) Muscles of vertebrates—comparative anatomy, evolution, homologies and development. CRC Press; Science Publisher, Enfield, New Hampshire

Diogo R, Tanaka EM (2014) Development of fore- and hindlimb muscles in GFP-transgenic axolotls: Morphogenesis, the tetrapod bauplan, and new insights on the Forelimb-Hindlimb Enigma. J Exp Zool Part B 322:106–127

Diogo R, Ziermann JM (2014) Development of fore- and hindlimb muscles in frogs: Morphogenesis, homeotic transformations, digit reduction, and the forelimb-hindlimb enigma. J Exp Zool Part B 322B:86–105

Diogo R, Ziermann JM (2015) Development, metamorphosis, morphology, and diversity: The evolution of chordate muscles and the origin of vertebrates. Dev Dyn 244:1046–1057

Diogo R, Linde-Medina M, Abdala V, Ashley-Ross MA (2013) New, puzzling insights from comparative myological studies on the old and unsolved forelimb/hindlimb enigma. Biol Rev 88:196–214

Diogo R, Kelly RG, Christiaen L et al (2015) A new heart for a new head in vertebrate cardiopharyngeal evolution. Nature 520:466–473

Edgeworth FH (1935) The cranial muscles of vertebrates. Cambridge at the University Press, London

Flood PR (1966) A peculiar mode of muscular innervation in amphioxus. Light and electron microscopic studies of the so-called ventral roots. J Comp Neurol 126:181–217

Fritzsch B, Northcutt RG (1993) Cranial and spinal nerve organization in Amphioxus and Lampreys: Evidence for an ancestral craniate pattern. Cells Tissues Organs 148:96–109

Gans C (1989) Stages in the origin of vertebrates: Analysis by means of scenarios. Biol Rev 64:221–268

Gans C, Northcutt RG (1983) Neural crest and the origin of vertebrates: a new head. Science 220:268–274

Garcia-Fernàndez J, Holland PWH (1994) Archetypal organization of the amphioxus Hox gen cluster. Nature 370:563–566

Garstang W (1928) Memoirs: The morphology of Tunicata, and its bearing on the phylogeny of the Chordata. Q J Micro Sci 2:51–187

Gegenbaur C (1878) Elements of comparative anatomy. Macmillan and Company, New York

Gill T (1895) The lowest of the vertebrates and their origin. Science:645–649

Gillis JA, Dahn RD, Shubin NH (2009) Shared developmental mechanisms pattern the vertebrate gill arch and paired fin skeletons. PNAS 106:5720–5724

Gillis JA, Fritzenwanker JH, Lowe CJ (2012) A stem-deuterostome origin of the vertebrate pharyngeal transcriptional network. Proc Roy Soc London B 279:237–246

Gopalakrishnan S, Comai G, Sambasivan R et al (2015) A cranial mesoderm origin for esophagus striated muscles. Dev Cell 34:694–704

Graham A, Shimeld SM (2013) The origin and evolution of the ectodermal placodes. J Anat 222:32–40

Gregory WK (1935) On the evolution of the skulls of vertebrates with special reference to heritable changes in proportional diameters (anisomerism). PNAS 21:1–8

Hall BK, Gillis JA (2013) Incremental evolution of the neural crest, neural crest cells and neural crest-derived skeletal tissues. J Anat 222:19–31

Holland LZ (2015) Genomics, evolution and development of amphioxus and tunicates: the goldilocks principle. J Exp Zool Part B 324:342–352

Holland LZ, Holland ND (1998) Developmental gene expression in amphioxus: new insights into the evolutionary origin of vertebrate brain regions, neural crest, and rostrocaudal segmentation. Am Zool 38:647–658

Holland LZ, Holland ND (2001) Evolution of neural crest and placodes: amphioxus as a model for the ancestral vertebrate? J Anat 199:85–98

Holland ND, Venkatesh TV, Holland LZ, Jacobs DK, Bodmer R (2003) AmphiNk2-tin, an amphioxus homeobox gene expressed in myocardial progenitors: insights into evolution of the vertebrate heart. Dev Biol 255:128–137

Holland LZ, Holland ND, Gilland E (2008) Amphioxus and the evolution of head segmentation. Integr Comp Biol 48:630–646

Holland ND, Holland LZ, Holland PW (2015) Scenarios for the making of vertebrates. Nature 520:450–455

Holmes W (1953) The atrial nervous system of amphioxus (Branchiostoma). Q J Micro Sci 3:523–535

Kaji T, Aizawa S, Uemura M, Yasui K (2001) Establishment of left-right asymmetric innervation in the lancelet oral region. J Comp Neurol 435:394–405

Kaji T, Keiji S, Artinger KB, Yasui K (2009) Dynamic modification of oral innervation during metamorphosis in Branchiostoma belcheri, the oriental lancelet. Biol Bull 217:151–160

Kaplan N, Razy-Krajka F, Christiaen L (2015) Regulation and evolution of cardiopharyngeal cell identity and behavior: insights from simple chordates. Curr Opin Genet Dev 32:119–128

Knight RD, Mebus K, d'Angelo A et al (2011) Ret signalling integrates a craniofacial muscle module during development. Development 138:2015–2024

Köntges G, Lumsden A (1996) Rhombencephalic neural crest segmentation is preserved throughout craniofacial ontogeny. Development 122:3229–3242

Koop D, Holland LZ (2008) The basal chordate amphioxus as a simple model for elucidating developmental mechanisms in vertebrates. Birth Def Res C Embryo Today 84:175–187

Koop D, Chen J, Theodosiou M et al (2014) Roles of retinoic acid and Tbx1/10 in pharyngeal segmentation: amphioxus and the ancestral chordate condition. EvoDevo 5:1

Kuratani S (1997) Spatial distribution of postotic crest cells defines the head/trunk interface of the vertebrate body: embryological interpretation of peripheral nerve morphology and evolution of the vertebrate head. Anat Embryol 195:1–13

Kuratani S, Adachi N, Wada N, Oisi Y, Sugahara F (2013) Developmental and evolutionary significance of the mandibular arch and prechordal/premandibular cranium in vertebrates: revising the heterotopy scenario of gnathostome jaw evolution. J Anat 222:41–55

Kusakabe R, Kuraku S, Kuratani S (2011) Expression and interaction of muscle-related genes in the lamprey imply the evolutionary scenario for vertebrate skeletal muscle, in association with the acquisition of the neck and fins. Dev Biol 350:217–227

Lescroart F, Kelly RG, Le Garrec J-F et al (2010) Clonal analysis reveals common lineage relationships between head muscles and second heart field derivatives in the mouse embryo. Development 137:3269–3279

Lescroart F, Hamou W, Francou A et al (2015) Clonal analysis reveals a common origin between nonsomite-derived neck muscles and heart myocardium. PNAS 112:1446–1451

Lowe CJ, Clarke DN, Medeiros DM, Rokhsar DS, Gerhart J (2015) The deuterostome context of chordate origins. Nature 520:456–465

Mackenzie S, Walsh FS, Graham A (1998) Migration of hypoglossal myoblast precursors. Dev Dyn 213:349–358

Mahadevan NR, Horton AC, Gibson-Brown JJ (2004) Developmental expression of the amphioxus Tbx1/10 gene illuminates the evolution of vertebrate branchial arches and sclerotome. Dev Genes Evol 214:559–566

Mallatt J (2008) The origin of the vertebrate jaw: neoclassical ideas versus newer, development-based ideas. Zoo Sci 25:990–998

Mallatt J, Chen JY (2003) Fossil sister group of craniates: predicted and found. J Morph 258:1–31

Marinelli W, Strenger A (1954) Vergleichende Anatomie und Morphologie der Wirbeltiere: von W. Marinelli und A. Strenger. *Lampetra fluviatilis* (L.). Franz Deuticke, Austria. 80

Matsuoka T, Ahlberg PE, Kessaris N et al (2005) Neural crest origins of the neck and shoulder. Nature 436:347–355

Minchin JE, Williams VC, Hinits Y et al (2013) Oesophageal and sternohyal muscle fibres are novel Pax3-dependent migratory somite derivatives essential for ingestion. Development 140:2972–2984

Minot CS (1897) Cephalic homologies. A contribution to the determination of the ancestry of vertebrates. Am Nat 31:927–943

Moreno TR, Rocha RM (2008) Phylogeny of the Aplousobranchia (Tunicata: Ascidiacea). Rev Brasil Zool 25:269–298

Niederreither K, Vermot J, Le Roux I et al (2003) The regional pattern of retinoic acid synthesis by RALDH2 is essential for the development of posterior pharyngeal arches and the enteric nervous system. Development 130:2525–2534

Noden DM (1983) The embryonic origins of avian cephalic and cervical muscles and associated connective tissues. Am J Anat 168:257–276

Noden DM (1986) Patterning of avian craniofacial muscles. Dev Biol 116:347–356

Noden DM, Francis-West P (2006) The differentiation and morphogenesis of craniofacial muscles. Dev Dyn 235:1194–1218
Noden DM, Schneider RA (2006) Neural crest cells and the community of plan for craniofacial development: historical debates and current perspectives. In: Saint-Jeannet J (ed) Neural crest induction and differentiation. Landes Bioscience, San Francisco, pp 1–31
Northcutt RG (2005) The new head hypothesis revisited. J Exp Zool Part B 304:274–297
Oisi Y, Fujimoto S, Ota KG, Kuratani S (2015) On the peculiar morphology and development of the hypoglossal, glossopharyngeal and vagus nerves and hypobranchial muscles in the hagfish. Zool Lett 1:1
Piekarski N, Olsson L (2007) Muscular derivatives of the cranialmost somites revealed by long-term fate mapping in the *Mexican axolotl* (Ambystoma mexicanum). Evol Dev 9:566–578
Piotrowski T, Nüsslein-Volhard C (2000) The endoderm plays an important role in patterning the segmented pharyngeal region in zebrafish (*Danio rerio*). Dev Biol 225:339–356
Presley R, Horder T, Slipka J (1996) Lancelet development as evidence of ancestral chordate structure. Israel J Zool 42:S97–S116
Prunotto C, Crepaldi T, Forni PE et al (2004) Analysis of *Mlc-lacZ* Met mutants highlights the essential function of Met for migratory precursors of hypaxial muscles and reveals a role for Met in the development of hyoid arch-derived facial muscles. Dev Dyn 231:582–591
Putnam NH, Butts T, Ferrier DE et al (2008) The amphioxus genome and the evolution of the chordate karyotype. Nature 453:1064–1071
Razy-Krajka F, Lam K, Wang W et al (2014) Collier/OLF/EBF-dependent transcriptional dynamics control pharyngeal muscle specification from primed cardiopharyngeal progenitors. Dev Cell 29:263–276
Sambasivan R, Kuratani S, Tajbakhsh S (2011) An eye on the head: the development and evolution of craniofacial muscles. Development 138:2401–2415
Sasakura Y, Kanda M, Ikeda T et al (2012) Retinoic acid-driven Hox1 is required in the epidermis for forming the otic/atrial placodes during ascidian metamorphosis. Development 139:2156–2160
Sauka-Spengler T, Le Mentec C, Lepage M, Mazan S (2002) Embryonic expression of Tbx1, a DiGeorge syndrome candidate gene, in the lamprey Lampetra fluviatilis. GEP 2:99–103
Schubert M, Yu J-K, Holland ND et al (2005) Retinoic acid signaling acts via Hox1 to establish the posterior limit of the pharynx in the chordate amphioxus. Development 132:61–73
Shearman RM, Burke AC (2009) The lateral somitic frontier in ontogeny and phylogeny. J Exp Zool Part B 312:603–612
Shimeld SM, Holland PW (2000) Vertebrate innovations. PNAS 97:4449–4452
Simões-Costa MS, Vasconcelos M, Sampaio AC et al (2005) The evolutionary origin of cardiac chambers. Dev Biol 277:1–15
Stokes M, Holland N (1995) Ciliary hovering in larval lancelets (= Amphioxus). Biol Bull 188:231–233
Stolfi A, Gainous TB, Young JJ et al (2010) Early chordate origins of the vertebrate second heart field. Science 329:565–568
Tirosh-Finkel L, Elhanany H, Rinon A, Tzahor E (2006) Mesoderm progenitor cells of common origin contribute to the head musculature and the cardiac outflow tract. Development 133:1943–1953
Tzahor E (2009) Heart and craniofacial muscle development: a new developmental theme of distinct myogenic fields. Dev Biol 327:273–279
Valentine JW (2004) On the origin of phyla. University of Chicago Press, Chicago
Wendling O, Dennefeld C, Chambon P, Mark M (2000) Retinoid signaling is essential for patterning the endoderm of the third and fourth pharyngeal arches. Development 127:1553–1562
Willey A (1894) Amphioxus and the ancestry of the vertebrates. MacMillan & Co., New York
Yasui K, Kaji T, Morov AR, Yonemura S (2014) Development of oral and branchial muscles in lancelet larvae of *Branchiostoma japonicum*. J Morph 275:465–477
Yoshida K, Nakahata A, Treen N et al (2017) Hox-mediated endodermal identity patterns the pharyngeal muscle formation in the chordate pharynx. Development 44(9):1629–1634. https://doi.org/10.1242/dev.144436
Zhu M, Yu X, Ahlberg PE et al (2013) A Silurian placoderm with osteichthyan-like marginal jaw bones. Nature 502:188–193
Ziermann JM, Diogo R (2013) Cranial muscle development in the model organism *Ambystoma mexicanum*: implications for tetrapod and vertebrate comparative and evolutionary morphology and notes on ontogeny and phylogeny. Anat Rec 296:1031–1048
Ziermann JM, Diogo R (2014) Cranial muscle development in frogs with different developmental modes: direct development vs. biphasic development. J Morph 275:398–413
Ziermann JM, Miyashita T, Diogo R (2014) Cephalic muscles of cyclostomes (hagfishes and lampreys) and Chondrichthyes (sharks, rays and holocephalans): comparative anatomy and early evolution of the vertebrate head muscles. Zool J Lin Soc 172:771–802

2 Early Vertebrates and the Emergence of Jaws

Zerina Johanson, Catherine A. Boisvert, and Kate Trinajstic

2.1 Introduction

The evolution of the vertebrate head, with placode-derived sensory organs such as the eyes and nasal capsules, was crucial in developing the ability to sense and interact with the surrounding environment, while a cartilaginous and mineralized braincase provided a support for these structures and an increasingly complex brain. Jaws were an important evolutionary innovation enabling **gnathostomes** to become efficient food processors. Dental structures enabled gnathostomes to adopt a predatory lifestyle, while the jaws themselves acted as an effective buccal pump, not only to bring food into the oral cavity but also increasing the volume of water passing over the gill arches. The functional effectiveness of the jaws depends on the associated **musculature**, with the evolution of a moveable neck (separation of the skull from the pectoral girdle) allowing the head to be raised to increase the size of the oral cavity.

Z. Johanson (✉)
Department of Earth Sciences, Natural History Museum, London, UK
e-mail: z.johanson@nhm.ac.uk

C. A. Boisvert · K. Trinajstic
Department of Environment and Agriculture, School of Molecular and Life Sciences,
Curtin University, Bentley, WA, Australia
e-mail: catherine.boisvert@curtin.edu.au;
K.Trinajstic@curtin.edu.au

The 'New Head' Hypothesis (Gans and Northcutt 1983; Northcutt 2005; see also Chap. 1) suggested that neural crest cell-derived cartilages and bones in the skull (the facial skeleton), and sensory placodes, were vertebrate innovations associated with the new development of a predatory lifestyle. However, as outlined below, certain features of both neural crest cells and placodes evolved prior to the origin of the vertebrates, questioning whether these features were related to the evolution of predation.

Vertebrate jaws comprise the **palatoquadrate** dorsally and **Meckel's cartilage** ventrally, with the palatoquadrate supported on the **braincase** in various ways, including by the more posterior hyoid arch. Dorsal and ventral elements of the jaws (mandibular arch) and hyoid arch meet at functional joints, as do the more posterior branchial or gill arches, and it was this similarity that led to early suggestions of serial homology of these elements and transformation of an unmodified anteriormost branchial arch into the jaws (Gegenbaur 1859, 1878). Evidence from the fossil record for this type of transformation may be best preserved in the recently described stem group vertebrate ***Metaspriggina***, which has a series of opposing, jointed branchial arches, with the most anterior said to be slightly larger and representing the mandibular arch [Cambrian Period, Burgess Shale (508 mya—million years ago) and other Burgess Shale-type deposits; Conway Morris and Caron 2014]. The jawless

J. M. Ziermann et al. (eds.), *Heads, Jaws, and Muscles*, Fascinating Life Sciences,
https://doi.org/10.1007/978-3-319-93560-7_2

vertebrate group **Euphaneropidae** possesses a large number of antero-posteriorly arranged arches (Janvier and Arsenault 2007), although an anterior size difference comparable to that in *Metaspriggina* is not apparent. In fact, most major clades of fossil jawless vertebrates preserve some evidence for the branchial arches even if the arches themselves are not preserved, including a series of branchial openings in the **Anaspida** and **Thelodonti** (Ritchie 1980; Wilson and Caldwell 1993; Blom 2012), impressions of the arches on the bony plates of the **Heterostraci** (Janvier 1993) and Osteostraci (Janvier 1985) and sediment impressions in the Thelodonti (Donoghue and Smith 2001). Information regarding arch muscle innervation and attachment can also be inferred in **Osteostraci** and **Galeaspida** based on the location of relevant foramina and attachment surfaces on the internal surface of the head shield (reviewed in Miyashita 2016).

The evolutionary origins of the jaws have also been based on feeding and pumping structures in extant jawless vertebrates (**Cyclostomata**: **lampreys** and **hagfish**; see Chap. 3), but this is problematic due not only to the differing morphology and development of these structures relative to vertebrate jaws (e.g., Miyashita 2016) but also the increasing appreciation that many cyclostome characters are derived, rather than primitive, and so less relevant to interpretation of jawed vertebrate morphology. For cxample, the dorsoventrally opposing branchial arches in ***Metaspriggina*** are very different from the unjointed cyclostome branchial basket (Marinelli and Strenger 1954, 1956; Conway Morris and Caron 2014). As well, important new information has come from the fossil jawless group Galeaspida, relevant to the **Heterotopy Hypothesis** of jaw evolution, involving the separation and lateral positioning of the nasal capsules (Shigetani et al. 2002; Gai and Zhu 2012; Kuratani 2012), recently identified in galeaspid *Shuyu zhejiangensis* (Gai et al. 2011). Fossil jawless vertebrates have also played an important role in new hypotheses of jaw evolution, such as the **Mandibular Confinement Hypothesis**, discussed further below (Miyashita 2016; see also Chap. 3).

Within the jawed vertebrates themselves, important new information on jaw evolution comes from the phylogenetically basal placoderms (Fig. 2.1), including innovations related to

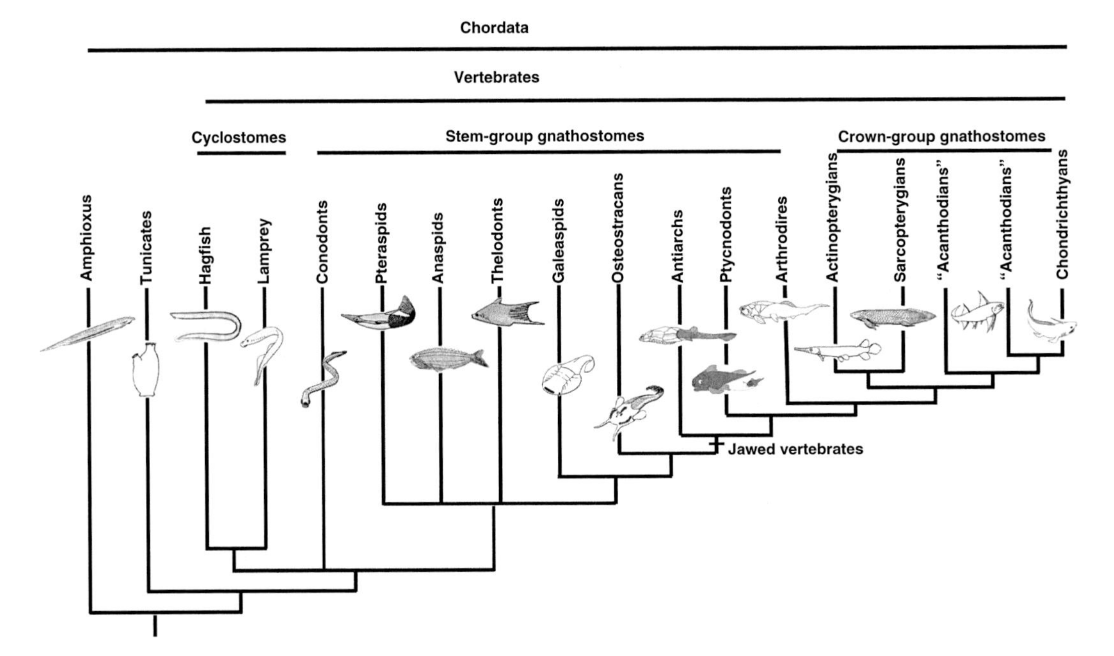

Fig. 2.1 Chordate phylogeny, including crown and stem group gnathostomes (adapted from Donoghue and Purnell 2005; Brazeau 2009; Zhu et al. 2013, with images modified from these, also Janvier 1996, Hurley et al. 2007, Kriwet et al. 2008, Tree of Life http://tolweb.org/tree/)

the jaws, including a functional neck and related musculature, dermal bones such as the maxillary and dentary bones associated with the cartilaginous jaws and dentition. It is these innovations, along with developmental underpinnings of jaw evolution itself, which are the focus of this chapter.

2.2 Innovations

The innovations that will be described in this chapter include:

1. Origin of the New Head
2. Evolution and development of jaws (palatoquadrate, Meckel's cartilage) and the hyoid arch (hyomandibular, ceratohyal)
3. Separated nasal sacs and the Heterotopy and Mandibular Confinement Hypotheses
4. The first jaws in phylogenetically basal jawed vertebrates: the placoderms
5. Mineralization associated with the jaws, including teeth and organized dentitions associated with feeding, and dermal bones such as the premaxilla, maxilla and dentary
6. Musculature associated with the jaws and hyoid arch
7. A moveable head, with functional separation from the postcranial skeleton (neck) and associated musculature: epaxial muscles and cucullaris muscle

2.2.1 Origin of the New Head

What features of the vertebrate head were innovations, and which had evolved prior to the evolution of the vertebrates? Gans and Northcutt (1983) proposed the '**New Head Hypothesis**', where these innovations, related to a more active and predatory lifestyle, were derived in large part from **neural crest cells** and **cranial placodes**. Neural crest cells contribute to the bones of the facial skeleton and to the cartilage of the pharyngeal arches and form early in development, during the folding and closure of the dorsal neural plate, which creates the neural tube. Neural crest cells develop at the border between the neural plate and ectodermal tissue. These cells are migratory and can differentiate into a range of tissues. These sequential stages of neural crest cell development can be associated with a crest gene regulatory network, including transcription factors such as *Snail1/2*, *FoxD3* and *SoxE*, with *SoxE* being a particularly important upstream regulator (Green et al. 2015).

Although neural crest cells were said to be a vertebrate innovation, certain components of crest development have evolved outside the group. For example, in the **Tunicata** (sister group to the Vertebrata), certain cells originate near the neural tube, migrate and give rise to pigmented cell types (Stolfi et al. 2015); pigment cells are a neural crest derivative in vertebrates. Neural crest cell homologs may also be present in other chordate group, the **Cephalochordata**: in the **amphioxus** genome, *Snail1/2*, *FoxD3* and *SoxE* are present, although only *Snail1/2* is expressed in the relevant region (Hall and Gillis 2013; see Green et al. 2015 for review).

Placodes, the other structures implicated in the evolution of the 'New Head' in vertebrates, are thickenings or invaginations of the cranial epithelium for which development within the head is dependent on reciprocal interactions with neural crest cells (reviewed in Schlosser et al. 2014; Steventon et al. 2014). These include olfactory, lens, trigeminal, adenohypophysial, otic and lateral line placodes (e.g., Schlosser 2010; Patthey et al. 2014; Fig. 2.2). But, as with the neural crest cells, there are certain features of placodal development that are present before the evolution of the vertebrates. For example, anterior and posterior regionalization of the head ectoderm, and the genes involved, have a deep phylogenetic history, with this regionalization forming into proto-placodal domains in tunicates, from which the anterior and posterior siphons develop (Schlosser et al. 2014). Thus, it appears that components of the vertebrate 'New Head' were well established before the origin of the group.

Although all vertebrates are characterized by placode-derived sensory structures such as eyes (lens placode) and inner ears (otic placode),

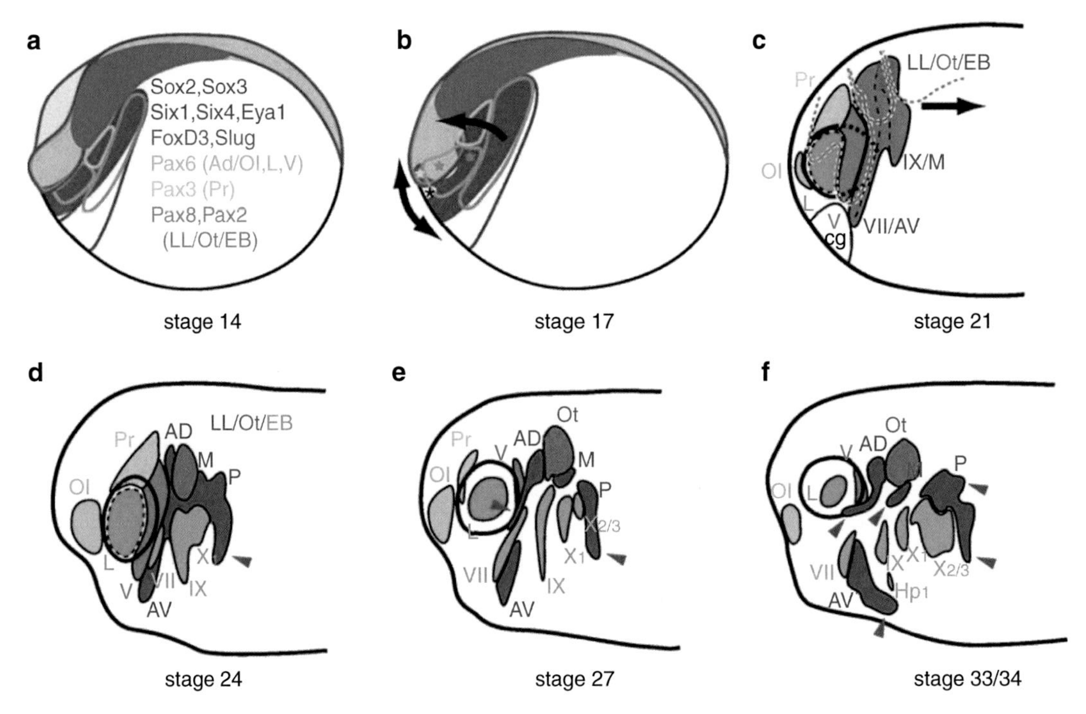

Fig. 2.2 Development of vertebrate sensory placodes in *Xenopus*, from Schlosser (2006, used with permission; see original text for more detail), including (**a**, **b**) gene expression domains (coloured outlines) during earlier neural plate (**a**) and neural fold stages (**b**). (**c–f**) Later stages of placode development. Brown arrowheads in (**d–f**), developing lateral line primordia. *Ad/Ol* anterior placodal area, from which adenohypophyseal (Ad) and olfactory placodes (Ol) develop; *AV* anteroventral lateral line placode; *L* prospective lens placode (hatched outline); *LL/Ot/EB* posterior placodal area, from which lateral line (LL), otic (Ot) and epibranchial (EB) placodes develop; *Ol* olfactory placode; *Ot* otic placode; *P* posterior lateral line placode; *Pr* profundal placode; *V* trigeminal placode; *VII* facial epibranchial placode; *IX* glossopharyngeal epibranchial placode; X_1 first vagal epibranchial placode; $X_{2/3}$ second and third vagal epibranchial placodes (fused)

lampreys and hagfish (Cyclostomata), and most fossil jawless vertebrates, possessed a single, and combined, **nasohypophysial** organ combining olfactory and adenohypophysial placodes (see also Chap. 3). Notably though, fossil taxa representing stem vertebrates (phylogenetically outside the crown group vertebrates, which include the cyclostomes + living jawed vertebrates and all related taxa) such as *Metaspriggina walcotti* (Conway Morris and Caron 2014) and ***Haikouichthys*** *ercaicunensis* (lower Cambrian Chengjiang Lagerstätte; Shu et al. 2003) possess eyes and paired nasal sacs (lens and olfactory placode derivatives), while *Haikouichthys* may possess otic capsules.

Vertebrates appear to have evolved a cartilaginous or mineralized structure that non-vertebrates lack, the braincase. This serves to protect and support the brain as well as the placode-derived sensory structures. Hagfishes have a very poorly developed braincase, compared to lampreys (Marinelli and Strenger 1954, 1956), while the braincase is thought to be cartilaginous and unpreserved in a range of fossil jawless vertebrates. With the evolution of **perichondral bone** (deposited on the cartilage surface), the braincase in the fossil jawless group Osteostraci was preserved. Imprints of the braincase are preserved on the internal head shield surface in galeaspids (Janvier 1996; Gai et al. 2011) and osteostracans (Stensiö 1927; Janvier 1981), where clear divisions of the telencephalon, mesencephalon, metencephalon and medulla oblongata can be observed. The location of the olfactory (nasal) and hypophyseal organs can also be observed; they are described below in the context of the Mandibular Confinement Hypothesis for the evolution of jaws (Miyashita 2016). The acanthothoracid placoderms (stem gnathostomes) retain a number of primitive characters, such as a short nasal capsule between the eyes and a short forebrain, which are more characteristic of jawless fishes. Throughout the evolution of the jawed vertebrates, the forebrain elongates as does the nose, and the upper lip shortens to be positioned under the nose (Dupret et al. 2014).

2.2.2 Evolution and Development of Jaws and the Hyoid Arch

2.2.2.1 Gill Arch Theory/Serial Hypothesis

There are multiple hypotheses regarding the evolution of the jaws, with the most recent summary of these provided by Miyashita (2016). The classical hypothesis for the origin of jaws is the **gill arch theory**, also called the **hypothesis of serial homology**, which suggests that jaws evolved from the modification of agnathan gill arches. It was proposed at the end of the 19th century by Gegenbaur (1859) based on an idealized ancestor constructed from anatomical observations of vertebrate embryos. This ancestor would have had a pharynx supported by serially repeated (and identical) segments. Embryonic **pharyngeal arches** (PAs), structures temporarily seen during development, can be subdivided into three categories: **mandibular arches** which develop into jaws in gnathostomes, **hyoid arches** which develop into the jaw suspension and a series of **branchial arches** which develop into skeletal branchial bars. They are formed in part by neural crest cells (as noted above, these are migrating cells emanating from the dorsal part of the neural tube and that differentiate into cartilages; Fig. 2.3), as well as mesodermal mesenchymal cells (also migratory cells but of mesodermal origin, forming the muscles of the pharyngeal region). The hypothesis of serial homology states that, in the hypothetical ancestor, the mandibular and hyoid arches originally developed into gill-bearing structures comparable to more posterior arches, which evolved into jaws in an ancestral gnathostome once freed from supporting the gills. As well, following from this hypothesis, the domain anterior to the mandibular domain (premandibular domain consisting of neural crest cells without any mesodermal component; Le Douarin 1982; Couly et al. 1993) would be either incorporated into the neurocranium or lost (De Beer 1937; Gegenbaur 1859).

This hypothesis for jaw evolution has been hotly debated for over a century and is still presented in textbooks, but there is little support from comparative anatomy, palaeontology, and embryology. One recent study (Gillis et al. 2013) has shown that a conserved pattern of nested gene expression for ***Dlx*** exists in all gnathostomes studied. This nested expression specifies the morphological identities of upper and lower jaws in mammals, teleosts and chondrichthyans (Depew et al. 2002; Gillis et al. 2013), and the three regions of **pharyngeal arches (mandibular, hyoid and gill arches)** are derived from equivalent domains specified by the genes (see also Chap. 3). Dorsoventral *Dlx* expression patterns are also present in the hyoid and more posterior branchial arches, with genes not only expressing dorsally and ventrally but also in an intermediate region in the arch (e.g., Square et al. 2015: Fig. 4), suggesting that these arches are serial homologues (e.g., Gegenbaur 1859, 1878; Gillis et al. 2013), forming the common pattern in jawed vertebrates. *Dlx* gene patterning has also been recognized in the cyclostome pharyngeal arches (Fujimoto et al. 2013), along with nested expression of the related *Hand* and *Msx* genes that characterizes jawed vertebrates (Cerny et al. 2010; Kuraku et al. 2010; see review in Medeiros and Crump 2012). The shared gene regulatory networks of pharyngeal arches in jawless vertebrates and basal gnathostomes were taken as support for the serial hypothesis (Gillis et al. 2013). However, *Dlx* expression in cyclostomes is not nested as in gnathostomes (Kuraku et al. 2010). In addition, there is still uncertainty regarding the orthology of lamprey *Dlx* genes (Gillis et al. 2013; Square et al. 2015), with differences in the number of orthologues expressed between lampreys and chondrichthyans, expression patterns

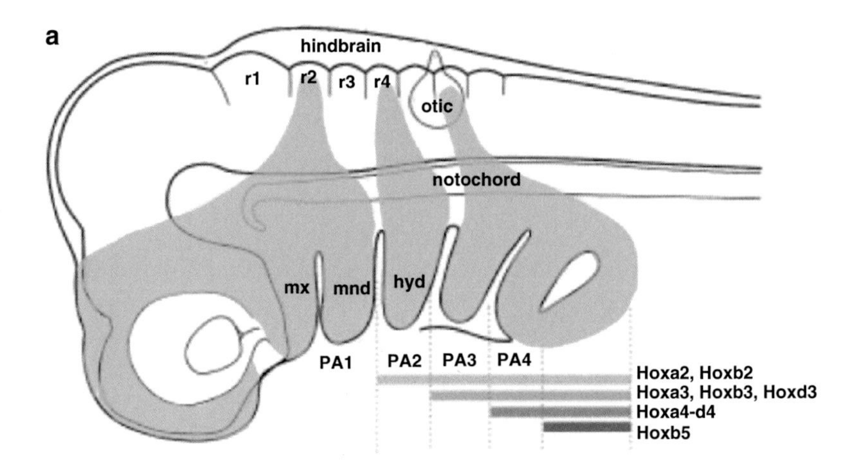

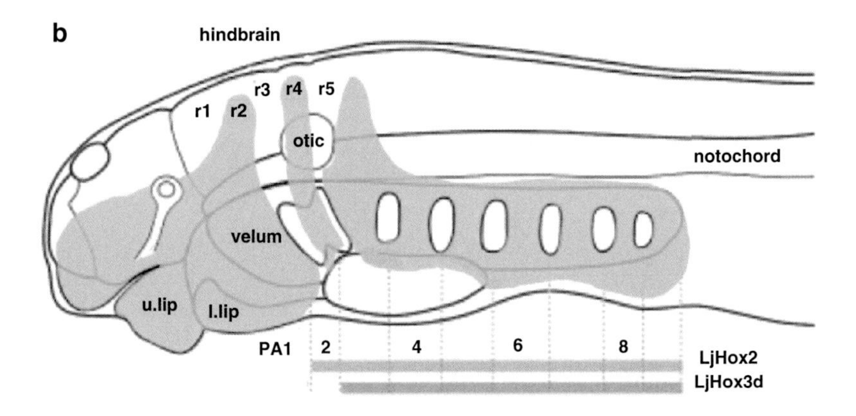

Fig. 2.3 Comparative development of the pharyngeal region in (**a**) shark and (**b**) lamprey, including neural crest cell streaming and anterior expression boundaries of *Hox* genes (modified from Kuratani 2004)

not being equivalent, and other differences also noted by Medeiros and Crump (2012). Overall, this research shows that some dorsoventral patterning of pharyngeal arches is ancient but provides limited evidence for the serial hypothesis.

The main problems with the serial hypothesis of homology are that no fossil or living vertebrates have respiratory gills on the mandibular and hyoid arches (but see *Metaspriggina*, below), or fully formed arches in a premandibular position, meaning that there is no evidence for a 'transitional' jawed vertebrate caught in the act of transforming gill arches into jaws. Additionally, the hypothesis would be supported if the pharyngeal skeleton of agnathans and gnathostomes were homologous. However, the **branchial** skeleton supporting the gill region in cyclostomes is located outside (lateral) relative to the pharyngeal muscles, but in sharks, branchial bars are on the inside of the chamber, medial to the pharyngeal muscles (Sewertzoff 1911; Schaeffer and Thomson 1980; Mallatt 1996). This different conformation results from underlying developmental differences. Other classical hypotheses for the origin of jaws such as the transformation of the velum of cyclostomes (the ventilation structure of cyclostomes which extends into the pharynx from the mandibular arch) into jaws (Ayers 1921; Janvier 1993; Forey 1995) are based on the assumption that the pharynx is a structure composed of repeating identical elements, but this is not reconcilable with the fact that the cyclostome pharyngeal arches are patterned differently from the gnathostome pharyngeal arches (Barreiro-Iglesias et al. 2011).

A major difference in development between cyclostome and gnathostome pharyngeal arch

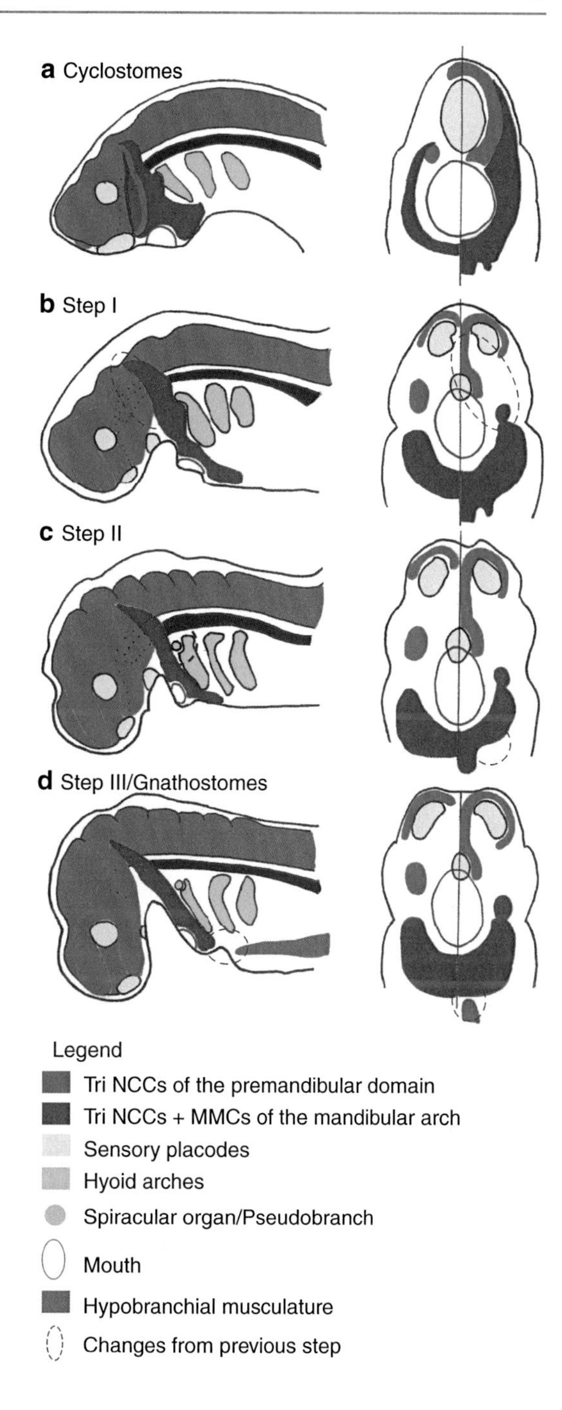

Fig. 2.4 Mandibular Confinement Hypothesis (Miyashita 2016, based in part on the Heterotopy Hypothesis, Kuratani 2012). Four hypothetical evolutionary grades showing the transformation from an agnathan velum to a gnathostome mouth. Adapted from Miyashita 2016, Kuratani 2004. Left column lateral view and right column, ventral view. For the right column, the left side represents an earlier developmental stage than the right side. (**a**) Development and anatomy of cyclostomes showing that the nasohypophyseal placode is single and medial. The premandibular neural crest cells (NCCs) proliferate with the NCCs and the mesodermal mesenchymal cells (MMCs) to form the upper lip. (**b**) The anterior boundary is established in Step II. The nasal placodes are paired, allowing the premandibular stream to proliferate and form the trabecula. The mandibular stream is now confined anteriorly (purple dotted ellipse). (**c**) The posterior boundary is established by the loss of posterior extension of the mandibular stream leading to the loss of a velum and the creation of a jaw joint. (**d**) Establishment of a ventral boundary as in extant chondrichthyans where the hypobranchial musculature extends anteriorly in the pharyngeal region, allowing for the development of the shoulder and neck musculature

development concerns how the different domains of neural crest cells and mesenchymal mesodermal cells are restricted (Fig. 2.4; Miyashita 2016). In gnathostomes, mesenchymal cells contributing to the mandibular arches (that will later form the jaw) become sandwiched between premandibular derivatives (which include the nasal capsule and hypophyseal canal) anteriorly and the hyomandibular pouch (which becomes the spiracle in some sharks). In cyclostomes and at least in **osteostracans** and **galeaspids** (see below), the premandibular and hyoid domains overlap and are not segregated. This shows that, rather than segmented structures being reorganized as per the serial hypothesis, the domain boundaries are absent in jawless vertebrates, with the origin of jaws occurring when these boundaries appear (see below, **Mandibular Confinement Hypothesis**).

Another important developmental feature disclaiming the hypothesis of serial homology is that the mandibular domain of gnathostomes and cyclostomes is distinct amongst pharyngeal arches in its gene expression and patterning requirements, as outlined further in the following sections. Although pharyngeal arches are patterned antero-posteriorly by a unique combination of collinearly expressed *Hox* genes, mandibular arches lack *Hox* gene expression, and jawlike cartilages develop on *Hoxa*-deficient gnathostome embryos (Takio et al. 2004; Minoux et al. 2009).

In conclusion, despite the fact that pharyngeal arches might look similar initially in the embryos of gnathostomes and cyclostomes and are regulated by similar genes early in development (*Tbx1, Wnt11 and Ffg3/8*) (Crump et al. 2004; Choe and Crump 2014; Shone and Graham 2014), the pharyngeal arches have different effectors (neural crest cells, mesodermal mesenchyme cells) and other differences outlined above. They are therefore not developmentally serially homologous.

2.2.2.2 Evolution and Development of the Jaws

In gnathostomes, there are three distinct developmental domains: **Premandibular**, **mandibular**, and **hyoid**. The first domain does not contribute to the jaws directly but is the domain for the migration of the **preoptic** and **postoptic** streams of **trigeminal neural crest cells** (Kuratani 2012; green streams, Fig. 2.4), discussed later below. The jaws themselves include a modified anterior **mandibular arch** (PA1), along with a supporting **hyoid arch** (PA2). The cartilages of all arches derive from particular neural crest cell streams related to the midbrain and hindbrain: the mandibular from the mandibular stream of the trigeminal neural crest cells, the hyoid from the stream just posterior to the mandibular stream, and more posterior arches from the branchial (circumpharyngeal) crest streams (Fig. 2.3; see Kuratani 2012; Meideros and Crump 2012 for recent reviews). Notably, comparable streams are present in the lamprey (McCauley and Bronner-Fraser 2003; Cerny et al. 2004) but they interact with the mesoderm in a different manner from gnathostomes (see next section). In gnathostomes, both the palatoquadrate and Meckel's cartilage are derived from the more ventral maxillomandibular component of the mandibular stream, with the dorsal component (preoptic, postoptic; see below) forming the trabeculae of the braincase (Cerny et al. 2004).

As discussed further below, the **olfactory** (nasal) and **adenohypophysial placodes** also play an important role in jaw evolution. According to the **Heterotophy Hypothesis** (Kuratani et al. 2001, 2013; Shigetani et al. 2002, 2005; Kuratani 2004, 2005, 2012), the single medial **nasohypophyseal** placode of cyclostomes forms a physical barrier to the most anterior neural crest cells moving towards the mandibular area. When the nasal and adenohypophyseal placodes become separated during vertebrate evolution, non-*Dlx* expressing trigeminal neural crest cells are able to move posteriorly to form the ethmoid, trabecula, and the nasal septum of the braincase (Kuratani 2004). The **trabecula** develops without contribution of the mesodermal mesenchyme cells in gnathostomes (Kuratani 2012). Due to the **heterotopic shift** of epithelial-mesenchymal interaction, the *Dlx* expression domain becomes shifted posteriorly in gnathostomes and becomes restricted to the jaws (Depew et al. 2005). The mandibular/mesodermal mesenchyme domain of gnathostomes is more restricted than in cyclostomes (red, Fig. 2.4) and does not interact with the postoptic trigeminal neural crest cells (green, behind eye placode, Fig. 2.4). The maxillary process (mx) is formed when the mandibular mesodermal mesenchyme cells undergo a secondary forward extension (Kuratani 2012).

As noted, in jawed vertebrates, the arches themselves are composed of neural crest cell-derived cellular cartilage, controlled by a conserved **gene regulatory network (GRN**;

Cattell et al. 2011) that includes genes in the SoxE family, for example *Sox9*, often known as the master regulator of cartilage development. Expression of *Sox9* is dependent on FGF-signalling through FGF receptors, while the SOX9 transcription factor binds to the fibrillar collagen gene *Col2a1*, a critical component of cartilage (Jandzik et al. 2015). SoxE genes are also important in the development of the cartilaginous arches in lampreys (Zhang et al. 2006; reviewed by Lakiza et al. 2011). Recently, Jandzik et al. (2015) investigated the cartilaginous structures around the amphioxus mouth (cirri), finding the amphioxus gene for fibrillar collagen (*ColA*) in these structures, and also detected *SoxE* and a FGF receptor in this region. Migratory neural crest cells are absent in amphioxus, so Jandzik et al. (2015) suggested that during vertebrate evolution, neural crest cells recruited these amphioxus genes, with changes in SoxE *cis*-regulatory sequences producing new transcription factor binding sites, which resulted in novel expression of SoxE in neural crest cells, driving vertebrate cellular cartilage development.

Nested *Hox* genes provide anteroposterior pharyngeal arch identity, including the jaw and hyoid arch, with these being absent from the jaw and the lamprey's velum and lower lip (Fig. 2.3; Hunt et al. 1991; Kuratani 2004; Takio et al. 2007; reviewed in Knight and Schilling 2013; Square et al. 2015). These differences resulting from *Hox* absence and nesting may be related to the different morphologies of the jaw and hyoid arch relative to more posterior arches (Square et al. 2015). New material from the stem vertebrate *Metaspriggina walcotti* (Burgess Shale) included serially arranged, bipartite, opposing dorsal and ventral arch elements with gills, but no distinct hyoid arch, nor jaws (Conway Morris and Caron 2014). Nevertheless, the anterior-most arch was said to be slightly thicker than the more posterior arches, and to lack gills, which may represent one of the first indications, in an evolutionary sense, of anterior arch differentiation, relative to the more posterior arches, and supportive of the hypothesis of serial homology above. This may suggest that *Hox* expression was absent from this anterior arch in *Metaspriggina*, but nested *Hox* expression more posteriorly had not yet occurred. As well, one important gene expressed in jawed vertebrates, but not lampreys (but see Miyashita 2016), is *Bapx1*, related to joint formation, including the joint between the opposing jaws, hyoid and more posterior arches (Cerny et al. 2010). The bipartite and opposing branchial arches in *Metaspriggina* (Conway Morris and Caron 2014) presumably also possessed a functional joint, indicating that *Bapx1* expression may have evolved early in vertebrate history. As mentioned above, one important implication is that the morphology (and putative gene expression) of *Metaspriggina* is primitive for vertebrates, and that the continuous, unjointed branchial baskets of lampreys and hagfish, are not homologous to arches of other vertebrates.

2.2.2.3 Mandibular Confinement Hypothesis

One of the more recent hypotheses of jaw evolution is the **Mandibular Confinement Hypothesis** proposed by Miyashita (2016), including discussion of several other hypotheses of jaw evolution (Fig. 2.4; see also Chap. 3). In this hypothesis, jaws can only evolve when the mandibular arches (the subset of pharyngeal arches that will become the jaws) become confined by clear anterior and posterior boundaries and do not overlap with the anterior premandibular domain and the posterior hyoid domain (the latter forming gill arches; Fig. 2.4). Compared to the serial hypothesis, which suggests that a metameric pharynx (made of identical pharyngeal arches) needs to be remodelled, this hypothesis suggests that jaws evolve when an unsegmented pharyngeal region becomes segmented. As well, initially distinct structures become more similar during jaw evolution according to this hypothesis.

For example, in cyclostomes and fossil jawless vertebrates, the 'mandibular' structures (muscular upper lip and **velum**) originally differ substantially from the more posterior branchial arches because the mandibular mesodermal mesenchyme cells interact with the trigeminal neural crest cells of the postoptic stream to form the upper lip (Kuratani et al. 2001, 2004). Cyclostome-like elements can be recognized in fossil jawless vertebrates: An **upper lip** can be identified in **osteostracans** and **galeaspids**, based on the foramina in the anterior head shield for the trigeminal nerve and associated muscle scars (Kuratani and Ahlberg 2018), while the euphaneropids have a ventral skeletal rod comparable to the cyclostome piston cartilage, and the conodont feeding apparatus may have moved in a manner similar to the lingual apparatus in cyclostomes (Keating and Donoghue 2016; Miyashita 2016). It is worth noting that in the stem vertebrate *Metaspriggina*, the anterior-most arch is said to be larger, and so differing, from the more posterior arches and lack gills, which may support the Mandibular Confinement Hypothesis, although overall, arches in *Metaspriggina* appear similar, particularly compared to the differences described in other jawless vertebrates (Conway Morris and Caron 2014). Miyashita's summary of the presence of cyclostome 'mandibular' elements in a variety of fossil jawless vertebrates suggests this, rather than the *Metaspriggina* morphology, may be characteristic for the vertebrate clade. The commonality of cyclostome mandibular morphology in extant and extinct taxa also suggests that the mandibular domain interacted with neural crest cells and mesodermal mesenchyme cells to form those specialized structures in early vertebrates.

The Mandibular Confinement Hypothesis suggests that for jaws to emerge, the neural crest cells and mesenchymal mesodermal cells must be constrained anteriorly and posteriorly to avoid interactions between each other in the three different compartments. In the embryo, many tissues and genes can constrain the migration streams of neural crest cells and mesodermal mesenchyme cells. Those include **sensory placodes** (embryonic structures from the head ectoderm giving rise to sense organs, as discussed above), the epithelium, cells from surrounding domains as well as BMP4 signalling centres (Holzschuh et al. 2005; Hall 2009; Steventon et al. 2014). The first event in the formation of jaws according to the Mandibular Confinement Hypothesis is the creation of an anterior barrier confining the mandibular arches (Step II, Fig. 2.4). In cyclostomes, there is a single nasohypophyseal placode, and the most anterior stream of neural crest cells (trigeminal neural crest cells) is prevented from moving posteriorly by the nasohypophyseal placode. Instead, the trigeminal neural crest cells migrate with mesodermal mesenchymal cells, forming a posthypophyseal process which develops into the upper lip (Oisi et al. 2013; Miyashita 2016). By comparison, in gnathostomes, the nasal placodes are paired anterolaterally and the hypophyseal placode is situated more caudoventrally (in Rathke's pouch). This reorganization of the placodes allows for the proliferation of the preoptic and postoptic trigeminal neural crest cells, forming the composite trabecula (anterior floor of the braincase), and largely separates the neural crest cells from the mesodermal mesenchymal cells in the premandibular region (Step I, Fig. 2.4), leading to the loss of the posthypophyseal process and the cyclostome upper lip (Kuratani 2012). The mesodermal mesenchymal cells (and some neural crest cells) undergo a secondary anteriorly migration to form the maxillary process of the jaws (Kuratani 2012; Miyashita 2016). Along with this, at the base of gnathostomes, the neural crest cells expressing *Dlx* became shifted and restricted posteriorly to the mandibular domain (Kuratani 2004, 2005, 2012; Shigetani et al. 2005; Kuratani et al. 2013). This created a clear non-*Dlx*-expressing anterior domain (premandibular) and a mandibular *Dlx*-expressing domain. In addition, *Dlx* genes in the mandibular region became dorsoventrally patterned at the base of the gnathostomes (see below).

The fossil record offers support for this first step, as the **Galeaspida** have separated **nasal sacs** (discussed in the following section), allowing the rearrangement of neural crest and mesodermal tissue, but in the absence of jaws. The next steps in the evolving jaws are the establishment of a **joint** (*Bapx1*, above) and a shift from 'hyomandibular ventilation' via a velum-like structure and of a posterior boundary to the mandibular arches. In cyclostomes, the mandibular neural crest cells and mesodermal mesenchymal cells can migrate ventrally to form the velum (Oisi et al. 2013). During these next evolutionary steps, the posterior hyomandibular extension of these cells is lost (Step II, Fig. 2.4), and the mandibular area becomes restricted posteriorly. *Barx1* inhibits joint formation (Nichols et al. 2013) in the ventral part of the mandibular arch but is broadly expressed in cyclostomes (Cerny et al. 2010); its restriction allows *Bapx1* expression and the formation of a mandibular arch joint. The loss of the hyomandibular extension leads to the establishment of a hyoid interface and to hyomandibular structures (spiracle, pseudobranch), while the new joint allows for new mechanisms of **buccal pumping**. The dorsoventral *Dlx* patterning of the hyoid arches acquired in the previous step now allows for the epi-(dorsal) and cerato-(ventral) branchial hinges to prevent flow reversal and allow for jawed vertebrate-like buccal respiration.

The final step in this evolutionary scenario is the establishment of a ventral boundary. In the previous step, the branchial bars move medially, which allow hypobranchial muscle precursor cells from the anterior rostral somites (paired balls of mesenchymal cells in the postotic region giving rise to the vertebral skeleton and musculature) to migrate rostrally. These develop into hypobranchial muscles extending from the separate pectoral region to the branchial arches. In osteostracans and galeaspids, these arches are enclosed in an expanded head shield that, in certain osteostracans, also encloses the pectoral fin. These arches would be associated with a musculature, but enclosure of the arches suggests that somite-derived hypobranchial muscle cells (and ventrally confined neural crest cells) would have been blocked from migrating into the branchial region; absence of a separate pectoral region precludes a musculature extending from the girdle to the branchial arches and forming the ventral boundary necessary for jaw evolution in these sister taxa to the jawed vertebrates (Miyashita 2016). Therefore, the last step of this evolutionary sequence is thought to first occur in the placoderms, the most basal gnathostomes (see below).

2.2.3 Separated Nasal Sacs and the Heterotopy and Mandibular Confinement Hypotheses

In cyclostomes, a combined nasohypophyseal organ is present (olfactory sac and a blind duct including the hypophysial pouch in lampreys; this extends into the pharynx in hagfish; Oisi et al. 2013, Kuratani and Ahlberg 2018), opening via a single median nasal opening, either more anteriorly in hagfish or dorsally in lampreys (Marinelli and Strenger 1954, 1956). A single dorsal, median opening also characterizes some fossil jawless vertebrates, for example, the **anaspids** (Ritchie 1980), osteostracans (Ritchie 1967; Janvier 1985) and galeaspids (Gai et al. 2011). A single nasal sac has been demonstrated for the osteostracans (Janvier 1985); by comparison, stem vertebrates such as *Metaspriggina* and *Haikouichthys* possess paired but closely approximated nasal sacs (Shu et al. 2003; Conway Morris and Caron 2014). Notably, paired nasal openings also appear to be present in the jawless fossil group, the **Arandaspida** (*Sacabambaspis*, Gagnier 1989; Gagnier and Blieck 1992).

Recently, application of synchrotron computed tomography has allowed the head shield of the galeaspids to be studied in considerable detail, in particular *Shuyu zhejiangensis* (Gai et al. 2011). In *Shuyu*, not only were the nasal sacs paired (compared to the closely related osteostracans and presumably most other fossil taxa with a single median opening), but they were separated laterally, on either side of a large median opening in the head shield. This morphology is relevant to the Mandibular Confinement Hypothesis, as described above, and the **Heterotopy Hypothesis** of jaw evolution presented by Kuratani and his colleagues (Shigetani et al. 2002; Kuratani 2004, 2012). In the latter hypothesis, there is a caudal heterotopic shift and restriction of *Dlx*-expressing neural crest cells to the mandibular arch, relative to their broader distribution in the oral region of lampreys. In lampreys, these include premandibular and mandibular cells, differentiating into the upper lip and lower lip and velum, respectively

(Gai and Zhu 2012). With the separation of the now-paired median nasal capsules, these tissues can be rearranged, migrating between the capsules to develop into the trabecula, and rostrally to form a maxillary process. Gai and Zhu (2012) noted the presence of a small trabecular component in the head shield of *Shuyu*, although this was subsequently questioned by Kuratani and Ahlberg (2018), as the 'trabecula' did not underlie the brain. Thus, the separation of the nasal capsules is only the first step in jaw evolution, which correlates well with this separation in the Galeaspida, but in the absence of jaws, and potentially the trabecula (Miyashita 2016).

As noted, paired nasal capsules occur in stem vertebrates, although in *Metaspriggina* and *Haikouichthys*, these are proximate to one another and positioned between the eye capsules, rather than more widely separated (e.g., Conway Morris and Caron 2014: Fig. 1j). This may have served to also block the tissue migration and differentiation described above, and if this is the case, one implication of the Heterotopy and Mandibular Confinement hypotheses is that the modified first branchial arch in *Metaspriggina* is not homologous to the jaws, as these would not have formed until midline space is made available by the separation of the nasal capsules. Also worth noting is that in arandaspids such as *Sacabambaspis*, a T-shaped bone appears to separate the nasal capsules, suggesting that these were more separate capsules, again in the absence of jaws.

2.2.4 The First Jaws in Phylogenetically Basal Jawed Vertebrates: The Placoderms

The placoderms are, phylogenetically, the first taxa with clearly functional jaws (Fig. 2.5a, c, d), although the nature of the hyoid arch has been more controversial (Brazeau et al. 2017). In placoderms, the **palatoquadrate** represents the upper jaw, articulating to the braincase and attached to the inner surface of the bony plates of the cheek in a range of groups including the **Acanthothoracida** (Ørvig 1975), Rhenanida (reviewed in Young 1986) and the **Arthrodira** (Miles 1969; Goujet 1984; Young 1986; Hu et al. 2017; Fig. 2.5d). In the **Antiarchi**, the palatoquadrate only articulates with the braincase (Young 1984), and this is also the case in the **Ptyctodontida**, where the cheek bones are substantially reduced (e.g., Trinajstic et al. 2012). The palatoquadrate also supports the dental plates in the Antiarchi (homologue of the suborbital cheek plate; Young 1984), Ptyctodontida and Arthrodira (the posterior dental plate; Fig. 2.5d). In the **Rhenanida**, upper dental plates appear to be absent (e.g., Young 1986; Lelièvre et al. 1995), while in the Acanthothoracida, the dental plates are currently only known in one specimen, where they articulate to the braincase and are also attached to dermal bones of the head shield (Ørvig 1975; Smith and Johanson 2003; Smith et al. 2017).

By comparison to this upper jaw diversity, the lower jaw in placoderms is very similar among the groups, with a cartilaginous Meckel's cartilage supporting a dermal element, the infragnathal (Miles 1969; Young 1984, 1986; Trinajstic et al. 2012; Hu et al. 2017); the lower jaws are unknown in the Acanthothoracida, and the jaws as a whole are unknown in the Petalichthyida.

As noted, by comparison to the jaws, identification of the **hyoid arch** is more problematic in the placoderms. In particular, the cartilage supporting the bony plate covering the gill arches, the **submarginal**, has been identified by different authors as either an **epihyal** (hyomandibular; Fig. 2.5d) or an **opercular** cartilage. The submarginal has a groove on the internal surface, with preserved perichondral bone in the Rhenanida (*Jagorina*), the Petalichthyida (Young 1986: Fig. 16), the Arthrodira (Young 1986: Fig. 14), the Acanthoracida (Young 1980: Fig. 17) and also *Entelognathus*, a placoderm with osteichthyan-type dermal jaw bones (Zhu et al. 2013; discussed further below). A groove is also present on the internal surface of the submarginal in the Antiarchi (Johanson and Young 1995: Fig. 3L). Goujet (1972, 1973, 1975, 1984; also Miles 1971; Forey and Gardiner 1986) identified

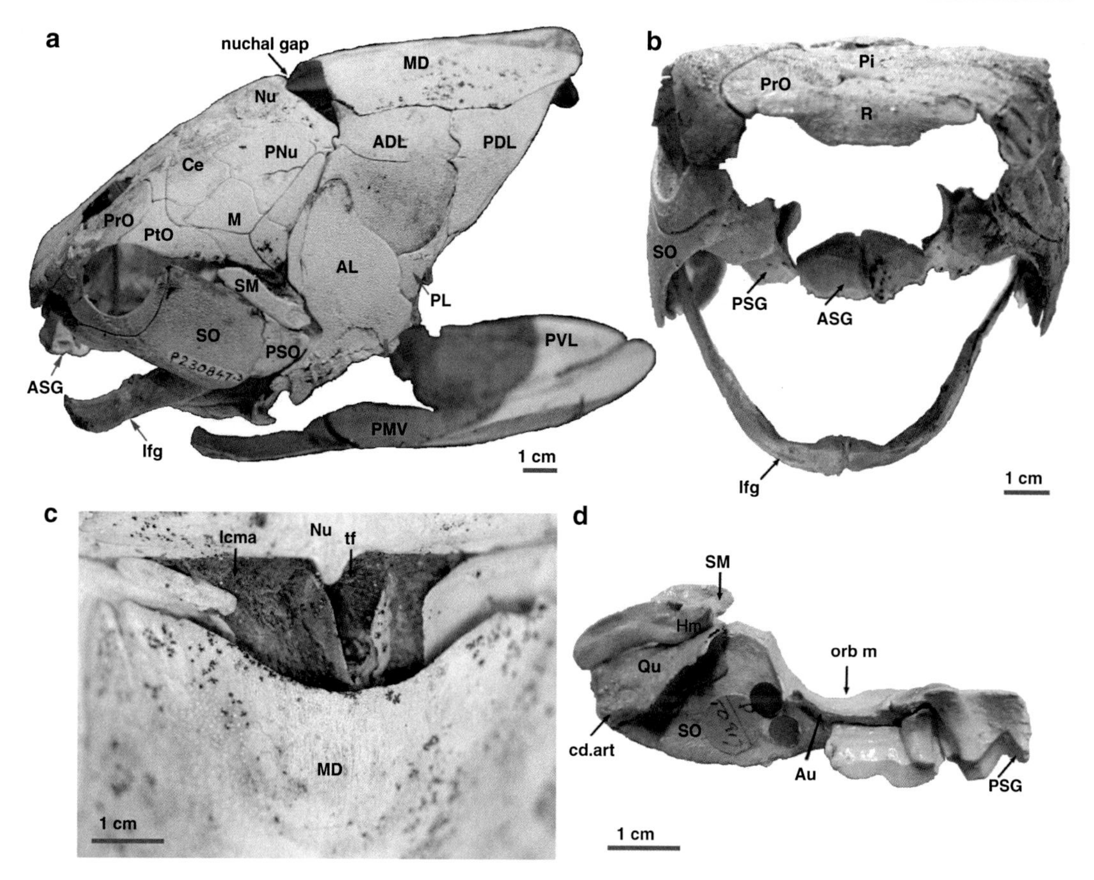

Fig. 2.5 (**a–c**) *Eastmanosteus calliaspis*—a coccosteomorph arthrodire from the Gogo Formation, Western Australia. (**a, b**) Museum of Victoria P 2308473, (**a**) lateral view; (**b**) anterior view; (**c**) Australian National University V2582, dorsal view showing the nuchal gap muscles; (**d**) *Torosteus pulchellus* (Western Australian Museum 88.2.7 = BMNH P50917), internal view of the cheek unit showing the hyomandibula and quadrate articulation. *ADL* anterior dorsolateral plate, *AL* anterior lateral plate, *ASG* anterior superognathal plate, *Au* autopalatine, *cd. art* articular condyle, *Ce* central plate, *Hm* hyomandibula, *Ifg* inferognathal plate, *lcma* levator capitis major, *M* marginal plate, *MD* median dorsal plate, *Nu* nuchal plate, *orb m* orbital margin, *PDL* posterior dorsolateral plate, *Pi* pineal plate, *PL* posterior lateral plate, *PMV* posterior median ventral plate, *PNu* paranuchal plate, *PrO* preorbital plate, *PSG* posterior superognathal, *PtO* postorbital plate, *PVL* posterior ventrolateral plate, *Qu* quadrate, *R* rostral plate, *SO* suborbital plate, *tf* transverse muscle fibres

submarginals with perichondral bone in a number of primitive arthrodires (Phlyctaenida) and interpreted the articulation of this bone on the braincase [anterior to a foramina for the branch of the facial nerve (VII) innervating the hyomandibular] as indicating that this was the epihyal rather than the opercular cartilage. The submarginal covered the gill arches, so there should also be an associated opercular cartilage, but in this reconstruction, a second braincase articulation was absent. Thus, the identity of this bone attached to the submarginal was problematic. Young (1986) instead suggested this perichondral bone represented the opercular cartilage in all placoderms, with the epihyal associated with, and positioned medial to, the opercular cartilage. However, the general absence of this epihyal and the putative lack of space for this epihyal (Gardiner and Miles 1990) meant the identity of the bone articulated to the submarginal remained open, including in recent phylogenetic analyses (Giles et al. 2015; King et al. 2016).

The well-preserved placoderms from the Late Devonian Gogo Formation (Western Australia) provided important new information in this regard. In the ptyctodont *Materpiscis*, Trinajstic et al. (2012: Fig. 2H) identified a narrow submarginal plate (typical for the reduced cheek in ptyctodonts) with a perichondral bone, associated with a second articulating bone, with these representing the opercular and epihyal bones, respectively. Most recently, Hu et al. (2017: Fig. 5b, c) provided a reexamination of a well-preserved specimen of a buchanosteid arthrodire and, via 3D prints of CT-scanned material, were able to demonstrate that when the submarginal with accompanying perichondral bone was placed in articulation with the braincase, space remained for a separate epihyal, along with a second braincase articulation. Thus, a separate epihyal/hyomandibula, not articulating to the submarginal, may be present in placoderms. Along with this, Hu et al. (2017) identified an **interhyal** associated with the hyoid arch in the buchanosteid, a bone otherwise only found in osteichthyans.

2.2.5 Mineralization Associated with Jaws

2.2.5.1 Organized Teeth and Dentitions in Early Vertebrates

In jawless vertebrates, mineralized elements in the oral and branchial cavities have been recognized in **conodonts** (Purnell and Donoghue 1997), heterostracans (Purnell 2002) and thelodonts (Van der Bruggen and Janvier 1993; Smith and Coates 1998; Märss and Wilson 2008), the latter being organized into whorl-like structures. Phylogenetically, these elements are convergent with respect to **teeth** in jawed vertebrates (Donoghue and Rücklin 2016), particularly in the conodonts, where taxa possessing oropharyngeal elements similar in composition to jawed vertebrate teeth were demonstrated to be phylogenetically derived within the group (Murdock et al. 2013).

Nevertheless, one feature of dentitions in crown group gnathostomes (Osteichthyes + Chondrichthyes; bony fishes + sharks, rays and holocephalans) is that the oral mineralized elements, or teeth, are organized in a spatiotemporal manner along the jaw and replaced. These characteristics allow the dentition to be highly functional through the life of the individual. This organization and replacement is controlled by a series of genes, as part of a **gene regulatory network,** including *ß-catenin*, *bmp2*, *bmp4*, *dlx2*, *fgf3*, *fgf10*, *notch2*, *pitx2*, *runx2* and *shh* (e.g., Fraser et al. 2009, 2010: Fig. 3), which is broadly shared with other oral structures (taste buds) and those externally, such as **scales** (Martin et al. 2016). These elements are known collectively as **odontodes**, deriving from interactions between the epithelial and mesenchymal tissues. One important feature of the teeth that external odontodes appear to lack is ongoing replacement, associated with the expression of the gene *Sox2* (Martin et al. 2016). Recent work on the shark dentition, where teeth are replaced in a continuous manner within a dental lamina, is highlighting the importance of the taste buds in tooth replacement, as the source of stem cells responsible for tooth regeneration (Martin et al. 2016). These cells are transferred from the taste bud niche to the deep successional lamina, where they contribute to the development of new teeth. The replacing dentition of extant chondrichthyans evolved by combining the regenerative ability of Sox2+ progenitors from the taste bud epithelium and the gene regulatory networks of odontodes (Martin et al. 2016). However, an ongoing question is when teeth and functional dentitions evolved, between the origins of the jawed vertebrate clade and crown group gnathostomes.

As mentioned, placoderms are a group of phylogenetically basal jawed vertebrates that may be either a monophyletic group (reviewed in Brazeau and Friedman 2014; also King et al. 2016) or a paraphyletic series of taxa at the base of the gnathostome clade (e.g., Brazeau 2009; Zhu et al. 2013; Coates et al. 2017; Fig. 2.1). All placoderms have jaws and dermal dentitions of some type (unknown in Petalichthyida), but only certain taxa within the Arthrodira have organized teeth comparable to the crown gnathostomes. For example, in *Compagopiscis croucheri*, new teeth are spatiotemporally organized, being sequentially added posteriorly to the ends of a restricted number of pre-existing functional rows of teeth (Smith and Johanson 2003; Johanson and Smith 2005; Rücklin et al. 2012). Temporal addition of teeth is also indicated by the progressive infilling of the pulp cavity of teeth, from the teeth closer to the jaw symphysis to those more posteriorly. However, these teeth do not appear to be replaced in the same manner as in crown gnathostome dentitions (e.g., Johanson and Trinajstic 2014), which suggests that tooth regeneration, as outlined by Martin et al. (2016), is a feature of the crown group, including chondrichthyans and bony fishes. Moreover, the discovery of two new Silurian placoderms from China (Zhu et al. 2013, 2016; see below) calls into question the homology of these arthrodiran dentitions with those of crown gnathostomes.

2.2.5.2 Homology of Placoderm Tooth Plates

Previously, the dorsal **tooth plates** in placoderms were thought to be homologous to internal bones of the roof of the mouth in bony fishes, including the vomers, coronoids, entopterygoids and dermopalatines. This was based on similarities in position and presence of teeth or toothlike structures. For example, the anterior tooth plates (**supragnathals**) were thought to be homologous to the anterior, paired vomers of osteichthyans (Stensiö 1963, 1969; Young 1986; Zhu et al. 2016).

Recently, two new placoderms, ***Entelognathus primordialis*** and ***Qilinyu rostrata,*** have been described, possessing typical placoderm head shield plates, but with the addition of bones previously associated with the osteichthyan skull and jaw, including the premaxilla, maxilla and dentary (Zhu et al. 2013, 2016). Phylogenetic analyses resolve these taxa more crownward relative to other placoderms, including the arthrodires.

The **maxilla** and **premaxillae** of *Entelognathus* and *Qilinyu* have an external component known as the facial lamina, as well as an internal, or palatal, lamina, the latter being broad. Other internal bones, such as the placoderm tooth plates described above, are absent. The palatal lamina itself is absent from the maxilla and premaxilla of osteichthyans. The dentary of *Qilinyu* has a mesial lamina, although this is absent in *Entelognathus*; the dentary of *Entelognathus* is described as possessing a 'biting ridge' as in bony fishes (Zhu et al. 2016). This suggests that *Entelognathus* is more closely related to crown group gnathostomes than *Qilinyu.*

Observing that the tooth plates in more phylogenetically basal placoderms such as the arthrodires lack the external facial lamina, Zhu et al. (2016) proposed that these tooth plates are not homologous to the internal roofing bones of bony fishes (the dental arcade; another difference is that there are more bones in this arcade than in the placoderm tooth plates) but instead are homologous to the internal palatal lamina of *Entelognathus* and *Qilinyu.* Otherwise the inner dental arcade would be present in basal placoderms, lost in *Entelognathus* and *Qilinyu* and then regained in osteichthyans, which they suggest is less parsimonious.

One implication of this interpretation is that the osteichthyan-type dermal jaw bones evolved even earlier than indicated by *Entelognathus* and *Qilinyu.* More importantly, the palatal laminae of *Entelognathus* and *Qilinyu* lack teeth, which suggests that teeth seen in the arthrodiran placoderms were lost at this point phylogenetically (or evolved separately in these arthrodires) and are not homologous to teeth in crown gnathostomes.

2.2.6 Musculature Associated with the Jaws, Hyoid Arch and Branchial Arches

An effective ventral musculature can rapidly drop the lower jaw and branchial arches, increasing suction, while ability to raise the lower jaw rapidly increases the probability of prey capture. Muscles raising the head can also affect an increase in the size of the gape. However, the ability to determine the musculature in fossils has been limited and largely dependent on the presence of muscle scars and comparison to extant relatives. Studies in the evolution of the jaw musculature have been particularly difficult because of the lack of extant comparative taxa for the agnathans and the earliest jawed vertebrates, the placoderms.

There is no evidence for **hypobranchial** or jaw-depressing musculature within jawless osteostracans, the group considered the most closely related to the jawed vertebrates (e.g., Fig. 2.1; Keating and Donoghue 2016). This assumption was based on the lack of muscle scars on the posterior wall of the osteostracan orobranchial chamber, the only site where these muscles could have attached. However, synchrotron microtomography has successfully identified muscle attachment sites where no muscle scars occur on the bone (Sanchez et al. 2013); so further analyses of osteostracans is warranted, particularly for penetration of Sharpey's fibres, anchoring musculature to the bone.

In addition to synchrotron microtomography, the preservation of mineralized soft tissues in the head and neck of some placoderms has greatly enhanced our understanding of the musculature of the head and jaws within arthrodires, and so basal jawed vertebrates. Synchrotron scans of anteroventral trunk shield plates (interolateral plate) in the arthrodire *Compagopiscis* demonstrated the insertion area for the **coracobranchialis** (a ventral branchial muscle) and the **coracomandibularis** and **coracohyoideus** muscles, both hypobranchial muscles (Sanchez et al. 2013). In extant chondrichthyans, the gill arch-depressing muscles originate on the midline coracoid bar of the scapulocoracoid; however, in placoderms, the origin is more lateral, on the upper part of the postbranchial lamina (Johanson 2003; Sanchez et al. 2013). It is proposed that, as in sharks, the coracomandibularis and coracohyoideus muscles acted independently of each other (Wilga et al. 2000), the coracomandibularis lowering the jaw (Heintz 1932; Miles and Westoll 1968; Johanson 2003) and the coracohyoideus depressing the hyoid arch (Johanson 2003). The position of the **coracobranchialis** is dorsal to the coracomandibularis and coracohyoideus; however it also originates on the postbranchial laminae (Sanchez et al. 2013). Thus, the hypobranchial musculature within placoderms is similar to that found in extant chondrichthyans.

The **levator arcus palatini**, a muscle involved in feeding, is attached to the ventral postocular process on the postorbital dermal plate in arthrodires and functions to expand the buccal cavity. Only the upper portion of the muscle is preserved, but as in other fishes, it is thought to insert on the hyoid arch (hyomandibula); as noted above, among placoderms, a separate hyoid bone was confirmed in ptyctodonts (Trinajstic et al. 2012) and arthrodires (Hu et al. 2017), suggesting this feeding muscle also evolved early in jawed vertebrate history.

The **mandibular adductor muscle** is the most powerful of the jaw muscles and acts to close the jaw. Within placoderms the infragnathal (lower jaw) is divided into an anterior biting division and a posterior blade, with the adductor mandibulae muscle inserting on a ridge on the latter; this is confirmed by the preservation of muscle on this region of the infragnathal of an arthrodire (pers. obs. KT, Western Australian Museum specimen WAM 10.1.1). Young et al. (2001) have suggested that the division of the infragnathal including a separate posterior section for muscle attachment may represent the ancestral condition for gnathostomes, and Maisey (1989) has noted a similar ridge for the adductor muscle insertion in some chondrichthyans. In placoderms, the **palatoquadrate** is cleaver-shaped and attaches to the dermal bones of the cheek, with the adductor muscle passing medially to it, whereas in extant gnathostomes, the adductor muscle is lateral to the palatoquadrate (e.g., Hu et al. 2017). Similarities between the jaw bones in extant osteichthyans and *Entelognathus* suggest that changes in muscle organization preceded the development of the cleaver-shaped palatoquadrate, common to placoderms at or immediately prior to the emergence of the crown gnathostomes (Zhu et al. 2013).

Constriction of the brachial muscles forces water across the gill chamber. Within arthrodires part of the dorsal branchial constrictor muscle is identified on the visceral surface of the paranuchal plate, lateral to the **cucullaris** muscle (Trinajstic et al. 2013: Fig. S1 I-J). In extant sharks this muscle originates on the cucullaris and inserts on the branchial arch and to the pectoral girdle through a tendon (Trinajstic et al. 2013: SI). The portion of the branchial constrictor muscle preserved in placoderms suggests the origin and insertion were similar in placoderms and chondrichthyans.

2.2.7 Evolution of a Neck and a Moveable Head

The evolution of a head that is moveable on the vertebral column has numerous benefits, including the ability to produce a larger oral cavity for food and water intake. Although a series of bones joins the pectoral girdle to the bones at the rear of the skull in bony fishes, there is movement between these bones, along their overlap surfaces. In bony fishes the separation of the skull and the pectoral girdle is associated with muscles that raise the head, counteracted by muscles that draw the head downward. Along with muscularized jaws and dentitions, this movement at a functional **'neck'** make crown group gnathostomes highly effective feeders.

When considering the evolution of a **moveable neck** in a phylogenetic context, it appears that the neck evolved along with jaws at the base of the jawed vertebrate clade, in the placoderms (Ericsson et al. 2013; Trinajstic et al. 2013; Kuratani et al. 2018). The sister group to the jawed vertebrates is currently debatable (e.g., Zhu et al. 2016), with the galeaspids sharing widely separated nasal sacs with jawed vertebrates, as noted above, but lacking a pectoral fin supported by a fin radial articulating to a pectoral girdle. The latter characterizes the Osteostraci (Janvier 1985; Janvier et al. 2004) and is shared with jawed vertebrates. However, the pectoral girdle is surrounded by perichondral and dermal bone that is continuous with the braincase and large bony head shield. In other words, the neck is absent. The heart is also enclosed posteriorly within the **head shield** (Janvier 1981, 1985).

In the placoderms, separate trunk shield bony plates surround the anterior part of the body, including the pectoral girdle. The trunk shield articulates with the bony plates of the head shield, such that the head shield is moveable on the trunk shield (Fig. 2.5a, b). Along with this, musculature is required to move the head shield. Using the exquisitely preserved placoderm specimens from the Upper Devonian Gogo Formation, Trinajstic et al. (2013) identified levator muscles running dorsally between the head and trunk shield (Fig. 2.5c), along with a cucullaris muscle extending from the lateral head shield to the anterolateral trunk shield. The latter muscle actively depresses the head shield, returning it to its original position.

New research is providing a better understanding of the development of the cucullaris. Some of the earliest studies (e.g., Edgeworth 1935) identified the cucullaris as developing from mesoderm intercalated between postotic somites and the branchial arches. More recently, the cucullaris was believed to derive from anterior paraxial mesoderm (somites; Noden 1983; Couly et al. 1993; Piekarski and Olsson 2007), but as summarized by Ericsson et al. (2013), and most recently demonstrated for the axolotl by Sefton et al. (2016), the cucullaris derives from cranial lateral plate mesoderm adjacent and ventral to somites 1–3 (see also Chap. 7). Lateral plate mesoderm is not well differentiated from the paraxial mesoderm in this region (Noden 1988), but the cucullaris also shows genetic pathways more similar to the head, than trunk, musculature (Theis et al. 2010). Somites 1–3 represent the occipital somites, with more posterior somites differentiating to form the vertebral column and the majority of the postcranial musculature. Being derived from the cranial mesoderm, the cucullaris was said to be similar to posterior branchial arch levators, representing part of the serial arch musculature (Sefton et al. 2016). By comparison, in chondrichthyans, the cucullaris was said to develop in the dorsal region of the branchial arches, with no contribution from somites (Ziermann et al. 2017). In the mouse, the muscles homologous to the cucullaris, the **trapezius** and **sternocleidomastoideus**, are also non-somitic and derived from branchial muscles and share a gene regulatory network with cardiac muscle progenitor cells (Lescroart et al. 2015).

The levator and cucullaris muscles in placoderms differ considerably from all known extant gnathostomes. Paired **levator *capitis* major** and **minor** muscles span the neck joint and function as head elevators (Trinajstic et al. 2013; Fig. 2.5c). Prior to the discovery of arthrodire specimens with the musculature preserved, a single levator muscle was proposed (Miles 1969), and the cucullaris muscle was reconstructed as in a shark with origin depicted at the dorsal longitudinal bundle fascia and the insertions depicted at the epibranchial cartilage and scapular process. However, the preserved cucullaris muscle in placoderms showed that it spanned the neck joint, the origin being in the cucullaris fossa on the visceral surface of the paranuchal plate at the posterior margin of the head and the insertion dorsal to the neck joint on the anterior dorsolateral plate on the trunk shield. The position of the cucullaris muscle indicated its function in placoderms is to depress the head at the unique hinge joint between the head and trunk armour (Miles 1969). In chondrichthyans the cucullaris muscle elevates both the gill arches and the pectoral girdle.

2.3 Conclusions

The origin of jaws was fundamental to the evolutionary success of the gnathostomes and has been the subject of numerous hypotheses involving both palaeontological and developmental evidence. Also to be considered is the evolution of a moveable and functional neck region, allowing the head to be moved upward and the jaw dropped, increasing the size of the oral opening and so the size of food that could be ingested. How this separation occurred from jawless vertebrates such as the Galeaspida and Osteostraci, with continuous bony head shields and pectoral fin girdles, is an area of exciting current research, associated with the origin of the cucullaris muscle. Another area of current research, both palaeontological and developmental, is the origins of teeth and dentitions, crucial for food processing. New fossil discoveries from China make a reevaluation of tooth homologies in early vertebrates necessary, while molecular studies on shark dentitions are assembling the tissue and cellular interactions and the genes involved in tooth development in model sharks.

Acknowledgments CB is supported by the Curtin Research Fellowship and the Australian Research Council grant DP 160104427 and KT by DP140104161. We thank the Western Australian Museum and the Natural History Museum (UK) for access to specimens. We wish to thank Janine Ziermann, Rui Diogo and Raul Diaz Jr for inviting us to contribute to this volume. We would like to thank Tatsuya Hirasawa and Carole Burrow for their constructive reviews which improved the chapter.

Further Reading

Origin of jaws: https://sites.ualberta.ca/~tetsuto/Trajectory_of_a_Curious_Mind/Fishing_for_Jaws.html

Long JA (2011) The rise of fishes: 500 million years of evolution. The John Hopkins University Press, Baltimore

References

Ayers H (1921) Vertebrate Cephalogenesis V. Origin of jaw apparatus and trigeminus complex—Amphioxus, Ammocoetes, *Bdellostoma*, *Callorhynchus*. J Comp Neurol 33:339–404

Barreiro-Iglesias A, Romaus-Sanjurjo D, Senra-Martínez P, Anadón R, Rodicio MC (2011) Doublecortin is expressed in trigeminal motoneurons that innervate the velar musculature of lampreys: considerations on the evolution and development of the trigeminal system. Evol Dev 13:149–158

Blom H (2012) New Birkeniid anaspid from the Lower Devonian of Scotland, and its phylogenetic implications. Palaeontology 55:641–652

Brazeau M (2009) The braincase and jaws of a Devonian 'acanthodian' and modern gnathostome origins. Nature 457:305–308

Brazeau M, Friedman M (2014) The characters of Palaeozoic jawed vertebrates. Zool J Linn Soc 170:779–821

Brazeau MD, Friedman M, Jerve A, Atwood RC (2017) A three-dimensional placoderm (stem-group gnathostome) pharyngeal skeleton and its implications for primitive gnathostome pharyngeal architecture. J Morphol 278:1220–1228

Cattell M, Lai S, Cerny R, Medeiros DM (2011) A new mechanistic scenario for the origin and evolution of vertebrate cartilage. PLoS One 6(**7**):e22474

Cerny R, Lwigale P, Ericsson R, Meulemans D, Epperlein HH, Bronner-Fraser M (2004) Developmental origins and evolution of jaws: new interpretation of "maxillary" and "mandibular". Dev Biol 276:225–236

Cerny R, Cattell M, Sauka-Spengler T, Bronner-Fraser M, Yu F, Medeiros DM (2010) Evidence for the

prepattern/cooption model of vertebrate jaw evolution. Proc Natl Acad Sci U S A 107:17262–17267
Choe CP, Crump JG (2014) Tbx1 controls the morphogenesis of pharyngeal pouch epithelia through mesodermal Wnt11r and Fgf8a. Development 141:3583–3593
Coates MI, Gess RW, Finarelli JA, Criswell KE, Tietjen K (2017) A symmoriiform chondrichthyan braincase and the origin of chimaeroid fishes. Nature 541:208–211
Conway Morris S, Caron J-B (2014) A primitive fish from the Cambrian of North America. Nature 512:419–422
Couly GF, Coltey PM, Le Douarin NM (1993) The triple origin of skull in higher vertebrates: a study in quail-chick chimeras. Development 117:409–429
Crump JG, Maves L, Lawson ND, Weinstein BM, Kimmel CB (2004) An essential role for Fgfs in endodermal pouch formation influences later craniofacial skeletal patterning. Development 131:5703–5716
De Beer GR (1937) The development of the vertebrate skull. Oxford University Press, London
Depew MJ, Lufkin T, Rubenstein JL (2002) Specification of jaw subdivisions by Dlx genes. Science 298:381–385
Depew MJ, Simpson CA, Morasso M, Rubenstein JLR (2005) Reassessing the *Dlx* code: the genetic regulation of branchial arch skeletal pattern and development. J Anat 207:501–561
Donoghue PCJ, Purnell MA (2005) Genome duplication, extinction and vertebrate evolution. Trends Ecol Evol 20:312–319
Donoghue PCJ, Rücklin M (2016) The ins and outs of the evolutionary origin of teeth. Evol Dev 18:19–30
Donoghue PCJ, Smith MP (2001) The anatomy of *Turinia pagei* (Powrie), and the phylogenetic status of the Thelodonti. Trans R Soc Edinburgh: Earth Sci 92:15–37
Dupret V, Sanchez S, Goujet D, Tafforeau R, Ahlberg PE (2014) A primitive placoderm sheds light on the origin of the jawed vertebrate face. Nature 507:500–503
Edgeworth FH (1935) The cranial muscles of vertebrates. Cambridge University Press, Cambridge
Ericsson R, Knight R, Johanson Z (2013) Evolution and development of the vertebrate neck. J Anat 222:67–78
Forey PL (1995) Agnathans recent and fossil, and the origin of jawed vertebrates. Rev Fish Biol Fishe 5:267–303
Forey PL, Gardiner BG (1986) Observations on *Ctenurella* (Ptyctodontida) and the classification of placoderm fishes. Zool J Linn Soc 86:43–74
Fraser GJ, Hulsey CD, Bloomquist RF, Uyesugi K, Manley NR, Streelman JT (2009) An ancient gene network is co-opted for teeth on old and new jaws. PLoS Biol **7**(**2**):e1000031
Fraser GJ, Cerny R, Soukup V, Bronner Fraser M, Streelman JT (2010) The odontode explosion: The origin of tooth-like structures in vertebrates. Bioessays 32:808–817
Fujimoto S, Oisi Y, Kuraku S, Ota KG, Kuratani S (2013) Non-parsimonious evolution of hagfish Dlx genes. BMC Evol Biol 13:15
Gagnier PY (1989) The oldest vertebrate: a 470-million-year-old jawless fish, *Sacabambaspis janvieri*, from the Ordovician of South America. Nat Geog Res 5:250–253
Gagnier PY, Blieck A (1992) On *Sacabambaspis janvieri* and the vertebrate diversity in Ordovician Seas. In: Kurik E-M (ed) Fossil fishes as living animals. Academy of Sciences of Estonia, Tallinn, pp 9–20
Gai Z, Zhu M (2012) The origin of the vertebrate jaw: Intersection between developmental biology-based model and fossil evidence. Chin Sci Bull 57:1–10
Gai Z, Donoghue PCJ, Zhu M, Janvier P, Stampanoni M (2011) Fossil jawless fish from China foreshadows early jawed vertebrate anatomy. Nature 476:324–327
Gans C, Northcutt RG (1983) Neural crest and the origin of vertebrates: a new head. Science 220:268–274
Gardiner BG, Miles RS (1990) A new genus of eubrachythoracid arthrodire from Gogo, Western Australia. Zool J Linn Soc 99:59–204
Gegenbaur C (1859) Grundzüge der vergleichenden Anatomie. Wilhelm Engelmann, Leipzig
Gegenbaur C (1878) Elements of comparative anatomy. Macmillan, London, UK. [English translation of Gegenbaur 1859]
Giles S, Friedman M, Brazeau MD (2015) Osteichthyan-like cranial conditions in an Early Devonian stem gnathostome. Nature 520:82–85
Gillis JA, Modrell MS, Baker CVH (2013) Developmental evidence for serial homology of the vertebrate jaw and gill arch skeleton. Nat Commun 4:1436
Goujet D (1972) Nouvelles observations sur la joue d'*Arctolepis* (Eastman) et d'autres Dolichothoraci. Ann Paleontol 58:1–10
Goujet D (1973) *Sigaspis*, un nouvel arthrodire du Dévonien inférieur du Spitsberg. Palaeontograph A 143:73–88
Goujet D (1975) *Dicksonosteus*, un nouvel arthrodire du Dévonien du Spitsberg. Remarques sur le squelette viscéral des Dolichothoraci. Colloq Int Centre Natl Recher Sci 218:81–99
Goujet D (1984) Les Poissons Placoderms du Spitsberg. Centre National de la Recherche Scientifique, Paris, p 284
Green SA, Simoes-Costa M, Bronner ME (2015) Evolution of vertebrates: a view from the crest. Nature 520:474–482
Hall BK (2009) The neural crest and neural crest cells in vertebrate development and evolution. Springer, New York
Hall BK, Gillis A (2013) Incremental evolution of the neural crest, neural crest cells and neural crest-derived skeletal tissues. J Anat 222:19–31
Heintz A (1932) The structure of *Dinichthys* contribution to our knowledge of the Arthrodira. In: Gudger EW (ed) Archaic fishes. bashford dean memorial volume. Article 4, pp 115–224

Holzschuh J, Wada N, Wada C et al (2005) Requirements for endoderm and BMP signalling in sensory neurogenesis in zebrafish. Development 132:3731–3742

Hu Y, Lu J, Young GC (2017) New findings in a 400 million-year-old Devonian placoderm shed light on jaw structure and function in basal gnathostomes. Sci Rep **7**:7813

Hunt P, Gulisano M, Cook M, Sham M-H, Faiella A, Wilkinson D, Boncinelli E, Krumlauf R (1991) A distinct *Hox* code for the branchial region of the vertebrate head. Nature 353:861–864

Hurley IA et al (2007) A new time-scale for ray-finned fish evolution. Proc R Soc Lond B 274:489–498

Jandzik D, Garnett AT, Square TA, Cattell MV, J-Km Y, Medeiros DM (2015) Evolution of the new vertebrate head by co-option of an ancient chordate skeletal tissue. Nature 518:534–537

Janvier P (1981) *Norselaspis glacialis* n.g., n.sp, et les relations phylogénétiques entre les Kiaeraspidiens (Osteostraci) du Dévonien inférieur du Spitsberg. Palaeovertebrata 11:19–131

Janvier P (1985) Les Céphalaspides du Spitsberg. Anatomie, phylogénie et systemématique des Ostéostracés siluro-dévoniens. Révision des Ostéostracés de la Formation de Wood Bay (Dévonian inférieur du Spitsberg). Centre National de la Recherche Scientifique, Paris, p 252

Janvier P (1993) Patterns of diversity in the skull of jawless fishes. In: Hanken J, Hall BK (eds) The skull: patterns of structural and systematic diversity, vol Vol. II. The University of Chicago Press, Chicago, pp 131–188

Janvier P (1996) Early vertebrates. Oxford monographs in geology. Oxford University Press, Oxford

Janvier P, Arsenault M (2007) The anatomy of *Euphanerops longaevus* Woodward, 1900, an anaspid-like jawless vertebrate from the Upper Devonian of Miguasha, Quebec, Canada. Geodiversitas 29:143–216

Janvier P, Arsenault M, Desbiens S (2004) Calcified cartilage in the paired fins of the osteostracan *Escuminaspis laticeps* (Traquair 1880), from the Late Devonian of Miguasha (Quebec, Canada), with a consideration of the early evolution of the pectoral fin endoskeleton in vertebrates. J Vert Paleontol 24:773–779

Johanson Z (2003) Placoderm branchial and hypobranchial muscles and origins in jawed vertebrates. J Vert Paleo 23:735–749

Johanson Z, Smith MM (2005) Origin and evolution of gnathostome dentitions: a question of teeth and pharyngeal denticles in placoderms. Biol Rev Camb Philos Soc 80:303–345

Johanson Z, Trinajstic K (2014) Fossilized ontogenies: the contribution of placoderm ontogeny to our understanding of the evolution of early gnathostomes. Palaeontology 57:505–516

Johanson Z, Young GC (1995) New (Antiarchi: Placodermi) from the Braidwood region, New South Wales, Australia (Middle-Late Devonian). Rec West Austral Mus Suppl 57:55–75

Keating JN, Donoghue PCJ (2016) Histology and affinity of anaspids, and the early evolution of the vertebrate dermal skeleton. Proc R Soc B 283:20152917

King B, Qiao T, Lee MSY, Zhu M, Long JA (2016) Bayesian morphological clock methods resurrect placoderm monophyly and reveal rapid early evolution in jawed vertebrates. Syst Biol 66:499–516

Knight RD, Schilling TF (2013) Cranial neural crest and development of the head skeleton. Madame Curie Bioscience Database

Kriwet J, Witzmann F, Klug S, Heidtke UHJ (2008) First direct evidence of a vertebrate three-level trophic chain in the fossil record. Proc R Soc B 275:181–186

Kuraku S, Takio Y, Sugahara F, Takechi M, Kuratani S (2010) Evolution of oropharyngeal patterning mechanisms involving *Dlx* and endothelins in vertebrates. Dev Biol 341:315–323

Kuratani S (2004) Evolution of the vertebrate jaw: comparative embryology and molecular developmental biology reveal the factors behind evolutionary novelty. J Anat 205:335–347

Kuratani S (2005) Developmental studies of the lamprey and hierarchical evolutionary steps towards the acquisition of the jaw. J Anat 207:489–499

Kuratani S (2012) Evolution of the vertebrate jaw from developmental perspectives. Evol Dev 14:76–92

Kuratani S, Ahlberg PE (2018) Evolution of the vertebrate neurocranium: problems of the premandibular domain and the origin of the trabecula. Zool Lett 4:1

Kuratani S, Nobusada Y, Horigome N, Shigetani Y (2001) Embryology of the lamprey and evolution of the vertebrate jaw: insights from molecular and developmental perspectives. Philos Trans R Soc Lond B Biol Sci 356:15–32

Kuratani S, Murakami Y, Nobusada Y, Kusakabe R, Hirano S (2004) Developmental fate of the mandibular mesoderm in the lamprey, Lethenteronjaponicum: Comparative morphology and development of the gnathostome jaw with special reference to the nature of the trabecula cranii. J Exp Zool B Mol Dev Evol 302:458–468

Kuratani S, Adachi N, Wada N, Oisi Y, Sugahara F (2013) Developmental and evolutionary significance of the mandibular arch and prechordal/premandibular cranium in vertebrates: revising the heterotopy scenario of gnathostome jaw evolution. J Anat 222:41–55

Kuratani S, Kusakabe R, Hirasawa T (2018) The neural crest and evolution of the head/trunk interface in vertebrates. Dev Biol. https://doi.org/10.1016/j.ydbio.2018.01.017

Lakiza O, Miller S, Bunce A, Myung-Jae Lee E, McCauley DW (2011) SoxE gene duplication and development of the lamprey branchial skeleton: insights into development and evolution of the neural crest. Dev Biol 359:149–161

Le Douarin N (1982) The neural crest. Cambridge University Press, Cambridge

Lelièvre H, Janvier P, Janjou D, Halawani M (1995) *Nefudina qalibahensis* nov. gen., nov. sp. un rhenanide (Vertebrata,

Placodermi) du Dévonien inférieur de la Formation Jauf (Emsien) d'Arabie Saoudite. Geobios 28:109–115

Lescroart F, Hamou W, Francou A, Théveniau-Ruissy M, Kelly RG, Buckingham M (2015) Clonal analysis reveals a common origin between nonsomite-derived neck muscles and heart myocardium. Proc Natl Acad Sci U S A 112:1446–1451

Maisey JG (1989) Visceral skeleton and musculature of a Late Devonian shark. J Vert Paleo 9:174–190

Mallatt J (1996) Ventilation and the origin of jawed vertebrates: a new mouth. Zool J Linn Soc 117:329–404

Marinelli W, Strenger A (1954) Vergleichende Anatomie und Morphologie der Wirbeltiere. 1. *Lampetra fluviatilis*. Franz Deuticke, Wien

Marinelli W, Strenger A (1956) Vergleichende Anatomie und Morphologie der Wirbeltiere. 2. *Myxine glutinosa*. Franz Deuticke, Wien

Märss T, Wilson MVH (2008) Buccopharyngo-branchial denticles of *Phlebolepis elegans* Pander (Thelodonti, Agnatha). J Vert Paleo 28:601–612

Martin KJ, Rasch LJ, Cooper RL, Metscher BD, Johanson Z, Fraser GJ (2016) Sox2+ progenitors in sharks link taste development with the evolution of regenerative teeth from denticles. Proc Natl Acad Sci U S A 113:14769–14774

McCauley DW, Bronner-Fraser M (2003) Neural crest contributions to the lamprey head. Development 130:2317–2327

Medeiros DM, Crump JG (2012) New perspectives on pharyngeal dorsoventral patterning in development and evolution of the vertebrate jaw. Dev Biol 371:121–135

Miles RS (1969) Features of placoderm diversification and the evolution of the arthrodire feeding mechanism. Trans R Soc Edinburgh 68:123–170

Miles RS (1971) The Holonematidae (placoderm fishes): a review based on new specimens of *Holonema* from the Upper Devonian of Western Australia. Philos Trans R Soc Lond B 263:101–234

Miles RS, Westoll TS (1968) The placoderm fish *Coccosteus cuspidatus* Miller ex Agassiz from the Middle Old Red Sandstone of Scotland. Part I. Descriptive morphology. Trans R Soc Edinburgh 67:373–476

Minoux M, Antonarakis GS, Kmita M, Duboule D, Rijli FM (2009) Rostral and caudal pharyngeal arches share a common neural crest ground pattern. Development 136:637–645

Miyashita T (2016) Fishing for jaws in early vertebrate evolution: a new hypothesis of mandibular confinement. Biol Rev 91:611–657

Murdock DJE, Dong XP, Repetski JE, Marone F, Stampanoni M, Donoghue PCJ (2013) The origin of conodonts and of vertebrate mineralized skeletons. Nature 502:546–549

Nichols JT, Pan L, Moens CB, Kimmel CB (2013) *barx1* represses joints and promotes cartilage in the craniofacial skeleton. Development 140:2765–2775

Noden DM (1983) The role of the neural crest in patterning of avian cranial skeletal, connective, and muscle tissues. Dev Biol 96:144–165

Noden DM (1988) Interactions and fates of avian craniofacial mesenchyme. Development 103: 121–140

Northcutt RG (2005) The new head hypothesis revisited. J Exp Zool B Mol Dev Evol 304B:274–297

Oisi Y, Ota KG, Kuraku S, Fujimoto S, Kuratani S (2013) Craniofacial development of hagfishes and the evolution of vertebrates. Nature 493:175–180

Ørvig T (1975) Description, with special reference to the dermal skeleton, of a new radotinid arthrodire from the Gedinnian of Arctic Canada. Colloq Internat Centre Nat Rech Sci 218:41–71

Patthey C, Schlosser G, Shimeld SM (2014) The evolutionary history of vertebrate cranial placodes—I: Cell type evolution. Dev Biol 389:82–97

Piekarski N, Olsson L (2007) Muscular derivatives of the cranialmost somites revealed by long-term fate mapping in the Mexican axolotl (*Ambystoma mexicanum*). Evol Dev 9:566–578

Purnell MA (2002) Feeding in extinct jawless heterostracan fishes and testing scenarios of early vertebrate evolution. Proc R Soc Lond B 269:83–88

Purnell MA, Donoghue PCJ (1997) Architecture and functional morphology of the skeletal apparatus of ozarkodinid conodonts. Proc R Soc Lond B 352:1545–1564

Ritchie A (1967) *Ateleaspis tessellata* Traquair, a non-cornuate cephalaspid from the Upper Silurian of Scotland. Zool J Linn Soc 47:69–81

Ritchie A (1980) The late Silurian anaspid genus *Rhyncholepis* from Oesel, Estonia, and Ringerike, Norway. Am Mus Novit 2699:1–18

Rücklin M, Donoghue PCJ, Johanson Z, Trinajstic K, Marone F, Stampanoni M (2012) Development of teeth and jaws in the earliest jawed vertebrates. Nature 491:748–751

Sanchez S, Dupret V, Tafforeau P, Trinajstic KM, Ryll B, Gouttenoire PJ, Wretman L, Zylberberg L, Peyrin F, Ahlberg PE (2013) 3D microstructural architecture of muscle attachments in extant and fossil vertebrates revealed by synchrotron microtomography. PLoS One 8:e56992

Schaeffer B, Thomson KS (1980) Reflections on Agnathan-Gnathostome relationships. In: Aspects of Vertebrate History, Essays in Honor of Edwin Harris Colbert. Museum of Northern Arizona Press, Flagstaff, pp 19–34

Schlosser G (2006) Induction and specification of cranial placodes. Dev Biol 294:303–351

Schlosser G (2010) Making senses: development of vertebrate cranial placodes. Int Rev Cell Mol Biol 283:129–234

Schlosser G, Patthey C, Shimeld SM (2014) The evolutionary history of vertebrate cranial placodes II. Evolution of ectodermal patterning. Dev Biol 389:98–119

Sefton EM, Bhullar BA, Modaddes Z, Hanken J (2016) Evolution of the head-trunk interface in tetrapod vertebrates. Elife 5:e09972

Sewertzoff AN (1911) Die Kiemenbogennerven der Fische. Anat Anz 38:487–495

Shigetani Y, Sugahara F, Kawakami Y, Murakami Y, Hirano S, Kuratani S (2002) Heterotopic shift of epithelial–mesenchymal interactions for vertebrate jaw evolution. Science 296:1319–1321

Shigetani Y, Sugahara F, Kuratani S (2005) A new evolutionary scenario for the vertebrate jaw. Bioessays 27:331–338

Shone V, Graham A (2014) Endodermal/ectodermal interfaces during pharyngeal segmentation in vertebrates. J Anat 225:479–491

Shu D-G, Conway Morris S, Han J, Zhang Z-F, Yasui K, Janvier P, Chen L, Zhang X-L, Liu J-N, Li Y, Liu H-Q (2003) Head and backbone of the Early Cambrian vertebrate *Haikouichthys*. Nature 421:526–529

Smith MM, Coates MI (1998) Evolutionary origins of the vertebrate dentition: phylogenetic patterns and developmental evolution. Eur J Oral Sci 106:482–500

Smith MM, Johanson Z (2003) Separate evolutionary origins of teeth from evidence in fossil jawed vertebrates. Science 299:1235–1236

Smith MM, Clark B, Goujet D, Johanson Z (2017) Evolutionary origins of teeth in jawed vertebrates: conflicting data from acanthothoracid dental plates ('Placodermi'). Palaeontology 60:829–836

Square T, Jandzik D, Romášek M, Cerny R, Medeiros DM (2015) The origin and diversification of the developmental mechanisms that pattern the vertebrate head skeleton. Dev Biol 427:219–229

Stensiö EA (1927) The Devonian and Downtonian vertebrates of Spitsbergen. 1. Family Cephalaspidae. Skrift Svalbard Ish 12:1–391

Stensiö EA (1963) Anatomical studies on the arthrodiran head. Part 1. Preface, geological and geographical distribution, the organization of the head in the Dolichothoraci, Coccosteomorphi and Pachyosteomorphi. Kungl Svenska Vetenskaps Handl 9:1–419

Stensiö EA (1969) *Traité de Paléontologie*, vol Vol. 4(2). Masson, Paris, pp 71–692

Steventon B, Mayor R, Streit A (2014) Neural crest and placode interaction during the development of the cranial sensory system. Dev Biol 389:28–38

Stolfi A, Ryan K, Meinertzhagen IA, Christiaen L (2015) Migratory neuronal progenitors arise from the neural plate borders in tunicates. Nature 527:371

Takio Y, Pasqualetti M, Kuraku S, Hirano S, Rijlii FM, Kuratani S (2004) Evolutionary biology: lamprey Hox genes and the evolution of jaws. Nature 429:262

Takio Y, Kuraku S, Murakami Y, Pasqualetti M, Rijli FM, Narita Y, Kuratani S, Kusakabe R (2007) Hox gene expression patterns in *Lethenteron japonicum* embryos: insights into the evolution of the vertebrate Hox code. Dev Biol 308:606–620

Theis S, Patel K, Valasek P et al (2010) The occipital lateral plate mesoderm is a novel source for vertebrate neck musculature. Development 137:2961–2971

Trinajstic K, Long J, Johanson Z, Young G, Senden T (2012) New morphological information on the ptyctodontid fishes (Placodermi, Ptyctodontida) from Western Australia. J Vert Paleo 32:757–780

Trinajstic K, Sanchez S, Dupret V et al (2013) Fossil musculature of the most primitive jawed vertebrates. Science 341:160–164

Van der Bruggen W, Janvier P (1993) Denticles in thelodonts. Nature 364:107

Wilga CD, Wainwright PC, Motta PJ (2000) Evolution of jaw depression mechanics in aquatic vertebrates: insights from Chondrichthyes. Biol J Linn Soc 71:165–185

Wilson MVH, Caldwell MW (1993) New Silurian and Devonian fork-tailed 'thelodonts' are jawless vertebrates with stomachs and deep bodies. Nature 361:442–444

Young GC (1980) A new early Devonian placoderm from New South Wales, Australia, with a discussion of placoderm phylogeny. Palaeont Abt A 167:10–76

Young GC (1984) Reconstruction of the jaws and braincase in the Devonian placoderm fish *Bothriolepis*. Palaeontology 27:635–661

Young GC (1986) The relationships of placoderm fishes. Zool J Linn Soc 88:1–57

Young GC, Lelievre H, Goujet D (2001) Primitive jaw structure in an articulated brachythoracid arthrodire (placoderm fish; early Devonian) from southeastern Australia. J Vert Paleo 21:670–678

Zhang GJ, Miyamito MM, Cohn MJ (2006) Lamprey type II collagen and *Sox9* reveal an ancient origin of the vertebrate collagenous skeleton. Proc Natl Acad Sci 103:3180–3185

Zhu M, Ahlberg PE, Pan Z, Zhu Y, Qiao T, Zhao W, Jia L, Lu J (2013) A Silurian maxillate placoderm illuminates jaw evolution. Science 354:334–336

Zhu Y-A, Zhu M, Wang J-Q (2016) Redescription of *Yinostius major* (Arthrodira: Heterostiidae) from the Lower Devonian of China, and the interrelationships of Brachythoraci. Zool J Linn Soc 176:806–834

Ziermann JM, Freitas R, Diogo R (2017) Muscle development in the shark Scyliorhinus canicula: implications for the evolution of the gnathostome head and paired appendage musculature. Front Zool 14:31

Cranium, Cephalic Muscles, and Homologies in Cyclostomes

3

Janine M. Ziermann

3.1 Introduction

There are two major extant vertebrate groups: jawed and jawless vertebrates ("gnathostomes" and cyclostomes, respectively). The former includes jawless fossil taxa. The "ostracoderms" (i.e., arandaspids, heterostracans, thelodonts, galeaspids, osteostracans, pituriaspids; see Chap. 2) are currently regarded as a paraphyletic group which are characterized by having an array of bone- and dentine-producing tissues and are therefore viewed as jawless stem gnathostomes (e.g., Janvier 2008). Here, I continue to use the term gnathostome throughout the chapter as jawed vertebrates are almost identical with modern gnathostomes. **Cyclostomes** are animals that on first sight resemble giant worms (Fig. 3.1) and comprise hagfishes (**Myxiniformes**) and lampreys (**Petromyzontiformes**) (Heimberg et al. 2010). Their name indicates the presence of a round mouth (Fig. 3.1e, f), and they are often grouped with other jawless extinct vertebrates in the paraphyletic group **agnathans** (see Chap. 2), i.e., vertebrates without jaws, from which jawed vertebrates diverged 430–520 million years ago. The jawless vertebrates were diverse during the mid-Paleozoic, but only lampreys and hagfishes are still extant (Potter 1980). It is not the intention of this chapter to analyze the relationship between fossil and/or extant hagfishes and lampreys. Information about the fossil record can be found in diverse literature (e.g., Gess et al. 2006 and citations within).

There are currently 38 extant **lamprey** species known, which live in the sea but spawn in rivers (Gee 2018). Larval lampreys are commonly known as "ammocoetes" (Fig. 3.1d) because they were erroneously regarded as adult forms (Leach 1944). Lampreys are distributed antitropical and the distribution is dependent on the lethal temperature of the ammocoetes which lies between 28° C and 32° C (Potter 1980). As adults, lampreys are parasitic or nonparasitic, with the latter being marked by an extended larval live, reduced postmetamorphic time, and smaller adult size (Potter 1980). Furthermore, nonparasitic forms do not feed during postlarval life (Potter 1980). Lamprey larvae live burrowed in river mud with their front end exposed to the water from which they filter particles. Larval and adult lampreys are often characterized by their mouthparts (dentition, tentacles), length, coloration, and "tongue" precursor/lingual apparatus. The body proportions are also important to distinguish different life stages from larval to adult specimens (Potter 1980). Adult parasitic lampreys have a circular sucker with many teeth (Fig. 3.1e) and a tongue that also contains teeth, which can be protruded from the mouth to grab onto passing fishes to rib chunks out of them. They only have a single,

J. M. Ziermann
Department of Anatomy, Howard University College of Medicine, Washington, DC, USA
e-mail: janine.ziermann@howard.edu

J. M. Ziermann et al. (eds.), *Heads, Jaws, and Muscles*, Fascinating Life Sciences,
https://doi.org/10.1007/978-3-319-93560-7_3

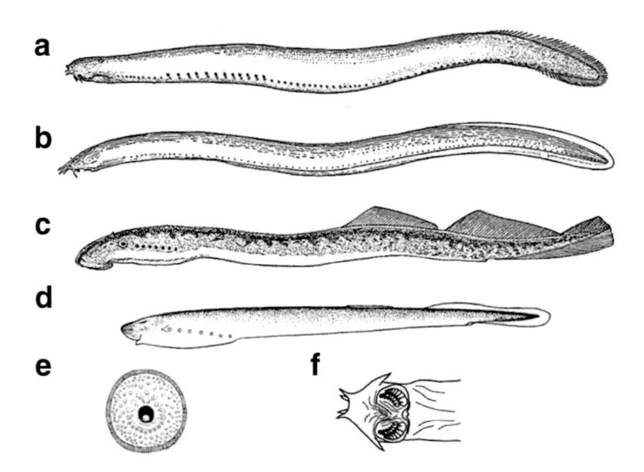

Fig. 3.1 Cyclostomes: (**a**) slime hag, *Eptatretus* sp. (formerly Bdellostoma); (**b**) hagfish, *Myxine* sp.; (**c**) lamprey, *Petromyzon* sp.; (**a–c**) from Romer (1950). (**d**) Ammocoete (larval lamprey), from Hardisty et al. (1989). (**e**) Mouthpart from *Petromyzon marinus*, from Potter (1980). (**f**) Ventral view of the mouthpart of a hagfish during maximum gape, from Clark et al. (2010)

medial nostril which is connected to the olfactory capsule. Seven gill slits are located behind the eye; in a historical description, the unpaired nostril, the lateral eye, and those seven gill slits together led to the misleading German name "Neunaugen" (nine eyes; Fig. 3.1c).

All extant **hagfishes** are benthic, opportunistic scavengers of marine invertebrates and vertebrates (Auster and Barber 2006). Their feeding apparatus has teeth (Fig. 3.1f) and cartilage but is dominated by muscles; the proportions of the feeding apparatus and the number of horny teeth are used to distinguish different species (Clark and Summers 2007). Hagfishes are able to forcefully remove tissue from carcasses and to ingest large pieces of food, despite having no jaws (Clark and Summers 2007). The use of gape cycles to grasp, ingest, and intraorally transport food was described by Clark and Summers (2007) and Clark et al. (2010).

It is often assumed that hagfishes are the more basal taxon in cyclostomes, partly because of the secondary loss of structures (Forey and Janvier 1993; Gess et al. 2006), which even included traits that are used to define vertebrates, as, for example, eyes and eye-related structures. However, developmental studies in hagfish have shown the presence of neural crest, somites, and even the appearance of putative vertebrae in the most caudal trunk region (Ota et al. 2007, 2011). There are also hagfish-specific traits such as the secondary opening of the nasohypophyseal duct into the pharynx (Oisi et al. 2013b) and the posterior shift of the caudal branchial arches (Holmgren 1946).

The monophyly of cyclostomes, i.e., that lampreys and hagfishes belong to the same taxon, was long questioned, but more recent molecular and developmental studies support this view (e.g., Kuraku et al. 1999; Delsuc et al. 2006; Heimberg et al. 2010). Their phylogenetic position as sister taxon to extant jawed vertebrates, Gnathostomata (Heimberg et al. 2010), makes them the most interesting group to study the origin and evolution of vertebrate structures (e.g., Janvier 1996, 2007; Kuratani et al. 2001; Kuratani 2004, 2005a, b, 2008a, b,). The comparative analysis of traits in those groups enables the uncovering of evolutionary patterns across early vertebrate lineages. In particular ammocoetes are often studied to understand vertebrate evolution as they resemble closer to the ancestral vertebrate (see below).

For example, cyclostomes are studied to understand the evolution of hypophyses and thyroid gland development (Leach 1944), thyroid hormone receptors (Holzer and Laudet 2017), adaptive immune system (Poole et al. 2017), heart physiology (Augustinsson et al. 1956), *Tbx1/10* gene expression (Sauka-Spengler et al. 2002; Tiecke et al. 2007), oxygen transport with hemoglobin (Hoffmann et al. 2010), telencephalon (Sugahara et al. 2013), hindbrain segmentation (Parker et al. 2014), neural crest gene regulatory network (Ota et al. 2007; Green et al. 2015), vertebrate paired fins (Tulenko et al. 2013), and many other structures.

Comparing the two major taxa of living vertebrates, the cyclostomes and gnathostomes, revealed many shared traits that had to be present in their last common ancestor (LCA), the vertebrates (Oisi et al. 2013b). The LCA of vertebrates had a musculoskeletal body plan that only consisted of branchial and axial structures, including skeletal arches that supported the gills, segmental myotomes, vertebrae, and median fins (Janvier 1996; Ota et al. 2011). Recently, it was shown that gills in cyclostomes and gnathostomes are homologous (Gillis and Tidswell 2017), a subject discussed for several decades (Mallatt 1984). There are also cyclostome-specific developmental and morphological traits that cannot be identified in gnathostomes.

This book focuses on the evolution of heads, jaws, and associated muscles in vertebrates. Therefore, I focus in this chapter on characters shared between jawless and jawed vertebrates. Those characters include extraocular muscles (e.g., Suzuki et al. 2016), branchiomeric muscles (e.g., Ziermann et al. 2014), and neural crest cells (Horigome et al. 1999; Ota et al. 2007). Importantly, neural crest cells interact with other tissues and influence not only the craniofacial but also the cranial musculoskeletal development (Green et al. 2015).

3.2 Skull and Jaw Evolution

The importance to compare cyclostomes with other vertebrates and even cephalochordates to understand the evolution of the cranium was already recognized in the nineteenth century (Huxley 1876). The chondrocrania of gnathostomes and cyclostomes are very difficult to compare, even at the modular level, but the results of comparative studies shed light onto the evolution of vertebrate crania. Furthermore, the origin of the vertebrate jaw has fascinated scientists for centuries. Recent advances in the ability to study cyclostomes lead to an abundance of studies that try to shed light on the emergence of jaws (Kuratani et al. 2001; Shigetani et al. 2002, 2005; Kuratani 2004, 2012; Mallatt 2008; Cerny et al. 2010; Medeiros and Crump 2012; Gillis et al. 2013; Miyashita 2016).

In order to enable a comparison of cyclostomes and gnathostomes, it is important to understand the homology of cranial elements between the taxa. Most studies compared each skeletal element, including the relation to cranial muscles and cranial nerves, in order to establish homology (e.g., Holmgren 1946; Yalden 1985). However, even the comparison between hagfish and lamprey crania is difficult because their anatomical pattern differ substantially (Fig. 3.2; Fürbringer 1875; Oisi et al. 2013a). Therefore, the evaluation of the development of the crania is essential for the homologization of the skeletal elements in cyclostomes (e.g., Johnels 1948).

Lampreys are the better accessible extant jawless vertebrates, and therefore more studies are published about them than about hagfishes. The larval and adult crania in lampreys are well studied (Fig. 3.2b, c; e.g., Huxley 1876; Marinelli and Strenger 1954; Oisi et al. 2013a and citations within). The embryonic development and metamorphosis of the lamprey cranium was described by Johnels (1948). In hagfish the adult cranium was described by several researchers (e.g., Fig. 3.2a; Marinelli and Strenger 1956b; Miyashita 2012; Oisi et al. 2013a and citations within), but only few developmental descriptions exist (Holmgren 1946; Ota et al. 2007; Ota and Kuratani 2008; Oisi et al. 2013a and references within; Oisi et al. 2013b; Miyashita and Coates 2015 and citations within). The most detailed description up to today of the development of the chondrocranium in hagfishes (*Eptatretus burgeri*, *E. atami*) is by Oisi et al. (2013a).

Hagfish embryos and lamprey larvae share similar ontogenetic skeletal features (Fig. 3.2; Oisi et al. 2013a, b), but lamprey adults have structures they share with hagfish but that are underdeveloped in larvae. The lingual apparatus, for example, is present in both adult taxa but only appears after metamorphosis in lampreys (Yalden 1985). Larval lampreys (ammocoetes) are filter feeders, which are found usually in the soft sediment of streams (Moore and Mallatt 1980); they possess some apparently plesiomorphic ("primitive," ancestral) characters like an endostyle. Hagfish do not have an endostyle, and the homology of the lamprey's endostyle with that of amphioxus or ascidians is also still questioned by some, as is the homology of thyroid gland and endostyle (e.g., Holland and Chen 2001). The early embryonic pattern of lampreys is similar to that of hagfishes but their oral apparatus, including the lips, resembles those of some adult fossil heterostracans or osteostracans better than does the lips of adult lampreys (Kuratani et al. 2002).

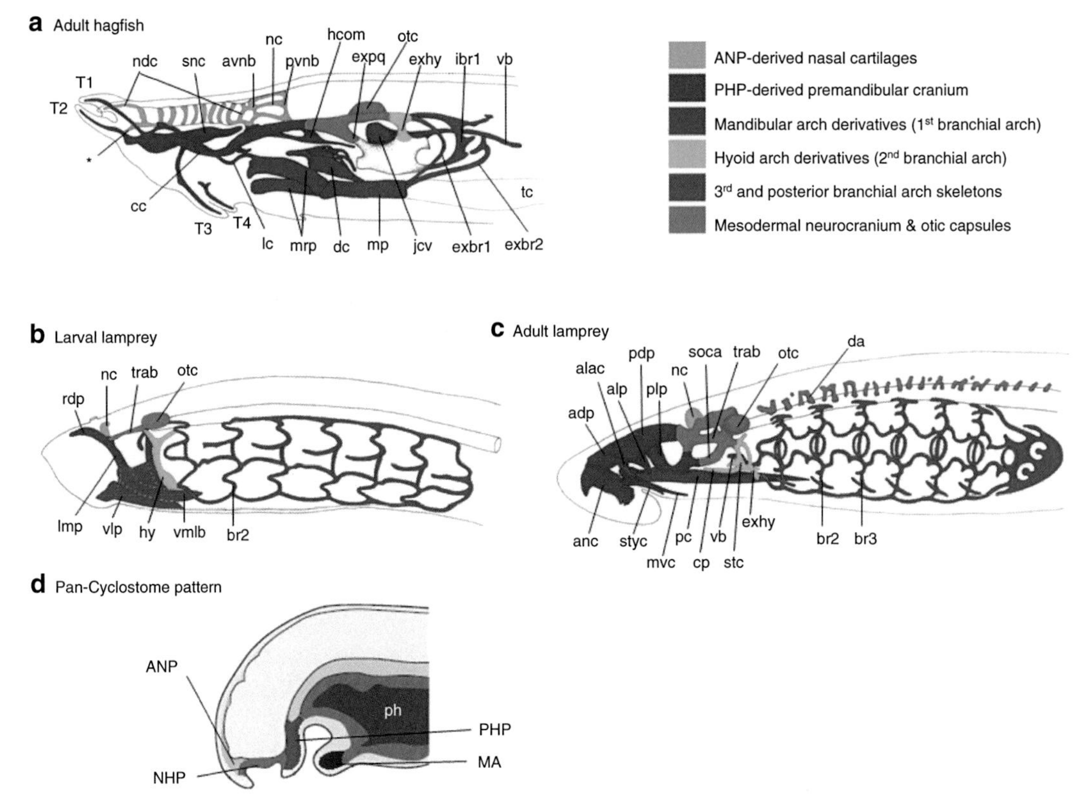

Fig. 3.2 Cranial skeleton in cyclostomes. (**a**) Adult hagfish, (**b**) larval lamprey (ammocoetes), (**c**) adult lamprey, (**d**) hypothetical pan-cyclostome embryonic pattern. (**a**) Asterisk (*)—this cartilage is indicated as PHP derivative by Oisi et al. (2013a) but should be part of the nasal duct cartilages and was recolored as ANP derivative; (**c**) Oisi et al. (2013a) mark the branchial cartilages in lampreys with the same abbreviation as the internal branchial arch in the adult hagfish. However, in their Table 1, they correctly homologize the extrabranchiale of hagfishes with the branchiale of lampreys. The styliform cartilage (stc) is listed as PHP-derivative in their Table 1, but their Fig. 1 and Fig. 10 compared would conclude it is hyoid derivative as also shown here. (**a–c**) Recolored and modified from Oisi et al. (2013a), terminology follows Oisi et al. (2013a) if not otherwise mentioned; (**d**) recolored from Oisi et al. (2013b). *adp* anterior dorsal plate, *alac* anterior lateral apical cartilage, *alp* anterior lateral plate, *anc* annular cartilage, *ANP* anterior nasal process, *avnb* anterior vertical nasal bar, *br 3* branchiale 3 (extrabranchiale *sensu* Marinelli and Strenger 1954; intrabranchiale *sensu* Oisi et al. 2013a), *cc* cornual cartilage, *da* dorsal arcualia, *dc* dental cartilage (after Marinelli and Strenger 1956b), *exbr1/2* extrabranchiale 1/2, *exhy* extrahyal, *exqp* extra palatoquadrate, *hcom* hypophyseal commissure, *hy* hyoid, *ibr1* internal branchial arch 1, *jcv* joint caput for velum, *lc* labial cartilage, *lmp* lateral mouth plate, MA Mandibular arch *mp* medial part of basal plate, *mrp* medio-rostral part of basal plate, *mvc* medioventral cartilage, *nc* nasal capsule, *ndc* nasal duct cartilages, *NHP* nasohypophyseal plate, *otc* otic capsule, *pc* piston cartilage, *pdp* posterior dorsal plate, *ph* pharynx, *PHP* posthypophyseal process, *plp* posterior lateral plate, *pvnb* posterior vertical nasal bar, *rdp* rostrodorsal plate, *snc* subnasal cartilage, *soca* subocular arch, *stc* styliform cartilage, *styc* stylet cartilage, *T1–T4* tentacles with supporting cartilages, *tc* tongue cartilage, *trab* trabecula, *vb* velar bar, *vlp* ventrolateral plate, vmlb ventromedial longitudinal bar

3.2.1 The Cyclostome Chondrocranium

The hagfish **chondrocranium** includes the **nasal capsule** cartilages, **otic capsule**, neurocranial base (mesodermal neurocranium), **lingual cartilages**, other branchial arch cartilages, and premandibular cartilages (Fig. 3.2a; Oisi et al. 2013a). The latter includes also the cartilages that support the tentacles. The otic capsule, the trabecula, and the dorsal longitudinal bar present likely the entire mesodermal-derived neurocranial elements (Oisi et al. 2013a). Elements derived from the **anterior nasal process** (ANP in Fig. 3.2) are in hagfishes the cartilages of the supranasal region (nasal duct cartilages and cartilaginous elements of the nasal capsule) and in lampreys the dorsal wall posterior to the nostril. Elements derived from the **posterior hypophyseal process** (PHP in Fig. 3.2) include also cartilages derived from both the premandibular crest and mandibular arch cells: tentacular (T1–4) cartilages (perhaps with the exception of the T4 cartilage) and the subnasal cartilage of hagfishes, along with the palatine bar and the hypophyseal commissure, and perhaps the dorsal longitudinal bar and the trabecula (Oisi et al. 2013a). The mucocartilage in the upper lip of lampreys and possibly the rostral trabeculae parts appear to develop from the equivalent anlage (Fig. 3.2c). Therefore, all posterior hypophyseal process derived cartilages of hagfishes should be homologous to the rostral dorsal plate and lateral wall of the upper lip, the trabecula, and part of the nasal capsule of lampreys (Oisi et al. 2013a). Based on the innervation pattern in hagfishes and lampreys, the lateral wall in lampreys may correspond to the tentacular cartilages T1, T3, and T4 in hagfishes, while T2 seems to be more similar to the dorsal roof (Oisi et al. 2013a).

Based on development, innervation, and gene expression pattern, Oisi et al. (2013a) summarized the homologous relationship of cyclostome crania (Fig. 3.2); but see Kuratani et al. (2016) for an updated interpretation of a cartilaginous element at the level of the hyoid arch. Several cyclostome-specific characters were identified (Oisi et al. 2013a): differentiation of the lingual apparatus and the velum in the ventral and middle mandibular arch region, respectively, and lateral and posterior hypophyseal process-derived cartilages. The (external) branchial arch skeleton is also thought to be cyclostome specific (Mallatt 1984), but a recent cell lineage tracing study demonstrated an endodermal origin of gills in both gnathostomes and cyclostomes which supports the homology of the gills in both taxa (Gillis and Tidswell 2017). Compared to gnathostomes, cyclostomes lack homologues to the intertrabecula and have no occipital vertebrae (Oisi et al. 2013a). It is currently not clear which of those characters are plesiomorphic (retained from the LCA of vertebrates and lost in gnathostomes) or synapomorph (newly developed in the LCA of cyclostomes).

3.2.2 Development of the Chondrocranium

Oisi et al. (2013a, b) compared the development of the chondrocranium in *Eptatretus* with the development in a lamprey (*Lethenteron reissneri*). They not only showed that there is a conserved embryonic pattern of head development in cyclostomes but also that chondrocrania of lampreys and hagfishes can be compared at least at the module level (Fig. 3.2); the latter corresponds to the craniofacial primordia that build up the cyclostome morphotype (Oisi et al. 2013a). The most conserved stage during cyclostome development is the pharyngula stage, which is before the chondrification of the cranium. However, as adults, the crania are very different from each other (Fig. 3.2), and homology establishment is difficult because of the adaptations in both taxa and because of the highly apomorphic nature of the hagfish cranium.

Both hagfishes and lampreys have a neural crest development comparable to that of gnathostomes, but the **nasohypophyseal process/plate** is unique to cyclostomes (Fig. 3.2d; Ota et al. 2007; Oisi et al. 2013b; Kuratani et al. 2016), but it was suggested that this process might even be plesiomorphic present for all vertebrates. However, the similarity between the nasohypophyseal complex in lampreys and osteostracans is likely due to convergent evolution (Gai et al. 2011). **Neural crest** cells give rise to numerous skeletal elements of the head, connective tissue, tendons, etc., but they do not from head muscles (Noden 1983; Noden and Francis-West 2006). However, muscle fibers form within cranial neural crest-derived connective tissue in a coordinated manner (Ziermann et al. 2018). That leads to the association of muscles with the proper skeletal region; i.e., muscles of a certain branchial arch are associated with connective tissue and through this with skeletal elements from the same arch (Köntges and Lumsden 1996).

The observation of the growth and transformation of the **posthypophyseal process** in larval lampreys showed that the nostril (**nasohypophyseal opening**) is moved to the dorsal side of the larval head (Damas 1944; Kuratani et al. 2016). In hagfishes the process enlarges anteriorly and forms a septum that divides the oronasal cavity dorsoventrally, as well as the ventral margin of the nostril rostrally; the posterior root of this process disappears during further development which leads to the formation of a continuous connection between the nasohypophyseal duct and the pharynx (Oisi et al. 2013b). Due to those developmental processes, the hagfish and lamprey heads are less comparable during later developmental stages, while during early development, the set of craniofacial primordia are identical in cyclostomes (Kuratani et al. 2016). This is just one example of how embryological studies can help to identify similarities and even homologies based on the assumption that all included taxa share the same ancestral developmental plan. However, when studying the emergence of the jaw it is important to keep in mind that the jaw elements (derivatives from the mandibular arch) evolved likely after the divergence of cyclostomes and gnathostomes. Therefore, Oisi et al. (2013a, b) used deeper levels of homology to establish their homology hypotheses. In their first study (here "2013b" because of sorting in alphabetical order), they established a **pan-cyclostome embryonic pattern** (Fig. 3.2d; see below), which is shared by cyclostomes but not by crown gnathostomes. They did not only compare the embryological development of lampreys and hagfishes but also gene expression patterns in different tissues. In later studies it was suggested that this embryonic pattern might even represent the ancestral vertebrate embryonic pattern.

The **pan-cyclostome embryonic pattern** includes the presence of a **nasohypophyseal plate** (a single median placode that yields the nasal epithelium and adenohypophysis), which is bordered by an **anterior nasal process** and a **posthypophyseal process** (Fig. 3.2d). Both of those processes and the ventral part of the mandibular arch serve as craniofacial primordia in cyclostomes—similar to the nasal prominences and maxillomandibular processes in jawed vertebrates (Oisi et al. 2013a, b). The anterior nasal process of cyclostomes differentiates into the posterodorsal margin of the nasohypophyseal duct, and the posthypophyseal process differentiates into the upper lip of lamprey larva (or the oral funnel in adult lampreys) and the oronasohypophyseal septum in hagfishes (Oisi et al. 2013b).

The mandibular arch mesoderm gives rise to three parts: the dorsal one shifts rostrally to reside in the posthypophyseal process and its derivatives (Kuratani et al. 2004), the mid-part transforms into the velum, and the ventral part differentiates into the tongue apparatus (Kuratani 2012). The described pattern is not present in gnathostomes (Oisi et al. 2013b). Even with this knowledge, the comparison with the gnathostome pattern is still difficult, e.g., comparing the undifferentiated mandibular arch mesoderm before the taxon-specific compartmentalization. It was furthermore shown that the trigeminal nerve divisions and pattern of innervation is comparable within cyclostomes but not between them and gnathostomes (Oisi et al. 2013b; Higashiyama and Kuratani 2014) (see below: trigeminal innervation).

Cephalic **neural crest**-derived ectomesenchyme contributes to craniofacial components in cyclostome embryos and to craniofacial primordia in gnathostomes (Horigome et al. 1999; Kuratani et al. 1999; Shigetani et al. 2002; Oisi et al. 2013b). Furthermore, the initial migration pattern and anteroposterior specification of neural crests as well as the expression patterns of regulatory genes are similar between those taxa (Horigome et al. 1999; McCauley and Bronner-Fraser 2003; Green et al. 2015). However, the otocyst is slightly more rostral in cyclostome embryos than in gnathostome embryos with respect to the hyoid arch, and the hyoid neural crest stream is found medial to the otocyst in cyclostomes (Horigome et al. 1999; Oisi et al. 2013a, b).

Besides some cyclostome-specific traits, it is likely that basic ectomesenchymal (neural crest) distribution and skeletogenic properties are very similar in cyclostomes and gnathostomes. This in turn suggests that a craniofacial skeleton with pharyngeal arch components and prechordal neurocranial elements can also be identified in cyclostomes. In fact, the mesodermal cranial elements in hagfishes and lampreys (Fig. 3.2) are similar to gnathostomes, and the head mesoderm distribution in early lamprey embryos resembles that of gnathostome embryos (Kuratani et al. 1999; Adachi and Kuratani 2012).

The so-called **trabeculae** are described in hagfishes, lampreys, and gnathostomes. However, as detailed in (Oisi et al. 2013a), they are likely not homologous, because they develop differently in all three taxa. Trabeculae in gnathostomes are neural crest-derived prechordal cranial elements (Couly et al. 1993; Wada et al. 2011), in lampreys they are mesodermal elements (Kuratani et al. 2004), and in hagfishes they seem to be composites of the trabecula and the dorsal longitudinal bar (Oisi et al. 2013a) (Fig. 3.2). However, the anterior portion of the trabeculae of hagfishes might be homologous to the gnathostome trabeculae, which appears to be supported by its position within the posthypophyseal process (Oisi et al. 2013a).

3.2.3 The Evolution of Jaws

As the name indicates, extant gnathostomes possess an upper and lower jaw that derives from the mandibular arch (see also Chap. 2 for discussion on the origin of the jaw). The cartilaginous primordia are usually called palatoquadrate (which is the main part of the upper jaw; the latter also includes premandibular components, e.g., trabecula) and Meckel's cartilage (lower jaw) (e.g., Goodrich 1930). The mandibular arch is characterized by the absence of *Hox* gene expression, while all posterior arches have a specific *Hox* gene patterning (Rijli et al. 1993, 1998). This is also shared by lampreys (Takio et al. 2004, 2007). The homologizations of caudally located branchial arch skeletons between hagfishes and lampreys are usually done by branchial muscle distribution (Marinelli and Strenger 1954, 1956b; Oisi et al. 2013b) and their cranial nerve innervation patterns (e.g., Song and Boord 1993; Oisi et al. 2013b), because each cranial nerve can be associated with a specific branchial arch (1st arch = mandibular arch = trigeminal nerve, cranial nerve V; 2nd arch= hyoid arch = facial nerve, cranial nerve VII; 3rd and following arches = caudal branchial arches = cranial nerves IX, X; Edgeworth 1935).

However, the cyclostome mandibular arch cannot easily be divided into upper and lower jaw elements. The dorsoventral patterning of branchial arches is regulated by ***Dlx* gene** expression in the ectomesenchyme (Depew et al. 2002, 2005; Minoux and Rijli 2010; Gillis et al. 2013). In mouse, *Dlx5* and *Dlx6* are specifically expressed in the ventral (lower) half of the mandibular arch (Depew et al. 2002). The simultaneous disruption of those genes leads to an upper jaw morphology instead of a lower jaw morphology. If in turn their upstream regulator *Ednra* is activated in the upper jaw domain, lower jaw morphology develops (Sato et al. 2008).

Lampreys have at least six ***Dlx* genes** (A–F); five of them are expressed in the branchial arch ectomesenchyme including the mandibular arch (Kuraku et al. 2010). However, there seems to be no dorsoventrally nested expression (see Chap. 2); but, a dorsoventrally symmetrical nested expression pattern around the gill pores was suggested (Cerny et al. 2010). *Bapx1* specifies the jaw joint in gnathostomes (Miller et al. 2003), but its lamprey homologue is not expressed in the lamprey's mandibular arch (Cerny et al. 2010; Kuraku et al. 2010). Yet, *dHand* cognate, a ventral pole specifier, is expressed in a way supporting a dorsoventral patterning in lampreys, while the unpolarized *Dlx* expression is consistent with the dorsoventrally symmetrical morphology of its posterior branchial arches. Hagfishes also do not have an apparent dorsoventral polarity in the preliminary analyses of Oisi et al. (2013a). The **lingual apparatus** in hagfishes derives from the ventral portion of the mandibular arch, and the homology of this structure to the lingual apparatus in lampreys is well established (Yalden 1985). The musculoskeletal structure, however, seems to develop through a different mechanism as it is independent of *Dlx* expression, and it is therefore not homologous to Meckel's cartilage (lower jaw) of gnathostomes (Oisi et al. 2013a).

The dorsal half of the crown gnathostome mandibular arch is developmentally patterned as the default state of the Dlx code (Depew et al. 2002; Gillis et al. 2013; see also Chap. 2). As those genes are ubiquitously expressed in the branchial arch ectomesenchyme (neural crest), it is assumed that the "upper jaw" is the default state of the Dlx code in gnathostomes (Kuratani et al. 2013; Oisi et al. 2013a). Therefore, *Dlx* expression in ectomesenchyme of branchial arches seems to be a common vertebrate character. However, the patterning in those arches changes during the evolution of gnathostomes, which may play an important role in the establishment of the lower jaw (for a detailed discussion, see Miyashita 2016) but also leads to questioning of the homology between palatoquadrate (upper jaw) and dorsal mandibular arch skeletal derivates in gnathostomes (Oisi et al. 2013a).

3.3 Muscle Evolution

Cephalic muscles, that is, muscles associated with the head, can be grouped based on their developmental origin into eye (extraocular), mandibular, hyoid, branchial (including epibranchial and laryngeal muscles), and hypobranchial muscles (Edgeworth 1935; Diogo and Abdala 2010). The muscles, except the extraocular and hypobranchial muscles, originate from mesodermal anlagen associated with the same named branchial arches (aka pharyngeal arches) and are innervated by nerves associated with their respective region of origin (Edgeworth 1935; Diogo and Abdala 2010; Harel and Tzahor 2013). For example, the first branchial arch is called the **mandibular arch**, and the tissues associated with this arch give rise to the upper and lower jaw elements, the associated mandibular muscles (e.g., muscles of mastication: adductor mandibulae), and the connective tissue. The nerve innervating the muscles and receiving sensory information from the mandibular region is the trigeminal nerve (cranial nerve V). It is not the aim of this chapter to review the complete development of all the cephalic muscles; for more detailed information, see, for example, Diogo and Abdala (2010), Harel and Tzahor (2013), and references within.

Anatomical descriptions of cyclostome musculature were performed several times during the past 150 years, but without a comparison to gnathostomes or between cyclostomes (Fürbringer 1875; Cole 1907; Tretjakoff 1926; Marinelli and Strenger 1954, 1956b). More recent publications, however, compare the morphology and development of the head muscles between cyclostomes and gnathostomes (Miyashita 2012; Ziermann et al. 2014; Diogo and Ziermann 2015). Functional analyses of feeding in hagfishes and lampreys reveal the underlying kinematics (Moore and Mallatt 1980; Rovainen 1996; Clark and Summers 2007; Clark et al. 2010).

In addition to the above-mentioned differences in the head skeleton between cyclostomes and extant gnathostomes, the associated musculature seems also to be quite different. Yet, comparing the morphology (attachments, number of bellies), innervation, overall position, and development of muscles associated with the head in cyclostomes and jawed fishes can provide insights into the homology and evolution of those muscles. Such a comparative study was performed, for example, by Ziermann et al. (2014) and Diogo and Ziermann (2015). The most intriguing observation, besides the obvious different morphology of the head muscles, is the difference in number of cranial muscles (Fig. 3.3). While cyclostomes have over 20 **mandibular arch muscles**, gnathostome fishes possess less than 10. The adult hagfish has four **hyoid arch muscles** which is similar to most of the gnathostome fishes (2–3), but the larval lamprey has only one hyoid muscle that is absent in adult lampreys. The **branchial arch muscles** are largely reduced in hagfishes (3); the numbers are increased in adult lampreys (76) but similar in larval lampreys (32) and cartilaginous fishes (16–28).

Based on the dissection and comparison of cyclostomes with chondrichthyans (cartilaginous fishes like sharks, skates, chimera), Ziermann et al.

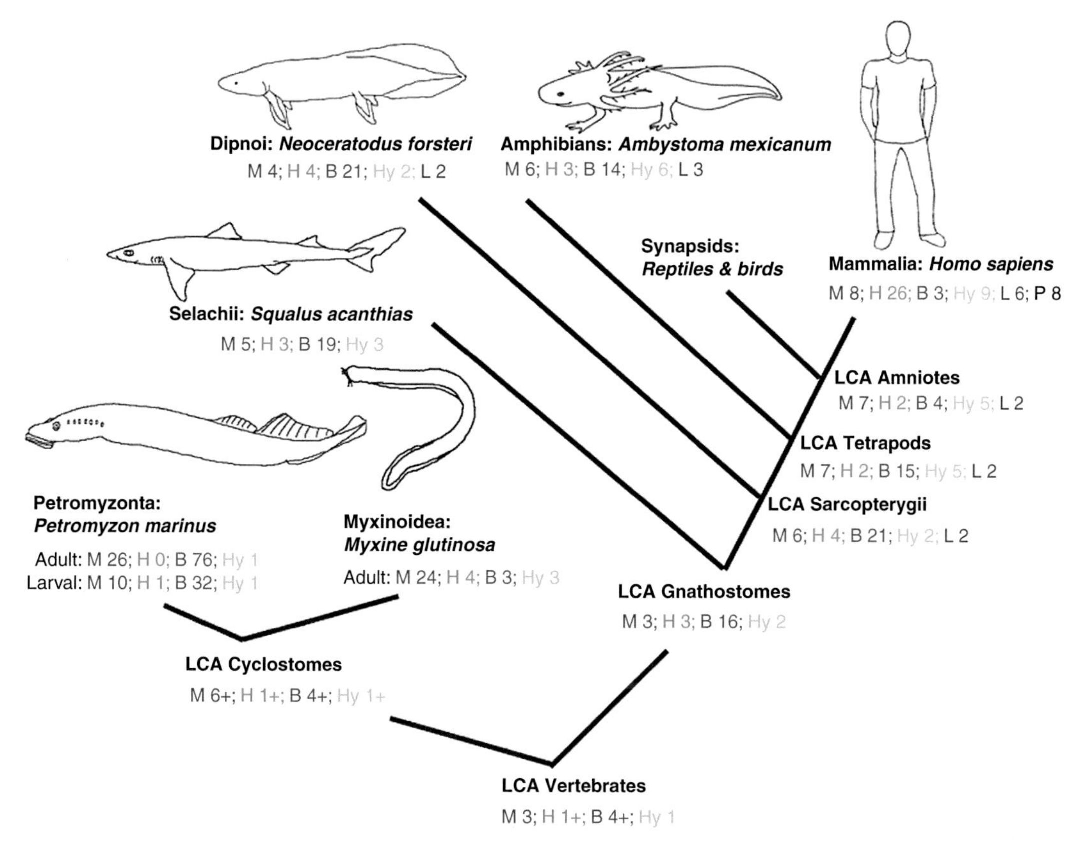

Fig. 3.3 Number of cephalic muscles in vertebrates (based on results from: Ziermann et al. 2014; Diogo and Ziermann 2015; Diogo et al. 2015b). Cephalic muscles are colored according to their developmental origin: mandibular arch muscles (red); hyoid arch muscles (green); branchial arch muscles (blue); hypobranchial muscles (yellow); laryngeal muscles (purple); pharyngeal muscles (black). Comparing and homologizing the muscles of diverse vertebrates, it is possible to infer the muscles in the last common ancestor (LCA) of extant taxon

(2014) inferred the ancestral condition of cephalic muscles in cyclostomes, gnathostomes, and even vertebrates (Fig. 3.3). In order to study the ancestral condition in cyclostomes, they studied and reviewed the literature of embryonic, larval, and adult hagfishes (Fig. 3.4) and lampreys (Fig. 3.5). The last common ancestor (LCA) of vertebrates had a single intermandibularis (i.e., a ventral muscle sheet) and other **mandibular muscles** (e.g., labial muscles), some constrictores hyoidei and branchiales, and epibranchial and hypobranchial muscle sheets (Ziermann et al. 2014). From this condition, the number of mandibular arch muscles increased toward the LCA of cyclostomes (synapomorphy) and then further in the different lineages of hagfishes and lampreys with 24 and 26 mandibular muscles in adult hagfishes and lampreys, respectively (red in Figs. 3.3, 3.4, and 3.5). Alternatively, the increase in mandibular muscles could have evolved independently from each other as the amount of branchial muscles also differs significantly in both taxa (see below).

The number of **hyoid muscles** stays almost constant throughout vertebrates until the diversification of amniotes (reptiles, birds, mammals; Fig. 3.3; see also Table 11.2 in Chap. 11). Interestingly, adult lampreys do not have any hyoid muscles, but the associated nerve (facial nerve, cranial nerve VII) can clearly be identified (Figs. 3.3 and 3.5). However, hagfishes have four hyoid muscles and larval lampreys also have one (constrictor prebranchialis, Fig. 3.5b). Therefore, the LCA of cyclostomes had at least one hyoid arch muscle. The number of **branchial muscles**

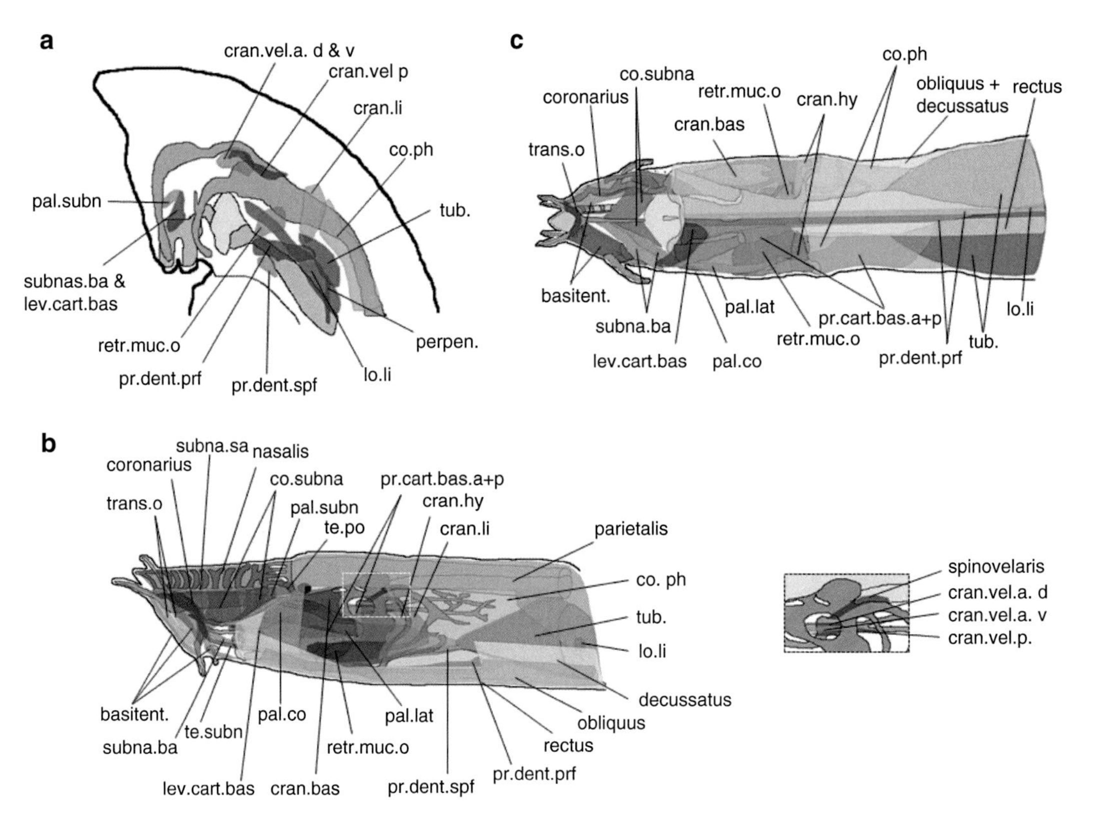

Fig. 3.4 Muscles of the Atlantic hagfish, *Myxine glutinosa*. Specimens not to scale. Not all muscles are shown on both sides in ventral view. Cephalic muscles are colored according to their developmental origin: mandibular arch muscles (red, orange, pink); hyoid arch muscles (green); true branchial arch muscles (blue); epibranchial muscles (brown); hypobranchial muscles (yellow). (**a**) Embryo, left lateral view, redrawn from Miyashita (2012); somites (not shown) extend to just behind the otic capsule. (**b**) Adult, left lateral view; (**c**) Adult, ventral view. (**b**) Parietalis and decussatus cut to enable view of deeper layers; white box—velar muscles in window on the right side of the animal. (**c**) Basitentacularis (basitent.) cut on right side. (**b**, **c**) Modified from Ziermann et al. (2014) and Diogo and Ziermann (2015). *basitent.* basitentacularis, *co.ph* constrictor pharynges, *co.subna* cornuosubnasalis, *cran.bas* craniobasalis, *cran.hy* craniohyoideus, *cran.li* carniolingualis, *lev.cart.bas* levator cartilagines basalis, *lo.li* longitudinalis lingua, *cran.vel.a. d & v* craniovelaris anterior dorsalis & ventralis, *cran.vel.p.* craniovelaris posterior, *pal.co* palatocoronarius, *pal.lat* palatinalis lateralis, *pal.subn* palatosubnasalis, *perpen.* perpendicularis, *pr. cart.bas.a + p* protractor cartilagines basalis anterior and posterior, *pr.dent.prf* protractor dentium profundus, *pr. dent.spf* protractor dentium superficialis, *retr.muc.o* retractor mucosae oris, *subna.ba* subnasobasalis, *subn.sa* subnasonasalis, *te.po* tentacularis posterior, *te.subn* tentaculosubnasalis, *trans.o* transversus oris, *tub.* tubuatus

in hagfishes and lampreys are quite different from each other, while adult lampreys have 76, larval lampreys possess 32, and adult hagfishes only 3. Cartilaginous fishes have on average about 19 branchial arch muscles. Ziermann et al. (2014) inferred from those numbers and their studies of vertebrate muscles that the LCA of cyclostomes had at least four branchial arch muscles, and with the evolution and adaptation to their specific lifestyles, lampreys increased and hagfishes reduced the number of branchial arch muscles. The number of hypobranchial muscles is almost constant throughout vertebrates and only slightly increases in tetrapods (Fig. 3.3).

Based on the comparison of morphology, innervation, overall position, and development, Ziermann et al. (2014) suggested to group the **mandibular muscles** of cyclostomes into five groups (red, orange, pink in Figs. 3.4 and 3.5):

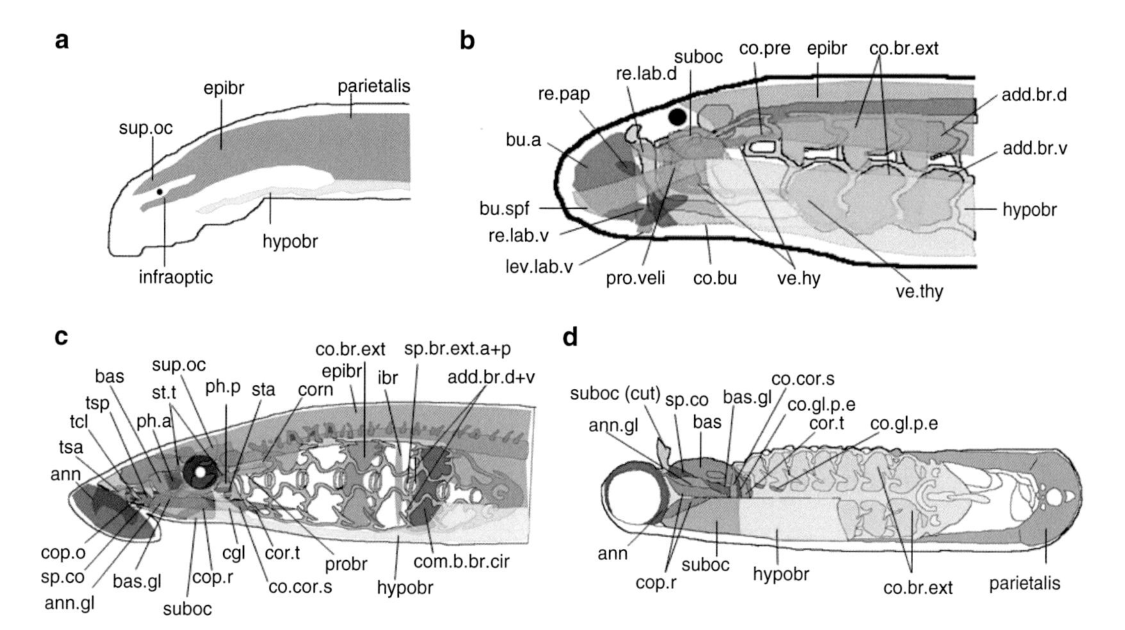

Fig. 3.5 Muscles of the sea lamprey, *Petromyzon marinus*. Specimens not to scale. Cephalic muscles are colored according to their developmental origin: mandibular arch muscles (red, orange, pink); hyoid arch muscles (green); true branchial arch muscles (blue); epibranchial muscles (brown); hypobranchial muscles (yellow). (**a**) Ammocoete larva, left lateral view; redrawn from Tulenko et al. (2013). (**b**) Ammocoete larva, left lateral view; redrawn from Miyashita (2012). (**c**) Adult, left lateral view; all branchial muscles (blue) are present in each segment but not shown in each segment. (**d**) Adult, ventral view; not all muscles shown on both sides; the hypobranchialis (hypobr) extends backward but was cut to show the branchial basket; the subocularis (suboc) on the left of the animals was cut and reflected. Modified from Ziermann et al. (2014) and Diogo and Ziermann (2015). *add.br.d* + *v* adductores branchiales dorsales + ventrales, *ann* annularis, *ann.gl* annuloglossus, *bas* basilaris, *bas.gl* basilariglossus, *bu.a* buccalis anterior, *bu.spf* buccalis superficialis, *cgl* cornuoglossus, *co.br.ext* constrictores branchiales externi, *co.bu* constrictor buccalis, *co.cor.s* constrictor cornualis superficialis, *co.gl.p.e* constrictor glossae profundus externus, *co.pre* constrictor prebranchialis, *com.b.br.cir* compressores bursae branchiales circulares, *cop.r* copuloglossus rectus, *cop.o* copuloglossus obliquus, *cor.t* cornuotaenalis, *corn* cornealis, *epibr* epibranchialis, *hypobr* hypobranchialis, *ibr* interbranchiales, *lev.lab.v* levator labialis ventralis, *ph.a* pharyngicus anterior, *ph.p* pharyngicus posterior, *probr* probranchialis, *sp.br.ext.a* + *p* sphincters branchiales anteriores + posteriores, *pro.veli* protractor veli, *re.lab.d* retractor labialis dorsalis, *re.lab.v* retractor labialis ventralis, *re.pap* retractor papillaris, *sp.co* spinosocopularis, *sta* styloapicalis, *st.t* stylotectalis, *sub.oc* subocularis, *sup.oc* supraocularis, *tcl* tectolateralis, *tsa* tectospinosus anterior, *tsp* tectospinosus posterior, *ve.hy* velohyoideus, *ve.thy* velothyroideus

derivatives of the intermandibularis muscle sheet, labial muscles, nasal muscles, lingual and dental muscles, and velar muscles. Importantly, it is argued that because there are significant differences in the developmental patterning of mandibular arch derivatives (muscles, cartilages, nerves), no homologies can be established between mandibular muscles of cyclostomes and gnathostomes; this is because if there are homologies, then it would have to be concluded that there are upper and lower jaws in cyclostomes (Kuratani pers. com.). As also pointed out with respect to the velar muscles (see below), gene expression does not always support homology. However, mandibular arch muscles were present before the split of the cyclostomes from the stem gnathostomes. Therefore, even if there is no one-to-one homology and if there is an increase of mandibular muscles in cyclostomes (addition), this does not exclude that at least some muscle groups are (as group) homologous to gnathostome mandibular muscles, as are the mammalian sternocleidomastoid and the trapezius together homologous to the protractor pectoralis of reptiles and adult amphibians (see Chap. 7).

The **intermandibularis muscle sheet** in gnathostomes derives from the ventral part of the mandibular muscle plate (Edgeworth 1935). The tubulatus muscle in hagfishes and the constrictor cornualis superficialis and constrictor glossae profundus internus of lampreys are suggested to belong to the intermandibularis group. Therefore, the LCA of vertebrates likely also possessed an intermandibularis muscle sheet.

The **labial muscles** were suggested to be conserved across vertebrates (Mallatt 1996, 1997b, 2008). However, the "upper lips" of cyclostomes and gnathostomes seem not to be homologous as they differ fundamentally in their developmental process (Horigome et al. 1999; Noden and Francis-West 2006; Kuratani 2012; Kuratani et al. 2013; Oisi et al. 2013b). Based on their anatomical comparisons, Ziermann et al. (2014) tentatively suggested that the constrictor buccalis in larval lampreys (Fig. 3.5b), which seems to develop from the "mandibular branchiomere" (Mallatt 1996), is homologous to the labial muscles in gnathostomes. The labial muscles in holocephalans (e.g., ratfish) are innervated by cranial nerve V2 (CNV2; maxillary branch of trigeminal nerve) (Song and Boord 1993; Mallatt 1996). The hagfish *Myxine glutinosa* has six muscles that are innervated by CNV2 (retractor mucosae oris, longitudinalis linguae, protractor dentium profundus, protractor dentium superficialis, tubulatus, and perpendicularis; Fig. 3.4). Corresponding muscles in the lamprey *Petromyzon marinus* are all but one innervated by CNV2 ramus mandibularis (*sensu* Marinelli and Strenger 1954) according to the analyzes of Miyashita (2012) and (Ziermann et al. 2014) (CNV2 ramus velaris: pharyngicus posterior; CNV2 ramus mandibularis: levator valvulae velaris, cardioapicalis, annuloglossus, copuloglossus rectus, constrictor cornualis superficialis, constrictor glossae profundus internus). Those observations can be interpreted in two ways: (1) holocephalans retained a cyclostome-like innervation of labial muscles, which was lost in other gnathostomes and (2) cyclostomes and holocephalans independently developed an innervation of labial muscles by CNV2. Currently, the latter hypothesis is supported as it is more parsimonious (two steps of independent gain of innervation, as compared to three steps: one gain and two losses of innervation by CNV2). Furthermore, it is questionable if the branches of the trigeminal nerve are homologues as currently assumed (Miyashita 2012; Higashiyama and Kuratani 2014; Modrell et al. 2014).

Velar muscles are suggested to derive from the same anlage that also gives rise to the gnathostome levator arcus palatini. The levator arcus palatini and spiracularis are gnathostome muscles that derive from the dorsal part (aka constrictor dorsalis) of the mandibular muscle plate (Edgeworth 1935). The expression of *engrailed* in the velothyroideus muscle of lamprey larvae (Fig. 3.5b) and in the gnathostome levator arcus palatini was used to infer homology between these muscles (Holland et al. 1993). However, similar gene expression does not support unambiguously homology, and it is not clear if the larval lamprey muscles degenerate during metamorphosis entirely or if they give rise to adult muscles; NB: the levator arcus palatini would corresponds to an "adult" muscle in gnathostomes (Miyashita 2012; Ziermann et al. 2014). With respect to the velothyroideus, two hypotheses exist: (1) the muscle is reduced at metamorphosis together with the larval velum and (2) the muscle becomes incorporated in adult velar muscles (e.g., depressor veli). Ziermann et al. (2014) favored the latter hypothesis and suggested further that velar muscles of hagfishes and lampreys derive from the same anlage as the dorsal mandibular ("constrictor dorsalis") muscles of gnathostomes such as the levator arcus palatini. The description that in both hagfishes and lampreys, the velum arises from the middle portion of the mandibular arch, between the rostral endodermal wall of the first branchial pouch and oral ectoderm, supports the homology of the velum in cyclostomes (Oisi et al. 2013b).

The **adductor mandibulae** of gnathostomes is related to "biting" of the jaw and derives from the transversely medial and dorsoventrally intermediate part of the mandibular muscle plate (Edgeworth 1935). However, it is currently not possible to identify a clear homologue to the adductor mandibulae in cyclostomes. Furthermore, it was suggested that the mandibular muscle development in gnathostomes is tightly linked to the patterning of the jaw skeleton (Noden 1983; Rinon et al. 2007; Medeiros and Crump 2012), which is supported by developmental defects observed in knockdown mutants (Schilling et al. 1996; Heude et al. 2010; Hinits et al. 2011).

Hagfishes have four **hyoid muscles** (green in Fig. 3.4), from which only two (craniolingualis, craniohyoideus) are likely homologous to the constrictor hyoideus dorsalis of gnathostome fishes (Ziermann et al. 2014). Adult lampreys have no hyoid muscle, but larval lampreys possess the constrictor prebranchialis (Miyashita 2012; Diogo and Ziermann 2015).

True **branchial muscles** include branchial muscles *sensu stricto* (*sensu* Diogo and Abdala 2010) and the cucullaris muscle and its derivatives. Laryngeal and epibranchial muscles are included as "other" branchial muscles, but laryngeal muscles do not evolve until gnathostomes (Ziermann et al. 2014). Interestingly, larval lampreys and sharks share two functions of their branchial muscles *sensu stricto*, as in both (1) the expiration is due to peristaltic action of superficial branchial constrictors and interbranchiales, and (2) the inspiration is caused by a passive recoil of the branchial arches (Mallatt 1996). This might indicate that this type of ventilation could be ancestral for the LCA of vertebrates. However, while the superficial branchial constrictors seem to be homologous throughout vertebrates, the interbranchiales could not be identified in cyclostomes and osteichthyans (bony fishes and tetrapods), which weakens this idea as both or one of those muscles could have been present in the LCA of vertebrates or gnathostomes (Ziermann et al. 2014).

The **cucullaris** muscle was discussed intensively in recent literature (e.g., Diogo and Abdala 2010; Diogo and Ziermann 2015) and basically two hypothesis regarding its developmental origin are most common: (1) the cucullaris is a true branchial muscle (Diogo and Abdala 2010; Ziermann et al. 2014), and (2) it is from somitic origin (Kusakabe and Kuratani 2005; Kusakabe et al. 2011; Sambasivan et al. 2011) (see also Chaps. 2 and 7 for discussion on the evolution of the cucullaris and its derivatives). Due to topographic similarities of the "infraoptic" muscles in lampreys (Fig. 3.5a) and the gnathostome muscle cucullaris, and based on the observation that the **infraoptic muscles** derive from anterior somites, it was suggested that those muscles are homologous and that they resemble epibranchial muscles (Kusakabe and Kuratani 2005; Kusakabe et al. 2011; Sambasivan et al. 2011). However, there are three infraoptic muscles in lampreys (subocularis, cornealis, and probranchialis) (Kusakabe et al. 2011), and developmentally the cucullaris resembles closer the hypobranchial migratory muscles of lampreys (Matsuoka et al. 2005). The cucullaris and its derivatives are in most gnathostomes innervated by cranial nerve XI (CNXI; accessory nerve; spinal accessory nerve) but might also be innervated by spinal nerves (Edgeworth 1935). Based on ontogenetic studies and comparative anatomical studies, it was suggested that the cucullaris derives from the same anlage as true branchial muscles do, followed by an extension toward the pectoral region (Diogo and Abdala 2010). This is furthermore supported by genetic studies in mice, where branchiomeric muscle differentiation is regulated by *Pitx2* and *Tbx1*, while trunk muscles (somitic origin) are regulated by *Pax3*. *Pax3* mutant mice lack somitic-derived muscles, but the cucullaris derivatives trapezius and sternocleidomastoid are still present (Tajbakhsh et al. 1997; Ericsson et al. 2013; Minchin et al. 2013). Further supporting the branchial identity of cucullaris and derivates is that *Tbx1* mutant mice lack branchial muscles, including trapezius and sternocleidomastoid (Theis et al. 2010). As the presence of a cucullaris homologous muscles in cyclostomes is not proven yet, I infer that the muscle evolved in the LCA of gnathostomes and was not present in the LCA in vertebrates (Ziermann et al. 2014).

Epibranchial and hypobranchial muscles arise from anterior myotomes, migrate into the head below the pharynx, and retain spinal innervation

(Edgeworth 1935) (brown, yellow in Figs. 3.4 and 3.5). **Epibranchial muscles** derive from anterior parts of somites, whereas hypobranchial muscles derive from ventral parts of somites (Edgeworth 1935; Lours-Calet et al. 2014). Cyclostomes have one (hagfish) or more (lamprey) epibranchial muscles and three (hagfish) or one (lamprey) **hypobranchial muscles**. As those muscles are present in cyclostomes and gnathostomes, the LCA of vertebrates also possessed an undifferentiated epibranchial muscle sheet and an undifferentiated hypobranchial muscle sheet (Ziermann et al. 2014).

Extraocular muscles (EOMs) are highly conserved throughout vertebrates, and all vertebrates have six, with the exception of (some) placoderms that have seven (Burrow et al. 2005). While cell lineage studies in mice suggest that EOMs are not branchiomeric muscles (Harel and Tzahor 2013), clonal studies also performed in mice suggest that they are branchiomeric muscles that are related to mandibular arch and right ventricle (heart) musculature (Lescroart et al. 2010). The latter scenario is supported by cell-labeling studies in lampreys, which showed that mandibular mesodermal cells migrate near the eye; however, it is unclear if those cells differentiate to EOMs (Kuratani et al. 2004). Interestingly, disruption of the *Pitx2* gene during mesoderm differentiation disrupts the morphogenesis of all EOMs and some mandibular arch muscles, but also the myogenesis of the body wall and appendicular muscles (Shih et al. 2007; Sambasivan et al. 2011).

Overall it seems that the adult basal gnathostomes share more similarities with hagfish embryos and larval lampreys than with adult cyclostomes due to **peramorphic events** that occurred in the evolutionary history of cyclostomes (Diogo and Ziermann 2015). Peramorphic events describe the appearance of ancestral adult characters in descendant juveniles due to additions to terminal somatic developmental stages (McNamara 1990). As described above, Ziermann et al. (2014) and Diogo and Ziermann (2015) presented hypotheses about the homology and evolution of adult and larval muscles in cyclostomes and gnathostomes. According to them, the LCA of extant vertebrates had an undifferentiated intermandibularis muscle sheet, labial muscles, and some other mandibular muscles, at least one hyoid muscle (constrictor hyoideus = constrictor prebranchialis), at least some constrictores branchiales (branchial muscles), and undifferentiated epibranchial and hypobranchial muscle sheets. The adductores branchiales (branchial muscles) were likely independently acquired in lampreys and chondrichthyans. Furthermore, lamprey larvae seem to be a better model for cranial muscles of adults of the LCA of gnathostomes and LCA of vertebrates, than is the adult lamprey, because the inferred adult muscles in LCA of vertebrates is amazingly similar to lamprey larva. The absence of a hyoid muscle in adult lampreys as compared to the presence of one larval hyoid muscle is the most striking supporting this hypothesis. Another example is the presence of two muscles within the labial and intermandibularis group in lamprey larvae and LCA of vertebrates, while the adult lamprey has many more labial muscles. At least one velar and/or dorsal mandibular muscle was inferred for the LCA of vertebrates, and lamprey larvae have two of these muscles, while adult lampreys seem to have no muscles that can be easily put into this group. Also, branchial arch muscles are more similar in larval lampreys to the LCA of vertebrates, because the adult branchial muscles are far more complex. This is also true for epibranchial muscles. Therefore, larval lampreys are more similar to adult members of the LCA of extant gnathostomes, supporting the idea that peramorphic events occurred in the history of cyclostomes.

Metamorphosis is a process in which larval structures are remodeled into an adult form; the adult form differs from the larval form in morphology and ecology (see Chap. 7). In lampreys, the mouth, eyes, gut epithelium, larval kidney, and endostyle (thyroid gland in adults) are remodeled during metamorphosis (Youson 1980, 1997). Hagfishes have direct development. The adult cephalic muscles in lampreys develop from blastema as larval cephalic muscles degenerate during metamorphosis. Currently, it is not clear if the larval muscles are direct precursors of adult muscles. The rebuilding process during metamorphosis

(Marinelli and Strenger 1954) is so dramatic that it is hard to make any assumptions about correspondence of larval and adult muscles. However, if one would accept the homology as proposed by Holland et al. (1993) that the larval velothyroideus in lampreys is homologous to the levator arcus palatini and dilatator operculi of embryonic and adult teleosts, then the velum of lampreys could be homologous to the palatoquadrate (upper jaw) of gnathostomes, or at least homogenic (Miyashita 2012). The velum is reduced during metamorphosis of lampreys, and instead a lingual apparatus develops; this would be another example where larval lampreys represent better adult gnathostomes than adult lampreys do.

3.3.1 Evolution of the Gnathostome Jaw and Mandibular Arch Muscles

Some species develop their mandibular (first) arch muscles before or simultaneously with hyoid (second) arch muscles; however, others develop first hyoid arch muscles (see Table 2 in Ziermann et al. 2017). Furthermore, it appears that the most anterior (first) arch in basal chordates (cephalochordates and fossils like *Haikouella*) corresponds to the hyoid (second) arch of vertebrates (Mallatt and Chen 2003; Mallatt 2008). This in turn would imply that the mandibular arch of vertebrates was secondarily incorporated into the branchial arch series in derived chordates; the similarity between the jaws and the patterning of branchial arches are suggested to have evolved due to functional reasons (Janvier 1996). Supported is this view by the expression of *Hox* genes in all branchial arches, except the mandibular arch (see above). Miyashita (2016) followed up on this idea, and based on his studies and an extensive literature review, he suggested the "**mandibular confinement theory**" (see Chap. 2 for more details). Specifically, this theory proposes that the jaw in gnathostomes evolved through a developmental spatial confinement of an ancestral oral (anterior) chordate structure. This confinement lead to a co-option of genetics and patterning that is normally found in more posterior arches which lead to the evolution of the mandibular arch and its derivatives (in particular the gnathostome jaw).

A problem with the mandibular confinement theory is that it depends on the assumption that the common ancestor of cyclostomes and gnathostomes would be similar to modern cyclostomes, which has yet to be proven. For example, Miyashita (2016) assumes that LCA of vertebrates would have possessed a lingual apparatus and a velum (anteroposteriorly elongated mandibular arch) like cyclostomes; however, an equally likely hypothesis is that this is a cyclostome synapomorphy (Oisi et al. 2013a). Furthermore, the mandibular confinement theory is dependent on refuting several previous "branchial arch theories" for the origin of the jaw, including those of Goodrich (1930), Mallatt (1996, 1997a, 2008), Kuratani (2004, 2005b, 2012), and many others. Therefore, carefully designed experiments (e.g., gene manipulation), detailed comparative developmental and anatomical studies, etc., will be necessary in the upcoming years to test the proposed hypotheses regarding the evolution of the vertebrate jaw.

The findings that the **cardiopharyngeal field** (Diogo et al. 2015a; see Chap. 1 for more details) gives rise to branchiomeric (head) muscles and to the myocardium (heart musculature) partially supports the hypothesis from Miyashita (2016), because the oral siphon muscles in ascidians, which are urochordates, derive not from the cardiopharyngeal field, while the atrial siphon muscles do. However, the oral siphon is said to correspond to the mandibular region of gnathostomes, where all branchiomeric muscles including the mandibular muscles derive from the cardiopharyngeal field (Diogo and Ziermann 2015). Therefore, the mandibular arch was integrated secondarily into this field. As mentioned above *Tbx1* is a branchiomeric muscle marker, and in lamprey development, *Tbx1/10* is expressed first in the mesodermal core of the branchial and pharyngeal region below the otic vesicle and only later in the labial/oral and velar muscle-deriving mesoderm that corresponds to the mandibular mesoderm (Sauka-Spengler et al. 2002). This would furthermore explain why it is so difficult to homologize the mandibular arch muscles of cyclostomes with those of gnathostomes (see above: adductor mandibulae).

3.4 Summary

Cyclostomes are very peculiar and fascinating animals. Both the larval chondrocrania (Oisi et al. 2013a, b) and the larval muscles of cyclostomes (Diogo and Ziermann 2015) resemble the adult plesiomorphic vertebrate and gnathostome condition better than adult cyclostome structures do. This is currently best explained by peramorphic events during the evolution of cyclostomes. The upper jaw development is the default developmental mode in vertebrates. The lower jaw, however, is a novelty that evolved in the LCA of gnathostomes. The musculature associated with the different branchial arches can be homologized in vertebrates based on gene expression patterns, attachments, innervation, and overall position.

Acknowledgments I would like to thank Shigeru Kuratani and Philippe Janvier for their constructive reviews which improved the chapter.

Further Reading

From the extensive literature list in this chapter, I suggest for further reading on the relevance of cyclostomes to understand vertebrate and gnathostome evolution: Janvier (2008), Kuratani (2008a), and Miyashita (2016).

References

Adachi N, Kuratani S (2012) Development of head and trunk mesoderm in the dogfish, *Scyliorhinus torazame*: I. Embryology and morphology of the head cavities and related structures. Evol Dev 14:234–258

Augustinsson K-B, Fänge R, Johnels A, Östlund E (1956) Histological, physiological and biochemical studies on the heart of two cyclostomes, hagfish (*Myxine*) and lamprey (*Lampetra*). J Physiol 131:257–276

Auster P, Barber K (2006) Atlantic hagfish exploit prey captured by other taxa. J Fish Biol 68:618–621

Burrow CJ, Jones AS, Young GC (2005) X-ray microtomography of 410 million-year-old optic capsules from placoderm fishes. Micron 36:551–557

Cerny R, Cattell M, Sauka-Spengler T, Bronner-Fraser M, Yu F, Medeiros DM (2010) Evidence for the prepattern/cooption model of vertebrate jaw evolution. Proc Nat Acad Sci 107:17262–17267

Clark AJ, Maravilla EJ, Summers AP (2010) A soft origin for a forceful bite: motorpatterns of the feeding musculature in Atlantic hagfish, *Myxine glutinosa*. Zoology 113:259–268

Clark AJ, Summers AP (2007) Morphology and kinematics of feeding in hagfish: possible functional advantages of jaws. J Exp Biol 210:3897–3909

Cole FJ (1907) XXVI.—a monograph on the general morphology of the Myxinoid fishes, based on a study of Myxine. Part II. The anatomy of the muscles. Earth Env Sci Trans R Soc Edinb 45:683–757

Couly GF, Coltey PM, Douarin NML (1993) The triple origin of skull in higher vertebrates: a study in quail-chick chimeras. Development 117:409–429

Damas H (1944) Recherches sur la dcveloppement de *Lampetra fluviatilis* L. Contribution a l'etude de la cephalogenese des vertebres. Arch Biol Paris 55:1–289

Delsuc F, Brinkmann H, Chourrout D, Philippe H (2006) Tunicates and not cephalochordates are the closest living relatives of vertebrates. Nature 439:965–968

Depew MJ, Lufkin T, Rubenstein JL (2002) Specification of jaw subdivisions by *Dlx* genes. Science 298:381–385

Depew MJ, Simpson CA, Morasso M, Rubenstein JLR (2005) Reassessing the *Dlx* code: the genetic regulation of branchial arch skeletal pattern and development. J Anat 207:501–561

Diogo R, Abdala V (2010) Muscles of vertebrates—comparative anatomy, evolution, homologies and development. CRC Press; Science Publisher, Enfield, New Hampshire

Diogo R, Ziermann JM (2015) Development, metamorphosis, morphology, and diversity: the evolution of chordate muscles and the origin of vertebrates. Dev Dyn 244:1046–1057

Diogo R, Kelly RG, Christiaen L, Levine M, Ziermann JM, Molnar JL, Noden DM, Tzahor E (2015a) A new heart for a new head in vertebrate cardiopharyngeal evolution. Nature 520:466–473

Diogo R, Ziermann JM, Linde-Medina M (2015b) Specialize or risk disappearance - empirical evidence of anisomerism based on comparative and developmental studies of gnathostome head and limb musculature. Biol Rev 90:964–978

Edgeworth FH (1935) The cranial muscles of vertebrates. Cambridge at the University Press, London

Ericsson R, Knight R, Johanson Z (2013) Evolution and development of the vertebrate neck. J Anat 222:67–78

Forey P, Janvier P (1993) Agnathans and the origin of jawed vertebrates. Nature 361:129–134

Fürbringer P (1875) Untersuchungen zur vergleichenden Anatomie der Muskulatur des Kopfskelets der Cyclostomen. Jenaer Zeitschriften

Gai Z, Donoghue PC, Zhu M, Janvier P, Stampanoni M (2011) Fossil jawless fish from China foreshadows early jawed vertebrate anatomy. Nature 476:324

Gee H (2018) Across the bridge: understanding the origin of the vertebrates. University of Chicago Press, Chicago
Gess RW, Coates MI, Rubidge BS (2006) A lamprey from the Devonian period of South Africa. Nature 443:981–984
Gillis JA, Modrell MS, Baker CV (2013) Developmental evidence for serial homology of the vertebrate jaw and gill arch skeleton. Nat Commun 4:1436
Gillis JA, Tidswell OR (2017) The origin of vertebrate gills. Curr Biol 27:729–732
Goodrich ES (1930) Studies on the structure and development of vertebrates. Dover Publications Inc. P, USA
Green SA, Simoes-Costa M, Bronner ME (2015) Evolution of vertebrates as viewed from the crest. Nature 520:474–482
Hardisty M, Potter I, Hilliard R (1989) Physiological adaptations of the living agnathans. Earth Env Sci Trans R Soc Edinb 80:241–254
Harel I, Tzahor E (2013) Head muscle development. In: McLoon LK, Andrade FH (eds) Craniofacial muscles. Springer, New York, pp 11–28
Heimberg AM, Cowper-Sal lari R, Sémon M, Donoghue PCJ, Peterson KJ (2010) microRNAs reveal the interrelationships of hagfish, lampreys, and gnathostomes and the nature of the ancestral vertebrate. Proc Nat Acad Sci 107:19379–19383
Heude É, Bouhali K, Kurihara Y, Kurihara H, Couly G, Janvier P, Levi G (2010) Jaw muscularization requires *Dlx* expression by cranial neural crest cells. Proc Nat Acad Sci 107:11441–11446
Higashiyama H, Kuratani S (2014) On the maxillary nerve. J Morphol 275:17–38
Hinits Y, Williams VC, Sweetman D, Donn TM, Ma TP, Moens CB, Hughes SM (2011) Defective cranial skeletal development, larval lethality and haploinsufficiency in Myod mutant zebrafish. Dev Biol 358:102–112
Hoffmann FG, Opazo JC, Storz JF (2010) Gene cooption and convergent evolution of oxygen transport hemoglobins in jawed and jawless vertebrates. Proc Nat Acad Sci 107:14274–14279
Holland ND, Chen J (2001) Origin and early evolution of the vertebrates: new insights from advances in molecular biology, anatomy, and palaeontology. BioEssays 23:142–151
Holland ND, Holland LZ, Honma Y, Fujii T (1993) *Engrailed* expression during development of a lamprey, *Lampetra japonica*: a possible clue to homologies between agnathan and gnathostome muscles of the mandibular arch. Develop Growth Differ 35:153–160
Holmgren N (1946) On two embryos of *Myxine glutinosa*. Acta Zooligica XXVII:1–91
Holzer G, Laudet V (2017) New insights into vertebrate thyroid hormone receptor evolution. Nucl Recept Res 4
Horigome N, Myojin M, Ueki T, Hirano S, Aizawa S, Kuratani S (1999) Development of cephalic neural crest cells in embryos of *Lampetra japonica*, with special reference to the evolution of the jaw. Dev Biol 207:287–308
Huxley TH (1876) The nature of the craniofacial apparatus of *Petromyzon*. J Anat Physiol 10:412–429
Janvier P (1996) Early vertebrates. Oxford University Press, Clarendon
Janvier P (2007) Homologies and evolutionary transitions in early vertebrate history. In: Anderson JS, Sues HD (eds) Major transitions in vertebrate evolution. Indiana University Press, Bloomington, pp 57–121
Janvier P (2008) Early jawless vertebrates and cyclostome origins. Zool Sci 25:1045–1056
Johnels AG (1948) On the development and morphology of the skeleton of the head of *Petromyzon*. Acta Zool 29:139–279
Köntges G, Lumsden A (1996) Rhombencephalic neural crest segmentation is preserved throughout craniofacial ontogeny. Development 122:3229–3242
Kuraku S, Hoshiyama D, Katoh K, Suga H, Miyata T (1999) Monophyly of lampreys and hagfishes supported by nuclear DNA–coded genes. J Mol Evol 49:729–735
Kuraku S, Takio Y, Sugahara F, Takechi M, Kuratani S (2010) Evolution of oropharyngeal patterning mechanisms involving *Dlx* and *endothelins* in vertebrates. Dev Biol 341:315–323
Kuratani S (2004) Evolution of the vertebrate jaw: comparative embryology and molecular developmental biology reveal the factors behind evolutionary novelty. J Anat 205:335–347
Kuratani S (2005a) Craniofacial development and the evolution of the vertebrates: the old problems on a new background. Zool Sci 22:1–19
Kuratani S (2005b) Developmental studies of the lamprey and hierarchical evolutionary steps towards the acquisition of the jaw. J Anat 207:489–499
Kuratani S (2008a) Evolutionary developmental studies of cyclostomes and the origin of the vertebrate neck. Develop Growth Differ 50:S189–S194
Kuratani S (2008b) Is the vertebrate head segmented?—evolutionary and developmental considerations. Integr Comp Biol 48:647–657
Kuratani S (2012) Evolution of the vertebrate jaw from developmental perspectives. Evol Dev 14:76–92
Kuratani S, Adachi N, Wada N, Oisi Y, Sugahara F (2013) Developmental and evolutionary significance of the mandibular arch and prechordal/premandibular cranium in vertebrates: revising the heterotopy scenario of gnathostome jaw evolution. J Anat 222:41–55
Kuratani S, Horigome N, Hirano S (1999) Developmental morphology of the head mesoderm and reevaluation of segmental theories of the vertebrate head: evidence from embryos of an agnathan vertebrate, *Lampetra japonica*. Dev Biol 210:381–400
Kuratani S, Kuraku S, Murakami Y (2002) Lamprey as an evo-devo model: lessons from comparative embryology and molecular phylogenetics. Genesis 34:175–183
Kuratani S, Murakami Y, Nobusada Y, Kusakabe R, Hirona S (2004) Developmental fate of the mandibular mesoderm in the lamprey, *Lethenteron japonicum*: comparative morphology and development of the Gnathostome jaw with special reference to the nature

of the trabecula cranii. J Exp Zool (Mol Dev Evol) 302B:458–468
Kuratani S, Nobusada Y, Horigome N, Shigetani Y (2001) Embryology of the lamprey and evolution of the vertebrate jaw: insights from molecular and developmental perspectives. Philos Trans R Soc Lond Ser B Biol Sci 356:1615–1632
Kuratani S, Oisi Y, Ota KG (2016) Evolution of the vertebrate cranium: viewed from hagfish developmental studies. Zool Sci 33:229–238
Kusakabe R, Kuraku S, Kuratani S (2011) Expression and interaction of muscle-related genes in the lamprey imply the evolutionary scenario for vertebrate skeletal muscle, in association with the acquisition of the neck and fins. Dev Biol 350:217–227
Kusakabe R, Kuratani S (2005) Evolution and developmental patterning of the vertebrate skeletal muscles: perspectives from the lamprey. Dev Dyn 234:824–834
Leach WJ (1944) The archetypal position of amphioxus and ammocoetes and the role of endocrines in chordate evolution. American Naturalist, pp 341–357
Lescroart F, Kelly RG, Le Garrec J-F, Nicolas J-F, Meilhac SM, Buckingham M (2010) Clonal analysis reveals common lineage relationships between head muscles and second heart field derivatives in the mouse embryo. Development 137:3269–3279
Lours-Calet C, Alvares LE, El-Hanfy AS, Gandesha S, Walters EH, Sobreira DR, Wotton KR, Jorge EC, Lawson JA, Lewis AK (2014) Evolutionarily conserved morphogenetic movements at the vertebrate head–trunk interface coordinate the transport and assembly of hypopharyngeal structures. Dev Biol 390:231–246
Mallatt J (1984) Early vertebrate evolution: pharyngeal structure and the origin of gnathostomes. J Zool 204:169–183
Mallatt J (1996) Ventilation and the origin of jawed vertebrates: a new mouth. Zool J Linnean Soc 117:329–404
Mallatt J (1997a) Crossing a major morphological boundary: the origin of jaws in vertebrates. Zoology 100:128–140
Mallatt J (1997b) Shark pharyngeal muscles and early vertebrate evolution. Acta Zool 78:279–294
Mallatt J (2008) The origin of the vertebrate jaw: neoclassical ideas versus newer, development-based ideas. Zool Sci 25:990–998
Mallatt J, Chen J (2003) Fossil sister group of craniates: predicted and found. J Morphol 258:1–31
Marinelli W, Strenger A (1956a) Vergleichende Anatomie und Morphologie der Wirbeltiere: von W. Marinelli und A. Strenger. *Myxine glutinosa* (L.). Franz Deuticke:80
Marinelli W, Strenger A (1956b) Vergleichende Anatomie und Morphologie der Wirbeltiere: von W. Marinelli und A. Strenger. *Myxine glutinosa* (L.). Franz Deuticke:80
Matsuoka T, Ahlberg PE, Kessaris N, Iannarelli P, Dennehy U, Richardson WD, McMahon AP, Koentges G (2005) Neural crest origins of the neck and shoulder. Nature 436:347–355
McCauley DW, Bronner-Fraser M (2003) Neural crest contributions to the lamprey head. Development 130:2317–2327
McNamara KJ (1990) The role of heterochrony in evolutionary trends. Evolutionary Trends. K. J. McNamara, London
Medeiros DM, Crump JG (2012) New perspectives on pharyngeal dorsoventral patterning in development and evolution of the vertebrate jaw. Dev Biol 371:121–135
Miller CT, Yelon D, Stainier DYR, Kimmel CB (2003) Two *endothelin 1* effectors, *hand2* and *bapx1*, pattern ventral pharyngeal cartilage and the jaw joint. Development 130:1353–1365
Minchin JE, Williams VC, Hinits Y, Low S, Tandon P, Fan C-M, Rawls JF, Hughes SM (2013) Oesophageal and sternohyal muscle fibres are novel Pax3-dependent migratory somite derivatives essential for ingestion. Development 140:2972–2984
Minoux M, Rijli FM (2010) Molecular mechanisms of cranial neural crest cell migration and patterning in craniofacial development. Development 137:2605–2621
Miyashita T (2012) Comparative analysis of the anatomy of the Myxinoidea and the ancestry of early vertebrate lineages. M.Sc. In: University of Alberta
Miyashita T (2016) Fishing for jaws in early vertebrate evolution: a new hypothesis of mandibular confinement. Biol Rev 91:611–657
Miyashita T, Coates MI (2015) Hagfish embryology. Hagfish Biol 95
Modrell MS, Hockman D, Uy B, Buckley D, Sauka-Spengler T, Bronner ME, Baker CV (2014) A fate-map for cranial sensory ganglia in the sea lamprey. Dev Biol 385:405–416
Moore JW, Mallatt JM (1980) Feeding of larval lamprey. Can J Fish Aquat Sci 37:1658–1664
Noden DM (1983) The role of the neural crest in patterning of avian cranial skeletal, connective, and muscle tissues. Dev Biol 96:144–165
Noden DM, Francis-West P (2006) The differentiation and morphogenesis of craniofacial muscles. Dev Dyn 235:1194–1218
Oisi Y, Ota KG, Fujimoto S, Kuratani S (2013a) Development of the chondrocranium in hagfishes, with special reference to the early evolution of vertebrates. Zool Sci 30:944–961
Oisi Y, Ota KG, Kuraku S, Fujimoto S, Kuratani S (2013b) Craniofacial development of hagfishes and the evolution of vertebrates. Nature 493:175–180
Ota KG, Fujimoto S, Oisi Y, Kuratani S (2011) Identification of vertebra-like elements and their possible differentiation from sclerotomes in the hagfish. Nat Commun 2:373
Ota KG, Kuraku S, Kuratani S (2007) Hagfish embryology with reference to the evolution of the neural crest. Nature 446:672–675
Ota KG, Kuratani S (2008) Developmental biology of hagfishes, with a report on newly obtained embryos of the Japanese inshore hagfish, *Eptatretus burgeri*. Zool Sci 25:999–1011
Parker HJ, Bronner ME, Krumlauf R (2014) A *Hox* regulatory network of hindbrain segmentation is conserved to the base of vertebrates. Nature 514:490

Poole JRM, Paganini J, Pontarotti P (2017) Convergent evolution of the adaptive immune response in jawed vertebrates and cyclostomes: an evolutionary biology approach based study. Dev Com Immunol 75:120–126
Potter I (1980) The Petromyzoniformes with particular reference to paired species. Can J Fish Aquat Sci 37:1595–1615
Rijli FM, Gavalas A, Chambon P (1998) Segmentation and specification in the branchial region of the head: the role of the *Hox* selector genes. Int J Dev Biol 42:393–401
Rijli FM, Mark M, Lakkaraju S, Dierich A, Dollé P, Chambon P (1993) A homeotic transformation is generated in the rostral branchial region of the head by disruption of *Hoxa-2*, which acts as a selector gene. Cell 75:1333–1349
Rinon A, Lazar S, Marshall H, Büchmann-Møller S, Neufeld A, Elhanany-Tamir H, Taketo MM, Sommer L, Krumlauf R, Tzahor E (2007) Cranial neural crest cells regulate head muscle patterning and differentiation during vertebrate embryogenesis. Development 134:3065–3075
Romer AS (1950) The vertebrate body. WB Saunders, London
Rovainen CM (1996) Feeding and breathing in lampreys. Brain Behav Evol 48:297–305
Sambasivan R, Kuratani S, Tajbakhsh S (2011) An eye on the head: the development and evolution of craniofacial muscles. Development 138:2401–2415
Sato T, Kurihara Y, Asai R, Kawamura Y, Tonami K, Uchijima Y, Heude E, Ekker M, Levi G, Kurihara H (2008) An endothelin-1 switch specifies maxillomandibular identity. Proc Natl Acad Sci 105:18806–18811
Sauka-Spengler T, Le Mentec C, Lepage M, Mazan S (2002) Embryonic expression of *Tbx1*, a DiGeorge syndrome candidate gene, in the lamprey *Lampetra fluviatilis*. Gene Expr Patterns 2:99–103
Schilling TF, Piotrowski T, Grandel H, Brand M, Heisenberg C-P, Jiang Y-J, Beuchle D, Hammerschmidt M, Kane DA, Mullins MC, Eeden FJM, Kelsh RN, Furutani-Seiki M, Granato M, Haffter P, Odenthal J, Warga RM, Trowe T, Nüsslein-Volhard C (1996) Jaw and branchial arch mutants in zebrafish I: branchial arches. Development 123:329–344
Shigetani Y, Sugahara F, Kawakami Y, Murakami Y, Hirano S, Kuratani S (2002) Heterotopic shift of epithelial-Mesenchymal interactions in vertebrate jaw evolution. Science 296:1316–1319
Shigetani Y, Sugahara F, Kuratani S (2005) A new evolutionary scenario for the vertebrate jaw. BioEssays 27:331–338
Shih HP, Gross MK, Kioussi C (2007) Cranial muscle defects of Pitx2 mutants result from specification defects in the first branchial arch. Proc Natl Acad Sci 104:5907–5912
Song J, Boord R (1993) Motor components of the trigeminal nerve and organization of the mandibular arch muscles in vertebrates. Cells Tissues Organs 148:139–149
Sugahara F, Murakami Y, Adachi N, Kuratani S (2013) Evolution of the regionalization and patterning of the vertebrate telencephalon: what can we learn from cyclostomes? Curr Opin Genet Dev 23:475–483
Suzuki DG, Fukumoto Y, Yoshimura M, Yamazaki Y, Kosaka J, Kuratani S, Wada H (2016) Comparative morphology and development of extra-ocular muscles in the lamprey and gnathostomes reveal the ancestral state and developmental patterns of the vertebrate head. Zool Lett 2:1
Tajbakhsh S, Rocancourt D, Cossu G, Buckingham M (1997) Redefining the genetic hierarchies controlling skeletal myogenesis: *Pax-3* and *Myf-5* act upstream of *MyoD*. Cell 89:127–138
Takio Y, Kuraku S, Murakami Y, Pasqualetti M, Rijli FM, Narita Y, Kuratani S, Kusakabe R (2007) *Hox* gene expression patterns in *Lethenteron japonicum* embryos—insights into the evolution of the vertebrate *Hox* code. Dev Biol 308:606–620
Takio Y, Pasqualetti M, Kuraku S, Hirano S, Rijli FM, Kuratani S (2004) Lamprey *Hox* genes and the evolution of jaws. Nature:1–2
Theis S, Patel K, Valasek P, Otto A, Pu Q, Harel I, Tzahor E, Tajbakhsh S, Christ B, Huang R (2010) The occipital lateral plate mesoderm is a novel source for vertebrate neck musculature. Development 137:2961–2971
Tiecke E, Matsuura M, Kokubo N, Kuraku S, Kusakabe R, Kuratani S, Tanaka M (2007) Identification and developmental expression of two *Tbx1/10*-related genes in the agnathan *Lethenteron japonicum*. Dev Genes Evol 217:691–697
Tretjakoff D (1926) Das Skelett und die Muskulatur im Kopfe des Flussneunauges. Z Wiss Zool 128:267–304
Tulenko FJ, McCauley DW, MacKenzie EL, Mazan S, Kuratani S, Sugahara F, Kusakabe R, Burke AC (2013) Body wall development in lamprey and a new perspective on the origin of vertebrate paired fins. Proc Natl Acad Sci 110:11899–11904
Wada N, Nohno T, Kuratani S (2011) Dual origins of the prechordal cranium in the chicken embryo. Dev Biol 356:529–540
Yalden D (1985) Feeding mechanisms as evidence for cyclostome monophyly. Zool J Linnean Soc 84:291–300
Youson J (1980) Morphology and physiology of lamprey metamorphosis. Can J Fish Aquat Sci 37: 1687–1710
Youson JH (1997) Is lamprey metamorphosis regulated by thyroid hormones? Am Zool 37:441–460
Ziermann JM, Diogo R, Noden DM (2018) Neural crest and the patterning of vertebrate craniofacial muscles. Genesis:e23097
Ziermann JM, Freitas R, Diogo R (2017) Muscle development in the shark *Scyliorhinus canicula*: implications for the evolution of the gnathostome head and paired appendage musculature. Front Zool 14:1–17. https://doi.org/10.1186/s12983-12017-10216-y
Ziermann JM, Miyashita T, Diogo R (2014) Cephalic muscles of cyclostomes (hagfishes and lampreys) and Chondrichthyes (sharks, rays and holocephalans): comparative anatomy and early evolution of the vertebrate head muscles. Zool J Linnean Soc 172:771–802

4 Chondrichthyan Evolution, Diversity, and Senses

Catherine A. Boisvert, Peter Johnston, Kate Trinajstic, and Zerina Johanson

4.1 Introduction

4.1.1 What Are Chondrichthyes?

Modern Chondrichthyes are jawed vertebrates lacking bones, instead possessing an internal skeleton composed of **cartilage**, with differing patterns of calcification in the vertebral centra (known as areolar) versus the rest of the skeleton, where it includes a combination of **globular** and **prismatic** calcification (Dean and Summers 2006). With respect to the latter, mineralization forms on the surface of the cartilage, in small plate-like structures known as **tesserae** (Dean et al. 2015). There are two major chondrichthyan clades, including the **Holocephali** and **Elasmobranchii**. The Elasmobranchii include sharks (**Selachii**), which include the **Galeomorphii** and **Squalomorphii**, and skates and rays (**Batoidea**), representing 96% of described modern species. By comparison, the Holocephali (**Chimaeroidei:** chimaerids, rhinochimaerids, callorhinchids) make up the remaining 4% of modern chondrichthyans. The phylogenetic relationships of these groups are presented in Fig. 4.1.

Chondrichthyans have a rich fossil record, originating in the Ordovician period (Andreev et al. 2015, 2016). Much of this early record comprises external dermal denticles, or scales, with chondrichthyan body fossils occurring in the Lower Devonian (e.g., *Cladoselache*, *Doliodus*; Williams 2001; Miller et al. 2003). Any discussion of the chondrichthyan fossil record must also now be re-evaluated in light of recent phylogenies that resolve acanthodians as stem chondrichthyans (Brazeau 2009; Davis et al. 2012; Zhu et al. 2012; Brazeau and Friedman 2015; Coates et al. 2017). Acanthodians are often referred to as "spiny sharks," with small scales and spines in front of the paired and unpaired fins. New fossils from the MOTH fauna in northern Canada included taxa with acanthodian characteristics such as these fin spines but with scales that were more chondrichthyan-like (Hanke and Wilson 2010). Following this, a variety of phylogenetic analyses resolved certain acanthodians as stem chondrichthyans, with Zhu et al. (2012) producing the first analysis of jawed fishes that placed all acanthodians on this stem (see Fig. 2.1 in Chap. 2 and Fig. 4.2, purple and lilac).

C. A. Boisvert · K. Trinajstic
School of Molecular and Life Sciences, Curtin University, Bentley, WA, Australia
e-mail: catherine.boisvert@curtin.edu.au; K.Trinajstic@curtin.edu.au

P. Johnston
Department of Anatomy and Medical Imaging, University of Auckland, Auckland, New Zealand

Z. Johanson (✉)
Department of Earth Sciences, Natural History Museum, London, UK
e-mail: z.johanson@nhm.ac.uk

J. M. Ziermann et al. (eds.), *Heads, Jaws, and Muscles*, Fascinating Life Sciences, https://doi.org/10.1007/978-3-319-93560-7_4

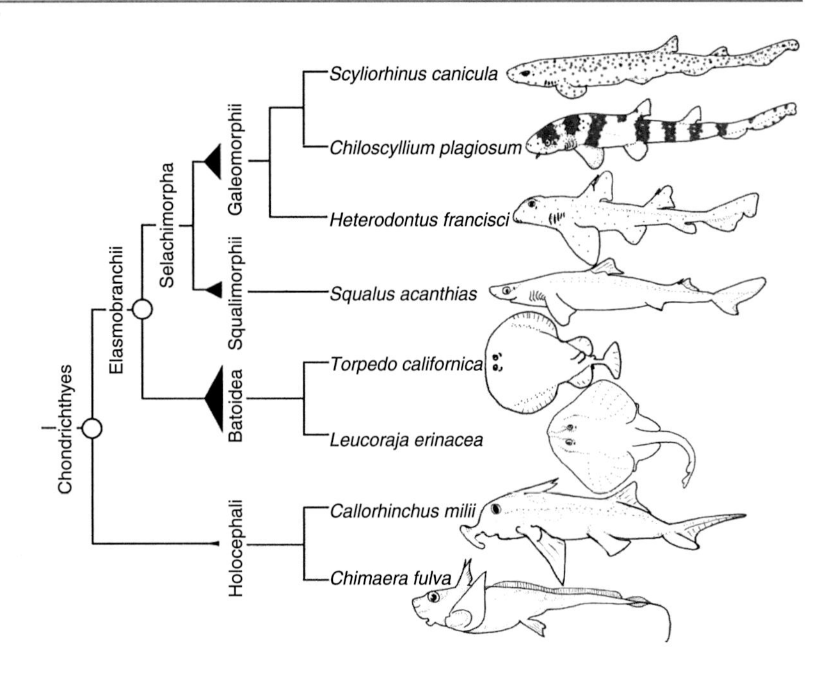

Fig. 4.1 Phylogeny of extant chondrichthyans, showing the major groups. The base of each node shows the relative number of species in each group. Modified from Renz et al. (2013). Specimen drawings by CB

As jawed vertebrates, chondrichthyans have opposing upper and lower **jaws**, with the upper jaw (**palatoquadrate**) articulating or fused with the **braincase** in a variety of ways, providing jaw support. Teeth are arranged on the upper and lower jaws into functional dentitions. One characteristic of chondrichthyans is that teeth are produced, lost, and replaced in a continuous manner. This is more apparent in sharks and rays, although the **tooth plates** of the holocephalan dentition also develop from their base as the biting surface is worn away (Stahl 1999). Other characteristic chondrichthyan features include the extensive covering of small scales (e.g., Reif 1982; but not the Holocephali), which are organized along the body to improve hydrodynamic function (Reif 1978; Dean and Bhushan 2010); internal fertilization involving modifications of the male pelvic fin to transfer sperm into the female; a connection between the inner ear and the outside environment via the endolymphatic duct; a spiral-shaped intestine and lipid-filled liver to aid in buoyancy; eyes supported by an eyestalk; and a solid cartilaginous braincase supporting the sensory capsules.

Although these features characterize chondrichthyans, there are notable differences with respect to fossil taxa, for example, various fossil species possess a braincase divided by various fissures, shared with bony fishes, and teeth that are not shed and lost, but are retained in often large **tooth whorls**. In this chapter we will review chondrichthyan features more specifically related to the cranium (braincase), brain and sensory organs, jaws, and related musculature. We will also review recent advances in chondrichthyan phylogeny, the framework upon which we base our evolutionary interpretations.

4.1.2 Historical Overview

The cartilaginous skeleton of chondrichthyans was traditionally thought to be a primitive (plesiomorphic) feature, but two lines of evidence refute this: recent phylogenetic analyses resolve the acanthodians (spiny sharks) as stem group Chondrichthyes (Zhu et al. 2012; Coates et al. 2017). This extinct group of fishes, first appearing in the fossil record in the Ordovician period (444 mya; Karatajūtė-Talimaa and Predtechenskyj 1995; Brazeau and Friedman 2015), had a superficial bony covering over parts of the head and the front (pectoral) fins. The majority of acanthodians possess bony spines supporting each fin as noted above and often a series of spines between

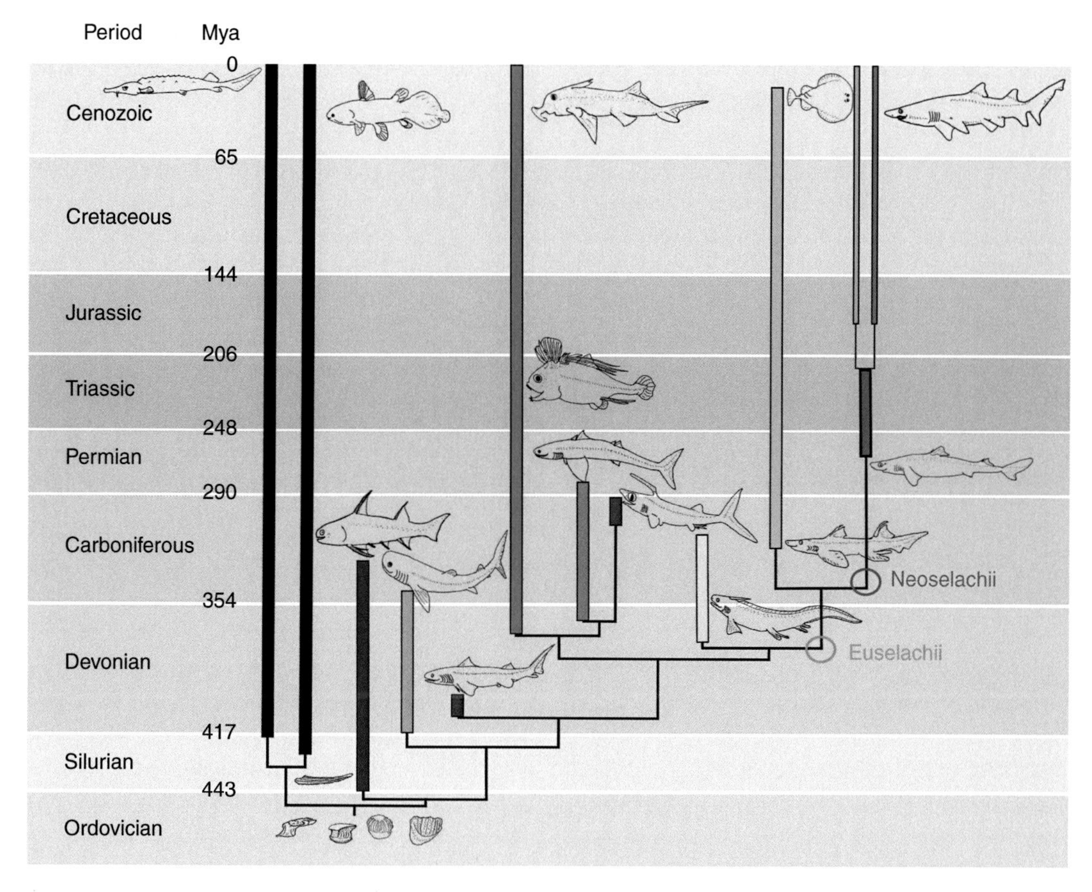

Fig. 4.2 Phylogeny of chondrichthyans showing the approximate time of first appearance of body fossils and extinction. Scales and spines of chondrichthyans and "acanthodians" are known from the Ordovician and Silurian and are illustrated by scales and fin at the base of the phylogeny. Topology and origination and extinction dates from Coates et al. (2018) and references in the text. Batoid and Selachimorpha split from Aschliman et al. (2012). All specimen drawings by CAB. **Black:** Osteichthyes, actinopterygians represented by a sturgeon (left) and sarcopterygian by a coelacanth (right). **Purple:** Acanthodid stem chondrichthyans represented by *Diplacanthus*. **Lilac**: Non-acanthodid stem chondrichthyans represented by *Gladbachus*. **Red**: *Doliodus*. **Dark Green**: Iniopterygians and holocephalans represented by *Rainerichthyes* (middle of dark green) and *Callorhinchus* (top of dark green). **Taupe**: Symmoriida represented by *Cladoselache*. **Brick red**: Paleoselachii represented by *Falcatus*. **Yellow**: Xenacanthiformes represented by *Triodus*. **Pale green**: Hybodontiformes represented by *Tristychius*. **Dark blue**: Neoselachii represented by *Squalus*. **Aqua**: Batoidea represented by *Torpedo*. **Pale blue**: Selachimorpha represented by *Carcharias taurus*

the paired fins (Janvier 1996), although some acanthodians lack these spines (Burrow and Young 1999; Hanke and Wilson 2010). The body is covered by small scales, while the internal skeleton is not preserved and so must have been made of unmineralized cartilage. Secondly, fossil sharks have been discovered with intermediate stages of evolution of the tessellated cartilages, a type of mineralized cartilage unique to chondrichthyans, and show it evolved from an ancestor possessing a bony skeleton (Long et al. 2015). Thus, chondrichthyans have lost the bone characteristic of the bony fishes (Actinopterygii; Ryll et al. 2014) and evolved surficial mineralization (tesserae) of their cartilaginous skeleton.

The concept of the great nineteenth-century anatomists of an ordered, linear progression of vertebrate evolution from cyclostomes (e.g., hagfish) to sharks and rays to bony fish and thence to tetrapods has dominated zoological thought and teaching ever since, and sharks had a key place in this progression, so much so that a skeptic among the classical anatomists coined the term "elasmobranch worship" (see Gee 2007). More recently this orderly sequence has been challenged with new fossil finds, particularly from the Silurian of China (Zhu et al. 2012, 2013) indicating extant chondrichthyans and osteichthyans exhibiting a mosaic of ancestral and derived characters. Nevertheless, elasmobranchs in particular do retain some primitive features not found in other extant fish, particularly in comparison with modern bony fish—notably in their skull, cardiovascular system, sensory organs, and fin structure—and are thus of great importance in understanding the evolution of organ systems.

4.1.3 Evolutionary History of Chondrichthyans

The first indication of chondrichthyan origins are "chondrichthyan-like" scales from the Middle Ordovician of North America, China, and Mongolia (e.g., Harding Sandstone; Smith and Sansom 1997; Andreev et al. 2015), some 50 million years before body fossils are known [with respect to the **acanthodians**, this gap is approximately 30 million years between the earliest scales, noted above, and the first acanthodian body fossil in the Late Silurian (Burrow and Rudkin 2014)]. Indeed, the early evolutionary origin of the group is known primarily from scales, suggesting early sharks were not very well mineralized, making them less likely to fossilize. It remains problematic to confidently identify chondrichthyans using only scale morphology and histology (Sansom et al. 2001, 2005) due to the plesiomorphic nature of early gnathostome scales. However, the earliest confirmed monodontoid chondrichthyan order, the **Elegestolepidida** (Lower Silurian to the Lower Devonian), is known only from scales that lack enamel and basal bone osteons and possess a distinctive neck canal formation (Andreev et al. 2015). A second order of mid-Ordovician chondrichthyans, the **Mongolepidida**, recognized by their complex polyodontoid scales, is also known from North America, China, and Mongolia (Andreev et al. 2016), indicating that a chondrichthyan radiation preceded the Lower Devonian "nektonic revolution" (*sensu* Klug et al. 2010).

Acanthodians first evolve in the Ordovician period, including body fossils, although these are often scales and spines with little other morphology preserved. Taxa now resolved as stem chondrichthyans (Fig. 4.2, purple and lilac; Coates et al. 2017) which are known from more than isolated scales include taxa such as *Poracanthodes*, *Acanthodes*, and *Kathemacanthus* (Miles 1973; Gagnier and Wilson 1996; Hanke and Wilson 2010), in addition to taxa that were previously recognized as "conventionally defined" chondrichthyans (Zhu et al. 2013), e.g., *Doliodus* (Lower Devonian; Miller et al. 2003; Maisey et al. 2009), *Pucapampella*, and *Gladbachus* (Middle Devonian; Janvier and Suarez-Riglos 1986; Heidtke and Krätschmer 2001; Maisey and Anderson 2001; Coates et al. 2018). These taxa have prismatic calcified cartilage, as a component of their skeleton, which is a diagnostic feature of chondrichthyans; by comparison, the primitive gnathostome endoskeleton comprises a core of globular cartilage surrounded by **perichondral bone** (Janvier 1996; Ørvig 1951). The endoskeleton of the early Devonian chondrichthyan *Gogoselachus* has endoskeletal elements comprising layers of **nonprismatic subpolygonal tesserae** which represent a transitional condition between globular calcified cartilage and prismatic calcified cartilage (Long et al. 2015).

Doliodus problematicus is recognized to possess a mosaic of acanthodian and shark characters (Fig. 4.2, red; Maisey et al. 2017). The squamation is shark-like, as is the dentition, braincase, jaws, and skeleton comprising prismatic cartilage (Maisey et al. 2014). However, paired pre-pectoral, pectoral, pre-pelvic, and pelvic **fin spines** are preserved along the body, which is a characteristic of acanthodians (Maisey et al. 2009, 2017). The presence of pectoral fin spines has been argued for *Antarctilamna*, another early Devonian chondrichthyan (Miller et al. 2003; Wilson et al. 2007; Gess and Coates 2015), which, along with the presence of fin spines in some placoderms and osteichthyans

(Zhu et al. 2009), indicates that paired pectoral fin spines are a gnathostome synapomorphy. ***Pucapampella*** has a **ventral cranial fissure** (Janvier and Suarez-Riglos 1986; Maisey 2001; Maisey and Anderson 2001; Maisey et al. 2009) which used to be a diagnostic feature for early osteichthyans and acanthodians (*Acanthodes*) but is now understood to be a shared characteristic of all crown group gnathostomes (Brazeau and Friedman 2015; absent in placoderms). Although ***Gladbachus*** (Fig. 4.2, lilac and reconstruction) is one the more complete stem chondrichthyans known (Burrow and Turner 2013; Coates et al. 2018), its phylogenetic position has been problematic (Coates 2005) due to the scale morphology and histology being plesiomorphic for gnathostomes (Burrow and Turner 2013). *Gladbachus* possesses teeth (Coates et al. 2018); otherwise the earliest teeth confirmed from chondrichthyans are from the Lower Devonian taxon *Leonodus* (Mader 1986).

Many other Early and Middle Devonian taxa are known only from isolated teeth (e.g., *Aztecodus*, *Celtiberina*, *Mcmurdodus*, *Portalodus*), which show a greater diversity in crown and base shape than in Late Paleozoic taxa (Ginter et al. 2010). However, in light of the combination of characters found in ***Doliodus*** which led to the description that it possessed an elasmobranch-like head and an acanthodian-like body (Maisey et al. 2017), many of the taxa established on the basis of tooth morphology alone may in fact represent stem chondrichthyans. The wide diversity of tooth morphology and the global distribution of many taxa do indicate that a diversification occurred in the Givetian (Ivanov et al. 2011). The age of the chondrichthyan crown group can be resolved to the Late Devonian, based on the appearance of taxa currently identified as stem group holocephalans, including those known from more complete specimens such as *Cladoselache* (Dean 1894), along with a variety of taxa based on teeth (Darras et al. 2008). Late Devonian elasmobranchs are also known from teeth (Ivanov et al. 2011) and fin spines (*Ctenacanthus*, Maisey 1984), but also braincases (*Cladodoides*, Maisey 2005), or some combination of these (*Tamiobatis*, Williams 1998).

By comparison, crown Elasmobranchii evolve much later, in the Mesozoic (Jurassic, Cretaceous; Maisey 2012; Janvier and Pradel 2015), while crown holocephalans also appear in the Mesozoic, but in the Triassic (Fig. 4.2; Stahl 1999; Janvier and Pradel 2015).

4.2 Early Origin of Holocephalans and Iniopterygians

Holocephalans or chimaeroids—elephant sharks, ratfish, or rabbitfish—originated in the Middle–Late Devonian (Darras et al. 2008; Janvier and Pradel 2015; Coates et al. 2017) and, as indicated by faunas such as Bear Gulch, Montana, were very morphologically diverse during the following period, the Carboniferous (Fig. 4.2, dark green; e.g., Stahl 1999; Grogan and Lund 2004; Lund and Grogan 2004). As mentioned above, the holocephalan crown group evolved in the Mesozoic, including callorhynchids (Jurassic), chimaeroids (Cretaceous), and rhinochimaerids (Triassic) (Stahl 1999; Janvier and Pradel 2015: Fig. 1.2). The composition of the holocephalan stem group has been changeable, with taxa such as *Cladoselache* and the Symmoriiformes (e.g., *Akmonistion*, *Cobelodus*) either resolved to the chondrichthyan stem (Pradel et al. 2011) or to the holocephalan stem (Coates and Sequeira 2001; Coates et al. 2017). The Iniopterygia has a more stable relationship as stem group holocephalans. The iniopterygians were an unusual and highly specialized group of stem group holocephalans, known only from a small number of genera, but with a range of body forms (Zangerl and Case 1973; Grogan and Lund 2009). All had stout pectoral fin projecting high up on the shoulder girdle (a synapomorphy of the group; Stahl 1980; Grogan and Lund 2009) with large pectoral spines. There are two families, the Iniopterygidae with upper jaws not fused to the braincase (**non-hyostylic**) and the Sibyrhynchidae, showing the hyostylic or **autostylic** condition (Zangerl and Case 1973; Stahl 1980; Pradel et al. 2010).

A new fossil from the Early Permian of the Karoo sandstone in South Africa, *Dwykaselachus*, has external anatomy of a group known as the Symmoriiformes (Fig. 4.2, taupe) but also chimaeroid specializations like the otic labyrinth arrangement and brain space configuration relative to large orbits (a potential adaptation to deep-water environments, a niche occupied by various holocephalans), showing a transitional phase to the characteristic chimaeroid cranium (Coates et al. 2017). Phylogenetic analyses establish the importance of the shared similarities between *Dwykaselachus* and chimaeroids, recovering Symmoriiformes as a stem holocephalan, sister clade to the iniopterygians and holocephalans. Notably, the Late Devonian taxon *Cladoselache* is resolved phylogenetically as a symmoriid, which implies a minimum age for the elasmobranch-holocephalan split within the Devonian (Coates et al. 2017).

4.3 Major Events in the Evolution of Chondrichthyans

Chondrichthyes and Osteichthyes are referred to as crown group Gnathostomata, but importantly, there are a range of fossil jawed vertebrates that are more closely related to this crown group than are jawless vertebrates. Therefore, any consideration of the evolution of major chondrichthyan characters needs to take these taxa into account, including the placoderms (covered in the Chap. 2), which are generally characterized by having the head and the anterior part of the body covered with thin bony plates, as well as *Ramirosuarezia* (Pradel et al. 2009, b), which is resolved as phylogenetically closer to the crown group than the placoderms (including *Entelognathus* and *Qilinyu*; Zhu et al. 2013, 2016). Important features that we discuss below are related to the skull, jaws, musculature, and gill arches, which are of particular interest because they are related to feeding, breathing, and the ability to sense the surrounding environment.

An additional consideration when discussing chondrichthyan evolution is the effects of major extinctions on the group. For example, the Late Devonian saw two extinctions, the later one associated with the disappearance of the placoderms and most acanthodians, presenting a major evolutionary opportunity for both chondrichthyans and osteichthyans. Most chondrichthyans evolving in the Carboniferous possessed flatter, presumably crushing dentitions (Sallan and Coates 2010; Sallan et al. 2011), including the wide diversity of holocephalans just mentioned (Grogan and Lund 2004; Lund and Grogan 2004). Many of these taxa were affected by later extinctions in the Permian, where 96% of marine life was lost (Sepkoski 1984), although others appear to have become extinct earlier in the Permian or in the Late Carboniferous (Friedman and Sallan 2012). However, *Hybodus* and the Hybodontidae survived this extinction and are currently resolved as the sister group to living sharks and rays (Fig. 4.2, pale green). These groups, along with the chimaeroids, originated in the Mesozoic, with sharks and rays splitting in the Upper Triassic (Aschliman et al. 2012) and diversifying in the Early–Middle Jurassic (Guinot and Cavin 2015). Recent research on lamniform sharks suggests that diversity and disparity decreased shortly after the end-Cretaceous extinction (75% of marine life; Sepkoski 1984), with sharks becoming smaller and teeth becoming less robust (Belbin et al. 2017).

4.3.1 What Makes Them Special?

Sharks and rays are a very diverse group that occupy many different ecological niches, including both fresh and saltwater environments, and at markedly different depths. Their feeding behavior ranges from stalking predation to ambush predation to plankton filtration, with different associated locomotor patterns (Motta and Huber 2012). Their main features include a mouth on the ventral surface of the head, mobility of the upper jaw, and separate gill slits. Their jaw suspension—hinged at the back of the skull—allows the shark snout to be elevated while taking prey and the jaws to protrude toward the prey or into the seafloor in the case of rays. Separate gill slits are visible externally, usually five but also six or seven in some species (e.g., the broadnose seven-gill shark *Notorynchus cepedianus*). The first gill slit is present in many species as the spiracle, a round opening on the dorsal side of the head, unlike bony fish in which the spiracle is found only in a few primitive types. Both inflow and outflow can occur via the spiracle. In rays, where the mouth is often directly in contact with the seafloor, the spiracle is particularly used for ventilation, as clean water can be taken in, rather than sand or mud (Summers and Ferry-Graham 2001). Some sharks need constant movement to force water over the gill surfaces, known as **ram ventilation**, while others can remain stationary and use suction from expansion of the mouth and pharynx to draw water in and over the gills. Feeding is similarly achieved by "ram" and "suction" methods in different groups of sharks (Wilga and Ferry 2016). Multiple rows of large sharp teeth are characteristic of large carnivorous sharks, but many other arrangements for biting, holding, and tearing smaller prey are found. Whale and basking sharks have large numbers of tiny teeth and depend on filtering plankton at the internal openings of the gills. Rays typically have hard tooth plates for crushing crustaceans and mollusks, although manta rays are **filter feeders**. Megamouth sharks are very large deepwater filter feeders that were only discovered in 1976 (*Megachasma pelagios*) and have bioluminescent tissue within the mouth to attract prey.

The holocephalans are unusual compared to the sharks and rays but share with them, for example, a superficially mineralized cartilaginous skeleton and claspers. Holocephalans are typically deepwater dwellers that feed along the bottom of the ocean. They have hard, mineralized tooth plates for crushing hard-bodied prey (Huber et al. 2008; Boisvert et al. 2015). The upper jaw (palatoquadrate) suspension involves fusion to the base of the skull (**holostylic** jaw suspension); however, as noted above, some Iniopterygia are non-holostylic, as are the symmoriiforms (*Cladoselache*, *Akmonistion*, and *Cobelodus*; Coates and Sequeira 2001; Maisey 2007), unlike the upper jaw attachment at two points on the cranium (amphistylic suspension) of primitive chondrichthyans or the single-point suspension of modern sharks and rays (**hyostylic**). Other key holocephalan features include an **operculum**, a flap-like covering over the external gill openings, similar to that present (but nonhomologous) in bony fish but absent in sharks, rays, and Symmoriiformes, gill arches positioned under the cranium, and a cranial clasper, which is a hook-like structure on top of the head of males of certain holocephalan taxa (but absent in Symmoriiformes) that is used to grip the pectoral fins of females during copulation, an adaptation for internal fertilization. Also, holocephalans generally lack scales, although sensory canals are lined with small, calcified rings, and embryos of taxa like *Callorhinchus milii* have rows of scales on the head that are lost during development (CB pers. obs.).

4.3.2 The Chondrichthyan Cranium

All craniates—animals with brains—have a cartilaginous braincase or **chondrocranium** during development, surrounding the brain (often open dorsally) and incorporating the cartilaginous supports or capsules of the eyes, inner ears, and nasal structures. In animals with bones, the osteichthyans, the chondrocranium becomes replaced by bone and covered by the bones of the outer skull. The persisting cartilaginous skull of chondrichthyans was, until quite recently, thought to be a primitive state. However, stem gnathostomes such as the placoderms had comparable outer bony plates, if not all of the chondrocranium ossified, so an entirely cartilaginous braincase is derived (Brazeau 2008). New fossils of early osteichthyans and chondrichthyans are emerging, and details of their cranial structure are used to understand the evolution of early gnathostomes.

4.3.3 Chondrichthyan Jaws and Jaw Suspension

As noted above, chondrichthyans are characterized by a range of types of jaw suspension, involving the attachment of the upper jaw (palatoquadrate) to the braincase or cranium (Maisey 1980, 2008) and the degree of support

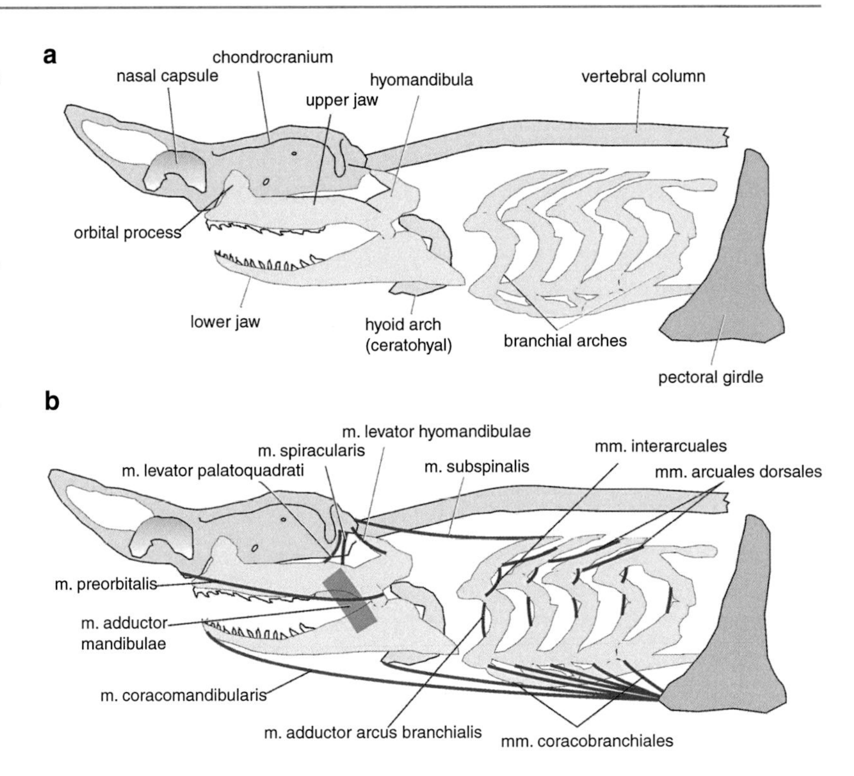

Fig. 4.3 Cranial and branchial anatomy of the school shark *Galeorhinus galeus*. (**a**) Cranial and branchial skeleton, schematic. Note the upper jaw (palatoquadrate cartilage) attached to the cranium ("suspension") by the hyomandibula (second branchial arch); there is a loose sliding articulation at the orbital process (hyostylic jaw suspension). (**b**) Locations of cranial and branchial muscles (after Hughes and Ballantijn 1965; Mikoleit 2004)

provided by the hyoid arch just posterior to the jaws, which includes the **hyomandibula** dorsally and the **ceratohyal** ventrally (Fig. 4.3). Although the hyomandibula provides support to the jaws, the ceratohyal forms an important part of the gill arch basket, relevant to suction feeding and how aerated water is drawn into the mouth and over the gills (suction versus ram). In sharks, the **palatoquadrate** can attach at a variety of points anteroposteriorly along the cranium (Maisey 1980: Figure 1; Wilga 2005: Figure 1), for example, on the postorbital process, and also more anteriorly, near the nasal capsules (ethmoid region). Several shark taxa also have a prominent orbital process of the palatoquadrate, including the Squalomorphii ("orbitostylic sharks"; Maisey 1980: Figure 4). These attachments can be at articular surfaces or ligamentous (Wilga 2005). The hyomandibular plays a role in jaw suspension and is relatively large, articulating to the otic region of the braincase (Maisey 1980). In skates and rays (Batoidea), palatoquadrate articulations are absent, with jaw support provided only by the hyomandibular. The batoid palatoquadrates are also shorter and do not extend anteriorly to the ethmoid region of the braincase. Among sharks, this is also the case in the galeomorphs and squalomorphs apart from *Chlamydoselachus* and hexanchoids (Lane and Maisey 2012).

With respect to stem chondrichthyans, jaws and jaw suspension are only known from *Acanthodes bronni* among the "acanthodians" (Miles 1973; Davis et al. 2012; Brazeau and de Winter 2015). Miles (1973) identified characters, including the position of the hyomandibula relative to the jugular vein that suggested a more osteichthyan-like condition. In previous phylogenetic analyses of the jawed vertebrates, the Acanthodii was resolved as a paraphyletic group, with *Acanthodes* resolved as more closely related to bony fishes, while other acanthodians were more closely related to the chondrichthyans (Brazeau 2009). Subsequent examination of the *Acanthodes* material demonstrated that the palatoquadrate articulated with the postorbital process, as in the stem chondrichthyans discussed below. The hyomandibula was thought to articulate on the otic region, another chondrichthyan character (Davis et al. 2012). Despite this, the analysis of Davis et al. (2012) resolved all acanthodians to the osteichthyan stem, contrary to Brazeau (2009). Later analyses, though, assigned acanthodians to the chondrichthyan stem, a largely stable result to this day (Zhu et al. 2013; Coates et al. 2017, 2018) (Fig. 4.2, purple and lilac). As well, Brazeau and de Winter (2015) confirmed that the hyomandibular position relative to the jugular groove was more similar to the chondrichthyan condition. Therefore, *Acanthodes* could act as a proxy for the acanthodian braincase, jaws, and jaw suspension and an outgroup condition for jaw suspension the rest of the chondrichthyan (non-acanthodian) lineage.

More phylogenetically derived stem group elasmobranchs and holocephalans generally have an elongate, cleaver-shaped palatoquadrate, including cladoselachians, *Cobelodus*, symmoriids, xenacanths, and ctenacanths, with the large flange located posterior to the orbit and articulating with the posterior margin of the postorbital process of the cranium (Maisey 1980; Lane and Maisey 2012). The jaw joint is posterior to the otic region of the braincase (Lane and Maisey 2012). An ethmoidal articulation is present, considered plesiomorphic for chondrichthyans along with the articulation to the postorbital process (Maisey 2008; Lane and Maisey 2012). Jaw musculature has been reconstructed in cladoselachians (Late Devonian; Maisey 1989) and *Cobelodus* (Maisey 2007).

The **Hybodontoidea** (Devonian to Miocene; Fig. 4.2, pale green) represent the sister group to extant sharks and rays, with their jaw suspension being recently reviewed by Lane and Maisey (2012). In many hybodont taxa, the palatoquadrate articulates anteriorly but not to the postorbital process, with a suspensory hyomandibular. Among extant sharks, this also characterizes the heterodontiforms, lamniforms (ligamentous), and the carcharhiniforms (articular facet; see also Wilga 2005).

4.4 Chondrichthyan Dentitions

Chondrichthyan dentitions are enormously varied, with the flatter, pavement-like dentitions of the skates and rays (e.g., Underwood et al. 2015) and the crushing tooth plates of the chimaeroids (Patterson 1965; Didier 1995; Stahl 1999), along with the range of dentitions in the shark, from the very small teeth in the filter-feeding basking shark (*Cetorhinus*) to the functional row of cutting blade dentitions in taxa such as the cookiecutter shark (*Isistius*; Underwood et al. 2016) to taxa with different teeth within their dentitions, such as the Port Jackson shark *Heterodontus*.

One major feature of the chondrichthyan dentition is the ongoing **regeneration** and replacement of teeth, which develop along the base of the jaw in a structure known as the **dental lamina**, and the organization or patterning of teeth along the jaw. Recent research has provided considerable insight into the genetic network involved in the development of chondrichthyan teeth and dentitions, focused on the catshark *Scyliorhinus* (Fraser et al. 2009, 2010; Smith et al. 2009; Debiais-Thibaud et al. 2011; Fraser and Smith 2011), and links to the external dermal denticles (Fraser et al. 2010). Most recently, focus has been on shark tooth regeneration and replacement and the role played by stem cells (Rasch et al. 2016; Martin et al. 2016). The first teeth develop in more superficial epithelium along the jaw, known as the odontogenic band (Smith et al. 2009, 2016). Intriguingly, this band is associated with taste buds in the mouth, to form an odonto-gustatory band (Rasch et al. 2016). Several gene families are expressed within this band, including *Hh*, *Wnt/β-catenin*, *Bmp*, *Pitx2*, and *Fgf*, important in tooth development in bony fishes (Fraser et al. 2006, 2012), suggesting a very deep evolutionary history. These gene families are also important in all stages of tooth regeneration within the dental lamina.

As the epithelial cells of the odonto-gustatory band proliferate, the dental lamina begins to develop, with cells expressing *β-catenin*, *Pitx1*, and *Sox2* (Rasch et al. 2016). Stem cells are held within the dental lamina (successional lamina) but, also more superficially, within taste buds proximate to the oral epithelium (Rasch et al. 2016). The genes *β-catenin*, *Pitx1*, and *Lef1* are expressed in conjunction with regenerating teeth, but not *shh*, which is only involved in tooth initiation.

4.5 Musculature

4.5.1 Muscles for Jaw Mechanics

Filter feeders aside, chondrichthyans use their jaws to bite, crush, or grasp their prey, with tooth shapes and sizes appropriate to these mechanisms. The jaws are developmentally separate to the rest of the skull, which houses the brain and sensory organs: the upper and lower jaws are developed from the first and second pharyngeal arches, a series of developmental structures that form jaws anteriorly and branchial arches posteriorly (origin of jaws covered in Chap. 2, but see also Chaps. 1 and 3). Branchial arches are skeletally supported arches that typically have openings between them, forming the gill slits. The jaws have an outer layer of mineralized (calcified) cartilage arranged in tiles over a non-mineralized core. This arrangement allows considerable force to be applied. In sharks and rays, two sets of muscles are used, one to protrude the upper jaw and the other to close the lower jaw against the upper (Fig. 4.4). Upper jaw protrusion by the **m. preorbitalis** (Fig. 4.4) is most obvious in large predatory sharks, which elevate the snout to allow the protruded jaw mechanism a more front-on approach to the prey. The jaw is closed by the **m. adductor mandibulae**, which can have several portions (see Ziermann et al. 2017).

4.5.2 Modification of the Jaw Musculature in Suction Feeders

Rays have developed a very mobile jaw mechanism that can be hinged ventrally to the body to extend into the substrate (seafloor) to take prey by biting or suction. The arrangement of muscle fibers in the jaw-closing muscle is effective at maximizing force or speed of closure, varying among different feeding patterns in diverse sharks. In fact, the jaw-closing muscles of sharks are mechanically more effective than those of mammals, which evolved many millions of years later.

4.5.3 Jaw Musculature in Prey Crushers

Holocephalans have the upper jaw fused along the base of the skull and cannot protrude the upper jaw; these fish have a complex pattern of muscles in the snout including the **m. levator anguli oris**, **m. labialis,** and **m. prelabialis** which are labial muscles, in addition to the **m. preorbitalis** found in elasmobranchs. Holocephalan snout muscles have been attempted to be homologized to the labial muscles of cyclostomes and the m. preorbitalis of elasmo-

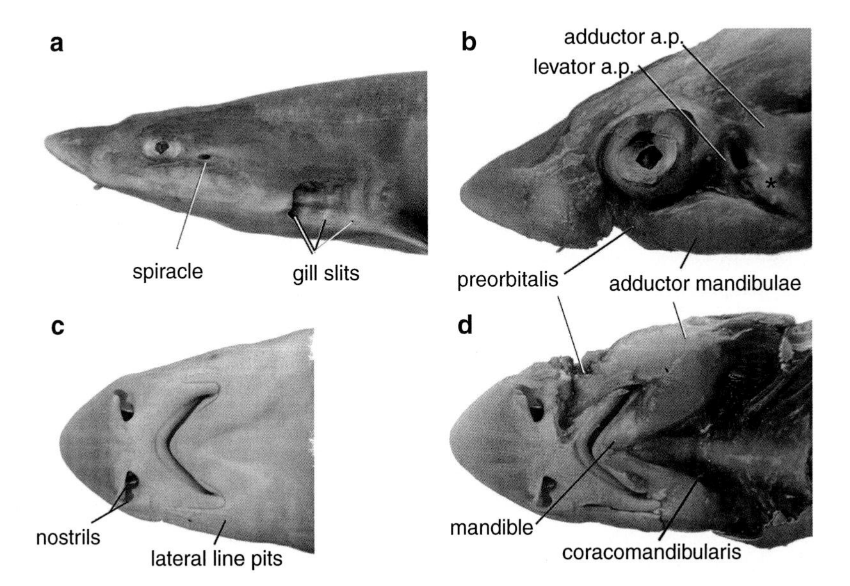

Fig. 4.4 External and muscular anatomy of the head of the school shark *Galeorhinus galeus*. (**a**, **b**) Lateral view; (**c**, **d**) ventral view. (**a**) External anatomy and (**b**) superficial dissection to show jaw muscles. (**c**) External anatomy and (**d**) superficial dissection to show jaw adductors (adductor mandibulae, preorbitalis) and depressor (coracomandibularis). Abbreviations: levator a.p., levator arcus palatini; adductor a.p., adductor arcus palatine

branchs (Ziermann et al. 2014), but as seen in the previous chapter (Chap. 2), the labial muscles of cyclostomes are unlikely to be homologous to those of chondrichthyans (see also Chap. 3). Homologizing all the labial muscles of holocephalans to only the m. preorbitalis of elasmobranchs may be an oversimplification as the complex labial musculature of holocephalans appears unrelated to jaw mechanics and probably is related to movements of the sensory apparatus concentrated in the snout.

4.5.4 Musculature Used for Breathing, and Spiracular Breathing

The other important muscle function in the head is that of expansion and contraction of the pharynx and gill chambers, for the mechanics of ventilation and suction feeding. Some sharks are "obligate ram ventilators" and need to be moving all the time to have a flow of water over the gills, and other slow-moving or largely stationary species use **suction ventilation** entirely, but most chondrichthyans use a combination of ram and suction ventilation (Brainerd and Ferry-Graham 2006). To generate suction, the gill chamber (pharynx) and to a lesser extent the mouth cavity need to expand rapidly. This is accomplished by the hinged nature of the branchial skeleton, which allows changes in volume of the enclosed chamber as the hinges are operated by muscles that mainly pull the floor of the mouth and pharynx downward (Wilga and Ferry 2016). These muscles originate from the pectoral (shoulder) girdle and are the **coracomandibularis** that inserts onto the mandible and the **coracobranchialis** that inserts onto the branchial arches (Fig. 4.3).

Lateral expansion is also possible in some species, depending on the orientation of the hinges, and is assisted by sheets of muscle that surround the gill chambers. Externally there is a valve mechanism to allow an effective pump with negative pressure being generated within the gill chambers and mouth—in sharks and rays, there are soft tissue valves over each of the separate gill slits, preventing water from entering but allowing escape during the compression part of the pump cycle. In holocephalans, the gill slits are covered externally by the operculum, and these fishes are similar to bony fishes, lungfish, and coelacanths in this respect. The operculum forms a flap valve over all the external gill openings and appears to assist suction ventilation and feeding in the way that the individual gill slits do in sharks, although the evidence in holocephalans is uncertain.

4.6 Gill Arch Evolution

As noted above, chondrichthyans have 5–7 gill arches posterior to the hyoid arch. Each jointed arch is formed from a number of dorsal (**pharyngo**- and **epibranchials**) and more ventral arch elements (**hypo**- and **ceratobranchials**) that articulate with a ventral midline series of **basibranchials**, forming a highly flexible unit with associated musculature (Miyake et al. 1992; Mallatt 1997; Wilga et al. 2001) to not only bring food into the oral cavity but aerated water as well. The arrangement of these arches has been reviewed by Nelson (1969), including denticles on the oropharyngeal arch surface. The general structure of the gill arches is conserved through jawed fishes, including the acanthodians (Miles 1964) and more phylogenetically basal placoderms (Carr et al. 2009; Stensiö 1963), recently reviewed by Pradel et al. (2014: Figure 3). There are differences in the number and arrangement of these arches; for example, in chondrichthyans, the pharyngo- and hypobranchials are directed posteriorly (but anteriorly in chimaeroids), while in osteichthyans these are oriented anteriorly. In keeping with previous ideas that chondrichthyan morphologies represented the primitive condition for jawed vertebrates, it was thought that the chondrichthyan arrangement of the arches, often described as a "∑," was primitive (Pradel et al. 2014). Although the ventral gill arches are known in placoderms, a full complement of gill arches is unknown for comparison to crown group gnathostomes (e.g., Carr et al. 2009; Brazeau et al. 2017).

Recently, a complete and associated series of gill arches was described in the symmoriiform

Ozarcus (Pradel et al. 2014). These showed multiple similarities to the arches of bony fishes, rather than other chondrichthyans, including having two pharyngobranchials (infra, supra), with the infrapharyngobranchials and more ventral hypobranchials having an anterior orientation, rather than posterior. As well, the last three ceratobranchials articulated to the posteriormost basibranchial, while in chondrichthyans, only the last ceratobranchial articulates with this basibranchial. When *Ozarcus* was first described, the symmoriiforms were considered to be stem group chondrichthyans (Pradel et al. 2011), and therefore the gill arches, and their similarity to bony fishes, were highly relevant to the evolution of chondrichthyan arches. However, symmoriiforms, including *Ozarcus*, are most recently resolved as stem group holocephalans (Coates et al. 2017). Important in this regard are the stem group chondrichthyans *Doliodus* and *Gladbachus*. Although branchial arches are present in *Doliodus* (Fig. 4.2, red), hypo- and pharyngobranchials were not described (Maisey et al. 2009). In *Gladbachus* (Fig. 4.2, lilac), the pharyngobranchials are oriented anteriorly, the bony fish condition (Coates et al. 2017). The retention of this state in the holocephalan stem in *Ozarcus* suggests that the posterior orientation of the pharyngobranchials was attained independently in crown group holocephalans and elasmobranchs.

4.7 Chondrichthyan Brains and Senses

Among the chondrichthyans, sharks are legendary for their sensory abilities. For example, we have all heard that sharks can smell a single drop of blood in the ocean, and movies have often exaggerated the sensory abilities of sharks. This section is an overview of the **brains** and **senses** of sharks, batoids, and holocephalans, keeping in mind that sensory abilities, brain size, and organization differ greatly between species and are highly associated with a given species' ecological niche (Yopak et al. 2007). Overall, chondrichthyans have large brains relative to body size when compared with other vertebrates, and galeomorph sharks and myliobatiform rays have similar brain/body ratios to those found in mammals (Bauchot et al. 1976; Northcutt 1978; Yopak et al. 2010). Phylogenetically more basal groups tend to have a smaller brain/body ratio, while brain size and cerebellar complexity (including foliation or folding) increase from phylogenetically more basal squalomorph sharks to more derived galeomorphs such as Carcharhinidae and Lamnidae (see phylogeny in Fig. 4.1; Yopak et al. 2007). Hammerhead sharks have the relatively largest brains, whereas whale sharks and the great white and gray nurse sharks have among the smallest brains relative to body size (Yopak et al. 2007; Yopak and Frank 2009). A similar pattern is found in batoids, with rajiforms, rhinopristiforms, and torpediniforms having smaller, less structurally complex brains than myliobatiforms (Northcutt 1978; Lisney et al. 2008). Among the rays, devil rays appear to have the largest and most complex brains (Lisney et al. 2008; Ari 2011; Yopak 2012). The brains of holocephalans are generally similar in size (and morphology) to those found in squalomorph sharks (Northcutt 1978; Yopak and Montgomery 2008). However, there is more to the brain than simply its size. Building on the pioneering work of Northcutt (1978), Yopak et al. (2007, 2010; Yopak and Frank 2009) and Lisney et al. (2008) have assessed chondrichthyan brains to evaluate the proportions of the different parts of the brain (Fig. 4.5) relative to each other and the degree to which the cerebellum is foliated. The five major brain areas are the forebrain, composed of the (1) **telencephalon** (Fig. 4.5b, pink), which as well as receiving primary **olfactory** input from the olfactory bulbs also receives multisensory input from the other modalities and is involved with multisensory processing and higher cognitive functions, and (2) the **diencephalon** (Fig. 4.5b, yellow), a multisensory relay center that acts as an interface between the brain and the endocrine systems and which plays an important role in homeostasis; (3) the midbrain or **mesencephalon** (Fig. 4.5c, blue), which is characterized by two prominent dorsal lobes, the optic tectum, which receives the majority of visual input from the retina (as well as input from other sensory modalities); and (4) the **hindbrain** (Fig. 4.5b, green), composed of the **medulla** and (5) the **cerebellum** (Fig. 4.5a). The medulla receives primary sensory input from the **octavolateralis** systems (acoustic, electroreceptive, and lateral line systems), while the cerebellum is a multimodal integration center that is important in muscle coordination and monitors the body's position in space. In some species, the cerebellum is **foliated**, which increases the surface area of the brain and is believed to increase cognitive ability as the cerebellum is involved in the integration of different stimuli (Walker and Homberger 1992; Demski and Northcutt 1996; Yopak 2012). Cerebellar foliation is phylogenetically relevant, with more basal elasmobranchs lacking foliation, holocephalans having a low **foliation index** while galeomorph sharks such as hammerheads and myliobatiform rays such as manta rays having high foliation indices (Yopak et al. 2007; Lisney et al. 2008) (see Fig. 4.1 for phylogeny). However, foliation is not entirely correlated with phylogeny and depends strongly on environment adaptability. Additionally, brain foliation often comes at the expense of brain size except in highly derived species such as the hammerhead.

The species with the most foliated cerebellum are those that are migratory and which hunt very active agile prey. In terms of brain proportions, there are cerebrotypes (brain configuration types) where species living in similar environments cluster (Yopak et al. 2007; Lisney et al. 2008). Bottom or near-bottom dwelling (demersal benthic) chondrichthyans such as batoids have an average-sized telencephalon, cerebellum, and medulla, with an enlarged mesencephalon. In contrast, holocephalans have very large cerebellums and an enlarged medulla but are below average relative to body size (Yopak et al. 2007). This suggests that holocephalans rely on electroreceptive, acoustic, and lateral line systems heavily. Demersal benthic species also have enlarged eyes, which may show a greater reliance on vision in these habitats. Wobbegongs and blind sharks are reef-associated bottom dwelling (benthic) species that have a reduced mesencephalon but enlarged medulla, the brain area that houses the primary sensory nuclei for the octavolateralis senses.

Many deepwater chondrichthyans have relatively small brains and a well-developed mesencephalon and medulla (Yopak and Montgomery 2008). This potentially reflects the fact that many

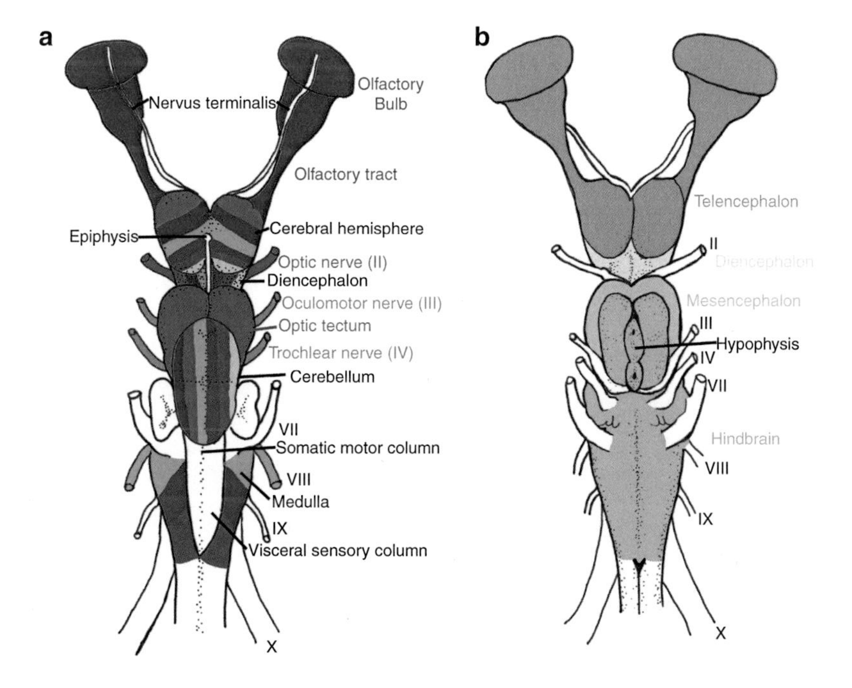

Fig. 4.5 Dogfish brain drawing in (**a**) dorsal view, color-coded to show the major sensory areas. Red: olfactory. Blue: visual. Green: hearing. Purple: electroreception. Yellow: taste. Brown: lateral line/mechanoreceptor. Areas with several colors are multimodal centers that integrate inputs from several senses. (**b**) Ventral view showing four main areas of the brain. Pink: telencephalon. Yellow: diencephalon. Pale blue: mesencephalon. Pale green: hindbrain. Modified from House and Pansky (1960) by CB

of these species prey on invertebrates, are slower moving, and inhabit an environment that is largely "two dimensional" (i.e., horizontal and above rather than above, below, and horizontal). In contrast, the largest brains are found in open water species associated with coastal and especially reef habitats. This might be due to the complexity of the reef environment, where animals have to learn the spatial organization of the habitat and its inhabitants (Bauchot et al. 1977; Northcutt 1978, 1979), as well as the complex social behaviors between conspecifics (members of the same species) and other species (Kotrschal et al. 1998), for example, when schooling. In the case of the thresher shark, brain size and morphology could be linked to its unique prey-capturing behavior using the extremely elongated upper tail fin lobe (Lisney and Collin 2006). Cerebellar foliation seems to be linked to locomotor abilities and sensory motor integration (New 2001) with slow-moving species using lateral undulation having less foliated brains and fast swimmers having more foliated cerebellum. Brain size is also correlated with the mode of reproduction, with viviparous species having the largest brain, but this could be a phylogenetic signal since the most derived species are all viviparous. Brain size and organization are therefore influenced by habitat, locomotion, and phylogeny, but having a more basic brain could make the animal more adaptable (Brabrand 1985; Lammens et al. 1987; Wagner 2002).

4.8 Sense Organ Development

Chondrichthyans have six well-developed senses: **vision**, **smell (olfaction)**, **taste (gustation)**, **mechanoreception** (touch and vibration through the **lateral line system**), **hearing,** and **electroreception**. Given the variety of habitats these animals live in, there is also a great variability in these senses, but this section focuses on the generalities of how these develop and function in chondrichthyans.

One of the greatest evolutionary novelties of vertebrates are sense organs developing from **migratory neurogenic placodes** and **neural crest cells** (Lipovsek et al. 2017; see also Chap. 2). **Cranial placodes** are patches of thickened ectoderm in the embryonic head that give rise to paired organs involved in hearing, olfaction, and detecting vibrations (through the lateral line which runs across the head and onto the body), lenses of the eyes, as well as neurons connecting them to the brain (O'Neill et al. 2007). Taste buds are not placode derived, but the neurons connecting them to the brain are derived from cranial placodes (O'Neill et al. 2007). The organs of electroreception derive from the lateral line placodes (Baker et al. 2013), which are themselves derived from neural crest cells (Juarez et al. 2013).

All placodes share a common developmental origin. In the early neurula stage of the embryo (when the nervous system develops), the folding neural plate is horseshoe-shaped (see Fig. 2.2 in Chap. 2), and the anterior domain is called the preplacodal or pan-placodal domain (Baker and Bronner-Fraser 2001). This domain is defined molecularly by the expression of the homeodomain transcription factors *Six1/2* and *Six4/5* which interact with the transcription cofactor *Eya1/2*. These genes are maintained in individual placodes, but different placodes are induced at different times during development by different tissues and molecules. The *Pax* (paired box) genes code for tissue-specific transcription factors and are upregulated later in cells fated to adopt different placodal fates (O'Neill et al. 2007): *Pax2* is expressed in the otic placode (hearing), *Pax3* in the ophthalmic placodes (vision), and *Pax6* in prospective lens and olfactory placodes (downregulated in olfactory placodes, Bhattacharyya et al. 2004). The **lateral line** and **electroreceptive ampullary organs** are linked developmentally by the expression of a novel chondrichthyan marker *Eya4*, and the lateral line ganglia initially express *Tbx3*. Overall, the expression of transcription factors underlying placode and cranial sensory ganglion development is highly conserved in all gnathostomes (O'Neill et al. 2007). The electrosensory ampullary organs are not unique to chondrichthyans, being present in larval lampreys, amphibians, and teleosts and are thought to be homologous in all non-teleosts (Baker et al. 2013), having re-evolved at least twice in teleosts.

4.8.1 Smell/Olfaction

Chondrichthyans rely on **olfaction** to detect prey, predators, and signal conspecifics (other sharks of the same species; Yopak et al. 2015; Theiss et al. 2009) as well as for navigation (Nosal et al. 2016) (Fig. 4.5 in red for areas of the brain and Fig. 4.6b). As the olfactory system is not connected to the respiratory system in sharks, water needs to be pumped into the nasal sacs to detect chemicals. In elasmobranchs, each nostril is divided by a flap of skin to separate incurrent from excurrent water flow (Walker and Homberger 1992) whereas, in chimaeroids, there is one pair of external nostrils but two channels diverting the water to the mouth, providing the same incurrent-excurrent flow-through system (Howard et al. 2013). In most chondrichthyans, water is pumped into the nasal sacs, but the forward motion of some continually swimming spe-

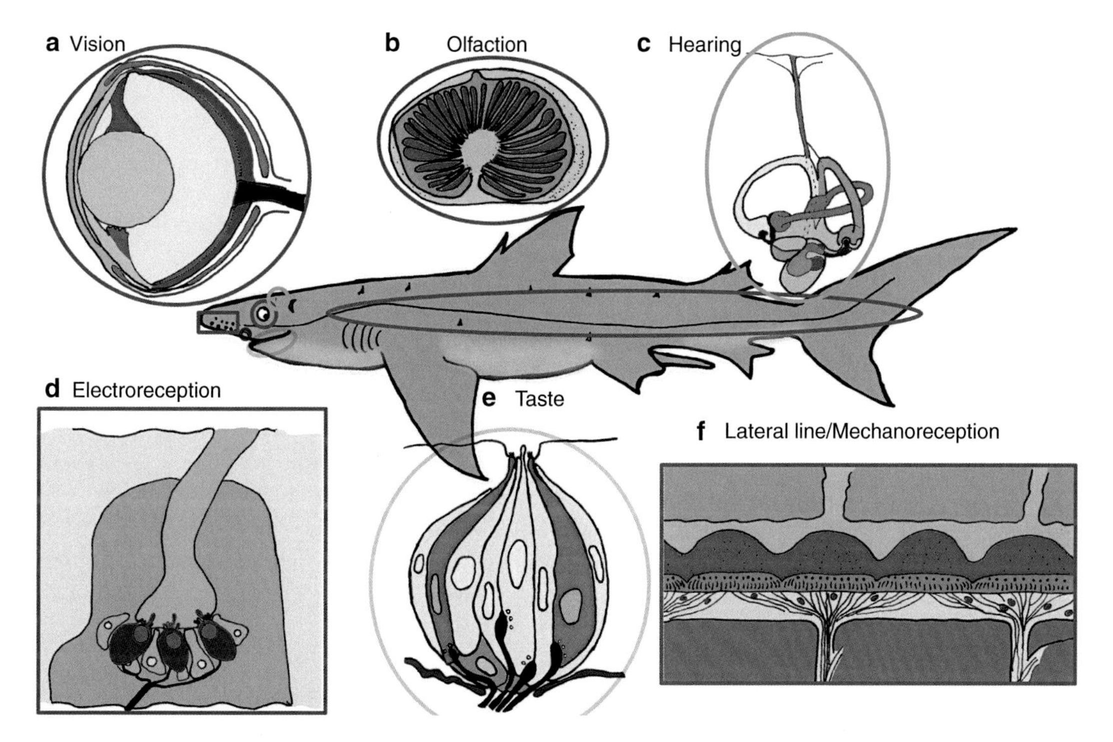

Fig. 4.6 The sensory system of chondrichthyans. Colored ellipse and rectangles on the shark drawing indicate the location of sensory organs. (**a**) Vision represented by a generalized shark eye redrawn from Lisney et al. (2012). Green: cornea. Pale orange: aqueous humor. Pale blue: iris. Dark blue: pseudocampanule, an intraocular muscle. Red: suspensory ligament. Pink: scleral cartilage. Yellow: vitreous humor. Gray: lens. Purple: retina. Brown: choroid and tapetum lucidum. Black: optic nerve. Dark green: sclera. (**b**) Linear olfactory lamella of a great white shark. (**c**) Hearing represented by the inner ear labyrinth from *Chimaera monstrosa* in lateral view. Yellow: anterior semicircular canal. Brown: endolymphatic duct. Blue: horizontal semicircular canal. Purple: lagenar macula (situated within the lagena). Red: macula neglecta. Green: posterior semicircular canal. Pink: saccular macular (situated within the sacculus). Orange: utricular macular (situated within the utriculus). Redrawn from Lisney (2010). (**d**) Electroreception showing the morphology of the ampullary electroreceptors in a skate redrawn from Baker et al. (2013). Purple: receptor cells. Pale blue: support cells. Aqua: conductive jelly. Orange: epidermal plug. Black: afferent nerves. (**e**) Generalized vertebrate taste bud showing light and dark sensory cells redrawn from Northcutt (2004). Yellow: light cells. Orange: dark cells. Purple: basal cells. Gray: basement membrane. Black: afferent nerves. (**f**) Lateral line system represented by a longitudinal section of the lateral canal in *Carcharhinus* redrawn from Tester and Kendall (1969). Purple: cupula. Green: neuromast zone. Yellow: fiber zone. Red: blood vessels. Black: nerves. Drawings by CB

cies (e.g., hexanchid sharks) may contribute to this as well (Howard et al. 2013). In order to detect the direction of the scent, the nostrils need to be well separated (Kajiura et al. 2005), and this depends on the taxa involved, with some holocephalans having closely placed nostrils but those of the Rhinochimaeridae (Bigelow and Schroeder 1953) being highly separated. When a scent is stronger in one nostril than the other, the animal will turn in that direction, following the scent trail in a zigzag pattern.

Odors are detected by the **sensory epithelium** (Vogel 1994) located in sensory channels made from the secondary folds of opposing **lamellae**, in the olfactory lamellar arrays in the nasal sacs (Fig. 4.6b; Howard et al. 2013). The arrangement of the lamellae differs between holocephalans and elasmobranchs. In holocephalans, there are fewer lamellae (25–36) arranged in a radial fashion around an elliptical central port, whereas neoselachians have linear arrays that can accommodate up to 700 lamellae (Fig. 4.6b; Howard et al. 2013). The total area of olfactory lamellar area is sometimes used as a proxy for olfactory sensitivity (Kajiura et al. 2005; Holmes et al. 2011), and Theiss et al. (2009) suggest it is a better estimate of olfactory sensitivity than lamella number. Neoselachians, especially ambush predators like wobbegongs (Theiss et al. 2009), would have better olfactory sensitivity than holocephalans (Howard et al. 2013), but other features are also important in olfaction including the proportion of the lamellae surface actually covered in sensory receptors (Hara 1992), the geometrical array of the olfactory lamellae which allow for different numbers of lamellae (differ in different species, e.g., Meng and Yin 1981; Theisen et al. 1986), the width of the sensory channels (Holmes et al. 2011), and the size of the olfactory bulb in the brain to process these stimuli (Yopak et al. 2015). The size of the olfactory bulb in chondrichthyans is tightly linked to their habitats rather than to phylogeny (Yopak et al. 2015). The largest olfactory bulbs occur in pelagic, coastal oceanic sharks such as the great white and tiger sharks, and this might be related to their reliance on olfaction for long distance migration and for the detection of food sources such as whale carcasses. Reef sharks of the same family (Carcharhinidae) as well as Hemiscylliidae and dasyatid batoids have the smallest olfactory bulbs of all species surveyed to date (Yopak et al. 2015). In the reef habitat, chondrichthyans rely on vision heavily. As for other senses, there is therefore a great variation of sensory abilities within the chondrichthyans.

4.8.2 Taste/Gustation

Taste buds are present in the mouth and neck (pharynx) of chondrichthyans, but their position and density vary between species (beige in Figs. 4.5 and 4.6; Northcutt 2004). In batoids, they are interspersed with **denticles**, which might protect against abrasion and parasites, reduce hydrodynamic drag from ram ventilation, and improve grasp and holding of prey (Rangel et al. 2016). The presence of denticles, however, reduces the surface area available for taste buds, which suggests that this generalist feeder does not have great taste capability, similar to other batoids (Atkinson and Collin 2012). In the spiny dogfish, the taste buds are most numerous on the roof of the mouth (Gardiner et al. 2012), while in the bamboo shark, they occur throughout the oral and pharyngeal region (Atkinson et al. 2016); a palatal organ was recently described for the rabbit fish (*Chimaera monstrosa*), common in all holocephalans (Ferrrando et al. 2016). This palatal organ has a low density of taste buds, so the primary use might be for general mechanical sensitivity involved in food sorting rather than tasting (Ferrrando et al. 2016).

The **taste buds** are pear-shaped multicellular chemoreceptors with apical (mouth) and basal (lying on the basement membrane) ends, oriented at a right angle to its position in the mouth (Fig. 4.6e; Northcutt 2004). The apical surface is the receptor area which has sensory cells consisting of large or small receptor villi (hair). At the apical surface are the light, main sensory cells (Fig. 4.6e, yellow) and dark, secondary sensory cells (Fig. 4.6e, orange), which are responsible for collecting the taste input. They communicate to the basal cells (Fig. 4.6e, purple) that lie directly on the basement membrane and contain vesicles rich in serotonin communicating with the nerves to relay messages to the brain. Stem cells are also present in the basal membrane, presumably for regeneration of the taste buds (Martin et al. 2016). The taste buds are innervated by branches of **facial**, **glossopharyngeal,** and **vagal** nerves. Although taste receptors in chondrichthyans closely resemble those of other vertebrates (Gardiner et al. 2012), there is a lot of variation in the oral papillae morphology with more than one type occurring in one species (Rangel et al. 2016).

4.8.3 Vision

In most chondrichthyan species, vision plays a role in spatial orientation, navigation, communication, and predatory and social behavior (Figs. 4.5 and 4.6a, blue; Lisney et al. 2012). Eye size relative to body size is an indicator of the relative importance of vision and is linked to habitat type (shallow vs. deep water), activity level, and prey type (Gardiner et al. 2012). The sharks with the largest eyes are thresher sharks that make deep vertical migrations (DVM), diving down more than 600 m within 10 min, and deep-sea sharks. Bigger eyes might be needed to adapt rapidly to light differences during DVM and for increased light sensitivity in species living in deep-sea habitats but also breeding in shallow water environments like the holocephalan *Callorhinchus milii* (Lisney 2010). The smallest eye size is found in benthic (bottom dwelling) sharks and batoids from coastal habitats where the water is turbid (Lisney and Collin 2007) and in some deep-sea batoids such as *Benthobatis* sp. and *Typhlonarke* sp., where the eyes are degenerate (Gruber 1977; Locket 1977).

The position of the eyes in elasmobranch depends on their habitats. Pelagic and benthopelagic species have laterally positioned eyes whereas benthic (bottom dwelling) batoids have dorsolateral eyes (Gardiner et al. 2012). There is only scarce data about the **visual field** of elasmobranchs, which is defined as the area seen without moving the eyes. It ranges from c. 104° to 196° in the vertical plane and 159° to 199° in the horizontal plane (Lisney et al. 2012) but is monocular in most species. There are areas of binocular overlap, but it varies from c. 7° to 48° in the horizontal plane and from 0° to 50° in the vertical plane and occurs at the expense of posterior blind areas (Lisney et al. 2012). The greatest anterior binocular overlap is in the hammerhead sharks and is linked with the lateral head expansion in these species. To overcome anterior blind spots, hammerheads move their heads side to side more than other species (greater yaw) when swimming (McComb et al. 2009). Hammerheads aside, batoids with dorsolaterally placed eyes have greater areas of binocular overlap in the horizontal plane compared to sharks which have lateral eyes. However, the dynamic visual field can be extended to 360° when swimming and moving the eyes (Gardiner et al. 2012).

Sharks have at least an upper and lower eyelid with some sharks, like lemon sharks, having the **nictitating membrane**, acting as a third eyelid to protect the eye from abrasion during feeding. Great white sharks and whale sharks rotate their eyeball while feeding to protect them from abrasion (Gardiner et al. 2012).

As in other vertebrates, light enters through the **cornea** (Fig. 4.6a, pale green) and **pupil** and is focused onto the **retina** (at the back of the eye, purple) by the **lens** (gray), which can be moved by the **intraocular muscles** such as the **pseudocampanule** (**protractor lentis**) (dark blue). The shape of the pupil varies according to habitat and behavior. It can be round as in deep-sea sharks, crescent-shaped in many skates and rays to camouflage the eye from predators and reduce the effects of spherical aberration (Lisney et al. 2012), or a slit in active predators like lemon sharks as it is the most mechanically efficient way to close down the pupil to a pinhole and allows the sharks to be active both during the day and night (Gardiner et al. 2012). The degree with which the pupil can be closed depends on light levels and activity pattern. It is almost immobile in low-light habitats and can constrict rapidly in sharks active both during night and day (Lisney et al. 2012).

4.8.4 Color Vision in Sharks?

There are visual and non-visual pigments in the eyes of chondrichthyans. The non-visual pigments are present in the **cornea** (Fig. 4.6a, pale green), **aqueous** (pale orange) and **vitreous humors** (yellow), and the **lens** (gray) to filter out wavelengths such as damaging UV light and to remove light prone to scatter. It is present mostly in surface dwellers. Additionally, chondrichthyans have a structure composed of mirror-like crystals (the **tapetum lucidum**) in the **choroid** (brown) behind the **retina** (purple). The tapetum lucidum is responsible for the eye shine in many vertebrates and is an adaptation for low-light conditions. It acts by reflecting light back to the photoreceptors layer of the retina and gives photons a second opportunity to excite the photoreceptors, hence increasing visual sensitivity (Ollivier et al. 2004). It is found in all chondrichthyan species (Lisney et al. 2012) and has different spectral properties depending on the width and spacing of the guanine crystals it contains, reflecting bluer light in deep-sea species and green blue in coastal species.

Visual pigments are located in photoreceptor cells classified according to their morphology and light sensitivity. **Rods** are highly sensitive to light but have low visual acuity (used in low-light conditions), and **cones** are used for bright light and color vision and are responsible for higher visual acuity (Lisney et al. 2012). The great majority of chondrichthyans have retinas containing both cones and rods, and the proportion of cones to rods in some species is relatively high. Species living in low-light conditions like deep-sea rays and the Port Jackson shark (living in turbid shallow water) have all-rod retinas indicating that they are possibly color blind (Bozzano et al. 2001; Bozzano 2004; Lisney et al. 2012). The presence of both cones and rods is not the only indicator of color vision, as at least two types of cones are required for color differentiation (Gardiner et al. 2012). The cones become specialized with visual pigments sensitive to specific wavelengths and intensity of light. Each visual pigment is composed of an **opsin** protein and a **chromophore,** and it is the properties of the chromophore that define the spectral sensitivity of the pigment as a whole. In vertebrates, there is only one class of medium-wavelength-sensitive pigments on rod photoreceptors for low-light vision, whereas there are four classes of pigments on cones with sensitivities ranging from long wavelength (λmax c. 500–575 nm) to UV and violet wavelength (λmax 355–445 nm) (Hunt et al. 2009). Deepwater species have photoreceptors shifted toward shorter wavelengths (blue) as ocean water becomes blue and monochromatic at depth, and these photoreceptors are better at detecting bioluminescence, the only source of light below 1000 m (Lisney et al. 2012). Sharks (selachians) appear to only have one cone type (Hart et al. 2011), so it is likely that sharks do not have **color vision**. In contrast, there is evidence that at least one deep-sea holocephalan *Callorhinchus milii* (Davies et al. 2009, 2012) and a number of species of rays (Hart et al. 2004; Theiss et al. 2007) do have multiple cone types suggesting that they have color vision. To date, very few behavioral experiments exist to demonstrate color detection when brightness is controlled as a factor. It has been found that the giant shovelnose ray could discriminate colors (Van-Eyk et al. 2011), but Schluessel et al. (2014) found that the bamboo shark was color blind but could distinguish different intensities very well. More behavioral experiments would be needed, but as many sharks are hard to keep in captivity or difficult to observe in the wild, most of the known data about color vision in elasmobranch is derived from anatomical observations (Lisney et al. 2012).

4.8.5 Hearing and Mechanosenses

Chondrichthyans are not known for making any sound (Gardiner et al. 2012) but can detect both sound through the **inner ear** (Fig. 4.6c) and vibrations through the **lateral line system** (Fig. 4.6f). Both senses are part of the **acousticolateralis system** which plays an important role in prey, predator, and conspecific detection as well as in orientation in relation to currents and hydrodynamic imaging (Lisney 2010). Chondrichthyan external ears consist of two small openings behind the eyes; they lack accessory organs like a swim bladder and a body connection between the swim bladder and the inner ear. They detect sound using the inner ear which is similar in structure to that of bony fishes (Lisney 2010). The inner ear consists of a **labyrinth** made up of several canals and sacs filled with a liquid (the **endolymph**). The three major canals are the anterior (ASC; Fig. 4.6c, yellow), posterior (PSC, green), and horizontal (HSC, blue) **semicircular canals** that are oriented in different planes and are responsible for detecting turning motion but are not involved in detecting sound. The three sac-like structures at the base of the labyrinth are the **lagena** (LM, Fig. 4.6c, purple), **sacculus** (SM, pink), and **utriculus** (UM, orange), which are involved in both balance and hearing (Lisney 2010). These structures contain **hair cells** called **maculae** which are associated with the eighth cranial (auditory) nerve (Fig. 4.5) and whose sensory hairs are covered by mineralized ear bones (**otoconia**) similar to bony fishes' otoliths (Lisney 2010). When sound enters the inner ear and hits the otoconia, the amplitude of the sound wave changes because of the different density of the otoconia relative to the water. This causes the hair cells to move, which is then transmitted as a nerve impulse to the auditory nerve. As well, chondrichthyans have an additional macula (area of neuromast-based sensory epithelium) called the **macula neglecta** (MN, Fig. 4.6c red). It is not covered by otoconial mass but is associated with the posterior semicircular canal (green) and is important for sound detection at least in elasmobranchs (Lowenstein and Roberts 1951; Fay et al. 1974; Corwin 1989). Free-swimming, piscivorous elasmobranchs tend to have a larger sacculus and posterior semicircular canal duct and a more complex, larger macula neglecta than bottom-dwelling, non-piscivorous species which suggests that the former have better hearing than the latter (Myrberg 2001). In chimaerids (the only holocephalans studied), there is a connection between the anterior and posterior semicircular canals which is lacking in elasmobranchs, and the saccular and utricular regions are not separated. This might represent specializations for sound detection, but the functional significance of these morphological differences is still unclear (Lisney 2010). Only a handful of elasmobranch species were tested for hearing, but their behavior and audiograms show that they can be attracted and detect low-frequency sounds (Myrberg 2001). In addition, lemon sharks can localize the source of a sound to around 10° (Nelson 1967), and blacktip sharks have been shown to be able to detect changes in barometric pressure as low as 5 mb to avoid storms. This is because the vestibular hair cells of their inner ear respond to changes in hydrostatic pressure (Heupel et al. 2003).

The second component of the acousticolateralis system is the **lateral line system** (Fig. 4.6f) which is a system of canals and superficial receptors around the snout and midline of all chondrichthyans involved in sensing water current, pressure waves, and, to a certain extent, sound (Gardiner et al. 2012). The functional units in the lateral line are the **neuromasts** (Fig. 4.6f, green), which are clusters of ciliated sensory cells as well as support cells encapsulated into a jelly-like sheath called the **cupula** (purple) which can be stimulated by water movement or pressure (Lisney 2010). They work in a similar way to the inner ear cells where water movement over the cupula and the sensory hairs transforms mechanical energy into a neuronal impulse transmitted to the **medulla** in the brain (Fig. 4.5a) (Maruska 2001). There are several types of mechanosensory lateral line organs: the superficial neuromasts (pit organs) which are located on the skin surface either in grooves or between modified scales to protect them from forward-swimming motion; canal neuromasts which are either connected to the outside environment by pores (pored) or isolated from it (non-pored) or located in a system of open grooves in chimaerids; spiracular organs which are stimulated by flexion of the cranial-hyomandibular joint and are situated in the diverticula of the first visceral pouch; and the **vesicles of Savi** which are present in some groups of rays and consist of neuromasts enclosed in subepidermal pouches (Maruska 2001). The distance range and sensitivity of the lateral line system are determined by the distribution and morphology of these mechanoreceptors. The large concentration of non-pored canals around the nose and mouth may function as specialized tactile receptors stimulated by prey contact and aid in feeding, and in batoids, the vesicles of Savi around the mouth would help in prey localization. The pored canals on the dorsal surface of the body and tail of elasmobranch could be used to detect water movement from conspecifics, predators, and currents (Maruska 2001). Although there is great diversity in the morphology of lateral line canals in chondrichthyans, much remains to be understood about its functional significance.

4.8.6 Electroreception (Ampullae of Lorenzini)

Chondrichthyans can detect small electric fields coming from other living organisms (biotic sources) as well as from physical sources such as geomagnetic induction of electric currents (abiotic sources) which aids in prey capture and orientation (Lisney 2010). Structures involved in this detection are called **ampullae of Lorenzini** in elasmobranch and **electroreceptive ampullae** in holocephalans and are homologous to each other (Fig. 4.6d). In marine elasmobranchs, ampullae are grouped together in bilateral head clusters that radiate in many directions and terminate in individual pores (Gardiner et al. 2012). This allows the electric potential of different **ampullae** within a cluster to be compared and the voltage difference between them to be measured. In holocephalans, the ampullae are also grouped in a number of distinct clusters but are associated with the lateral line canals (Lisney 2010). The **ampullary electroreceptors** (ampullae) consist of **sensory** (purple) and **support cells** (pale blue) located at the base of a canal filled with a low resistance **conductive jelly** (aqua). The tight junction of the canal wall and between the sensory and support cells serves as an electrical barrier. Current is detected as the difference between voltages at the top (apical) vs. base (basal) surface of the sensory cells. Ampullae are very sensitive, and elasmobranch can detect voltage gradients as low as 1–5 nV/cm (Tricas and Sisneros 2004). An animal's electroreceptive capabilities are likely to be determined by the density and distribution of the ampullary organ as well as the shape of the head (Lisney 2010). For example, the hammerhead sharks have an enlarged snout and have a larger number of ampullae and higher pore density than similarly sized carcharhinids, suggesting that they have better electroreceptive capability (Kajiura 2001). In most chondrichthyan species, electroreception is believed to be most important in prey detection and capture, but it has been shown to also be important in social communication such as mate and predator detection as well as the detection of magnetically induced fields involved in orientation behaviors (Lisney 2010). As for other senses, much is to be learned about the electrosensory abilities of chondrichthyans.

4.9 Conclusions

Chondrichthyans have evolved over 400 million years ago and have been incredibly morphologically diverse. They have survived large mass extinction events affecting vertebrates at the end of the Devonian and Permian geological periods, and although less diverse than at their peak in the Paleozoic, they are still ecologically and morphologically diverse today. Chondrichthyans are an excellent developmental model for understanding gnathostome evolution, possessing a suite of phylogenetically basal characters and long gestation periods in ovo to allow for developmental manipulation. The broad range of ecological adaptations has made chondrichthyans successful in the past, but many species are under increasing threat of extinction due to fishing, poor preservation status, habitat destruction, and human-induced climate change. Current conservation bodies are working hard to change the public perception, and it is hoped that we can make sure chondrichthyans thrive and continue to amaze us with their diversity and beauty.

Acknowledgments CB is supported by the Curtin Research Fellowship and the Australia Research Council grant DP 160104427 and KT by DP140104161. We thank the Western Australian Museum and the Natural History Museum for access to specimens. We wish to thank Alan Pradel, Tom Lisney, and an anonymous reviewer for improving the manuscript. We wish to thank Janine Ziermann, Rui Diogo, and Raul Diaz Jr for inviting us to contribute to this volume.

Further Readings

Carrier JC, Musick JA, Heithaus MR (2012) Biology of sharks and their relatives, 2nd edn. CRC Press, Boca Raton

Helfman G, Collette BB, Facey DE, Bowen BW (2009) The diversity of fishes: biology, evolution and ecology, 2nd edn. Wiley, London. Chapters 3, 6, 8, 11, 12

Janvier P (1996) Early vertebrates. Oxford University Press, New York

Long JA (2011) The rise of fishes: 500 million years of evolution. The John Hopkins University Press, Baltimore

Deban SM (2003) Constraint and convergence in the evolution of salamander feeding. In: Gasc JP, Casinos A, Bels VL (eds) Vertebrate biomechanics and evolution. BIOS Scientific Publishers, Oxford, pp 163–180

Deban SM, Wake DB (2000) Terrestrial feeding in salamanders. In: Schwenk K (ed) Feeding: form, function and evolution in tetrapod vertebrates. Academic Press, San Diego, pp 65–94

References

Andreev PS, Coates MI, Shelton RM, Cooper PR, Smith PM, Sansom IJ (2015) Upper Ordovician chondrichthyan-like scales from North America. Palaeontology 58:691–704

Andreev PS, Coates MI, Karatajūtė-Talimaa V et al (2016) The systematics of the Mongolepidida (Chondrichthyes) and the Ordovician origins of the clade. Peer J 4:e1850

Ari C (2011) Encephalization and brain organization in mobulid rays (Myliobatiformes, Elasmobranchii) with ecological perspectives. Open Anat J 3:1–13

Aschliman NC, Nishida M, Miya M, Inoue JG, Rosana KM, Naylor GJP (2012) Body plan convergence in the evolution of skates and rays (Chondrichthyes: Batoidea). Mol Phylogenet Evol 63:28–42

Atkinson CJL, Collin SP (2012) Structure and topographic distribution of oral denticles in elasmobranch fishes. Biol Bull 222:26–34

Atkinson CJL, Martin KJ, Fraser GJ, Collin SP (2016) Morphology and distribution of taste papillae and oral denticles in the developing oropharyngeal cavity of the bamboo shark, *Chiloscyllium punctatum*. Biol Open 5:1759–1769

Baker CVH, Bronner-Fraser M (2001) Vertebrate cranial placodes I. Embryonic induction. Dev Biol 232:1–61

Baker CVH, Modrell MS, Gillis JA (2013) The evolution and development of vertebrate lateral line electroreceptors. J Exp Biol 216:2515–2522

Bauchot R, Platel R, Ridet J-M (1976) Brain-body weight relationships in Selachii. Copeia 1976:305–310

Bauchot R, Bauchot ML, Platel R, Ridet JM (1977) The brains of Hawaiian tropical fishes: brain size and evolution. Copeia 1:42–46

Belbin RA, Underwood CJ, Johanson Z, Twitchett RJ (2017) Ecological impact of the end-Cretaceous extinction on lamniform sharks. PLoS One 12(6):e0178294

Bhattacharyya S, Bailey AP, Bronner-Fraser M, Streit A (2004) Segregation of lens and olfactory precursors from a common territory: cell sorting and reciprocity of *Dlx5* and *Pax6* expression. Dev Biol 271:403–414

Bigelow HB, Schroeder WC (1953) Fishes of the Gulf of Maine. Fish Bull 53:1–630

Boisvert CA, Martins CL, Edmunds AG, Cocks J, Currie P (2015) Capture, transport, and husbandry of elephant sharks (*Callorhinchus milii*) adults, eggs, and hatchlings for research and display. Zoo Biol 34:94–98

Bozzano A (2004) Retinal specialisations in the dogfish *Centroscymnus coelolepis* from the Mediterranean deep-sea. Scientia Mar 68:185–195

Bozzano A, Murgia R, Vallerga S, Hirano J, Archer S (2001) The photoreceptor system in the retinae of two dogfishes, *Scyliorhinus canicula* and *Galeus melastomus*: possible relationship with depth distribution and predatory lifestyle. J Fish Biol 59:1258–1278

Brabrand A (1985) Food of roach (*Rutilus rutilus*) and ide (*Leuciscus idus*): Significance of diet shifts for interspecific competition in omnivorous fishes. Oecologia 66:461–467

Brainerd EL, Ferry-Graham LA (2006) Mechanics of respiration. In: Shadwick R, Lauder G (eds) Biomechanics: A volume of the fish physiology series. Elsevier Science, New York, pp 1–29

Brazeau MD (2008) Early jaw and braincase morphologies with unorthodox implications for basal gnathostome interrelationships. J Vert Paleo 56a:28

Brazeau MD (2009) The braincase and jaws of a Devonian 'acanthodian' and modern gnathostome origins. Nature 457:305–308

Brazeau MD, de Winter V (2015) The hyoid arch and braincase anatomy of *Acanthodes* support chondrichthyan affinity of 'acanthodians'. Proc Biol Sci 282:20152210. https://doi.org/10.1098/rspb.2015.2210

Brazeau MD, Friedman M (2015) The origin and early phylogenetic history of jawed vertebrates. Nature 520:490–497

Brazeau MD, Friedman M, Jerve A, Atwood RC (2017) A three-dimensional placoderm (stem-group gnathostome) pharyngeal skeleton and its implications for primitive gnathostome pharyngeal architecture. J Morphol 278:1220–1228

Burrow CJ, Rudkin D (2014) Oldest near-complete acanthodian: the first vertebrate from the Silurian Bertie Formation Konservat-Lagerstätte, Ontario. PLoS One 9:e104171

Burrow C, Turner S (2013) Scale structure of putative chondrichthyan *Gladbachus adentatus* Heidtke & Krätschmer, 2001 from the Middle Devonian Rheinisches Schiefergebirge, Germany. Hist Biol 25:385–390

Burrow C, Young GC (1999) An articulated teleostome from the Late Silurian (Ludlow) of Victoria, Australia. Rec West Aust Mus 57:1–14

Carr R, Johanson Z, Ritchie A (2009) The phyllolepid placoderm *Cowralepis mclachlani*: Insights into the evolution of feeding mechanisms in jawed vertebrates. J Morphol 270:775–804

Coates MI (2005) *Gladbachus adentatus* Heidtke and Kratschmer: an awkward addition to the set of early jawed fishes. Abstract, Society of Vertebrate Palaeontology and Comparative Anatomy, Annual Symposium, London. Palaeontological Association

Coates MI, Sequeira SEK (2001) A new stethacanthid chondrichthyan from the Lower Carboniferous of Bearsden, Scotland. J Vert Paleo 21:438–459

Coates MI, Gess RW, Finarelli JA, Criswell KE, Tietjen K (2017) A symmoriiform chondrichthyan braincase and the origin of chimaeroid fishes. Nature 541:208–211

Coates MI, Finarelli JA, Sansom IJ, Andreev PS, Criswell KE, Tietjen K, Rivers ML, La Riviere PJ (2018) An early chondrichthyan and the evolutionary assembly of a shark body plan. Proc Biol Sci 285:20172418

Corwin JT (1989) Functional anatomy of the auditory system in sharks and rays. J Exp Zool A Ecol Int Phys 252:62–74

Darras L, Derycke C, Blieck AR, Vachard D (2008) The oldest holocephalan (Chondrichthyes) from the Middle Devonian of the Boulonnais (Pas-de-Calais, France). Comptes Rendus Palevol 7:297–304

Davies WL, Carvalho LS, Tay B-H, Brenner S, Hunt DM, Venkatesh B (2009) Into the blue: Gene duplication and loss underlie color vision adaptations in a deep-sea chimaera, the elephant shark *Callorhinchus milii*. Genome Res 19:415–426

Davies WL, Tay B-H, Zheng K et al (2012) Evolution and functional characterisation of melanopsins in a deep-Sea Chimaera (Elephant Shark, *Callorhinchus milii*). PLoS One 7:e51276

Davis SP, Finarelli JA, Coates MI (2012) *Acanthodes* and shark-like conditions in the last common ancestor of modern gnathostomes. Nature 486: 247–250

Dean B (1894) Contributions to the morphology of *Cladoselache* (*Cladodus*). J Morphol 9:87–114

Dean B, Bhushan B (2010) Shark-skin surfaces for fluid-drag reduction in turbulent flow: a review. Philos Trans A Math Phys Eng Sci 368:4775–4806

Dean MN, Summers AP (2006) Mineralized cartilage in the skeleton of chondrichthyan fishes. Zoology 109:164–168

Dean MN, Ekstrom L, Monsonego-Ornan E et al (2015) Mineral homeostasis and regulation of mineralization processes in the skeletons of sharks, rays and relatives (Elasmobranchii). Semin Cell Dev Biol 46:51–67

Debiais-Thibaud M, Oulion S, Bourat F, Laurenti P, Casane D, Borday-Birraux V (2011) The homology of odontodes in gnathostomes: insights from *Dlx* gene expression in the dogfish, *Scyliorhinus canicula*. BMC Evol Biol 11:307

Demski LS, Northcutt RG (1996) The brain and cranial nerves of the white shark: An evolutionary perspective. In: Klimley AP, Ainley DG (eds) Great white sharks. The Biology of *Carcharodon carcharias*. Academic Press, New York, pp 121–130

Didier DA (1995) Phylogenetic systematics of extant chimaeroid fishes (Holocephali, Chimaeroidei). Am Mus Novit 3119:1–86

Fay RR, Kendall JI, Popper AN, Tester AL (1974) Vibration detection by the macula neglecta of sharks. Comp Biochem Physiol 47A:1235–1240

Ferrrando S, Gallus L, Gambardella C, Croce D, Damiano G, Mazzarino C, Vacchi M (2016) First description of a palatal organ in *Chimaera monstrosa* (Chondrichthyes, Holocephali). Anat Rec 299:118–131

Fraser GJ, Smith MM (2011) Evolution of developmental pattern for vertebrate dentitions: an oro-pharyngeal specific mechanism. J Exp Zool B Mol Dev Evol 316B:99–112

Fraser GJ, Berkovitz BK, Graham A, Smith MM (2006) Gene deployment for tooth replacement in the rainbow trout (*Oncorhynchus mykiss*): a developmental model for evolution of the osteichthyan dentition. Evol Dev 8:446–457

Fraser GJ, Hulsey CD, Bloomquist RF, Uyesugi K, Manley NR, Streelman JT (2009) An ancient gene network is co-opted for teeth on old and new jaws. PLoS Biol 7(2):e31

Fraser GJ, Cerny R, Soukup V, Bronner-Fraser M, Streelman JT (2010) The odontode explosion: The origin of tooth-like structures in vertebrates. BioEssays 32:808–817

Fraser GJ, Britz R, Hall A, Johanson Z, Smith MM (2012) Replacing the first-generation dentition in pufferfish with a unique beak. Proc Natl Acad Sci 109:8179–8184

Friedman M, Sallan LC (2012) Five hundred million years of extinction and recovery: A Phanerozoic survey of large-scale diversity patterns in fishes. Palaeontology 55:707–742

Gagnier P-Y, Wilson MVH (1996) Early Devonian acanthodians from northern Canada. Palaeontology 39:241–258

Gardiner JM, Hueter RE, Maruska KP et al (2012) Sensory physiology and behavior of elasmobranchs. In: Carrier JC, Musick JA, Heithaus MR (eds) Biology of sharks and their relatives, 2nd edn. CRC Press, Boca Raton, Florida, pp 349–401

Gee H (2007) Before the Backbone. Springer, Berlin, p 346

Gess RW, Coates MI (2015) High-latitude chondrichthyans from the Late Devonian (Famennian) Witpoort formation of South Africa. Paläontol Z 89:147–169. https://doi.org/10.1007/s12542-014-0221-9

Ginter M, Hampe O, Duffin CS (2010) Paleozoic Elasmobranchii: Teeth. In: Schultze H-P (ed) Handbook of palaeoichthyology, vol 3D. Verlag Dr. Friedrich Pfeil, München, pp 1–168

Grogan E, Lund R (2004) The origin and relationships of early Chondrichthyes. In: Carrier JC, Musick JA, Heithaus MR (eds) Biology of sharks and their relatives. CRC Press, Boca Raton, pp 3–31

Grogan E, Lund R (2009) Two new iniopterygians (Chondrichthyes) from the Mississippian (Serpukhovian) Bear Gulch Limestone of Montana with evidence of a new form of chondrichthyan neurocranium. Acta Zool 90:134–151

Gruber SH (1977) The visual system of sharks; adaptations and capability. Am Zool 17:453–469

Guinot G, Cavin L (2015) Contrasting "fish" diversity dynamics between marine and freshwater environments. Curr Biol 25:2314–2318

Hanke G, Wilson M (2010) The putative stem-group chondrichthyans Kathemacanthus and Seretolepis from the Lower Devonian MOTH locality, Mackenzie Mountains, Canada. Morphology, phylogeny and paleobiogeography of fossil fishes. Verlag Dr. Friedrich Pfeil, Munich, pp 159–182

Hara TJ (ed) (1992) Fish chemoreception. Springer, Dordrecht

Hart NS, Lisney TJ, Marshall NJ, Collin SP (2004) Multiple cone visual pigments and the potential for trichromatic colour vision in two species of elasmobranch. J Exp Biol 207:4587–4594

Hart NS, Theiss SM, Harahush BK, Collin SP (2011) Microspectrophotometric evidence for cone monochromacy in sharks. Naturwissenschaften 98:193–201

Heidtke UHJ, Krätschmer K (2001) *Gladbachus adentatus* nov. gen et sp., ein primitiver Hai aus dem Oberen Givetium (Obers Mitteldevon) der Bergische Gladbach-Paffrath-Mulde (Rheinisches Schiefergebirge). Mainz Geowissen Mitteil 30:105–122

Heupel MR, Simpendorfer CA, Hueter RE (2003) Running before the storm: blacktip sharks respond to falling barometric pressure associated with Tropical Storm Gabrielle. J Fish Biol 63:1357–1363

Holmes WM, Cotton R, Xuan VB et al (2011) Three-dimensional structure of the nasal passageway of a hagfish and its implications for olfaction. Anat Rec 294:1045–1056

House EL, Pansky B (1960) A functional approach to neuroanatomy. Acad Med 35(11):1067–1068

Howard LE, Holmes WM, Ferrando S et al (2013) Functional nasal morphology of chimaerid fishes. J Morphol 274:987–1009

Huber DR, Dean MN, Summers AP (2008) Hard prey, soft jaws and the ontogeny of feeding mechanics in the spotted ratfish *Hydrolagus colliei*. J R Soc Interface 5:941–953

Hughes GM, Ballantijn CM (1965) The muscular basis of the respiratory pumps in the dogfish (*Scyliorhynus canicula*). J Exp Biol 43:363–383

Hunt DM, Carvalho LS, Cowing JA, Davies WL (2009) Evolution and spectral tuning of visual pigments in birds and mammals. Philos Trans R Soc B 364:2941–2955

Ivanov A, Märss T, Kleesment A (2011) A new elasmobranch *Karksiodus mirus* gen. et sp. nov. from the Burtnieki Regional Stage, Middle Devonian of Estonia. Estonian J Earth Sci 60:22–30

Janvier P (1996) Early vertebrates. Oxford monographs on geology and geophysics, vol Vol. 33. Clarendon Press, Oxford, p 393

Janvier P, Pradel A (2015) Elasmobranchs and their extinct relatives: Diversity, relationships, and adaptations through time. Fish Physio 34A:1–17

Janvier P, Suarez-Riglos M (1986) The Silurian and Devonian vertebrates of Bolivia. Bull l'Institut Français d'Etudes Andines 15:73–114

Juarez M, Reyes M, Colman T et al (2013) Characterization of the trunk neural crest in the bamboo shark, *Chiloscyllium punctatum*. J Comp Neurol 521:3303–3320

Kajiura SM (2001) Head morphology and electrosensory pore distribution of carcharhinid and sphyrnid sharks. Environ Biol Fish 61:125–133

Kajiura SM, Forni JB, Summers AP (2005) Olfactory morphology of carcharhinid and sphyrnid sharks: Does the cephalofoil confer a sensory advantage? J Morphol 264:253–263

Karatajūtė-Talimaa V, Predtechenskyj N (1995) The distribution of the vertebrates in the Late Ordovician and Early Silurian palaeobasins: Vertebrate microremains from the Lower Silurian of Siberia and Central Asia 105 of the Siberian Platform. Bull Museum Nat d'Hist Natur, Paris, Ser 4(17):39–56

Klug C, Kröger B, Kiessling W et al (2010) The Devonian nekton revolution. Lethaia 43:465–477

Kotrschal K, Van Staaden MJ, Huber R (1998) Fish brains: evolution and environmental relationships. Rev Fish Biol Fish 8:373–408
Lammens EHRR, Geursen J, McGillavry PJ (1987) Diet shifts, feeding efficiency and coexistence of bream (*Abramis brama*), roach (*Rutilus rutilus*) and white bream (*Blicca bjoerkna*) in eutrophicated lakes. Proc V Congr Europ Ichtyol, Stockholm 153–162
Lane JA, Maisey JG (2012) The visceral skeleton and jaw suspension in the durophagous hybodontid shark *Tribodus limae* from the Lower Cretaceous of Brazil. J Paleontol 86:886–905
Lipovsek M, Ledderose J, Butts T et al (2017) The emergence of mesencephalic trigeminal neurons. Neural Dev 12:11
Lisney TJ (2010) A review of the sensory biology of chimaeroid fishes (Chondrichthyes; Holocephali). Rev Fish Biol Fisheries 20:571–590
Lisney T, Collin S (2006) Brain morphology in large pelagic fishes: a comparison between sharks and teleosts. J Fish Biol 68:532–554
Lisney TJ, Collin SP (2007) Relative eye size in elasmobranchs. Brain Behav Evol 69:266–279
Lisney TJ, Yopak KE, Montgomery JC, Collin SP (2008) Variation in brain organization and cerebellar foliation in chondrichthyans: batoids. Brain Behav Evol 72:262–282
Lisney TJ, Theiss SM, Collin SP, Hart NS (2012) Vision in elasmobranchs and their relatives: 21st century advances. J Fish Biol 80:2024–2054
Locket NA (1977) Adaptations to the deep–sea environment. In: Cresitelli F (ed) Handbook of sensory physiology, vol Vol. VII. Springer–Verlag, Berlin, pp 67–192
Long JA, Burrow CJ, Ginter M (2015) First shark from the Late Devonian (Frasnian) Gogo Formation, Western Australia sheds new light on the development of tessellated calcified cartilage. PLoS One 10(5):e0126066
Lowenstein O, Roberts TDM (1951) The localization and analysis of the responses to vibration from the isolated elasmobranch labyrinth. A contribution to the problem of the evolution of hearing in vertebrates. J Physiol 114:471–489
Lund R, Grogan E (2004) Five new euchondrocephalan *Chondrichthyes* from the Bear Gulch Limestone (Serpukhovian, Namurian E2b) of Montana, USA. In: Arratia G, Wilson MVH, Cloutier R (eds) Recent advances in the origin and early radiation of vertebrates. Verlag Dr Pfeil, München, pp 505–531
Mader H (1986) Schuppen und Zähne von Acanthodien und Elasmobranchiern aus dem Unter-Devon Spaniens (Pisces). Gött Arbeit Geol Paläont 28:1–59
Maisey JG (1980) An evaluation of jaw suspension in sharks. Am Mus Novit 2706:1–17
Maisey JG (1984) Studies on the Paleozoic selachian genus *Ctenacanthus* Agassiz. No. 3, Nominal species referred to Ctenacanthus. Am Mus Novit 2774:1–20
Maisey JG (1989) Visceral skeleton and musculature of a Late Devonian shark. J Vert Paleo 9:174–190
Maisey JG (2001) A primitive chondrichthyan braincase from the Middle Devonian of Bolivia. In: Ahlberg PE (ed) Major events in early vertebrate evolution: paleontology, phylogeny, genetics, and development. Taylor and Francis, New York, pp 263–288
Maisey JG (2005) Braincase of the Upper Devonian shark *Cladodoides wildungensis* (Chondrichthyes, Elasmobranchii), with observations on the braincase in early chondrichthyans. Bull Am Mus Nat Hist 288:1–103
Maisey JG (2007) The braincase in Paleozoic symmoriiform and cladoselachian sharks. Bull Am Mus Nat Hist 307:1–122
Maisey JG (2008) The postorbital palatoquadrate articulation in elasmobranchs. J Morphol 269:1022–1040
Maisey JG (2012) What is an 'elasmobranch'? The impact of palaeontology in understanding elasmobranch phylogeny and evolution. J Fish Biol 80:918–951
Maisey JG, Anderson ME (2001) A primitive chondrichthyan braincase from the early Devonian of South Africa. J Vert Paleo 21:4702–4713
Maisey JG, Miller RF, Turner S (2009) The braincase of the chondrichthyan *Doliodus* from the Lower Devonian Campbellton Formation of New Brunswick, Canada. Acta Zool 90:109–122
Maisey JG, Turner S, Naylor GJP, Miller RF (2014) Dental patterning in the earliest sharks: implications for tooth evolution. J Morphol 275: 586–596
Maisey JG, Miller R, Pradel A, Denton JSS, Bronson A, Janvier P (2017) Pectoral morphology in Doliodus: Bridging the 'acanthodian'-chondrichthyan divide. Am Mus Novit 3875:1–15
Mallatt J (1997) Shark pharyngeal muscles and early vertebrate evolution. Acta Zool 78:279–294
Martin KJ, Rasch LJ, Cooper RL, Metscher BD, Johanson Z, Fraser GJ (2016) Sox2+ progenitors in sharks link taste development with the evolution of regenerative teeth from denticles. Proc Natl Acad Sci 113:14769–14774
Maruska KP (2001) Morphology of the mechanosensory lateral line system in elasmobranch fishes: ecological and behavioral considerations. In: Tricas TC, Gruber SH (eds) The behavior and sensory biology of elasmobranch fishes: an anthology in memory of Donald Richard Nelson. Springer Netherlands, Dordrecht, pp 47–75
McComb DM, Tricas TC, Kajiura SM (2009) Enhanced visual fields in hammerhead sharks. J Exp Biol 212:4010–4018
Meng Q, Yin M (1981) A study of the olfactory organ of the shark. Trans Chinese Ichthyol Soc 2:1–24
Mikoleit G (2004) Phylogenetische Systematik der Wirbeltiere, vol 671. Dr Friedrich Pfeil, Munich
Miles R (1964) A reinterpretation of the visceral skeleton of Acanthodes. Nature 204:457
Miles R (1973) Articulated acanthodian fishes from the Old Red Sandstone of England, with a review of the

structure and evolution of the acanthodian shoulder-girdle. Bull Brit Mus (Nat Hist) 24:111–213
Miller RF, Cloutier R, Turner S (2003) The oldest articulated chondrichthyan from the Early Devonian period. Nature 425:501–504
Miyake T, McEachran JD, Hall BK (1992) Edgeworth's legacy of cranial muscle development with an analysis of muscles in the ventral gill arch region of batoid fishes (Chondrichthyes: Batoidea). J Morphol 212:213–256
Motta PJ, Huber DR (2012) Prey capture behavior and feeding mechanics of Elasmobranchs. In: Carrier JC, Musick JA, Heithaus MR (eds) Biology of sharks and their relatives, 2nd edn. Taylor & Francis, Boca Raton, pp 153–209
Myrberg AA (2001) The acoustical biology of elasmobranchs. Environ Biol Fish 60:31–46
Nelson DR (1967) Hearing thresholds, frequency discrimination, and acoustic orientation in the Lemon Shark, *Negaprion Brevirostris* (Poey). Bull Mar Sci 17:741–768
Nelson GJ (1969) Gill arches and the phylogeny of fishes, with notes on the classification of vertebrates. Bull Amer Mus Nat Hist 141:475–552
New JG (2001) Comparative neurobiology of the elasmobranch cerebellum: Theme and variations on a sensorimotor interface. Environ Biol Fish 60:93–108
Northcutt RG (1978) Brain organization in the cartilaginous fishes. In: Hodgson ES, Mathewson RF (eds) Sensory biology of sharks, skates and rays. Office of Naval Research, Arlington, pp 117–193
Northcutt RG (1979) Central projections of the eight cranial nerve in lampreys. Brain Res 167:163–167
Northcutt RG (2004) Taste Buds: development and evolution. Brain Behav Evol 64:198–206
Nosal AP, Chao Y, Farrara JD, Chai F, Hastings PA (2016) Olfaction contributes to pelagic navigation in a coastal shark. PLoS One 11:e0143758
O'Neill P, McCole RB, Baker CVH (2007) A molecular analysis of neurogenic placode and cranial sensory ganglion development in the shark, *Scyliorhinus canicula*. Dev Biol 304:156–181
Ollivier FJ, Samuelson DA, Brooks DE, Lewis PA, Kallberg ME, Komaromy AM (2004) Comparative morphology of the tapetum lucidum (among selected species). Vet Ophthalmol 7:11–22
Ørvig T (1951) Histologic studies of ostracoderms, placoderms and fossil elasmobranchs 1. The endoskeleton, with remarks on the hard tissues of lower vertebrates in general. Ark Zool 2:321–454
Patterson C (1965) The phylogeny of the chimaeroids. Philos Trans R Soc B 249:101–219
Pradel A, Maisey JG, Tafforeau P, Janvier P (2009) An enigmatic gnathostome vertebrate skull from the Middle Devonian of Bolivia. Acta Zool 90:123–133
Pradel A, Tafforeau P, Janvier P (2010) Study of the pectoral girdle and fins of the Late Carboniferous sibyrhynchid iniopterygians (Vertebrata, Chondrichthyes, Iniopterygia) from Kansas and Oklahoma (USA) by means of microtomography, with comments on iniopterygian relationships. Comptes Rendus Palevol 9:377–387
Pradel A, Tafforeau P, Maisey JG, Janvier P (2011) A new Paleozoic Symmoriiformes (Chondrichthyes) from the Late Carboniferous of Kansas (USA) and cladistic analysis of early chondrichthyans. PLoS One 6(9):e24938
Pradel A, Maisey JG, Tafforeau P, Mapes RH, Mallatt J (2014) A Palaeozoic shark with osteichthyan-like branchial arches. Nature 509:608–611
Rangel BDS, Ciena AP, Wosnick N, De Amorim AF, Kfoury JA Jr, Rici REG (2016) Ecomorphology of oral papillae and denticles of *Zapteryx brevirostris* (Chondrichthyes, Rhinobatidae). Zoomorphology 135:189–195
Rasch LJ, Martin KJ, Cooper RL, Metscher BD, Underwood CJ, Fraser GJ (2016) An ancient dental gene set governs development and continuous regeneration of teeth in sharks. Dev Biol 415:347–370
Reif W-E (1978) Shark dentitions: Morphogenetic processes and evolution. Neues Jarb Geol Palaontol 157:107–115
Reif W (1982) Morphogenesis and function of the squamation in sharks. Neues Jahrb Geol Palaontol Abh 164:172–183
Renz AJ, Meyer A, Kuraku S (2013) Revealing less derived nature of cartilaginous fish genomes with their evolutionary time scale inferred with nuclear genes. PLoS One 8(6):e66400
Ryll B, Sanchez S, Haitina T, Tafforeau P, Ahlberg PE (2014) The genome of *Callorhinchus* and the fossil record: a new perspective on SCPP gene evolution in gnathostomes. Evol Dev 16:123–124
Sallan LS, Coates MI (2010) End-Devonian extinction and a bottleneck in the early evolution of modern jawed vertebrates. Proc Natl Acad Sci U S A 107:10131–10135
Sallan LS, Kammer TW, Ausich WI, Cook LA (2011) Persistent predator–prey dynamics revealed by mass extinction. Proc Natl Acad Sci 108:8335–8338
Sansom IJ, Smith MM, Smith MP (2001) The Ordovicianradiation of vertebrates. In: Ahlberg PE (ed) Major events in early vertebrate evolution. Taylor & Francis, London, pp 156–171
Sansom IJ, Wang N-Z, Smith MM (2005) The histology and affinities of sinacanthid fishes: Primitive gnathostomes from the Silurian of China. Zool J Linnean Soc 144:379–386
Schluessel V, Rick IP, Plischke K (2014) No rainbow for grey bamboo sharks: evidence for the absence of colour vision in sharks from behavioural discrimination experiments. J Comp Phys A 200:939–947
Sepkoski JJ (1984) A kinetic model of Phanerozoic taxonomic diversity. III. Post-Paleozoic families and mass extinctions. Paleobiology 10:246–247
Smith MM, Sansom I (1997) Exoskeletal microremains of an Ordovician fish from the Harding Sandstone of Colorado. Palaeontology 40:645–658

Smith MM, Fraser GJ, Chaplin N, Hobbs C, Graham A (2009) Reiterative pattern of *sonic hedgehog* expression in the catshark dentition reveals a phylogenetic template for jawed vertebrates. Proc Biol Sci 276:1225–1233

Smith MM, Fraser GJ, Johanson Z (2016) Origin of teeth in jawed vertebrates. Infocus Proc Roy Microscop Soc 42:4–17

Stahl BJ (1980) Non-autostylic Pennsylvanian iniopterygian fishes. Palaeontology 23:315–324

Stahl BJ (1999) Handbook of paleoichthyology. In: Chondrichthyes III. Holocephali, vol 4. Verlag Dr Friedrich Pfeil, München

Stensiö EA (1963) Anatomical studies on the arthrodiran head. Kungl Sven Vetenskap Handl 9:1–419

Summers AP, Ferry-Graham LA (2001) Ventilatory modes and mechanics of the hedgehog skate (*Leucoraja erinacea*): testing the continuous flow model. J Exp Biol 204:1577–1587

Tester AL, Kendall JI (1969) Morphology of the lateralis canal system in shark genus *Charcharhinus*. Pacific Sci 23:1–16

Theisen B, Zeiske E, Breucker H (1986) Functional morphology of the olfactory organs in the spiny dogfish (*Squalus acanthius* L.) and the small-spotted catshark (*Scyliorhinus canicula* (L.)). Acta Zool 67:73–86

Theiss SM, Lisney TJ, Collin SP, Hart NS (2007) Colour vision and visual ecology of the blue-spotted maskray, Dasyatis kuhlii Müller & Henle, 1814. J Comp Physiol A 193:67–79

Theiss SM, Hart NS, Collin SP (2009) Morphological Indicators of Olfactory Capability in Wobbegong Sharks (Orectolobidae, Elasmobranchii). Brain Behav Evol 73:91–101

Tricas T, Sisneros J (2004) Ecological functions and adaptations of the elasmobranch electrosense. In: von der Emde G, Mogdans J, Kapoor BG (eds) The senses of fish: adaptations for the reception of natural stimuli. Springer, Berlin, pp 308–329

Underwood CJ, Johanson Z, Welten M et al (2015) Development and evolution of Dentition pattern and tooth order in the skates and rays (Batoidea; Chondrichthyes). PLoS One 10(4):e0122553

Underwood CJ, Johanson Z, Smith MM (2016) Cutting blade dentitions in squaliform sharks form by modification of inherited alternate tooth ordering patterns. Roy Soc Open Sci 3:160385

Van-eyk SM, Siebeck UE, Champ CM, Marshall J, Hart NS (2011) Behavioural evidence for colour vision in an elasmobranch. J Exp Biol 214:4186–4192

Vogel S (1994) Life in moving fluids, 2nd edn. Princeton University Press, Princeton

Wagner HJ (2002) Sensory brain areas in three families of deep-sea fish (slickheads, eels and grenadiers): comparison of mesopelagic and demersal species. Mar Biol 141:807–817

Walker WF, Homberger DG (1992) Vertebrate dissection. Saunders College Publishing, Orlando

Wilga CD (2005) Morphology and evolution of the jaw suspension in lamniform shark. J Morphol 265:102–119

Wilga CD, Ferry LA (2016) Functional anatomy and biomechanics of feeding in elasmobranchs. In: Shadwick RE, Farrell AP, Brauner CJ (eds) Physiology of elasmobranch fishes. Academic Press, Cambridge, pp 153–187

Wilga CD, Hueter RE, Wainwright PC, Motta PJ (2001) Evolution of upper jaw protrusion mechanisms in elasmobranchs. Am Zool 41:1248–1257

Williams ME (1998) A new specimen of *Tamiobatis vetustus* (Chondrichthyes, Ctenacanthoidea) from the Late Devonian Cleveland Shale of Ohio. J Vert Paleo 18:251–260

Williams ME (2001) Tooth retention in cladodont sharks: with a comparison between primitive grasping and swallowing, and modern cutting and gouging feeding mechanisms. J Vert Paleo 21:214–226

Wilson MVH, Hanke GF, Märss T (2007) Paired fins of jawless vertebrates and their homologies across the "agnathan"-gnathostome transition. In: Anderson JS, Sues H-D (eds) Major transitions in vertebrate evolution. Indiana University Press, Bloomington, pp 122–149

Yopak KE (2012) Neuroecology of cartilaginous fishes: the functional implications of brain scaling. J Fish Biol 80:1968–2023

Yopak KE, Frank LR (2009) Brain size and brain organization of the whale shark, Rhincodon typus, using magnetic resonance imaging. Brain Behav Evol 74:121–142

Yopak KE, Montgomery JC (2008) Brain organization and specialization in deep-sea chondrichthyans. Brain Behav Evol 71:287–304

Yopak KE, Lisney TJ, Collin SP, Montgomery JC (2007) Variation in brain organization and cerebellar foliation in chondrichthyans: sharks and holocephalans. Brain Behav Evol 69:280–300

Yopak KE, Lisney TJ, Darlington RB, Collin SP, Montgomery JC, Finlay JC (2010) A conserved pattern of brain scaling from sharks to primates. Proc Natl Acad Sci 107:12946–12951

Yopak KE, Lisney TJ, Darlington RB, Collin SP (2015) Not all sharks are "swimming noses": variation in olfactory bulb size in cartilaginous fishes. Brain Struct Funct 220:1127–1143

Zangerl R, Case GR (1973) Iniopterygia: a new order of chondrichthyan fishes from the Pennsylvanian of North America. Fieldiana Geol Mem 6:1–67

Zhu M, Zhao W, Jia L, Lu J, Qiao T, Qu Q (2009) The oldest articulated osteichthyan reveals mosaic gnathostome characters. Nature 458:469–474

Zhu M, Yu XB, Choo B, Wang JQ, Jia LT (2012) An antiarch placoderm shows that pelvic girdles arose at the root of jawed vertebrates. Biol Lett 8(3):453–456

Zhu M, Yu X, Ahlberg PE et al (2013) A Silurian placoderm with osteichthyan-like marginal jaw bones. Nature 502:188–194

Zhu M, Ahlberg PE, Pan Z et al (2016) A Silurian maxillate placoderm illuminates jaw evolution. Science 354:334–336

Ziermann JM, Miyashita T, Diogo R (2014) Cephalic muscles of Cyclostomes (hagfishes and lampreys) and Chondrichthyes (sharks, rays and holocephalans): comparative anatomy and early evolution of the vertebrate head muscles. ZJLS 172:771–802

Ziermann JM, Freitas R, Diogo R (2017) Muscle development in the shark *Scyliorhinus canicula*: implications for the evolution of the gnathostome head and paired appendage musculature. Front Zool 14:31

Actinopterygians: Head, Jaws and Muscles

5

Alessia Huby and Eric Parmentier

5.1 Introduction

5.1.1 Osteichthyes

Next to the group of cartilaginous fishes (Chondrichthyes) comprising sharks, skates and rays (Elasmobranchii) and chimaeras (Holocephali), the clade **Osteichthyes** includes more than 50,000 living species with all **bony fishes** and tetrapods (Lecointre and Le Guyader 2001). The main feature of bony fishes is the presence of two types of bone in their skeleton: endochondral bones constitute the deep endoskeleton, whereas the dermal exoskeleton is made of dermal bones resulting from intramembranous ossification. The deep bones form from previously developed cartilage models which are then progressively replaced by the bone. The dermal bones involve the replacement of connective tissue membrane sheets with bone tissue (Lecointre and (Lecointre and Le Guyader 2001; Kardong 2012). Bony fishes also have other special features such as the body which is entirely covered by bony scales, the distal part of their fin membrane which is supported by lepidotrichia (i.e., double rows of small transformed scales) and the swim bladder which is an air sac connected to the digestive tract (i.e., oesophageal diverticulum) serving to regulate fish density relative to water density as well as many other species-specific characters.

Among these organisms with a bony endoskeleton, we generally distinguish two main subgroups: the **ray-finned fishes** (**Actinopterygii** > 30,000 extant species) from the **lobe-finned fishes** and tetrapods (**Sarcopterygii** > 24,000 living species) (Nelson 2006). The first subgroup of actinopterygians, also named "ray-finned fishes" because of the transformed scales on their fins forming their dermal rays, is the most diverse and extremely successful class of vertebrates. In terms of number, they group more than 30,000 species which provides an extraordinary basis for diversity.

A diversity which is equivalent to about a half of all living vertebrates and more than 95% of all living fish species which are gathered into 431 families and 42 orders (Nelson 2006; Helfman et al. 2009). However, the number of species in this taxon should be more impressive since it is expected many more species are still to be discovered and identified, including the strange species that inhabit the deep sea. Excluding the four-legged vertebrates (Tetrapoda), the subgroup of sarcopterygians (see Chap. 6) consists of a minority of lobe-finned fishes which are represented by only two extant species of coelacanths (Actinistia) and six living species of lungfishes (Dipnoi). Many different taxa of fossil actinopterygians (e.g., Palaeonisciformes, Pholi-

A. Huby · E. Parmentier (✉)
Laboratory of Functional and Evolutionary Morphology, University of Liège, Quartier Agora, Institut de Chimie, Liège, Belgium
e-mail: alessia.huby@doct.uliege.be; e.parmentier@uliege.be

J. M. Ziermann et al. (eds.), *Heads, Jaws, and Muscles*, Fascinating Life Sciences,
https://doi.org/10.1007/978-3-319-93560-7_5

dopleuriformes, Perleidiformes, Semionotidae, Pycnodontidae, Macrosemiidae) were also studied, but they are not discussed in this chapter. Readers can find information on these taxa in many different reviews (e.g., Blot 1966; Poplin 1984; Miller and McGovern 1996; Cloutier and Arratia 2004; Nelson et al. 2016).

5.1.2 Actinopterygii

Within the large class of **actinopterygian** fishes, five current separate lineages (Fig. 5.1) are encountered: the lineage of **Polypteriformes** with bichirs and reedfishes (**Cladistia**), the lineage of **Acipenseriformes** with sturgeons and paddlefishes (**Chondrostei**), the lineage of **Lepisosteiformes** with all living gars (**Ginglymodi**), the lineage of **Amiiformes** which contains the only species of bowfin *Amia calva* (**Halecomorphi**) and the lineage of **teleostean** fishes (**Teleostei**). The latter includes the amazing majority of living ray-finned fish species since it contains almost 99% of vertebrate species we can encounter in the aquatic environment (Nelson 2006). Polypteriformes are one of the earliest and basal clades of actinopterygians dating from the Devonian period but still have a debated phylogenetic position. This group is currently considered the sister group of the four other lineages (Venkatesh et al. 2001; Inoue et al. 2003;

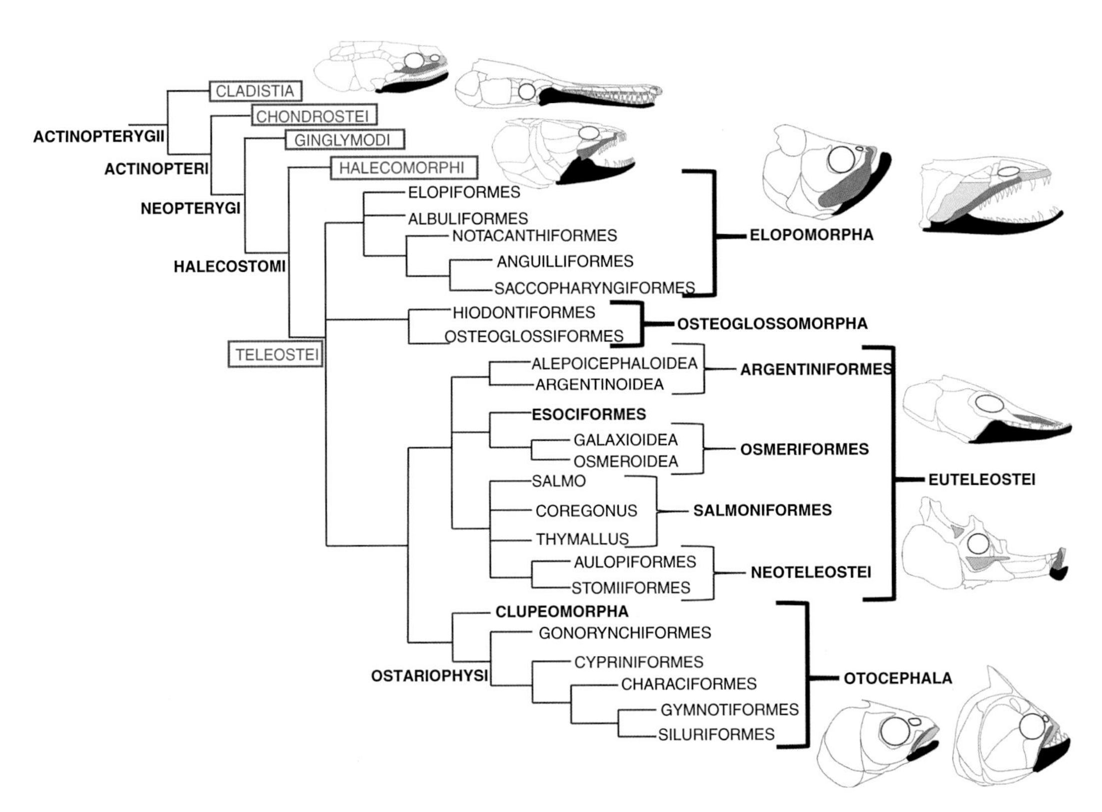

Fig. 5.1 Phylogenetic relationships among the major extant actinopterygian subgroups (modified from Diogo 2008) and illustrations of the head of some taxa. The five major lineages of actinopterygians are framed in red, and head schemata are those of *Polypterus senegalus* (Polypteriformes), *Lepisosteus platyrhincus* (Lepisosteiformes), *Amia calva* (Amiiformes), *Megalops atlanticus* (Elopiformes), *Gymnothorax favagineus* (Anguilliformes), *Esox lucius* (Esociformes), *Hippocampus* sp. (Syngnathiformes), *Alburnus alburnus* (Cypriniformes) and *Serrasalmus* sp. (Characiformes). In the skull illustrations, the black element highlights the lower jaw. The light and dark grey elements are, respectively, for the premaxilla and maxilla of the upper jaw. The eye and nares are circled in bold black. A note about the clade of Neoteleostei is that it includes many more orders than simply Aulopiformes and Stomiiformes, for example, Ateleopodiformes, Myctophiformes, Polymyxiiformes, Percopsiformes, Gadiformes, Zeiformes, Lampriformes, Perciformes, Beryciformes, etc.

Nelson 2006; Diogo 2008) where Acipenseriformes, Lepisosteiformes and Amiiformes are also thought to be basal actinopterygians. Teleosteans are regarded as the most modern and advanced ray-finned fishes (e.g., Lauder and Liem 1983; Nelson 1994, 2006; Patterson 1994; Janvier 1996; Bemis et al. 1997). Teleostean fishes are the most rich species and diversified vertebrate lineage since there are more teleost species than all the other vertebrate species combined (Peng et al. 2009). According to molecular and morphological phylogenetic analyses, this large lineage is subdivided into four major teleostean subgroups (Fig. 5.1): (1) **Elopomorpha** (e.g., Elopiformes, Albuliformes, Notacanthiformes, Anguilliformes and Saccopharyngiformes), (2) **Osteoglossomorpha** (e.g., Hiodontiformes and Osteoglossiformes), (3) **Otocephala** (e.g., Clupeomorpha and Ostariophysi which includes Gonorynchiformes, Cypriniformes, Characiformes, Gymnotiformes and Siluriformes) and (4) **Euteleostei** (e.g., Argentiniformes, Esociformes, Osmeriformes, Salmoniformes and Neoteleostei) (Diogo 2008). Formerly, the Euteleostei subgroup was subdivided into three "superorders": **Protacanthopterygii** (e.g., Esociformes, Osmeriformes and Salmoniformes), **Paracanthopterygii** (e.g., Batrachoidiformes, Gadiformes, Lophiiformes, Ophidiiformes and Percopsiformes) and **Acanthopterygii** (e.g., Atheriniformes, Beloniformes, Beryciformes, Cyprinodontiformes, Gasterosteiformes, Mugiliformes, Perciformes, Pleuronectiformes, Scorpaeniformes, Stephanoberyciformes, Synbranchiformes, Tetraodontiformes and Zeiformes) (Greenwood et al. 1966). Numerous studies have demonstrated that teleosts have extreme morphology and diversified heads, jaws and cranial muscles (e.g., Liem 1967; Osse 1969; Lauder and Liem 1981; Waltzek and Wainwright 2003; Hulsey and Garcia De Leon 2005; Geerinckx et al. 2007) which gives to the class a special position and a great importance for the study of evolutionary history.

In the framework of this chapter, it is therefore not possible to conduct an exhaustive description of all heads of bony fishes or actinopterygian species, their associated muscles and mechanisms. Although it is a truism, fishes are widespread **worldwide** and inhabit all aquatic biotopes as marine, brackish and freshwater systems. They can be encountered from the pelagic zone to the bottom of the ocean, as well as in lakes and rivers and in a variety of extreme environments including desert and thermal springs (e.g., pupfishes), sunless subterranean caves (e.g., cavefishes), torrential rivers (e.g., torrentfishes), hypersaline habitats (e.g., molly fishes), high-altitude lakes and streams (e.g., mountain carps), abyssal depths (e.g., anglerfishes), polar seas and arctic tundra (e.g., cods, flatfishes, salmons, trouts) (Helfman et al. 2009). Although this versatility of fishes to adapt to different environmental conditions is necessary to recall, they are all under the same basic constraint: in a dense and viscous **aquatic medium**, they all have to be able to ingest water at least to breathe and at best to feed. The **respiration** in bony fishes is mainly done by means of water flow entering through the mouth and flowing into the **buccal cavity** towards the **pharyngeal cavity** and the gills (i.e., pharyngeal arches and lamellae) in which respiratory gas exchanges occur. The water flow is created by the action of musculoskeletal pumps, which change the pressure and volume in the buccal and pharyngeal cavities. The existence of these successive suction-to-flowing pumps tends to streamline the water flow. In the same way, the most general way of **feeding** in actinopterygians corresponds to the ability to generate a strong pressure gradient inside the oral cavity by means of musculoskeletal pumps in order to draw a prey into the mouth (Lauder 1985; Wainwright et al. 2015). This mechanism has reached an important level of diversity (Westneat 1994; Liem 1978; Barel 1983; Ferry-Graham et al. 2001) because it is based on a high number of interconnected skeletal elements (e.g., up to 60 skeletal parts in adult teleosts) that are moved by an approximately equal number of muscles (Osse 1969; Aerts 1991). Although different innovations have been developed by ray-finned fishes and have been the subject of many different papers, the way to get food in an aquatic environment remains globally conserved. Its understanding requires however first the anatomical description of a generalized and simplified head musculoskeletal system.

5.2 Anatomy

The **anatomy** of the head in primitive and modern actinopterygian fishes is an area of vertebrate morphology that has a long and distinguished history (Ferry-Graham et al. 2001) and has been the subject of numerous comparative studies. Moreover, it is a research field that has been and is still much studied because of the **kinetics** and incredible movements executed by the **fish skull**. The skull of most actinopterygians is actually distinctive among vertebrates due to the presence of a large number of independent and mobile cartilaginous and bony elements. This unique cranial composition (Fig. 5.2a, b) makes it more complex and **kinetic** (i.e., the skeletal elements that compose the skull can move with respect to each other) than the skull of chondrichthyans, for example (Motta and Huber 2004). These various skull elements result from compromises between different functions as breathing, feeding, hydrodynamic movements, protecting the brain and supporting the sensory organs. In addition, the **cranial muscles** of actinopterygian fishes play a major role in respiration and feeding by moving skull components to control the opening and closing of the buccal and pharyngeal cavities. In the scientific literature, there are many descriptions and illustrations that explain these anatomical aspects for simpler or more complex skulls of ray-finned fish species (suggestions for: Polypteriformes: Traquair 1870; Allis 1919, 1922; Lauder 1980; Acipenseriformes: Carroll and Wainwright 2003; Miller 2004; Lepisosteiformes: Allis 1922; Lauder 1980; Kammerer et al. 2006; Konstantinidis et al. 2015;

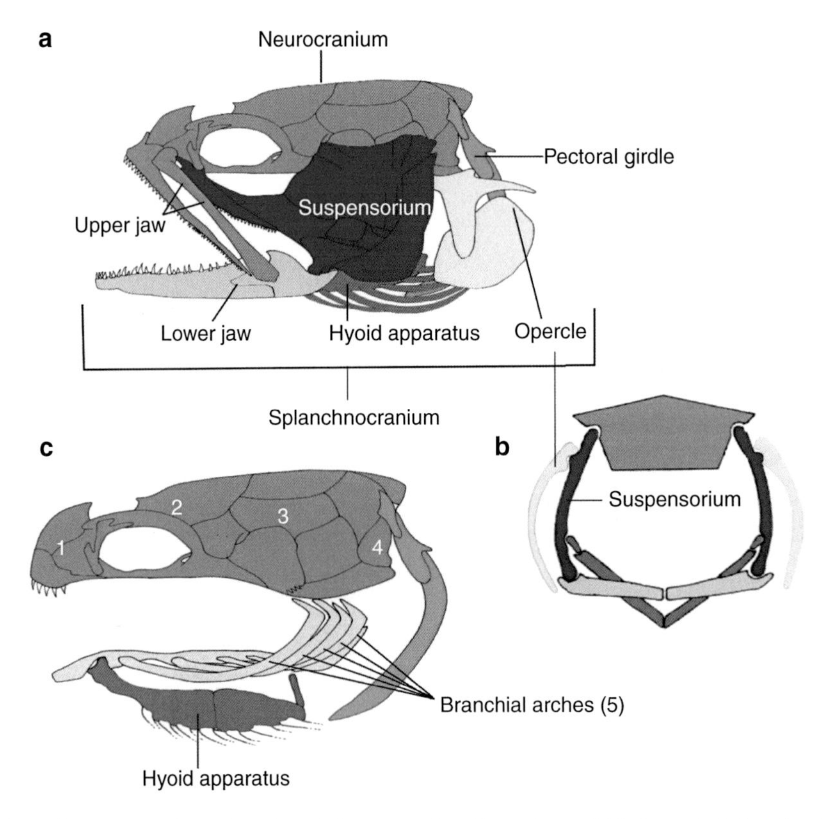

Fig. 5.2 Schematic representations of a teleost (*Carapus acus*) neurocranium and splanchnocranium in (**a**) lateral view and (**b**) frontal view. (**c**) Representation of the different regions of neurocranium and the hyoid and branchial regions of the splanchnocranium. The neurocranium is represented as a cranial box (blue) that includes four regions: (1) the ethmoid region, (2) the orbital region, (3) the otic region and (4) the occipital region. The splanchnocranium comprises the upper and lower jaws, the suspensorium, the hyoid apparatus, the opercular series and the branchial arches. Schematic representations (**a**) and (**c**) are redrawn from Parmentier (2003) and the schema (**b**) from Lauder (1985)

Amiiformes: Allis 1897; Lauder 1980; Elopiformes: Vrba 1968; Anguilliformes: De Schepper et al. 2005, 2007; Eagderi and Adriaens 2010; Osteoglossiformes: Sanford and Lauder 1989; Camp et al. 2009; Gadiformes: Herbing et al. 1996; Salmoniformes: Wilson and Veilleux 1982; Perciformes: Deary and Hilton 2016; Gidmark et al. 2015; Cypriniformes: Gosline 1973). Thereafter, we have tried to present an anatomical description of the skull, jaws and cranial muscles for a representative **actinopterygian model** of the group by doing a simplified summary of different studies.

5.2.1 Skull and Jaws

The **skull** (i.e., cranial skeleton) consists of two main parts: (1) the braincase called the **neurocranium** that protects the brain and sensory organs and (2) the **splanchnocranium** (i.e., visceral cranium) made of series of suspended skeletal elements supporting the jaws, cheeks and gills and offering attachment site for the respiratory and feeding muscles. Embryologically, the neurocranium is mainly formed from the cells having a mesodermal origin, whereas the splanchnocranium emerges from the cells of the **neural crest** (Kardong 2012). **Neural crest cells** migrate from the neural tube to the body wall where they contribute initially to **pharyngeal arches** and then give rise to a great variety of adult structures including the jaws and gill arches (e.g., in the zebrafish *Danio rerio*: Schilling and Kimmel 1994; Kimmel et al. 1995, 2001; Cubbage and Mabee 1996).

5.2.1.1 Neurocranium

Structurally, the **neurocranium** is divided into four regions (Fig. 5.2c): the olfactory region, the orbital region, the otic region and the occipital region (Helfman et al. 2009). The **olfactory (or ethmoid) region** is the most anterior region of the neurocranium that supports the nares related to smell (i.e., the ability to sense and detect odorous molecules) and consists mainly of the following bones: ethmoid, lateral ethmoids, vomer, preethmoids, mesethmoids, kinethmoid and nasals. The **orbital region** is the cavity of the skull in which the eye is located and is formed by several cartilaginous and bony elements: frontals, orbitosphenoid, pterosphenoids, sclerotic cartilage, suborbital series (i.e., lachrymal, jugal, postorbital, fourth orbital, fifth orbital and dermosphenoid) and the supraorbital series (i.e., supraorbital 1 and supraorbital 2). The **otic region** is the part of the skull delimited for the support of the hearing organs. It consists of numerous consolidated bones: sphenotics, pteroptics, prootics, epiotics, opisthotics, supratemporals, parietals, basisphenoid and parasphenoid. The **occipital (or basicranial) region** is at the back of the braincase and forms the cranial base. The region mainly consists of the following bones: exoccipital, basioccipital and supraoccipital (some reading suggestions: Liem 1967; Vandewalle et al. 1992; Diogo and Chardon 2000a, b; Bemis and Forey 2001; Parmentier et al. 2001).

5.2.1.2 Splanchnocranium

The **splanchnocranium** (Fig. 5.2a, c) is also divided into three regions or "functional units" in terms of feeding biomechanics: the oromandibular region, the hyoid region and the branchial region (Helfman et al. 2009; Kardong 2012). Each region is derived to a certain extent from an embryonic pharyngeal arch. The most anterior visceral arch gives rise to the oral jaws (i.e., mandibular arch), while the next arch becomes the hyoid apparatus and the main part of the suspensorium that support the jaws (i.e., hyoid arch). The other posterior pharyngeal arches contribute to the branchial basket which supports the gill arches and gill filaments (some reading suggestions: Vandewalle et al. 1997, 2000; Parmentier et al. 1998; Engeman et al. 2009; Carvalho and Vari 2015).

The **oromandibular region** is composed of the **upper jaw** (i.e., premaxilla, maxilla and supramaxilla) and **lower jaw** (i.e., dentary, anguloarticular, retroarticular, Meckel's cartilage and coronomeckelian bone), the **suspensorium** (i.e., palatine, entopterygoid, metapterygoid, quadrate, hyomandibula and symplectic) and the **opercular series** (i.e., opercle, preopercle, interopercle, subopercle and subtemporal) corresponding to the gill cover elements.

The upper and lower jaws constitute the buccal jaws whose function is to grab food, whereas the prey processing is realized deeper in the buccal cavity, at the level of the pharyngeal cavity, where there is a second set of **pharyngeal jaws** (Vandewalle et al. 2000). The premaxilla and maxilla constituting the upper jaws articulate on the olfactory region of the neurocranium. Their morphology, length and shape are highly variable in actinopterygians just as it is also the case for teeth that can be found on both bones, on one of the bones or are absent.

Behind the upper and lower jaws, the suspensorium complex possesses at least five articulations (Fig. 5.3) which are important for the understanding of the mechanical principles of the respiration and feeding: (1) the autopalatine anteriorly articulates with the neurocranium in front of the orbit (i.e., neurocranial-autopalatine joint); (2) the hyomandibula posteriorly articulates with the neurocranium on the otic region (i.e., neurocranial-hyomandibula joint); (3) the posterior margin of the hyomandibula articulates with the opercular series which is related by a ligament to the caudal part of the lower jaw (i.e., hyomandibula-opercle joint); (4) the medial ventral margin of the hyomandibula articulates with the interhyal of the hyoid apparatus allowing back and forth movements of the branchial basket and (5) the quadrate of the suspensorium ventrally articulates with the anguloarticular bone of the lower jaw allowing the pivoting of the mandible (i.e., anguloarticular-quadrate joint). The two articulations of the suspensorium with the neurocranium can be compared to door hinges allowing lateral movements of the "cheeks" of the fish. In some species, a fifth articulation (6) can be found between the palatine and the maxilla (see later).

The **hyoid region** includes the **hyoid apparatus** (i.e., **hyoid bar**) which is the primary element of the mouth floor generally comprising (Fig. 5.4a, b) urohyal, basihyal, hypohyal, ceratohyal, epihyal, interhyal and branchiostegal rays (Aerts 1991; Faustino and Power 2001; Helfman et al. 2009). On the medial side of the hyomandibula, the hyoid apparatus articulates with the suspensorium to the branchial basket (Fig. 5.3) allowing the back-and-forth movements of the buccal roof (Liem 1967; Osse 1969).

The **branchial region** corresponds to the region around the fish gills that includes the following skeletal elements (Fig. 5.4b): pharyngobranchials, pharyngeal plates, epibranchials, ceratobranchials, hypobranchials and basibranchials (e.g., Vandewalle et al. 2000; Faustino and

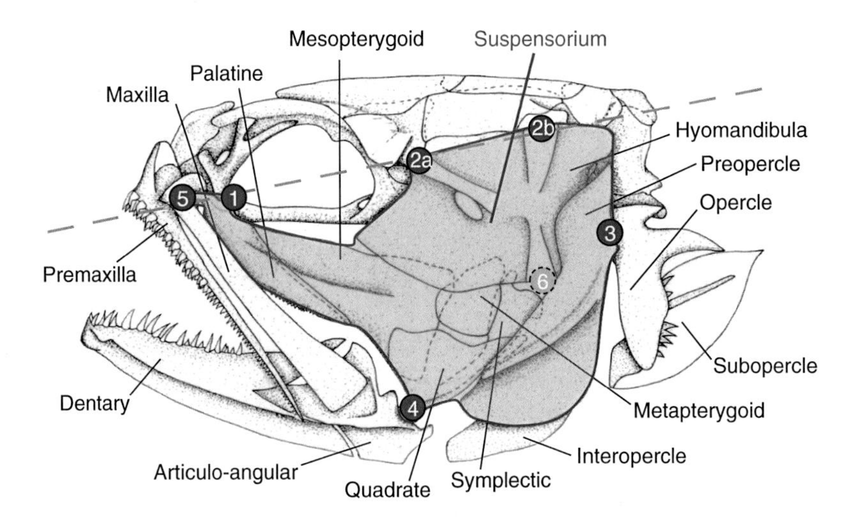

Fig. 5.3 Illustration of the main articulations of the suspensorium in a teleost species (*Carapus boraborensis*): (1) between the palatine and the neurocranium, (2a), (2b) between the hyomandibula and the neurocranium, (3) between the hyomandibula and the opercle, (4) between the quadrate of the suspensorium and the lower jaw, (5) between the maxilla and the palatine, and (6) between the lower part of the hyomandibula and the interhyal of the hyoid bar (this articulation is in light blue because the articulation takes places on the medial side of the suspensorium). The dotted line represents an axis passing through the articulations between the suspensorium and neurocranium and allowing lateral movements of the "cheeks" of the fish. The schema is redrawn from Parmentier (2003)

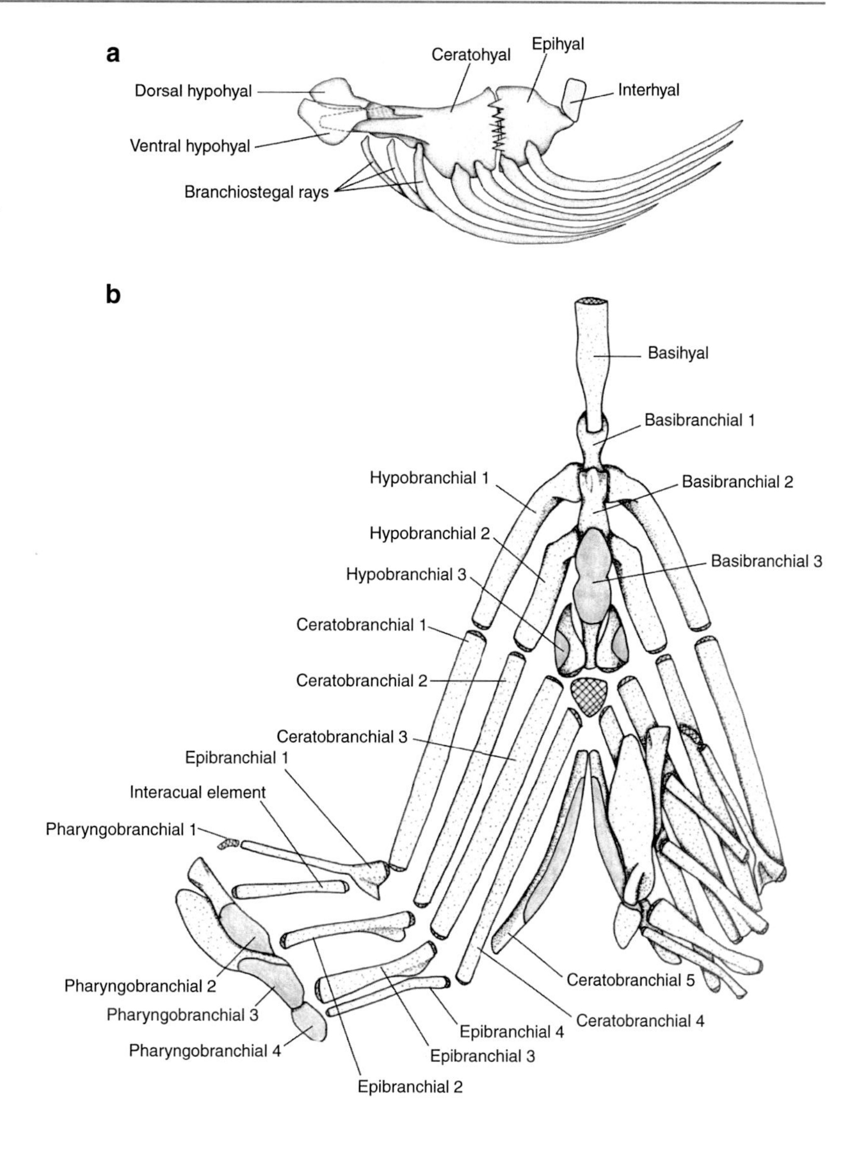

Fig. 5.4 Examples of (**a**) the hyoid apparatus in lateral external view and (**b**) the branchial basket in dorsal view in an actinopterygian (*Carapus boraborensis*). For the hyoid apparatus, urohyal is not shown (Parmentier 2003)

Power 2001; Helfman et al. 2009). **Pharyngeal jaws** are located in the branchial region and are used to process food (Fraser et al. 2009).

5.2.2 Cranial Musculature

The cranial musculature (i.e., the muscles associated with the skull, jaws and other skeletal components) is essential for the understanding of mechanisms and linkages involved in breathing and feeding movements as these muscles are the primary contributors involved in the opening and closing of the buccal and pharyngeal cavities. Besides, most of the cranial muscles are formed before yolk exhaustion to allow exogenous respiration and feeding (Herbing et al. 1996). The **cranial muscles** are divided into four main groups: mandibular muscles, hyoid muscles, branchial muscles and hypobranchial muscles (e.g., Edgeworth 1935; Diogo and Abdala 2010).

5.2.2.1 Mandibular Muscles

The **mandibular muscles** are directly or indirectly involved in movements of the lower jaw and are innervated by the trigeminal nerve (i.e., cranial nerve V). The main mandibular muscles are four in number: *adductor mandibulae, intermandibularis, levator arcus palatini and dilatator operculi*. They originate from an embryonic mandibular muscle plate that progressively contributes to the development of three structures: (1) the premyogenic condensation constrictor dorsalis that dorsally develops, (2) the adductor mandibulae that medially develops, and (3) the intermandibularis that ventrally develops. The premyogenic condensation constrictor dorsalis then gives rise to the levator arcus palatini and the dilatator operculi (Edgeworth 1935; Diogo et al. 2008). The studies of Edgeworth (1935) are a fundamental source of information about the development of the cranial muscles in actinopterygians.

The **adductor mandibulae** (i.e., jaw muscle, Fig. 5.5) is the more easily accessible cranial muscle and the largest superficial muscle complex of the fish cheek (Winterbottom 1974). It is specifically innervated by the ramus mandibularis nerve which is a motor branch of the trigeminal nerve. It is the more significant mandibular muscle in feeding biomechanics because it is responsible for the closing of the **lower jaw** and is consequently present in all actinopterygians. Structurally, the adductor mandibulae ranges from simple and undivided jaw muscle to a highly complex architecture incorporating up to ten discrete subdivisions. According to the new terminology of Datovo and Vari (2013), the adductor mandibulae muscle is composed of a large facial segment (i.e., segmentum facialis) and a smaller mandibular segment (i.e., segmentum mandibularis). The facial segment is positioned lateral to the suspensorium, whereas the mandibular segment is located medial to the lower jaw. These two muscle segments are usually connected by a tendinous complex (i.e., intersegmental aponeurosis) which is attached to the medial surface of the lower jaw. In many fishes, the facial segment can be also subdivided into three muscle sections, a ventrolateral section (i.e., pars rictalis), a dorsolateral section (i.e., pars malaris) and an anteromedial section (i.e., pars stegalis), and the mandibular segment can be separated into two muscle sections: a dorsal section (i.e., pars coronalis) and a ventral section (i.e., pars

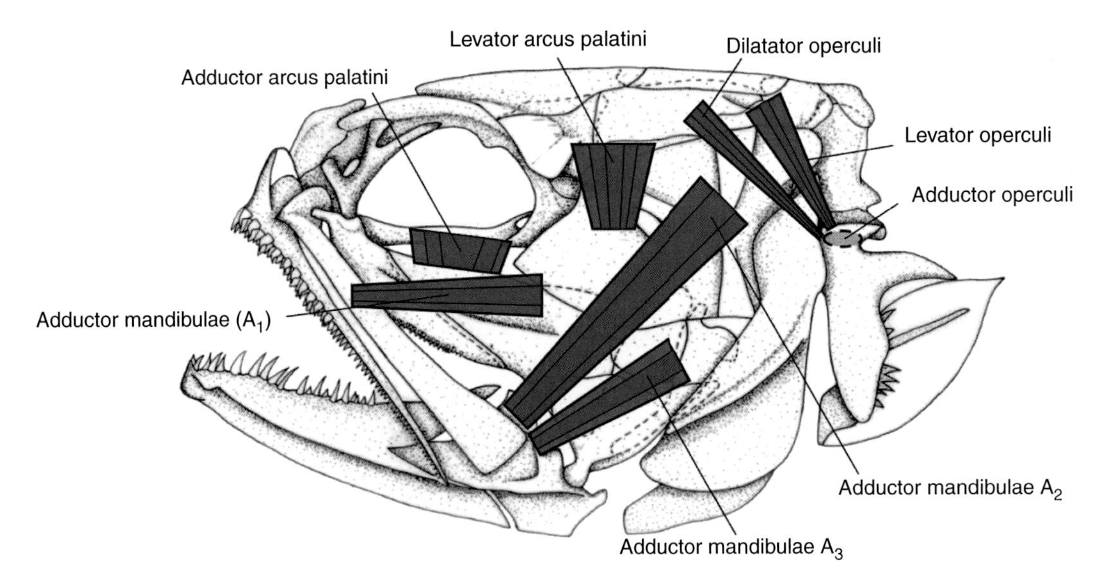

Fig. 5.5 Schematic representation of some cranial muscles in an actinopterygian (*Carapus boraborensis*). Mandibular muscles: adductor mandibulae (A1, A2, A3), levator arcus palatini and dilatator operculi. Hyoid muscles: adductor operculi, adductor arcus palatini and levator operculi. Protractor hyoideus (i.e., geniohyoideus) and sternohyoideus that can participate to the mouth opening are not shown. All the muscles and adductor arcus palatini and adductor operculi have at least one insertion on the lateral side of the suspensorium and opercle. In adductor arcus palatine, the insertion is on the medial side of the suspensorium, and in adductor operculi, the insertion is on the medial side of the opercle. The schema is redrawn from Parmentier (2003)

mentalis). Each muscle section can also be subdivided and differentiated into different subsections which are well explained and illustrated in the reference publications (Datovo and Vari 2013, 2014 for teleosteans).

The **intermandibularis** is a muscle which ventrally connects the two mandibles (i.e., dentaries) and is present in virtually all actinopterygians. It is unsubdivided in basal actinopterygians such as Cladistia, Chondrostei and Ginglymodi, but it is subdivided into intermandibularis anterior and posterior in the Halecomorphi *Amia calva* and Teleostei. The intermandibularis posterior combines with the interhyoid muscle and is involved in the mouth opening (see below).

The **levator arcus palatini** (Fig. 5.5) is also found in all actinopterygians apart from Chondrostei where there is instead the **protractor hyomandibulae** that is responsible for the protraction of the hyomandibula. The levator arcus palatini originates from the neurocranium and has an attachment site often along the hyomandibula on the suspensorium to lift the palatal arch.

The **dilatator operculi** (Fig. 5.5) is found in all actinopterygians but is also absent in Chondrostei. This muscle originates from the neurocranium and inserts along the dorsolateral faces of the opercula to move them apart and expand the pharyngeal cavity.

5.2.2.2 Hyoid Muscles

The **hyoid muscles** are closely related to movements occurring in the mouth opening and motions of the **hyoid apparatus**. They are generally innervated by the facialis nerve (i.e., cranial nerve VII). The four main hyoid muscles are interhyoideus, hyohyoideus, adductor operculi and adductor arcus palatini. Embryologically, they arise from the premyogenic condensation constrictor hyoideus that gives rise ventrally to the interhyoideus and hyohyoideus and dorsomedially to the adductor operculi and the adductor arcus palatini (Edgeworth 1935).

The **interhyoideus** operates in the opening of the mouth by having a site of origin from the basihyal and ceratohyal of the hyoid apparatus and an attachment site on the lower jaw. This hyoid muscle is found in all actinopterygians but is specifically fused in Teleostei with the intermandibularis posterior of the mandibular muscles to constitute the **protractor hyoideus** (i.e., **geniohyoideus**).

The **hyohyoideus** is a ventral muscle in contact with the hyoid apparatus, which is unsubdivided in basal actinopterygians such as Cladistia, Chondrostei and Ginglymodi but is subdivided into hyohyoideus inferior and superior in *Amia calva* and Teleostei. The hyohyoideus superior is also notably divided into one hyohyoideus abductor and two hyohyoidei adductors in *Amia calva* and Teleostei. The hyohyoideus abductor is responsible for the expansion of the branchiostegal membrane because of its origin from branchiostegal rays. The hyohyoidei adductors are in contrast responsible of the constriction of the branchiostegal membrane. This muscle originates from the opercle and subopercle and inserts on branchiostegal rays.

The **adductor operculi** (Fig. 5.5) is a dorsal hyoid muscle that has a site of origin from the neurocranium and an attachment site on the opercles causing their adduction. This muscle is present without exception in all actinopterygians.

The **adductor arcus palatini** (Fig. 5.5) is present in all actinopterygians (exclusive of Chondrostei where there is rather a **retractor hyomandibulae**) where it originates from the neurocranium and inserts on the medial side of several elements of the suspensorium such as hyomandibula, metapterygoid and entopterygoid in order to raise the suspensorium (i.e., suspensorial adduction). In addition to these major hyoid muscles, a levator operculi and an adductor hyomandibulae are, respectively, is found in Halecomorphi *Amia calva* and Teleostei and more advanced Teleostei such as Euteleostei, Otocephala and Clupeomorpha.

The **levator operculi** (Fig. 5.5) originates from the neurocranium and inserts on the opercles which moves essentially to the opercular series, which may interfere in lower jaw depression through the interoperculo-mandibular ligament.

The **adductor hyomandibulae** is a dorsal hyoid muscle that originates from the neurocranium to attach on the dorsomedial faces of the hyomandibula. Its function is to adduct the hyomandibula.

5.2.2.3 Branchial Muscles

The **branchial muscles** include the branchial muscles *sensu stricto* that are innervated by the glossopharyngeus and vagus nerves (i.e., cranial nerves IX and X, respectively) and the other branchial muscles such as the cucullaris, laryngeal, coracobranchialis and epibranchial muscles that are normally innervated by the spinal accessory nerve (i.e., cranial nerve XI). The development, organization, nomenclature and function of branchial muscles are complex and are not discussed herein. Research works such as those of Winterbottom (1974), Vandewalle et al. (2000) and others could be consulted for specific examples as well as better representation and understanding.

5.2.2.4 Hypobranchial Muscles

The **hypobranchial muscles** are usually innervated by spinal nerves. There is a single hypobranchial muscle in teleosteans such as in the zebrafish, the **sternohyoideus** (Schilling and Kimmel 1994, 1997; Diogo et al. 2008), while there are two hypobranchial muscles in basal actinopterygians such as Cladistia (e.g., *Polypterus senegalus*): the **coracomandibularis** (i.e., **branchiomandibularis**) and the sternohyoideus (Noda et al. 2017). The coracomandibularis connects the branchial arches to the lower jaw and is missing in living Lepisosteiformes and Teleosteans. The sternohyoideus is innervated by the anterior branches of the occipito-spinal nerves. It plays a major role in hyoid depression, and, through a series of mechanical linkages, in mouth opening and suspensorial abduction.

5.3 Breathing and Feeding Biomechanics

5.3.1 Breathing

Most actinopterygians breathe with gills that enable them to release carbon dioxide and to recover the oxygen that is dissolved in the aquatic environment (Brainerd and Ferry-Graham 2005). The **respiratory cycle** (Fig. 5.6) begins typically with the mouth opening that first implies the depression (i.e., ventral rotation) of the lower jaw. This mouth opening is directly followed by the depression of the hyoid apparatus and the lateral expansion of the suspensorium, which are

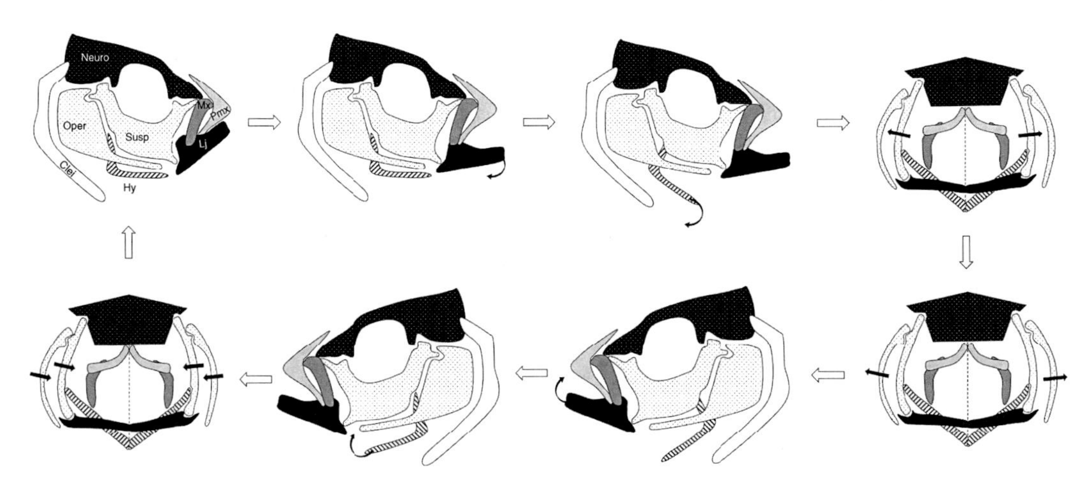

Fig. 5.6 Schematic representation of the respiratory cycle in an actinopterygian. The mouth opening begins with the depression (i.e., ventral rotation) of the lower jaw which is followed by the depression of the hyoid apparatus and the lateral expansion of the suspensorium leading to the opercular enlargement. The mouth closing results from the inverse movements which are the elevation (i.e., dorsal rotation) of the lower jaw and then the hyoid apparatus induced the adduction of the suspensorium and the opercular series. *Clei cleithrum* of the pectoral girdle, *Hy hyoid apparatus* (hatched in black), *Jj lower jaw* (black), *Mx maxillary* (dark grey), *Neuro neurocranium* (black), *Oper opercular series*, *Pmx premaxilla* (light grey), *Susp suspensorium*. The black arrows indicate the direction of movements

caused by the contraction of the sternohyoideus and levator arcus palatini muscles, respectively. The volume increase allows moving water from the buccal to the pharyngeal cavity. The spreading of the opercular series allows creating a more important volume on the lateral parts of the branchial basket. As a result, the water flow is directed towards the gill opening. The mouth closing increases the pressure in the buccal cavity forcing again the water to move in the branchial basket. The rising of the hyoid apparatus (i.e., dorsal rotation) and the adduction of the suspensorium complete the pressure increase. Once the water is ejected from the opercular cavity, the opercular series returns against the fish body, and the passive part of the branchiostegal membranes moves away from it. Finally, the mouth begins to open again in order to start a new respiratory cycle. The increase and decrease in volume is related to the decrease and increase in pressure, respectively, which results in the displacement of water towards the gills (e.g., Hughes and Shelton 1958; Ballintijn and Hughes 1965; Herbing et al. 1996).

5.3.2 Feeding

Feeding is actually more complex than simply opening the mouth and then closing it around a prey item (Shadwick and Lauder 2006). This action can be accomplished in two ways, depending of the fish movements. In the first case, the fish can swim with large gape allowing the water and potential prey to enter the mouth. The water leaves the fish through the gill openings, whereas food is directed towards the digestive tract. This mode of feeding is called the **ram feeding**. In the second option, the fish develops by means of its musculoskeletal system a large volume of the buccal cavity which in turn provokes a pressure decrease in the mouth cavity and results in entering of water (Lauder 1980). This mode of feeding is called the **suction feeding** and could be assimilated to an exaggeration of respiration movements. Powerful buccal expansion and rapid mouth opening are associated with extreme suction generation (Ferry-Graham et al. 2001). Ram and suction feeding were first considered as extremes of a continuum from pure ram to pure suction feeding, and it has been shown that many species of fish procure food using combinations of ram and suction feeding (Wainwright et al. 2001, 2007; Carroll 2004; Carroll et al. 2004; Day et al. 2005, 2007; Van Wassenbergh et al. 2005; Higham et al. 2006a, b; Staab et al. 2012). Even fish species that have abandoned capturing prey by suction feeding retain the mechanism during the processing and manipulation of prey (Wainwright et al. 2015). More recently, a third mode has been incorporated to create the ram-suction-biting domain, the action of biting being simply to close the jaws on the prey (Ferry et al. 2015). Adding this mode can provide more acute description of the feeding mechanism and give insight on the species ecology but does not change the basic fact that, after the biting, the fish has to find a way to move the prey into the mouth which requires suction and/or ram. In this way, there are three main feeding strategies that are encountered in fishes, ram feeding, suction feeding and **feeding with manipulation** (biting), but each mechanism relies on the use of the same musculoskeletal elements to capture prey.

The most common mode of prey capture in actinopterygian fishes is suction feeding, in particular, among teleosteans (Liem 1980; Lauder 1985) and could be understood as an exaggeration of the respiration movements, where some musculoskeletal elements can be modulated to modify the mouth gape or increase the feeding performance. The underlying mechanisms of suction feeding are complex and have been extensively studied. They could be divided into four phases: a preparatory phase, an expansive phase, a compressive phase and a recovery phase (Lauder 1980, 1985). The preparatory phase, which consists in buccal cavity compression and buccal volume decreasing, is absent in basal actinopterygians such as Polypteriformes, Lepisosteiformes and Amiiformes and can be observed only in acanthopterygian teleosteans. The most important phase of suction feeding is the expansive phase, which is defined by Lauder (1980, p. 294) as "the time from the start of the mouth opening to peak gape". During this phase, the mouth opens quickly with a rapid expansion of the buccal cavity, which occurs as a result of cranial elevation (i.e., dorsal rotation of the neurocranium) generated by epaxial muscles (Westneat and Olsen 2015), jaw opening (i.e., ventral rotation of lower jaws) that can be caused in different ways such as the contraction of either geniohyoideus, levator operculi, sternohyoideus (and related hyoid depression), hypaxial or epaxial muscles and at the same time lateral expansion of the suspensorium (Schaeffer and Rosen 1961; Lauder 1982; Grubich 2001) (Fig. 5.7). The rapid expansion creates a drop in pressure into the buccal cavity, which generates a flow of water directed towards the mouth (Higham et al. 2006b). The resulting water flow exerts a hydrodynamic force on the prey item and draws it towards the beginning of the digestive tract. The compressive phase, defined by Lauder (1980, p. 294) as "the time from the peak gape to complete closure of the jaws", involves the compression of buccal and pharyngeal cavities via hyoid protraction and suspensorium adduction and at the same time of the lower jaw closure via the adductor mandibulae (Grubich 2001). The last recovery phase results in the return in their original position of all the skeletal elements of the feeding system (Wainwright et al. 2001, 2007; Carroll 2004; Carroll et al. 2004; Day et al. 2005, 2007; Van Wassenbergh et al. 2005; Higham et al. 2006a, b).

The mechanical principles of suction feeding were mainly based on studies using high-speed camera and electromyography (e.g., Osse 1969). Actually three mechanisms can allow the mouth opening and correspond to the so-called expansive phase. According to the high amount of actinopterygian species and their related specificities, it is not possible to describe accurately all the different mechanisms encountered. They are voluntary simplified, and the reader has to keep in mind that they are not necessarily found in all species (Lauder 1982; Westneat 2005). Whatever the mechanism, the aim is basically to depress the lower jaw and to elevate the skull. The power required for suction expansion would be mainly generated by the **epaxial swimming muscles**, in which the body muscles just behind the head cause the skull to rotate upward during feeding (Camp et al. 2015).

1. The first basic mechanism implies from the back to the front the coupling of the ventral **hypaxial musculature**, the pectoral girdle, the sternohyoideus muscle, the hyoid bar, the geniohyoideus muscle and the lower jaw. Fundamentally, the contraction of the hypaxial musculature stabilizes at least the pectoral girdle and at best pulls it backward. Then, the contraction of the sternohyoideus muscle pulls the hyoid bar posteroventrally. This action is transferred to the geniohyoid muscle that depresses the lower jaw because it pivots around the articulation with the quadrate. Isolated or different combinations of contraction of the three muscles can modify the movement amplitude. In basal actinopterygians (Cladistia, Chondrostei, Ginglymodi and Halecomorphi), the geniohyoideus muscle is not found. In Teleostei, the hyoid apparatus can be related to the mandible by the mandibulo-hyoid ligament, while in other primitive species, this ligament is changed into interoperculo-hyoid ligament. This ligament connects the hyoid apparatus and the interoperculum which is connected by the interoperculo-mandibular ligament to the lower jaw (Lauder 1982).
2. The second mechanism consists in the elevation or dorsal rotation of the neurocranium. It has been modeled on the coupling between the skull, the epaxial musculature, the pectoral girdle, the urohyal from the hyoid apparatus and the lower jaw (Muller 1987). The contraction of the epaxial musculature inserting on the posterior part of the skull causes the neurocranium elevation because it pivots clockwise around the rostral end of the vertebral column (Schaeffer and Rosen 1961; Lauder 1982; Carroll et al. 2004). This skull movement induces the backward displacement of the pectoral

girdle. This movement is transferred to urohyal from the hyoid apparatus and by the first coupling explained above transmitted to the lower jaw.

3. The third mechanism implies the opercular series (i.e., operculum, suboperculum and interoperculum), the levator operculi muscle and the lower jaw. The contraction of the levator operculi muscle that connects the dorsal margin of the opercle to the neurocranium causes the elevation of the operculum that pivots around its articulation with the hyomandibula of the suspensorium. This motion pulls posteriorly the interoperculum which possesses on its anterior edge an interoperculo-mandibular ligament passing under the articulation between the quadrate and the lower jaw and inserting on the posterior part of the mandible. Consequently, the posterior movement of the interoperculum starts the depression of the lower jaw.

On another note while the mobility of the jaws and the shape of the opening of the mouth are modified in some species that have departed from a primary reliance on suction feeding, the anterior-to-posterior wave of expansion persists. The suction would be more efficient when the buccal cavity is shaped like a large cone with a small circular mouth opening (Liem 1990). The rate of expansion of the cone can change the shape of the cone, determining the water flow velocity and the resulting suction efficiency. Therefore, the buccal cavity may be modeled as an expanding cylinder with surrounding buccal pressure distributed across its internal surface (Muller et al. 1982). In addition, the action of the mouth opening, or the lower jaw depression, tends to pull on the upper jaw (maxilla and/or premaxilla) and protrude it due to **linkages** in most teleostean fishes between the upper and lower jaws (Westneat 2004). When the upper jaw protrudes, the descending arm of the premaxilla and the maxilla typically rotate forward and occlude the sides of the open mouth (Gibb 1996). Indeed, this helps create the round or planar opening of the mouth thought to be a key component of effective suction feeding.

5.4 Evolution Trends in Actinopterygians

The success of actinopterygians and mainly teleosteans has been associated with different **evolutionary trends**, but it remains to be shown. It would concern the repositioning and specialization of the dorsal fin, the change in placement and function of pectoral and pelvic fins, the elaboration of homocercal tail and the improvement of the swim-bladder function (Rosen 1982). At the skull level, there is fusion and reduction in a number of bony elements, such as dermal bones that originally constituted the exoskeleton of the braincase (Helfman et al. 2009). Dermal bones (i.e., exoskeleton) seem to have merged with deep bones (i.e., endoskeleton) to contribute to the development of a more laterally **kinetic** skull.

Nonetheless, it is rather difficult to generalize evolutionary trends within the skull, jaws and cranial muscles of actinopterygians because of the plethora of species from different taxa that were able to take advantage of different habitats and types of prey. The results are that **jaw mechanics** show numerous patterns of both diversification and convergence. A common large gap can be found, for example, in distant-phylogenetic species such as the Northern pike *Esox lucius* (Esociformes) and the grouper (Perciformes), but they are not phylogenetically related to meaning features which result from evolutionary convergence. Comparable observations concern herrings (Clupeiformes), minnows (Cypriniformes) or damselfishes (Perciformes) that have circular mouth to feed on plankton but use different mechanisms to do it. Although having different anatomy, all species are able to drop the lower jaw and then abduct the hyoid bar and the suspensorium before abducting the opercular series. This is because ray-finned fishes are characterized by an extremely large number of mobile bony elements in the skull allowing various **mouth opening** mechanisms.

Moreover, in relation to their actinopterygian Bauplan, the mandibular lever system of the mandible is present in virtually all ray-finned fishes. The lower jaw possesses however different shapes that directly impact both the force and velocity of the lower jaw closing abilities. Species vary from having high force transmission to those specialized for speed of jaw motion (Barel 1983; Alfaro et al. 2001; Westneat 2004; Wainwright et al. 2015). In parallel, there are important patterns concerning the number and shape of teeth on the jaws. Biters show rows of large conical teeth directed towards the buccal cavity, whereas many suction feeders can have minute teeth or are simply toothless (Schaeffer and Rosen 1961; Motta 1984).

Throughout actinopterygian phylogeny the increasing mobility of upper jaws from basal to more derived taxa is a subject of much interest. In the bichir *Polypterus* (Polypteriformes) and the gars *Lepisosteus* (Lepisosteiformes), both premaxilla and maxilla are firmly attached to the neurocranium and do not contribute to mouth opening. A first innovation is found in *Amia calva* (Amiiformes) where the maxilla is free from the cheek and is able to pivot anteriorly because it has gained a rotational joint with the neurocranium (Lauder 1980). The maxilla is attached by connective tissue to the palatine bone and is connected to the mandible via the maxillo-mandibular ligament. At the jaw opening, the lower jaw is dropped and pulls the posterior end of the maxilla that swings forward. As a result, maxilla and associated connective tissue form the lateral walls of the gape. This novelty has enhanced the control of fluid flow and has increased the velocity of water movement, both of which can improve suction feeding abilities (Lauder 1980).

In the next structural change that evolved in distantly related groups (e.g., Salmoniformes, Esociformes, Aulopiformes, Stomiiformes, Elopiformes, Clupeiformes), the proportionally small premaxilla acquires some mobility and can articulate with the maxilla (Gosline 1980; Wainwright et al. 1989; Grubich 2001). Both bones are joined on a butt joint meaning that an anterior swing of the maxilla causes (small) movements of the premaxilla (Rosen 1982). Although the fine structural organization between *Elops* and *Clupea* appears to be different, the result of the maxillary rotation is to rock the dental surface of the premaxilla forward and outward in both taxa (Gosline 1980). The maxilla has thus a propulsive function. These short movements of both bones (maxilla and premaxilla) could result in the **protrusion** of the premaxilla of higher teleosts (Alexander 1967; Motta 1984). Moreover, the maxillary articulation with the palatine is modified in these taxa since the ligamentous joint found in *Amia* is now replaced by a ball-and-socket joint articulation. As it was the case in *Amia*, the maxillary rotation forms a tubular mouth for suction feeding.

The next major structural specialization is encountered in teleosteans and is related to the increasing mobility of premaxilla and maxilla that are loosely connected by ligaments. It allowed many species to develop the **upper jaw protrusion** which is the ability to extend the premaxilla and maxilla towards the prey during feeding. Functional advantages of jaw protrusion include at least (1) the increase in the rate of approach of the predator to the prey (Westneat and Wainwright 1989; Ferry-Graham et al. 2001; Waltzek and Wainwright 2003), (2) the increase of the distance from which a prey may be sucked, (3) the decrease of lower jaw movements to close the mouth, (4) the reduction of energy expenditure during suction feeding (Osse 1985) and (5) the increase of the hydrodynamic force exerted on prey (Holzman et al. 2008; Staab et al. 2012). Morphologically, optimized anterior mouth opening for suction feeding also reduces the length of the toothed jaw edge to grasp, retain or bite a prey (Osse 1985). This mechanism would have evolved at least five times in distantly related phylogenetic groups and may help to explain the extraordinary diversity seen in ray-finned fish skulls (Westneat 2004, 2005; Wainwright et al. 2015) (Fig. 5.8). The upper jaw protrusion ability is found in taxa showing the fastest rates of speciation (Alfaro et al. 2009), and interestingly, three of these independent origins have occurred within Ostariophysi (e.g., Gonorynchiformes, Cypriniformes, Characiformes). Although the morphological and kinematical details of jaw protrusion appear to be quite variable (Liem 1980; Motta 1984), a particular system of premaxilla projection appears to be basic to modern teleosts (Osse 1985). Whatever the detailed mechanism, the **upper jaw protrusion** is always related to the rotation of the maxillary. The shift from the single premaxilla rotation to protrusion seems due to different features.

5.4.1 Acipenseriformes

Sturgeons and paddlefishes (**Acipenseriformes** or **Chondrostei**) constitute a basal group in Actinopterygii. They are notably characterized by reduced ossification of the endoskeleton, but they have numerous dermal bones that are associated with the head and the body. They also have a hyostylic jaw suspension, meaning the upper jaw (or **palatoquadrate** in sturgeons) is not directly connected to the cranium but it is suspended through loose connective tissue between the upper jaw and ventral surface of the neurocranium (Carroll and Wainwright 2003). The palatoquadrate articulates with the lower jaw (i.e., Meckel's cartilage), both parts being supported caudally by the hyoid bar. This organization is similar to the jaw anatomy of sharks (Wilga and Motta 1998; Huber et al. 2005). The **protrusion** mechanism could be summarized as follows (Carroll and Wainwright 2003); (Fig. 5.8). (1) the retraction of the hyoid bar is associated with lower jaw depression; and (2) the contraction of the protractor hyomandibularis, connecting the anterior margin of the hyomandibula to the neurocranium, would provoke forward dorsal rotation of the hyomandibula. This forward displacement would push the symplectic bone rostrally resulting in the jaws being placed outside (i.e., moving anteriorly) of the oral cavity and thus protruded.

5.4.2 Acanthopterygii (e.g., Perciformes)

In acanthopterygian protrusion, the proximal part of the premaxilla moves forward relative to the skull; this kinesis involves different modifications at the level of the skull, the ligaments, the shapes of the maxilla and premaxilla (Fig. 5.8). There is, for example, the development of a sliding articulation between the premaxilla and the skull that corresponds to the development of an ascending process extending over the anterior part of the neurocranium. Another modification corresponds also to the elongation of the toothed process of the premaxilla, excluding the maxilla from the gape (Alexander 1967; Gosline 1980). It would prevent the formation of an angle between maxilla and premaxilla, favouring the development of a rounded mouth gape. The cylindrical shape of the mouth is due to many connective tissues between both bones of the upper and lower jaws. In this system, the twisting of the maxilla during mouth opening does no more have propulsive function because the membranous attachment that previously concerned only the maxilla and the lower jaw is now also found at the level of the premaxilla: the lowering of the mandible directly pulls the premaxilla downward (Schaeffer and Rosen 1961). The maxilla, connective tissue and ligaments (between the premaxilla and the skull) determine the premaxilla protrusion distance. According to the species, this basic system can have numerous adaptations at the level of the morphology (of the upper jaw, anterior part of the skull, etc.) and on the moving mechanism (Motta 1984). The system of levers formed by the lower jaw, maxilla and premaxilla has been modeled as a four-bar linkage (Westneat 2004).

5.4.3 Cypriniformes

In **Cypriniformes** (carps, minnows, loaches and relatives), an additional sesamoid and synapomorphic bone, called the **kinethmoid**, is involved in the jaw protrusion mechanism (Fig. 5.8). This bone is located at the rostral neurocranium and is entirely suspended by ligaments which names provide information about their attachments premaxilla-kinethmoid ligament, mesethmoid-kinethmoid ligament, palatine-kinethmoid ligament and maxilla-kinethmoid ligament (Hernandez et al. 2007; Staab and

Hernandez 2010). During mouth opening, the kinethmoid makes an anterior 90°–180° rotation that protrudes the premaxilla. The amplitude of the displacement is the function of the kinethmoid size and shape of the ligaments. In this case, the cypriniformes does not have a long ascending process of the premaxilla. The fine mechanism still is not fully understood because its complexity is more important than in the previous group. It implies more components, and there are more connections between the elements (Staab and Hernandez 2010). It was first thought the lower jaw depression drives the premaxilla protrusion as it is the case in Acanthopterygii (Alexander 1967; Motta 1984). However, a recent study on five different species of Cypriniformes has shown it was not the case since the timing of peak gape is not correlated with the timing of peak protrusion (Staab et al. 2012). It shows at least lower jaw movement is not the only force acting on upper jaws. In Cypriniformes, the adductor mandibulae A_1 complex (see hereafter) inserts on the maxilla. The A_1 bundle organization is more complex than in Acanthopterygii and seems to be implicated in jaw protrusion. Moreover, its high diversity in terms of insertion sites combined with diversity in jaw and kinethmoid shapes highlight specialization in different kinds of movements, increasing the ability of the fish to interact with its environment (Hernandez et al. 2007; Hernandez and Staab 2015). Electromyographic-based studies support the contraction of the A_1 bundles and can lower the maxilla (Ballintijn et al. 1972). As a result, the ventral displacement of the maxillae produces tension in the paired maxilla-kinethmoid ligament and the anterior rotation of the kinethmoid. The main functional difference between Cypriniformes and Acanthopterygii (e.g., Perciformes) would be in the flexibility of the movements relative to jaw protrusion (Hernandez and Staab 2015). In acanthopterygians, jaw protrusion takes place simultaneously with full mandible lowering. In cypriniform, the full lower jaw depression is not required to have jaw protrusion. **Upper jaw protrusion** is decoupled from lower jaw depression, meaning the production can take place with closed or open mouth. According to Gidmark et al. (2012), this functional difference could be related to the ability to feed (lowered mandible + protrusion) or to sort food (raised mandible + protrusion). Additional studies showed movements are more flexible in the relative timing of jaw protrusion and suction flows (Staab et al. 2012). These differences could be related to the feeding niches. Acanthopterygians are found in different feeding niche (Wainwright et al. 2007) but are preferentially feeding on elusive prey in the water column: correlated movements between upper and lower jaw are required to provoke powerful water flow. The Cypriniformes are mostly benthic feeders (López-Fernández et al. 2012; Hernandez and Staab 2015) and could be compared to a vacuum cleaner: the higher kinesis of the jaw allows positioning of a rounded mouth on the substrate that prolongs the sucking action.

5.4.4 Characiformes (*Bivibranchia protractila*)

The characiform *Bivibranchia protractila* (junior synonym of *Bivibranchia fowleri*) is so named due to its protrusible upper jaw (Vari 1985; Vari and Goulding 1985). This feature can be found in different species of Hemiodontidae, but some differences can be found among species (Alexander 1964; Roberts 1974; Vari 1985). In this clade, the small premaxilla is fused to the maxilla, both structures being S-shaped (Fig. 5.8). The upper jaw has lost is ligamentous attachment to the ethmoid and is not articulated to the palatine. However, a ligament can be found between the palatine and the premaxilla, and there is also a maxilla-mandibular ligament between the maxilla and the dentary (Géry 1962). At the level of the rostral part of the suspensorium, the palatine, ectopterygoid and entopterygoid appear to be firmly connected, supporting the neurocranium and most probably articulating with the quadrate (Regan 1911; Alexander 1964). During the mouth opening, the ligament between the dentary and the maxilla pulls the upper jaw downwards and forwards. The upper jaw then pulls the anterior margin of the palatine and rotates the rostral complex of the suspensorium downwards. This movement is feasible thanks to the loose connection between the complex and the quadrate. Therefore, the upper jaw is protracted (Géry 1962; Vari 1985; Vari and Goulding 1985).

5.4.5 Gonorynchiformes (*Phractolaemus ansorgii*) (Grande and Poyato-Ariza 1999)

To the best of our knowledge, the mechanism of the jaw protrusion in the gonorynchiform *Phractolaemus ansorgii* is not known but inferred from dissection and handly manipulations. At rest, the mouth is unusually positioned being dorsally directed (Fig. 5.8). In this situation, the raised lower jaw forms a semicircle with the upper jaw (Géry 1963). The upper jaw is located under the mesethmoid. It connects both the suspensorium (through the palatine) and neurocranium (through the prevomer) by a ligament (Grande et al. 2010), meaning the upper jaw is extremely movable. During the lowering of the mandible, the lower jaw rotates anteriorly around the quadrate to gain a horizontal position. In this situation, the maxilla-mandibular ligaments pull the upper jaws anteriorly, what results in the loss of connection of the jaw with the skull and facilitates the protrusion. When the mouth is totally protracted, the oral cavity is completely directed anteriorly or anteroventrally (Thys van den Audenaerde 1961).

It is worth mentioning that some species can also show protrusible lower jaws (Westneat and Wainwright 1989). In the sling-jaw wrasse, *Epibulus insidiator*, this unusual ability is mainly related to deep modifications at the level of the suspensorium and opercle. In Perciformes, the quadrate found at the lower part of the suspensorium has usually the role of a stationary support for the lower jaw because it is firmly attached to other bones (symplectic, metapterygoid, etc.) of the jaw. In *Epibulus*, the quadrate can articulate with the metapterygoid and make rotations that push the lower jaw rostrally (Delsman 1925; Westneat and Wainwright 1989). However, it is also important to bear in mind that most of the skeletal pieces of the skull and jaws are able to perform these incredible movements because of the contraction of **cranial muscles**.

During the evolution of actinopterygians and more generally those of vertebrates, the cranial muscles underwent enormous diversification that was crucial to the success of each clade (Goodrich 1958). Within the large class of Actinopterygii, it is important to understand that virtually all species have the cranial musculature described in the "anatomy" part of this chapter. Some of these cranial muscles have however differentiated by subdividing into several muscle sections to probably respond to the increasing complexity and **kinetics** of the teleostean skull. Nevertheless, the role of each muscle element remains fundamentally conserved in all ray-finned fishes, except from species which have early diverged such as sturgeons or paddlefishes (Acipenseriformes) which are having deeply anatomical and functional differences (Carroll and Wainwright 2003; Miller 2004).

In the cranial musculature of actinopterygians, the most studied and differentiated muscle is undoubtedly the **adductor mandibulae** muscle complex since it participates both in breathing and feeding movements by raising the lower jaw and closing the mouth. However, the evolution and nomenclature of the different muscle bundles of the adductor mandibulae has been the subject of many discussions and predominantly for teleostean fishes. There are many hypotheses about the early differentiation of the adductor mandibulae muscle in the course of evolution of actinopterygians and mainly in the teleostean lineage (Lauder 1980; Gosline 1989) and numerous publications are devoted to the terminology (e.g., Owen 1846, 1866; Winterbottom 1974; Diogo and Chardon 2000a, b; Wu and Shen 2004; Diogo et al. 2008; Datovo and Bockmann 2010; Datovo and Castro 2012). In the framework of this chapter, we have decided to bring to the fore the new terminology of Datovo and Vari (2013) instead of that proposed by Vetter in 1878 and subsequently used by Winterbottom (1974) and other authors with some misinterpretations. Table 5.1 highlights the main differences with regard to the way of naming and understanding the subdivisions of the adductor mandibulae muscle. The terminology of Datovo

Table 5.1 Comparative table between the nomenclature of the adductor mandibulae muscle for teleostean fishes used by Winterbottom (1974) and subsequent authors with the proposed terminology of Datovo and Vari (2013)

Literature	Winterbottom (1974)						
Muscle name	*Adductor mandibulae*						
Segment name							
Origin(s)	Suspensorium (lateral surfaces)						
Insertion(s)	Lower (and upper) jaws						
Section name	**A1** dorsal section		**A2** ventrolateral section		**A3** medial section		**Aw** intramandibularis section
Origin(s)	Suspensorium: quadrate, symplectic, hyomandibula, ectopterygoid, mesopterygoid, metapterygoid, palatine Others: preopercle, articular, infraorbital series, neurocranium (prefrontal, ethmoid, parasphenoid)		Suspensorium: quadrate, symplectic, mesopterygoid, metapterygoid, hyomandibula, Others: preopercle opercle, neurocranium (prootic, parietals, pterotic, ethmoid, sphenotic, supraoccipital, postorbital, frontal)		Suspensorium: quadrate, symplectic, metapterygoid, hyomandibula, Others: preopercle, neurocranium (parasphenoid, sphenotic, frontal, parietal, prootic, squamosal, opistothic, postfrontal		Tendon from other sections (anteroventral face) Quadrate (medial face) Preopercle (medial face)
Insertion(s)	Lower jaw (articular) Upper jaw (maxilla and premaxilla posterodorsal face)		Lower jaw (posterior face of the coronoid process of dentary; Meckelian fossa; articular) Upper jaw (maxilla)		Lower jaw (medial face of dentary; meckelian fossa; articular)		Lower jaw (Meckelian fossa of dentary, articular)
Subsection name	**A1$_{\alpha}$** ventrolateral subsection	**A1$_{\beta}$** posterodorsal subsection	**A2$_{\alpha}$** dorsal subsection	**A2$_{\beta}$** ventral subsection	**A3$_{\alpha}$**	**A3$_{\beta}$** part of Aw	
Origin(s)	Suspensorium	Metapterygoid Hyomandibular					
Insertion(s)	Lower jaw	Upper jaw (maxilla)					

Literature	Datovo & Vari (2013)								
Muscle name	***Adductor mandibulae* muscle complex**								
Segment name	**Segmentum mandibularis** mandibular subdivision			**Segmentum facialis** facial subdivision					
	Aw division of Winterbottom			A1, A2, A3 divisions of Winterbottom					
Origin(s)	Suspensorium (lateral surfaces) Faucal ligament Buccopharyngeal membrane			Suspensorium (lateral surfaces) Infraorbital series (medial face) Neurocranium					
Insertion(s)	Lower jaw			Lower (and upper) jaws					
Section name	**Pars coronalis** dorsal section	**Pars mentalis** ventral section		**Pars rictalis** ventrolateral section		**Pars malaris** dorsolateral section		**Pars stegalis** anteromedial section	
				A1 (-OST) and A2		A1 and A3		A1 and A3	
Origin(s)	Dorsal part of mandibular tendon (near the coronoid process)	Faucal ligament Buccopharyngeal membrane		Preopercle (horizontal arm) Quadrate (ventrolateral portion)		Preopercle (vertical arm) Hyomandibula (lateral surfaces)		Metapterygoid (lateral surfaces) Hyomandibula (anterodorsal region)	
Insertion(s)	Portion of the lower jaw dorsal to the Meckel's cartilage			Lower jaw (via intersegmental aponeurosis) Maxilla (via retrojugal lamina and ligaments)		Lower jaw (via intersegmental aponeurosis) Upper jaw (maxilla via retrojugal lamina and endomaxillar ligament)		Coronomeckelian bone of lower jaw (via ventral portion of meckelian tendon)	
Subsection name		**Prementalis** (anterodorsal subsection)	**Postmentalis** (posteroventral subsection)	**Ectorictalis** (external subsection)	**Endorictalis** (internal subsection)	**Promalaris** (anterodorsal subsection)	**Retromalaris** (posteroventral subsection)	**Epistegalis** (anterodorsal subsection)	**Substegalis** (posteroventral subsection)
				A1(-OST)		A1		A1	
Origin(s)		Association with intersegmental aponeurosis and pars coronalis	Faucal ligament Buccopharyngeal membrane					Medial to levator arcus palatini muscle	Ventral to levator arcus palatini muscle
Insertion(s)		Lower jaw (medial face)	Lower jaw (medial face)	Upper jaw (maxilla via lateral region of retrojugal lamina)	Lower jaw	Upper jaw (maxilla via anterodorsal region of the retrojugal lamina directly connect with endomaxillar and/or ectomaxillar ligaments)	Lower jaw and upper jaw (maxilla via posterodorsal or lateral region of retrojugal lamina and preangular and paramaxillar ligaments)	Upper jaw (maxilla)	Coronomeckelian bone of lower jaw via meckelian tendon

and Vari (2013) concerns only **teleostean fishes** and is more intuitive since it is possible to designate the different muscle subdivisions based on their position. Besides, it is easy to understand the instances of evolutionary subdivision and/or coalescence of muscle subdivisions (McCord and Westneat 2016). For further discussion, the following reference books should be consulted (Winterbottom 1974; Datovo and Vari 2013, 2014).

In **basal actinopterygians** (Polypteriformes, Lepisosteiformes and Amiiformes), the adductor mandibulae muscle is generally subdivided into three main portions: anterior, medial and posterolateral portions (Lauder 1980) that do not have the same nomenclature as teleostean fishes (see hereafter). In *Polypterus*, the anterior portion is absent, whereas it is subdivided in *Lepisosteus* and *Amia calva*. In any basal actinopterygian lineage, the medial portion is also separated into two subdivisions, but they have different pathways of differentiation and muscle terminology. The posterolateral portion is not subdivided, and the entire muscle has a unique attachment site on the medial face of the lower jaw (Lauder 1980).

In the **teleostean** lineage, the complex configuration of the adductor mandibulae muscle, coupled with the fact that its different muscle sections are found in diverse groups, has suggested several pathways of differentiation (Gosline 1989). It must be imagined that at the beginning, there was only a single muscle adductor mandibulae mass and that the first differentiation of the jaw muscle would have been in the segregation between the facial and mandibular adductor mandibulae segment where the latter would have separated as a distinct entity (Edgeworth 1935; Winterbottom 1974; Gosline 1989). In addition, this first differentiation would be an actinopterygian plesiomorphy (Lauder 1980) which means that all ray-finned fishes have this first subdivision for the adductor mandibulae. Secondly, the facial segment would have begun to differentiate even though an unsubdivided facial segment is observable, for example, in Elopiformes (*Elops*), Osteoglossiformes (*Hiodon*), Salmoniformes (*Salvelinus*) and Clupeiformes (*Clupea*) (Lauder and Liem 1980; Gosline 1989; Datovo and Vari 2014). According to Gosline (1989), two pathways of second differentiation can be observed in teleosteans. In the first pathway of differentiation found in **acanthopterygians** (e.g., Atheriniformes, Cyprinodontiformes, Gasterosteiformes, Perciformes, Scorpaeniformes, Tetraodontiformes), an antero-dorso-lateral part of the facial segment is differentiated and develops an attachment site on the upper jaw at the level of the maxilla [that can be the A1 section for Winterbottom (1974) and the pars malaris section or the pars promalaris subsection for Datovo and Vari (2013)]. Then, another part of the facial segment separates more medially, which is observed in most of teleosteans [that can be the A2/A3 section for Winterbottom (1974) and the pars malaris and pars stegalis sections for Datovo and Vari (2013)]. The second pathway of differentiation is encountered in most **Ostariophysi** (i.e., Gonorynchiformes, Cypriniformes, Characiformes, Gymnotiformes, Siluriformes) where an antero-ventro-lateral part of the facial segment appears and attaches on the medial face of the lower jaw [that can be the A2 section for Winterbottom (1974) and the pars rictalis section for Datovo and Vari (2013)]. In that case, this division of the adductor mandibulae seems to have developed as a "supplementary system for raising the mandible"(Diogo and Chardon 2000a, b). There is also another part of the facial segment that differentiates more medially, but it is followed by a more external differentiation of the first ventral section. This new subdivision develops, via the primordial ligament, an attachment site on the upper jaw at the level of the maxilla [that can be the A1-OST section for Winterbottom (1974) and the pars ectorictalis subsection for Datovo and Vari (2013)]. Among some Siluriformes, another special differentiation would appear externally with an attachment site on the maxilla and could be termed the retractor tentaculi muscle (Diogo and Chardon 2000a, b; Datovo and Vari 2013, 2014).

In this way, the evolution and differentiation of the adductor mandibulae muscle is one of the most notable within the cranial musculature because, in our opinion, this development may be mainly related to the parallel specialization of the buccal jaws which become able to protrude for suction feeding in more advanced ray-finned fishes (Westneat 2004, 2005; Wainwright et al. 2015).

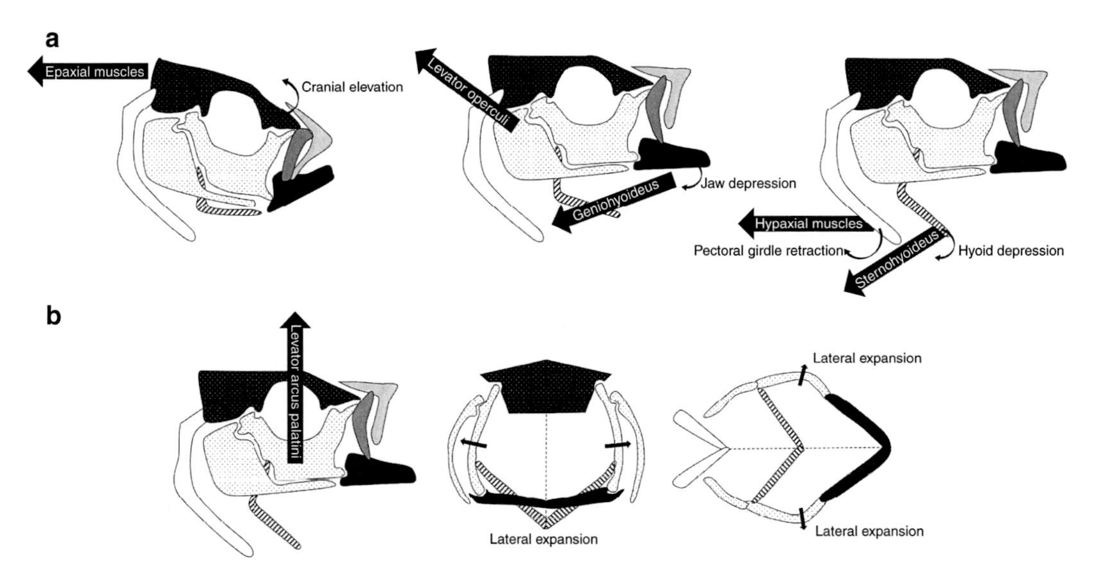

Fig. 5.7 Schematic representation of one of the different mechanisms that can be used for the expansive phase during suction feeding (**a**) and schematic representation of the lateral expansion (**b**). The mouth can open following the isolated or combined contraction of different muscles: (1) the contraction of the epaxial muscles causes the cranial elevation; (2) the contraction of the geniohyoideus (i.e., protractor hyoidei) and levator operculi involves the lower jaw depression; (3) the hyoid depression can be due to the isolated contraction of the sternohyoideus or to a combination with the contraction of the hypaxial muscles that lead to the pectoral girdle retraction; and (4) the contraction of the levator arcus palatini conducts to the lateral expansion of the suspensorium

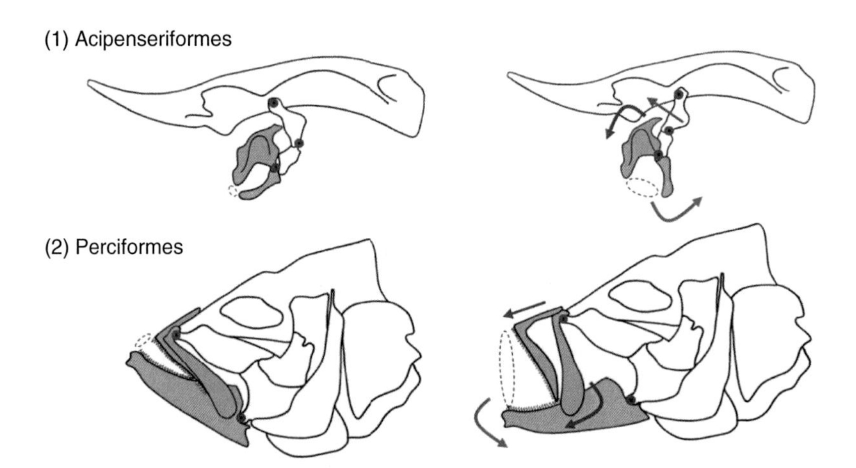

Fig. 5.8 Schematic representations in left lateral view of the different protraction mechanisms in Actinopterygii: (1) Acipenseriformes redrawn from Carroll and Wainwright (2003), (2) Perciformes redrawn from Motta (1984), (3) Cypriniformes redrawn from Staab et al. (2012), (4) Characiformes *Bivibranchia* sp. redrawn from Géry (1963) and (5) Gonorynchiformes *Phractolaemus* sp. redrawn from Géry (1963). Lower jaws are in orange, and the upper jaws are in blue. The kinethmoid bone of Cypriniformes is illustrated in green. The red circles localize the main articulations that are involved in the mechanism of the mouth opening. The dotted circles indicate the mouth opening, and the arrows show the direction of movements

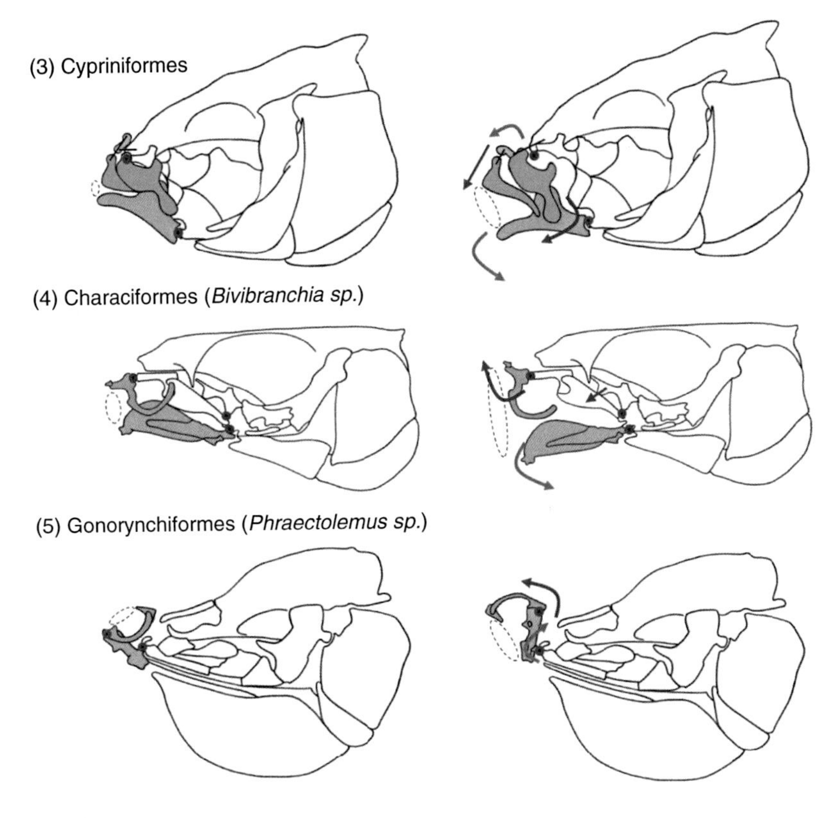

Fig. 5.8 (continued)

References

Aerts P (1991) Hyoid morphology and movements relative to abducting forces during feeding in Astatotilapia elegans (Teleostei: Cichlidae). J Morphol 208(3):323–345

Alexander RM (1964) Adaptation in the skulls and cranial muscles of south American characinoid fish. Zool J Linnean Soc 45(305):169–190

Alexander RM (1967) The functions and mechanisms of the protrusible upper jaws of some acanthopterygian fish. J Zool 151(1):43–64

Alfaro ME, Janovetz J, Westneat MW (2001) Motor control across trophic strategies: muscle activity of biting and suction feeding fishes. Am Zool 41(6):1266–1279

Alfaro ME et al (2009) Nine exceptional radiations plus high turnover explain species diversity in jawed vertebrates. Proc Natl Acad Sci U S A 106(32):13410–13414

Allis EP (1897) The cranial muscles and cranial and first spinal nerves in *Amia calva*. Ginn

Allis EP (1919) The homologies of the maxillary and vomer bones of Polypterus. Dev Dyn 25(4):348–394

Allis EP (1922) The cranial anatomy of Polypterus, with special reference to Polypterus bichir. J Anat 56(3-4):189–294

Ballintijn CM, Hughes GM (1965) The muscular basis of the respiratory pumps in the trout. J Exp Biol 43(2):349–362

Ballintijn CM, Van Den Burg A, Egberink BP (1972) An electromyographic study of the adductor mandibulae complex of a free-swimming carp (Cyprinus carpio L.) during feeding. J Exp Biol 57(1):261–283

Barel CDN (1983) Towards a constructional morphology of cichlid fishes (Teleostei, Perciformes). Neth J Zool 33(4):357–424

Bemis WE, Forey PL (2001) Occipital structure and the posterior limit of the skull in actinopterygians. In: Major events in early vertebrate evolution: palaeontology, phylogeny, genetics and development. Taylor & Francis, London, pp 41–62

Bemis WE, Findeis EK, Grande L (1997) An overview of Acipenseriformes. Environ Biol Fish 48(1-4):25–71

Blot J (1966) Étude des Palaeonisciformes du bassin houiller de Commentry. Allier, Paris

Brainerd EL, Ferry-Graham LA (2005) Mechanics of respiratory pumps. Fish Physiol 23:1–28

Camp AL, Konow N, Sanford CPJ (2009) Functional morphology and biomechanics of the tongue-bite apparatus in salmonid and osteoglossomorph fishes. J Anat 214(5):717–728

Camp AL, Roberts TJ, Brainerd EL (2015) Swimming muscles power suction feeding in largemouth bass. Proc Natl Acad Sci U S A 112(28):8690–8695

Carroll AM (2004) Muscle activation and strain during suction feeding in the largemouth bass Micropterus salmoides. J Exp Biol 207(6):983–991

Carroll AM, Wainwright PC (2003) Functional morphology of prey capture in the sturgeon, Scaphirhynchus albus. J Morphol 256(3):270–284

Carroll AM et al (2004) Morphology predicts suction feeding performance in centrarchid fishes. J Exp Biol 207(22):3873–3881

Carvalho M, Vari RP (2015) Development of the splanchnocranium in Prochilodus argenteus (Teleostei: Characiformes) with a discussion of the basal developmental patterns in the Otophysi. Zoology 118(1):34–50

Cloutier R, Arratia G (2004) Early diversification of actinopterygians. In: Recent advances in the origin and early radiation of vertebrates. Pfeil, Munich, pp 217–270

Cubbage CC, Mabee PM (1996) Development of the cranium and paired fins in the zebrafish Danio rerio (Ostariophysi, Cyprinidae). J Morphol 229(2):121–160

Datovo A, Bockmann FA (2010) Dorsolateral head muscles of the catfish families Nematogenyidae and Trichomycteridae (Siluriformes: Loricarioidei): comparative anatomy and phylogenetic analysis. Neotrop Ichthyol 8(2):193–246

Datovo A, Castro RMC (2012) Anatomy and evolution of the mandibular, hyopalatine, and opercular muscles in characiform fishes (Teleostei: Ostariophysi). Zoology 115(2):84–116

Datovo A, Vari RP (2013) The jaw adductor muscle complex in teleostean fishes: evolution, homologies and revised nomenclature (osteichthyes: actinopterygii). PLoS One 8(4):e60846

Datovo A, Vari RP (2014) The adductor mandibulae muscle complex in lower teleostean fishes (Osteichthyes: Actinopterygii): comparative anatomy, synonymy, and phylogenetic implications. Zool J Linnean Soc 171(3):552–622

Day SW et al (2005) Sucking while swimming: evaluating the effects of ram speed on suction generation in bluegill sunfish Lepomis macrochirus using digital particle image velocimetry. J Exp Biol 208(14):2653–2660

Day SW, Higham TE, Wainwright PC (2007) Time resolved measurements of the flow generated by suction feeding fish. Exp Fluids 43(5):713–724

De Schepper N, Adriaens D, De Kegel B (2005) Moringua edwardsi (Moringuidae: Anguilliformes): cranial specialization for head-first burrowing. J Morphol 266(3):356–368

De Schepper N, De Kegel B, Adriaens D (2007) Pisodonophis boro (Ophichthidae: Anguilliformes): specialization for head-first and tail-first burrowing. J Morphol 268(2):112–126

Deary AL, Hilton EJ (2016) Comparative ontogeny of the feeding apparatus of sympatric drums (Perciformes: Sciaenidae) in the Chesapeake Bay. J Morphol 277(2):183–195

Delsman HC (1925) Fishes with protrusile mouths. Treubia 6:98–106

Diogo R (2008) The origin of higher clades: osteology, myology, phylogeny and evolution of bony fishes and the rise of tetrapods. Science, New York

Diogo R, Abdala V (2010) Muscles of vertebrates: comparative anatomy, evolution, homologies and development. CRC Press, Boca Raton

Diogo R, Chardon M (2000a) Anatomie et fonction des structures céphaliques associées à la prise de nourriture chez le genre Chrysichthys (Teleostei: Siluriformes). Belg J Zool 130(1):21–37

Diogo R, Chardon M (2000b) Homologies among different adductor mandibuale sections of teleostan fishes, with special regard to catfishes (Teleostei: Siluriformes). J Morphol 243(2):193–208

Diogo R, Hinits Y, Hughes SM (2008) Development of mandibular, hyoid and hypobranchial muscles in the zebrafish: homologies and evolution of these muscles within bony fishes and tetrapods. BMC Dev Biol 8(1):24

Eagderi S, Adriaens D (2010) Cephalic morphology of Pythonichthys macrurus (Heterenchelyidae: Anguilliformes): specializations for head-first burrowing. J Morphol 271(9):1053–1065

Edgeworth FH (1935) The cranial muscles of vertebrates. Cambridge University Press, Cambridge

Engeman JM, Aspinwall N, Mabee PM (2009) Development of the pharyngeal arch skeleton in Catostomus commersonii (Teleostei: Cypriniformes). J Morphol 270(3):291–305

Faustino M, Power DM (2001) Osteologic development of the viscerocranial skeleton in sea bream: alternative ossification strategies in teleost fish. J Fish Biol 58(2):537–572

Ferry LA, Paig-Tran EM, Gibb AC (2015) Suction, ram, and biting: deviations and limitations to the capture of aquatic prey. Integr Comp Biol 55(1):97–109

Ferry-Graham LA, Lauder GV, Hulsey CD (2001) Aquatic prey capture in ray-finned fishes: a century of progress and new directions. J Morphol 248(2):99–119

Fraser GJ et al (2009) An ancient gene network is co-opted for teeth on old and new jaws. PLoS Biol 7(2):e1000031

Geerinckx T et al (2007) A head with a suckermouth: a functional-morphological study of the head of the suckermouth armoured catfish Ancistrus cf. triradiatus (Loricariidae, Siluriformes). Belg J Zool 137(1):47–66

Géry J (1962) Pterohemiodus luelingi sp. nov., un curieux poisson characoïde à nageoire dorsale filamenteuse, avec une clé des genres d'Hemiodontinae (Ostariophysi-Erythrinidae). Bonner zoologische Beiträge 59(12):332–342

Géry J (1963) L'appareil protracteur buccal de Bivibranchia (Characoidei) avec une note sur Phractolaemus (Chanoidei) (Pisces). Vie et Milieu 13(4):729–740

Gibb A (1996) The kinematics of prey capture in Xystreurys liolepis: do all flatfish feed asymmetrically? J Exp Biol 199(10):2269–2283
Gidmark NJ et al (2012) Flexibility in starting posture drives flexibility in kinematic behavior of the kinethmoid-mediated premaxillary protrusion mechanism in a cyprinid fish, Cyprinus carpio. J Exp Biol 215(13):2262–2272
Gidmark NJ et al (2015) Functional morphology of durophagy in black carp, Mylopharyngodon piceus. J Morphol 276(12):1422–1432
Goodrich ES (1958) Studies on the structure and development of vertebrates, vol II. Macmillan, London
Gosline WA (1973) Considerations regarding the phylogeny of cypriniform fishes, with special reference to structures associated with feeding. Copeia 1973(4):761–776
Gosline WA (1980) The evolution of some structural systems with reference to the interrelationships of modern lower teleostean fish groups. Japan J Ichthyol 27(1):1–28
Gosline WA (1989) Two patterns of differentiation in the jaw musculature of teleostean fishes. J Zool 218(4):649–661
Grande T, Poyato-Ariza FJ (1999) Phylogenetic relationships of fossil and recent gonorynchiform fishes (Teleostei: Ostariophysi). Zool J Linnean Soc 125(2):197–238
Grande T, Poyato-Ariza FJ, Diogo R (2010) Gonorynchiformes and Ostariophysan relationships: a comprehensive review. Science, New York
Greenwood PH et al (1966) Phyletic studies of teleostean fishes, with a provisional classification of living forms. Bulletin of the AMNH 131:4
Grubich JR (2001) Prey capture in actinopterygian fishes: a review of suction feeding motor patterns with new evidence from an elopomorph fish, Megalops atlanticus. Am Zool 41(6):1258–1265
Helfman GS et al (2009) The diversity of fishes: biology, evolution, and ecology. Wiley, Hoboken, NJ
Herbing IHV et al (1996) Ontogeny of feeding and respiration in larval Atlantic cod Gadus morhua (Teleostei, Gadiformes): I. Morphology. J Morphol 227(1):15–35
Hernandez LP, Staab KL (2015) Bottom feeding and beyond: how the premaxillary protrusion of cypriniforms allowed for a novel kind of suction feeding. Integr Comp Biol 55(1):74–84
Hernandez PL, Bird NC, Staab KL (2007) Using zebrafish to investigate cypriniform evolutionary novelties: functional development and evolutionary diversification of the kinethmoid. J Exp Zool B Mol Dev Evol 308(5):625–641
Higham TE, Day SW, Wainwright PC (2006a) Multidimensional analysis of suction feeding performance in fishes: fluid speed, acceleration, strike accuracy and the ingested volume of water. J Exp Biol 209(14):2713–2725
Higham TE, Day SW, Wainwright PC (2006b) The pressures of suction feeding: the relation between buccal pressure and induced fluid speed in centrarchid fishes. J Exp Biol 209(17):3281–3287
Holzman R et al (2008) Jaw protrusion enhances forces exerted on prey by suction feeding fishes. J R Soc Interface 5(29):1445–1457
Huber DR et al (2005) Analysis of the bite force and mechanical design of the feeding mechanism of the durophagous horn shark Heterodontus francisci. J Exp Biol 208(18):3553–3571
Hughes GM, Shelton G (1958) The mechanism of gill ventilation in three freshwater teleosts. J Exp Biol 35(4):807–823
Hulsey CD, Garcia De Leon FJ (2005) Cichlid jaw mechanics: linking morphology to feeding specialization. Funct Ecol 19(3):487–494
Inoue JG et al (2003) Basal actinopterygian relationships: a mitogenomic perspective on the phylogeny of the "ancient fish". Mol Phylogenet Evol 26(1):110–120
Janvier P (1996) Early vertebrates. Oxford University Press, New York, NY
Kammerer CF, Grande L, Westneat MW (2006) Comparative and developmental functional morphology of the jaws of living and fossil gars (Actinopterygii: Lepisosteidae). J Morphol 267(9):1017–1031
Kardong KV (2012) Vertebrates: comparative anatomy, function, evolution. McGraw-Hill Higher Education, New York
Kimmel CB et al (1995) Stages of embryonic development of the zebrafish. Dev Dyn 203(3):253–310
Kimmel CB et al (2001) Neural crest patterning and the evolution of the jaw. J Anat 199(1–2):105–119
Konstantinidis P et al (2015) The developmental pattern of the musculature associated with the mandibular and hyoid arches in the longnose gar, Lepisosteus osseus (Actinopterygii, Ginglymodi, Lepisosteiformes). Copeia 103(4):920–932
Lauder GV (1980) Evolution of the feeding mechanism in primitive actinopterygian fishes: a functional anatomical analysis of Polypterus, Lepisosteus, and Amia. J Morphol 163(3):283–317
Lauder GV (1982) Patterns of evolution in the feeding mechanism of actinopterygian fishes. Am Zool 22(2):275–285
Lauder GV (1985) Aquatic feeding in lower vertebrates. In: Hildebrand M, Bramble DM, Liem KF, Wake DB (eds) Functional vertebrate morphology. Harvard University Press, Cambridge, pp 210–229
Lauder GV, Liem KF (1980) The feeding mechanism and cephalic myology of Salvelinus fontinalis: form, function, and evolutionary significance. In: Charrs: Salomnids of the genus Salvelinus, pp 365–390
Lauder GV, Liem KF (1981) Prey capture by Luciocephalus pulcher: implications for models of jaw protrusion in teleost fishes. Environ Biol Fish 6(3):257–268
Lauder GV, Liem KF (1983) Patterns of diversity and evolution in ray-finned fishes. Fish Neurobiol 1:1–24
Lecointre G, Le Guyader H (2001) Classification phylogénétique du vivant, vol Vol. 2. Belin, Paris

Liem KF (1967) Functional morphology of the head of the anabantoid teleost fish Helostoma temmincki. J Morphol 121(2):135–157
Liem KF (1978) Modulatory multiplicity in the functional repertoire of the feeding mechanism in cichlids fishes. Part I. Piscivores. J Morphol 158(3):323–360
Liem KF (1980) Adaptive significance of intra-and interspecific differences in the feeding repertoires of cichlid fishes. Am Zool 20(1):295–314
Liem KF (1990) Aquatic versus terrestrial feeding modes: possible impacts on the trophic ecology of vertebrates. Am Zool 30(1):209–221
López-Fernández H et al (2012) Diet-morphology correlations in the radiation of South American geophagine cichlids (Perciformes: Cichlidae: Cichlinae). PLoS One 7(4):e33997
McCord CL, Westneat MW (2016) Evolutionary patterns of shape and functional diversification in the skull and jaw musculature of triggerfishes (Teleostei: Balistidae). J Morphol 277(6):737–752
Miller MJ (2004) The ecology and functional morphology of feeding of North American sturgeon and paddlefish. In: Sturgeons and paddlefish of North America. Springer, Dordrecht, pp 87–102
Miller RF, McGovern JH (1996) Preliminary report of fossil fish (Actinopterygii: Palaeonisciformes) from the Lower Carboniferous Albert Formation at Norton, New Brunswick (NTS 21 H/12). Current research, pp 97–104
Motta PJ (1984) Mechanics and functions of jaw protrusion in teleost fishes: a review. Copeia 1984(1):1–18
Motta PJ, Huber DR (2004) Prey capture behavior and feeding mechanics of elasmobranchs. In: Biology of sharks and their relatives, 2nd edn. Taylor & Francis, London, pp 153–197
Muller M (1987) Optimization principles applied to the mechanism of neurocranium levation and mouth bottom depression in bony fishes (Halecostomi). J Theor Biol 126(3):343–368
Muller M, Osse JWM, Verhagen JHG (1982) A quantitative hydrodynamical model of suction feeding in fish. J Theor Biol 95(1):49–79
Nelson JS (1994) Fishes of the world. Wiley, New York
Nelson JS (2006) Fishes of the world. Wiley, Hoboken
Nelson JS, Grande T, Wilson MVH (2016) Fishes of the world. Wiley, New York
Noda M, Miyake T, Okabe M (2017) Development of cranial muscles in the actinopterygian fish Senegal bichir, Polypterus senegalus Cuvier, 1829. J Morphol 278(4):450–463
Osse JWM (1969) Functional morphology of the head of the perch (Perca Fluviatilis L.): an electromyographic study. Neth J Zool 19(3):289–392
Osse JWM (1985) Jaw protrusion, an optimization of the feeding apparatus of teleosts? Acta Biotheor 34(2):219–232
Owen R (1846) Lectures on the comparative anatomy and physiology of the vertebrate animals: Delivered at the Royal College of Surgeons of England, in 1844 and 1846. Volume 2. Part I - Fishes. Longman, Brown, Green, and Longmans
Owen R (1866) Comparative anatomy and physiology of vertebrates: fishes and reptiles. Longman, Harlow
Parmentier E (2003) Contribution à l'étude des relations entre des poissons de la famille des Carapidae et leurs hôtes invertébrés: une approche mutidisciplinaire. University of Liège, Liège
Parmentier E et al (1998) Morphology of the buccal apparatus and related structures in four species of Carapidae. Aust J Zool 46(4):391–404
Parmentier E, Vandewalle P, Lagardere F (2001) Morphoanatomy of the otic region in carapid fishes: ecomorphological study of their otoliths. J Fish Biol 58(4):1046–1061
Patterson C (1994) Bony fishes. In: Prothero DR, Schoch RM (eds) Major features of vertebrate evolution, Short courses in paleontology, vol 7. Paleontological Society, University of Tennessee, Knoxville, pp 57–84
Peng Z et al (2009) Teleost fishes (Teleostei). The timetree of life. Oxford University Press, Oxford, pp 335–338
Poplin CM (1984) Lawrenciella schaefferi n.g., n.sp. (Pisces: Actinopterygii) and the use of endocranial characters in the classification of the Palaeonisciformes. J Vertebr Paleontol 4(3):413–421
Regan CT (1911) LXV. The classification of the teleostean fishes of the order Ostariophysi—2. Siluroidea. Ann Mag Nat Hist 8(47):553–577
Roberts TR (1974) Dental polymorphism and systematics in Saccodon, a neotropical genus of freshwater fishes (Parodontidae, Characoidei). J Zool 173(3):303–321
Rosen DE (1982) Teleostean interrelationships, morphological function and evolutionary inference. Am Zool 22(2):261–273
Sanford CP, Lauder GV (1989) Functional morphology of the tongue-bite in the osteoglossomorph fish Notopterus. J Morphol 202(3):379–408
Schaeffer B, Rosen BE (1961) Major adaptive levels in the evolution of the actinopterygian feeding mechanism. Am Zool 1(2):187–204
Schilling TF, Kimmel CB (1994) Segment and cell type lineage restrictions during pharyngeal arch development in the zebrafish embryo. Development 120(3):483–494
Schilling TF, Kimmel CB (1997) Musculoskeletal patterning in the pharyngeal segments of the zebrafish embryo. Development 124(15):2945–2960
Shadwick RE, Lauder GV (2006) Fish physiology: fish biomechanics, vol 23. Academic Press, Cambridge
Staab KL, Hernandez LP (2010) Development of the cypriniform protrusible jaw complex in Danio rerio: constructional insights for evolution. J Morphol 271(7):814–825
Staab KL et al (2012) Independently evolved upper jaw protrusion mechanisms show convergent hydrodynamic function in teleost fishes. J Exp Biol 215(9):1456–1463
Thys van den Audenaerde DFE (1961) L'anatomie de Phractolaemus ansorgei Blgr. et la position systéma-

tique des Phractolaemidae. Annales du Musée Royal de l'Afrique Centrale, Sciences Zoologiques, série 8(103):101–167
Traquair RH (1870) The cranial osteology of Polypterus. J Anat Physiol 5(Pt 1):166–184
Van Wassenbergh S, Aerts P, Herrel A (2005) Scaling of suction-feeding kinematics and dynamics in the African catfish, Clarias gariepinus. J Exp Biol 208(11):2103–2114
Vandewalle P et al (1992) Early development of the cephalic skeleton of Barbus barbus (Teleostei, Cyprinidae). J Fish Biol 41(1):43–62
Vandewalle P et al (1997) Postembryonic development of the cephalic region in Heterobranchus longifilis. J Fish Biol 50(2):227–253
Vandewalle P, Parmentier E, Chardon M (2000) The branchial basket in teleost feeding. Cybium 24(4):319–342
Vari RP (1985) A new species of Bivibranchia (Pisces: Characiformes) from Surinam, with comments on the genus. Proc Biol Soc Wash 98(2):511–522
Vari RP, Goulding M (1985) A new species of Bivibranchia (Pisces: Characiformes) from the Amazon River basin. Proc Biol Soc Wash 98(4):1054–1061
Venkatesh B, Erdmann MV, Brenner S (2001) Molecular synapomorphies resolve evolutionary relationships of extant jawed vertebrates. Proc Natl Acad Sci U S A 98(20):11382–11387
Vrba ES (1968) Contributions to the functional morphology of fishes. Part V The feeding mechanism of Elops taurus Linnaeus. Afr Zool 3(2):211–236
Wainwright PC et al (1989) Evolution of motor patterns: aquatic feeding in salamanders and ray-finned fishes. Brain Behav Evol 34(6):329–341
Wainwright PC et al (2001) Evaluating the use of ram and suction during prey capture by cichlid fishes. J Exp Biol 204(17):3039–3051
Wainwright P et al (2007) Suction feeding mechanics, performance, and diversity in fishes. Integr Comp Biol 47(1):96–106
Wainwright PC et al (2015) Origins, innovations, and diversification of suction feeding in vertebrates. Integr Comp Biol 55(1):134–145
Waltzek TB, Wainwright PC (2003) Functional morphology of extreme jaw protrusion in Neotropical cichlids. J Morphol 257(1):96–106
Westneat MW (1994) Transmission of force and velocity in the feeding mechanisms of labrid fishes (Teleostei, Perciformes). Zoomorphology 114(2):103–118
Westneat MW (2004) Evolution of levers and linkages in the feeding mechanisms of fishes. Integr Comp Biol 44(5):378–389
Westneat MW (2005) Skull biomechanics and suction feeding in fishes. Fish Physiol 23:29–75
Westneat MW, Olsen AM (2015) How fish power suction feeding. Proc Natl Acad Sci 112(28):8525–8526
Westneat MW, Wainwright PC (1989) Feeding mechanism of Epibulus insidiator (Labridae; Teleostei): evolution of a novel functional system. J Morphol 202(2):129–150
Wilga C, Motta P (1998) Conservation and variation in the feeding mechanism of the spiny dogfish Squalus acanthias. J Exp Biol 201(9):1345–1358
Wilson MVH, Veilleux P (1982) Comparative osteology and relationships of the Umbridae (Pisces: Salmoniformes). Zool J Linnean Soc 76(4):321–352
Winterbottom R (1974) A descriptive synonymy of the striated muscles of the Teleostei. Proc Acad Natl Sci Phila 125(125):225–317
Wu KY, Shen SC (2004) Review of the teleostean adductor mandibulae and its significance to the systematic positions of the Polymixiiformes, Lampridiformes, and Triacanthoidei. Zool Stud 43(4):712–736

6 Sarcopterygian Fishes, the "Lobe-Fins"

Alice M. Clement

6.1 Introduction

Sarcopterygian fishes ("lobe-fins") are distinctively different from the actinopterygians ("ray-finned" fish) we heard about in the previous chapter. Both of them belong to the Osteichthyes (bony fishes), but the sarcopterygians are characterised by having their fin anchored to the pectoral girdle by a single bone (i.e., humerus), rather than a series of bones as in actinopterygians. The Class includes enigmatic fish like "the most enduring vertebrate on the planet", the Australian lungfish *Neoceratodus* whose skeleton is indistinguishable from similar fossils living alongside the Cretaceous dinosaurs 100 million years ago, as well as the coelacanth "Lazarus taxon" *Latimeria*, rediscovered in the 1930s after it was thought its lineage had gone extinct some 70 million years ago! However surprising to some, strictly speaking we ourselves also belong to the Class Sarcopterygii (although you may not always feel like a "lobe-finned fish"). In fact all "tetrapods", that is first four-footed vertebrates (animals with a backbone), and all of their descendants are sarcopterygians. Thus "Sarcopterygii" also includes all amphibians, reptiles, birds and mammals. However, this chapter will deal exclusively with sarcopterygian *fishes*, from the primitive "ghost fish" *Guiyu* from the Silurian Period of China all the way up to the earliest tetrapods, as well as the lungfishes and coelacanths of today.

Special attention has been paid to the Devonian (359–419 million years ago) sarcopterygians as it was in this group and at that time that one of the greatest steps of evolution occurred, the transition by our ancestors from water to land. The first tetrapods would have had a number of obstacles to overcome in leaving the water, including having to develop limbs for body support and locomotion, acquiring aerial respiratory abilities, and learning new ways to feed, osmoregulate and reproduce in the terrestrial realm. Some of the earliest tetrapod trackways have been found in deposits from the Middle Devonian of Poland, suggesting that some sarcopterygians had already acquired limbs and digits by 390 million years ago (Niedźwiedzki et al. 2010).

There are just eight extant species of sarcopterygian fish, but during the Devonian Period (Age of Fishes), they were one of the most widespread, diverse and dominant groups. Although only the coelacanth and lungfish lineages survive, there were once many other groups of sarcopterygian fishes now known only from the fossil record. We will visit a number of these throughout this chapter, including primitive stem members and mysterious animals called "onychodonts", "porolepiforms" and "tetrapodomorphs". Due to the nature of

A. M. Clement
School of Biological Sciences, College of Science and Engineering, Flinders University, Adelaide, Australia
e-mail: alice.clement@flinders.edu.au

J. M. Ziermann et al. (eds.), *Heads, Jaws, and Muscles*, Fascinating Life Sciences,
https://doi.org/10.1007/978-3-319-93560-7_6

fossilisation (as soft tissue such as muscle rarely preserves), there will be more attention paid to the skull bones in these fishes, as well as their mandibles (lower jaws) and fascinating array of dentitions (teeth). Where living representatives exist (coelacanths and lungfishes), I will also provide information on cranial muscles and development.

One of the defining features of the sarcopterygian skull is that it is divided into two halves (Dzerzhinsky 2016). The front portion is called the ethmosphenoid, and the rear is called the otoccipital. The endocranial hinge splitting the two halves is termed the intracranial joint. The intracranial joint was later lost in some groups like the lungfishes and the tetrapods, but we can still see one today in the coelacanth. Another sarcopterygian feature in the skull is the presence of a hard tissue called cosmine. Cosmine is a mixture of true enamel and dentine and is closely associated with a network of pores and canals (Ørvig 1969). Again, this tissue was eventually lost in a number of sarcopterygians including all extant forms, so consequently we do not know what role it played. Some thought it may have had an electrosensory function (Thomson 1977), but the presence of a supporting vascular (blood) system was taken as evidence otherwise (Bemis and Northcutt 1992); some suggest it was related to the lymphatic system instead (Kemp 2017). Most recently, it was again suggested that cosmine and its associated clusters of pore-group pits are most likely electroreceptors after all (King et al. 2018). All extant and most extinct sarcopterygians possess enamel on their teeth (Qu et al. 2015).

There has been much debate over the interrelationships of sarcopterygians, both extinct and extant (Ahlberg 1991; Schultze 1994; Cloutier and Ahlberg 1996; Zardoya and Meyer 1997; Ahlberg and Johanson 1998; Friedman 2007; Yu et al. 2010; Lu et al. 2012; Amemiya et al. 2013; Betancur-r et al. 2013; Clack and Ahlberg 2016; Lu et al. 2016), but a generally accepted consensus supported by molecular and palaeontological evidence is presented in Fig. 6.1.

General Sarcopterygian Further Reading

Janvier, P. 1996. *Early Vertebrates,* New York, Oxford University Press.

Jarvik, E. 1980. *Basic Structure and Evolution of Vertebrates,* London, Academic Press.

Long, J. A. 2011. *The Rise of Fishes- 500 million years of evolution,* Sydney, University of New South Wales Press.

6.2 Stem Sarcopterygians

The earliest sarcopterygians appeared in what is now China during the Silurian Period, more than 425 million years ago. *Guiyu* (Zhu et al. 2009), or the "ghost fish", is one of the oldest near-complete gnathostome (jawed vertebrate) and gets its name due to its unusual ghostly or secretive combination of morphological characters (Fig. 6.2). *Guiyu* has a skull divided into two parts like other sarcopterygians (parietal and postparietal shields separated by the dermal intracranial joint), but its cheekbones were more like those in actinopterygians in being composed of just one large bone (preopercular), and it bears ridged ornamentation made of a tissue called ganoine (rather than the cosmine found in other sarcopterygians). Surprisingly, *Guiyu* also retains some features of the postcranial skeleton (a primitive shoulder girdle and a spine on the median fin) that are more similar to the condition in chondrichthyans, placoderms and acanthodians.

Other important stem sarcopterygians from China include *Psarolepis* (Yu 1998; Zhu et al. 1999), *Achoania* (Zhu et al. 2001) and *Megamastax* (Choo et al. 2014), as well as a number of early sarcopterygians that fall at the base of main subgroups of sarcopterygians, namely, *Styloichthys* (Zhu and Yu 2002) at the base of the coelacanth lineage and *Powichthys* (Jessen 1975, 1980) and *Youngolepis* (Chang 1982) at the base of the lungfish lineage (Fig. 6.1). Both *Psarolepis* (meaning "speckled scale") and *Achoania* (meaning "no choana/internal nostrils") had tooth whorls at the front of their lower jaw. They were both also covered by a layer of tissue called cosmine that carried large pits across its surface. The skull roof covering the front half of the skull (parietal shield) was longer than that covering at the back (postparietal shield). And while *Psarolepis* had

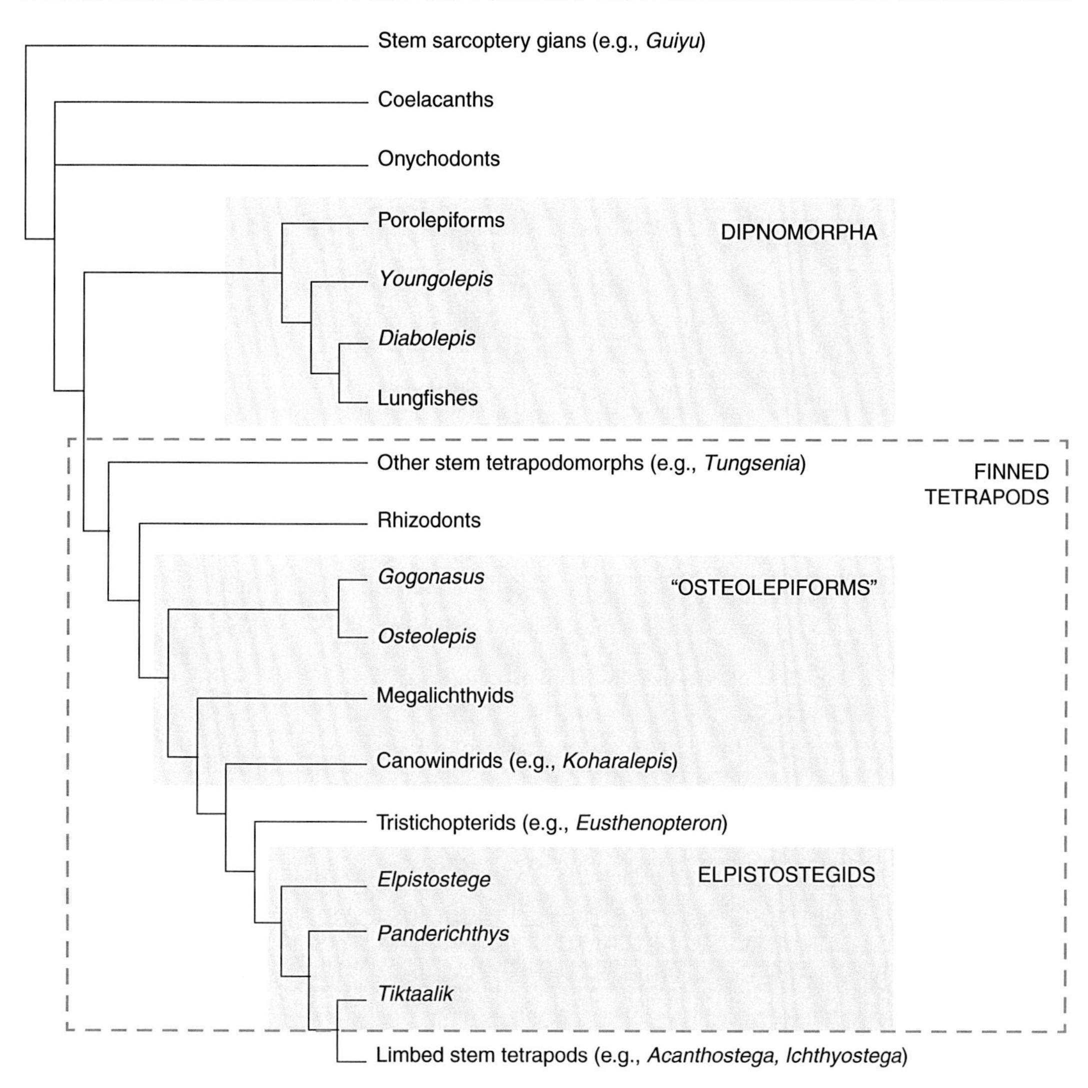

Fig. 6.1 Hypothesised phylogenetic relationships of major extinct and extant sarcopterygian fish and the first limbed tetrapods (phylogeny adapted from Lu et al. 2012, 2016)

nostrils situated high up on its snout, *Achoania* had a larger parasphenoid (palatal bone inside the roof of the mouth). *Megamastax* (meaning "big mouth") had rows of conical teeth and peculiar spiky "tooth cushions" along its jaws instead of a tooth whorl but was one of the few early fishes to grow to huge sizes.

Styloichthys ("pillar fish") lacks a maxilla (upper jaw bone) but has a very deep lower jaw and short dentary bones. *Styloichthys* has an eye stalk, a basal osteichthyan character also seen in *Psarolepis* and *Achoania*. Researchers still debate the exact phylogenetic position of *Styloichthys* but believe it must be close to the origin of the coelacanth lineage (Friedman 2007). One of the oldest sarcopterygians found outside of China is *Powichthys*, an animal very similar to the better-known *Youngolepis* (both "dipnomorphs"). *Youngolepis* and *Powichthys* are the first sarcopterygians to show a skull roof pattern with many small bones next to the main median bones of the skull roof, a pattern later established in lungfishes. Unexpectedly, the braincase is not cleanly divided in these fish unlike other sarcopterygians; however, they do both retain a cover of cosmine on their bones.

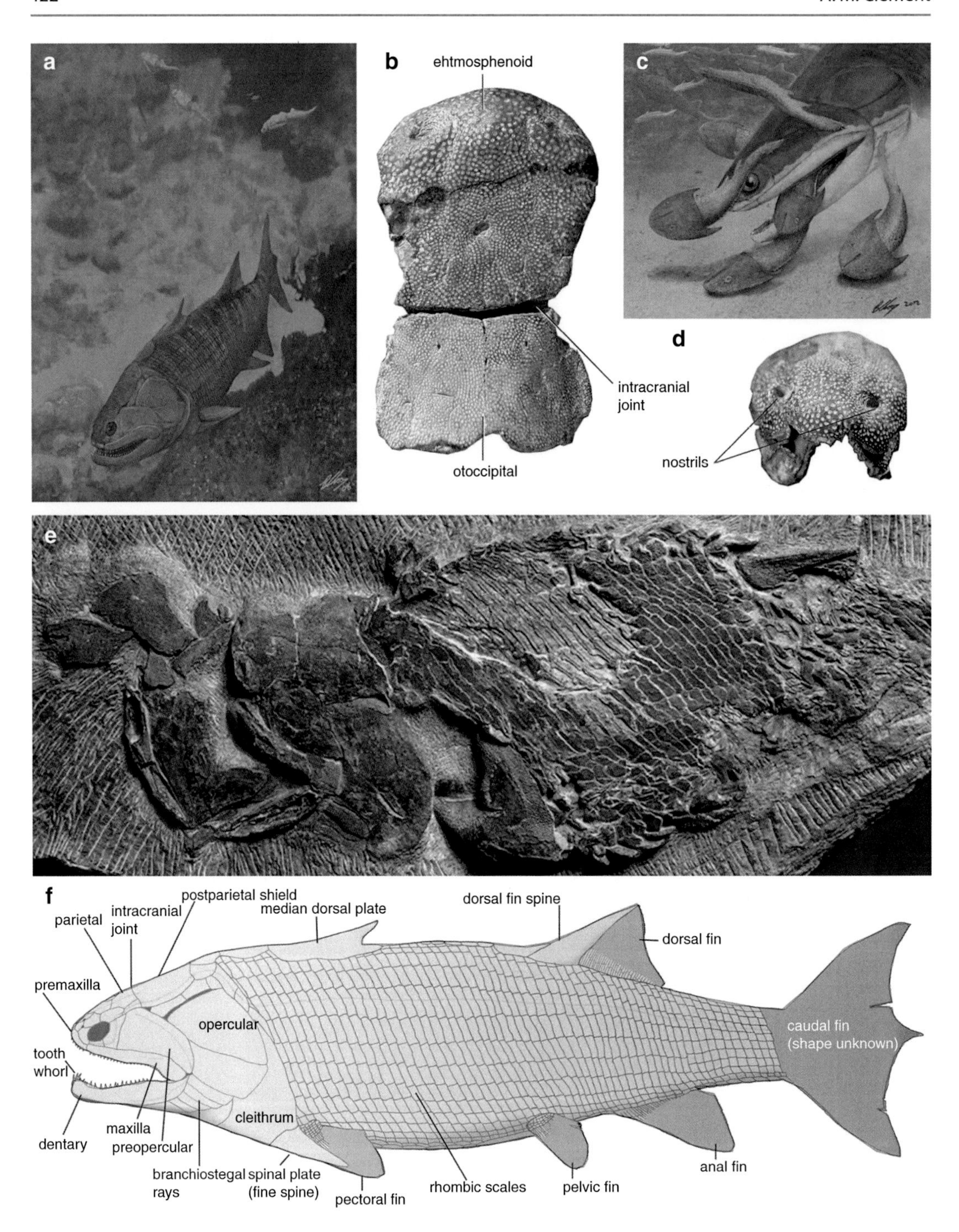

Fig. 6.2 Stem sarcopterygians. (**a**) *Guiyu oneiros* (life reconstruction artwork courtesy of Brian Choo); (**b**) *Psarolepis romeri* skull in dorsal view showing the front (ethmosphenoid) and rear (otoccipital) portions separated by the intracranial joint (photo courtesy of Zhu Min IVPP); (**c**) *Megamastax amblyodus* feeding on galeaspids (life reconstruction artwork courtesy of Brian Choo); (**d**) *Psarolepis romeri* snout in anterior view showing nostrils (photo courtesy of Zhu Min IVPP); (**e**) *Guiyu oneiros* specimen in lateral view (photo courtesy of Zhu Min IVPP); (**f**) *Guiyu oneiros* full-body reconstruction in lateral view (reconstruction artwork courtesy of John Long)

Stem Sarcopterygians Further Reading

Ahlberg, P. E. 1999. Something fishy in the family tree. *Nature,* 397, 564–565.

Yu, X., Zhu, M. & Zhao, W. 2010. The origin and diversification of osteichthyans and sarcopterygians: Rare Chinese fossil findings advance research on key issues of evolution. *Palaeoichthyology,* 24, 71–75.

6.3 Coelacanths

One of the most spectacular biological findings of our time was that of a living coelacanth discovered during the 1930s off the east coast of South Africa. Coelacanths (also known as "Actinistia") are an ancient group of fish with their origins dating back to the Early Devonian but were thought to have been extinct since the time of the dinosaurs some 70 million years ago. "Old four legs" was given the official name of *Latimeria chalumnae*, in honour of the curator of the local museum, Marjorie Courtenay-Latimer, who helped to identify the fish, and for the name of the river mouth (Chalumna) near where it was found (Fig. 6.3a). At the time there was much enthusiasm as coelacanths were thought to be the direct ancestors of tetrapods (and hence ultimately humankind). However, it is now accepted that lungfish are instead the closest extant group to the land vertebrates (Amemiya et al. 2013). A second species was discovered in 1997 in Indonesia. The "King of the Sea", *Latimeria menadoensis*, is brown (unlike the deep blue colour of *L. chalumnae*) and has been shown to be genetically distinct from the African populations, having split apart perhaps as long as 30 to 40 million years ago (Inoue et al. 2005), but otherwise the fish are very similar. The morphology of *Latimeria*, including details of the skull, muscles and nervous system, was described in exhaustive detail in a series of monographs (Millot and Anthony 1958a; b; Millot et al. 1978), but more recent work has further investigated the cranial and feeding morphology of these enigmatic fish (Dutel et al. 2013, 2015a; b).

There are many features of the postcranial skeleton of coelacanths that distinguish them from other sarcopterygian fish, their unusual fin shape and arrangement in particular. However, in the skull they also have some distinctive features such as a double jaw joint (one jaw joint at each quadrate, sometimes described as "tandem") for specialised feeding behaviour, and they lack an upper jaw bone (maxilla). In addition to this, *Latimeria* is the only extant sarcopterygian to retain an intracranial joint, it having been lost in lungfishes and tetrapods. The intracranial joint, separating the front (ethmosphenoid) and rear (otoccipital) portions of the skull, allows movement during prey capture. Relatedly, the basicranial muscle, supposedly like that in onychodonts, also allows for greater kinesis in the intracranial joint during feeding. Some researchers have suggested that the greater flexibility afforded by the intracranial joint and basicranial muscle elevates the snout and consequently enhances gape and contributes to powerful suction ability (Millot and Anthony 1958a). However there are other researchers who believe this flexibility has been overstated and highlights the need for further in vivo experiments with live animals (Dutel et al. 2013, 2015a). Moreover, coelacanths possess a large sensory organ in their snout (rostral organ) that uses electroreception to locate prey, much in the same way that sharks utilise ampullae of Lorenzini. Both groups of sarcopterygian fishes alive today (coelacanths and lungfishes) have electrosensory abilities (Watt et al. 1999), as well as some jawless fishes, cartilaginous fishes, amphibians and mammals.

The basicranial muscle in coelacanths is thought to be homologous with the retractor bulbi in tetrapods. The role of the coracomandibularis remains unclear; it might be involved in jaw opening or stabilising the hyoid (Dutel et al. 2015b). The adductor mandibular muscles are divided into three groups and composed of seven different muscle bundles (Dutel et al. 2013).

a
b
c
d
e
f
orbit
skull roof
tooth whorl
hyomandibular
lower jaw

Furthermore, *Latimeria* has five hyoid muscles, at least four true branchial muscles and two hypobranchial muscles (Diogo et al. 2008).

We were introduced to *Styloichthys* in the previous section discussing stem sarcopterygians, which is perhaps the oldest coelacanth known, from the Early Devonian of China. However, there are a number of other fossil coelacanths from the Devonian and even more so from the Carboniferous Period and throughout the Mesozoic Era; in total there are more than 130 fossil species described (Forey 1998). Some of the most primitive coelacanths include *Gavinia* from Australia (Long 1999) and *Miguashaia* from Canada (Cloutier 1996b). More distinctive forms that are considered "anatomically modern"—due to their possession of two pairs of parietals, large sensory pores in the skull and a long front portion (preorbital) of the skull—appeared later. This includes forms such as *Diplocercides* and *Holopterygius* and probably the best-known early coelacanth, *Euporosteus* (Zhu et al. 2012). Excitingly, there is also an as-yet undescribed fossil coelacanth from the Devonian Gogo Formation in Australia (Fig. 6.3b) that is preserved perfectly in three dimensions that closely resembles *Diplocercides* (Long and Trinajstic 2010). Even later still we find more derived coelacanths like *Allenypterus* from the Carboniferous of the USA (Fig. 6.3c) and *Coelacanthus* from the Permian Period (Forey 1998). We see a change over time from the possession of a broad parasphenoid (palatal bone inside the roof of the mouth) in primitive coelacanths like those just mentioned, to a narrower parasphenoid with a reduced area for teeth in the more modern forms such as and *Macropoma* and the giant *Megalocoelacanthus* from the Cretaceous Period (Dutel et al. 2012), as well as in the extant *Latimeria.* The reduction of the parasphenoid is thought to be related to the gradual forward movement of the basicranial muscle throughout coelacanth evolution.

Most coelacanth fossils are preserved as flattened specimens, revealing little about their internal anatomy. There has been just one taxon from the Devonian known substantially from three-dimensional (3D) remains to be able to reconstruct a cranial endocast, that is, a mould of the internal skull cavity that houses the brain. *Diplocercides* from the Late Devonian of Germany had its cranial anatomy painstakingly reconstructed using serial grinding and was rebuilt as a wax model (Stensiö 1963). The new coelacanth taxon currently being described from the Gogo Formation is similarly exceptionally preserved in 3D and is expected to reveal much about the internal cranial anatomy of early coelacanths (Clement et al. In prep.). The braincases and endocasts differ significantly from that of the extant coelacanth, *Latimeria* (Millot and Anthony 1958b). It is likely that the brains of the smaller Devonian taxa more closely resembled the size and shape of their endocasts than the large, deep-sea coelacanths of today.

Extant coelacanths are rare, deep-sea fishes, so consequently there is very little known about their development. They are ovoviviparous and give birth to between 5 and 25 fry, or live "pups", at a time. Their eggs are huge, at 9 cm in diameter, and they are the largest recorded for any fish (Forey 1998). Adult coelacanths are large fish that can grow more than 2 m in length and weigh 90 kg. Within the uterus each embryo is enclosed

Fig. 6.3 Coelacanths, onychodonts and porolepiforms. (**a**) A model of *Latimeria* on display in the Paris Natural History Museum (photo courtesy of John Long); (**b**) a new unnamed coelacanth from the Late Devonian Gogo Formation, Western Australia, skull in left lateral view (photo courtesy of John Long); (**c**) a Carboniferous coelacanth fossil from Bear Gulch (USA) *Allenypterus* (photo courtesy of John Long); (**d**) *Qingmenodus yui* life restoration (artwork courtesy of Brian Choo, from Lu et al. 2016); (**e**) skull of *Onychodus jandemarrai* from the Late Devonian Gogo Formation, Australia (photo courtesy of John Long); (**f**) skull of *Porolepis* in right lateral view (photo courtesy of John Long)

in its own compartment (Wourms et al. 1991). The unborn young are nourished by egg yolk (lecithotrophic), with some researchers suggesting they are also oophagous, which means they feed on eggs produced in the ovary while still inside the uterus (Musick et al. 1991).

Coelacanth Further Reading

Forey, P. L. 1998. *History of the Coelacanth Fishes,* London, Chapman and Hall.

Musick, J. A., Bruton, M. N. & Balon, E. K. 1991. *The biology of Latimeria chalumnae and evolution of coelacanths.*

6.4 Onychodonts

The next group we will look at are the onychodonts (Fig. 6.3d, e), also known as the "dagger-toothed" or "nail-toothed" fishes, and when you see a picture of one, it is easy to understand why. Onychodonts have a very distinctive tooth whorl at the front of their lower jaws and must have been fearsome predators. Some of these fish grew to be very large, perhaps as much as 4 m in length. Onychodonts went extinct at the end of the Devonian Period.

The first onychodonts appeared during the Early Devonian and are known from both Australia (*Bukkanodus*) and China (*Qingmenodus*), but they have also been found in Germany (*Strunius*), Spain (*Grossius*) as well as North America (*Onychodus*) (Long 2011). One of the best known and most complete onychodonts known is *Onychodus jandemarrai* from the Gogo Formation in Australia (Andrews et al. 2006), but details of the otoccipital (rear region of the skull) of this group have only more recently been brought to light by other specimens from China; *Qingmenodus* is the earliest onychodont to preserve a braincase (Lu and Zhu 2010; Lu et al. 2016). High-resolution computed tomographic scanning enabled a cranial endocast of *Qingmenodus* to be created. It revealed coelacanth-like features of the rear portion of the braincase, while the front portion appears to have retained more primitive characters similar to stem sarcopterygians (Lu et al. 2016).

The oldest onychodonts had a very long rear portion of the skull (otoccipital) compared to the short, broad ethmosphenoid (front region), but all onychodonts had a hinged braincase—a feature common to all primitive sarcopterygians. Onychodont skulls are all highly kinetic; they achieved this by incorporating more cartilage into the mandible and jaw joint for increased flexibility. Relatedly, the nasal capsules are reduced and situated wider in the skull than you might expect; again this is a result of having to make room for the tooth whorl to fit back inside the skull.

Onychodonts also have a different pattern of skull roof and cheekbones to those of other sarcopterygians (Long 2011). For example, the cheek has fewer bones than in other sarcopterygians, being comprised of only two bones (called the preoperculum and squamosal). Also, the upper jaw bone (maxilla) is very large and shaped like a long blade that extends far backwards from the orbit. The presence of the large, retractable tooth whorls is the most striking feature of this group and seems to have dominated changes in the skull bones of onychodonts from the general sarcopterygian condition; relatedly, this must have affected the cranial muscles also. Onychodonts are thought to have had a large basicranial muscle that attached far posteriorly underneath the braincase—similar to the condition in coelacanths today (Andrews et al. 2006). This muscle is active during movements involving the intracranial joint allowing greater flexibility of the skull when feeding, especially when biting down on prey. Somewhat surprisingly for these predators, the adductor muscles had only a small attachment area on the lower jaw, but biomechanical analysis shows that they would have maximum force when the mouth was wide open (Andrews et al. 2006).

Onychodont Further Reading

Andrews, S. M., Long, J., Ahlberg, P. E., Barwick, R. E. & Campbell, K. S. W. 2006. The structure of the sarcopterygian *Onychodus jandemarrai* n.sp. from Gogo, Western Australia: With a functional interpretation of the skeleton. *Transactions of the Royal Society of Edinburgh (Earth Science),* 96, 197–307.

Long, J. A. 1995. *The Rise of Fishes- 500 million years of evolution,* Sydney, University of new South Wales Press.

6.5 Porolepiformes

Another group of fish possessing fearsome tooth whorls are the Porolepiformes (Fig. 6.3f). These are also an exclusively Devonian group, most commonly found in nearshore/freshwater deposits and named for the rows of special pores in the cosmine covering the surface of their scales (Ahlberg 1991). Later members of this group lost their cosmine, and their scales became more rounded (Mondejar-Fernandez and Clément 2012). Porolepiform fossils are mostly known from Northern Hemisphere deposits (Long 2011), including Scotland (*Glyptolepis*, *Holoptychius*, *Duffichthys*), Spitsbergen (*Porolepis*, *Heimenia*), Latvia (*Laccognathus*) as well as Canada (*Nasogaluakus*, *Quebecius*).

Porolepiformes are characterised by their special teeth—termed dendrodont dentition—which show an unusually complex infolding of enamel and dentine in their large fangs. They also had small eyes, a short, broad head and a well-developed lateral line system on their skull. It is thought that porolepiforms were slow and sluggish predators and probably used their lateral line system to detect prey in the dark and murky waters where they lived. The shape of their body and tail also suggests that they were ambush predators.

Like other sarcopterygians, porolepiforms had their skull divided into two halves (ethmosphenoid and oticoccipital) by a joint running through the braincase. They possessed a robust palatoquadrate (region of the upper jaw and roof of the mouth) that could have withstood strong bite forces. The tooth whorls differ from those in onychodonts by having an additional series of large teeth in parallel rows; these were most likely used to grip struggling prey in their jaws.

The Porolepiformes are the sister group to a much larger and successful group, the lungfishes, or Dipnoi.

Porolepiform Further Reading

Downs, J. P., Daeschler, E. B., Jenkins, F. A. J. & Shubin, N. H. 2013. *Holoptychius bergmanni* sp. nov. (Sarcopterygii, Porolepiformes) from the Upper Devonian of Nunavut, Canada, and a review of *Holoptychius* taxonomy. *Proceedings of the Academy of Natural Sciences of Philadelphia,* 162, 47–59.
Jarvik, E. 1972. Middle and Upper Devonian Porolepiformes from East Greenland with Special Reference to *Glyptolepis groenlandica* n. sp. and a Discussion on the Structure of the Head in the Porolepiformes. *Meddelelser om Gronland,* 187, 1–307.

6.6 Lungfishes

The Dipnoi "twice breathers" are a group of sarcopterygian fish also known as lungfishes that first appeared in the Early Devonian and still survive to the present day (Fig. 6.4). Their name clearly derives from the fact that modern forms possess lungs as well as gills and can therefore obtain oxygen from both water and air. Living lungfish are what we call "sister taxa" to the tetrapods, meaning that they are the most closely related group of fishes to the land vertebrates (Amemiya et al. 2013). As you will learn from this chapter, there are other (now-extinct) groups more closely related to the tetrapods present in the fossil record.

There are just three extant lungfish genera: four species of *Protopterus* from Africa, *Lepidosiren* from South America and "the most enduring vertebrate on earth", *Neoceratodus* (Fig. 6.4g), from Australia, with a skeleton unchanged since the time of the dinosaurs (Kemp and Molnar 1981). Nonetheless, lungfishes were one of the most widespread and dominant vertebrate groups during the peak of their diversity in the Devonian, with close to 100 species described from this period alone. The earliest lungfishes were heavily ossified, fully marine animals such as *Dipnorhynchus* from Australia (Etheridge 1906) and Germany (Jaekel 1927), *Sorbitorhynchus* from China (Wang et al.

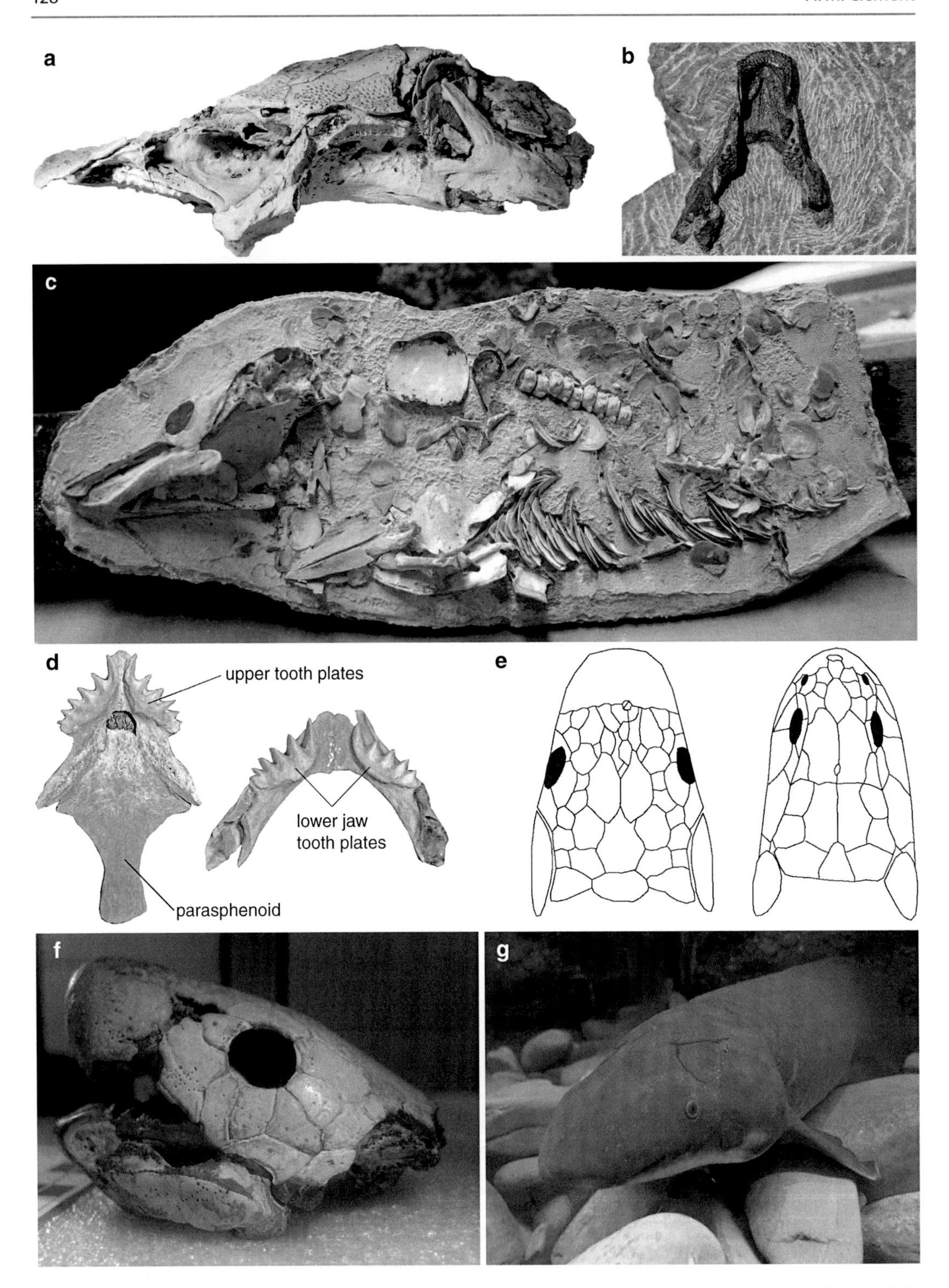

Fig. 6.4 Lungfishes. (**a**) Late Devonian lungfish, *Rhinodipterus kimberleyensis*, skull in left lateral view (photo courtesy of John Long); (**b**) *Rhinodipterus ulrichi*, lower jaw in dorsal view showing tooth plates; (**c**) *Griphognathus whitei*, a "duck-billed" lungfish from the Late Devonian Gogo Formation, being prepared out of the limestone rock (photo courtesy of John Long); (**d**) upper and lower tooth plates of the extant Australian lungfish, *Neoceratodus forsteri*; (**e**) outline drawing showing the difference between a Devonian lungfish skull roof (left) and a generalised contemporaneous sarcopterygian skull (right); (**f**) *Chirodipterus australis* skull and lower jaw in left lateral view; (**g**) the "most enduring vertebrate on earth", the Australian lungfish, *Neoceratodus forsteri*

1993), *Uranolophus* from the USA (Denison 1968), or *Tarachomylax* from Russia (Barwick et al. 1997). Over time, however, the skeletons of lungfishes became more cartilaginous, and the group eventually came to live exclusively in freshwater environments. Indeed, some even adapted to drastic drying conditions, developing the ability to aestivate in burrows in the mud during droughts (e.g., *Gnathorhiza* from the Permian, and *Protopterus* today).

Lungfish differ from most other sarcopterygian fish in their possession of an autostylic jaw, whereby the palatoquadrate has fused with the braincase (Schultze and Campbell 1986). Hence, their skull is no longer divided into two, and they have lost the intracranial joint. The jaw articulation joint shifts forward so that contact with the hyomandibular is lost, and then this bone is later lost (Clack and Ahlberg 2016). The oldest known and most basal lungfish is *Diabolepis* from the Early Devonian of China, which still retains traces of the intracranial joint (Chang and Yu 1984). The only other sarcopterygians to have lost the intracranial joint are the tetrapods, but this character is considered to be convergent. Lungfish also have a highly distinctive dentition, arrangement of their dermal skull bones, as well as the morphology of their postcranial skeleton. Primitive lungfishes also had bony supports between the braincase and skull roof (cristae) which created large vaults for the attachment of powerful jaw muscles, as well as an intricate network of small, bony tubules throughout the snout (rostral tubuli).

The arrangement of skull roof and cheekbones in lungfishes is mostly formed as a mosaic of small bones. It is so different to that of other fishes that a homology-independent lettering scheme was proposed (Forster-Cooper 1937). Thus, in the space where a sarcopterygian fish such as *Eusthenopteron* might have two large, paired median bones at the rear of the skull called postparietals, lungfishes would usually have a single bone demarked as the "B bone". Moreover, lungfishes undergo a great reduction in the number of dermal skull bones over time. The earliest lungfishes sometimes had more than 50 bones on the surface of their skull, whereas *Neoceratodus* has just 10 ossifications.

Perhaps the most distinctive feature of lungfishes is their unique dentition (Fig. 6.4b, d). Following changes in the generalised sarcopterygian skull to an autostylic condition, lungfishes additionally switched from marginal (teeth around the edges of the mouth) to palatal biting (Schultze and Campbell 1986). It was during the Devonian that lungfishes showed the greatest diversity of dentitions. These were grouped into three broad dentition categories: denticulated, dentine-plated and tooth-plated (Campbell and Barwick 1990). There were also some lungfishes (e.g., *Holodipterus*) which were able to resorb and remodel their dentition throughout life (Ahlberg et al. 2006; Long 2010) and others that showed different dentition types during different life stages (e.g., *Andreyevichthys*). Denticulated forms like *Griphognathus*, *Fleurantia* and *Soederberghia* possessed a covering of small, rounded protuberances called "denticles" spread across their palate (Miles 1977; Cloutier 1996a; Ahlberg et al. 2001). Dentine-plated lungfishes had large areas within their buccal (mouth) cavity covered in very hard but usually smooth hypermineralised dentine patches. This would have enabled these fish to crack and crush hard-shelled prey between their jaws (durophagy). Dentine-plated lungfish such as *Dipnorhynchus* and *Chirodipterus* usually had massive and robust jaws and large spaces for attachment of the adductor muscles and must have been capable of an extremely powerful bite (Miles 1977; Campbell and Barwick 1982, 1983, 2000). The final group, the tooth-plated lungfishes, had well-developed tooth ridges on the dental plates that only sometimes occluded with those on the opposite jaw. This group includes the best-known Devonian lungfish, *Dipterus* from Scotland, as well as other forms like *Rhinodipterus* and *Scaumenacia* (Ørvig 1961; Gross 1956; Cloutier 1996a; Clement 2012). After the Carboniferous, only tooth-plated lungfish remained. In the fossil record, Mesozoic and Cenozoic lungfishes are often only identified from isolated tooth plates. The fossilised tooth plates of lungfishes are the most commonly recovered component due to their great hardness of their hypermineralised dentine, especially in contrast with lungfish's ever increasingly cartilaginous skeletons.

Cranial ribs are paired structures that attached to the base of the braincase in extant and some fossil taxa (Long 1993). They are distinct in size and shape from the regular pleural (body) ribs. They have shown to be active during air gulping in extant lungfishes by anchoring the pectoral girdle and depressing the hyoid (Bishop and Foxon 1968). Cranial ribs are an important component of a lungfish-specific "two-stroke buccal-pump" mechanism thought to enable air-gulping behaviour. They are found in some fossil lungfishes from the Middle Devonian onwards suggesting lungfish acquired the ability to breathe air relatively early on in their history (Clement and Long 2010; Clement et al. 2016b).

The first cranial endocast of a lungfish published was that of *Chirodipterus wildungensis* from the Late Devonian of Germany, drawn from a number of shattered specimens (Säve-Söderbergh 1952). More recently, advances in modern scanning technology has enabled the reconstruction of virtual cranial endocasts that do not cause any damage to the specimens. *Rhinodipterus kimberleyensis* from the Late Devonian Gogo Formation was the first lungfish to have a virtual cranial endocast investigated (Clement and Ahlberg 2014). This was followed shortly by that of the Middle Devonian *Dipterus valenciennesi*, from Scotland (Challands 2015), and the Early Devonian lungfish, *Dipnorhynchus sussmilchi* from South-Eastern Australia (Clement et al. 2016a). The condition in the extant Australian lungfish, *Neoceratodus*, has also been explored using the same techniques (Clement et al. 2015). Comparison among these specimens suggests a trend in lungfishes for expansion of part of the forebrain involved in olfaction (telencephalon) and part of the labyrinth system of the inner ear (utriculus) over time.

Bemis (1986) grouped the cranial muscles of extant lungfishes into five functional groups: jaw closing, jaw opening, jaw and hyoid raising, hyoid depressing and head raising. The adductor mandibulae muscles are principle in jaw closing, with some additional lip retractor muscles found in the lepidosirenid lungfishes (*Lepidosiren* and *Protopterus*, but not *Neoceratodus*). *Neoceratodus* makes use of the geniocoracoideus muscle to open the jaws, whereas the lepidosirenids use geniothoracicus and depressor mandibulae instead. All three genera possess a rectus cervicis. For jaw and hyoid raising, lungfishes use their intermandibularis, interhyoideus and levator hyoideus. The rectus cervicis along with other hypaxials are employed during hyoid depression, and the epaxials muscles work to raise the head—most important when breathing at the water's surface.

In another approach, Diogo et al. (2008) listed the muscles topologically (according to their anatomical region) in lungfishes and discussed their hypothesised homologies with other sarcopterygians (including tetrapods). *Lepidosiren* is said to possess five mandibular muscles, three hyoid muscles, at least three branchial (pharyngeal/laryngeal) muscles and two hypobranchial muscles. Recent work by Ziermann et al. (2018) thoroughly investigated cephalic (cranial) muscle development in the Australian lungfish, *Neoceratodus*, in embryos, larvae and one juvenile specimen. It was found that lungfish show a number of developmental patterns also common to other vertebrate taxa. These include the tendency for muscles to first develop anteriorly and then later posteriorly, also the propensity to develop at their muscle origin and grow towards their insertion points. Similarly, the branchial arch muscles develop in an "outside-in" direction, from lateral to ventral/medial (Ziermann et al. 2018). The development and morphology of the cranial muscles lend support to the hypothesis that extant lungfishes have been shaped by paedomorphosis (see paragraph below).

Lungfishes have some of the largest genomes (the complete set of genetic material of an organism), comparable with those of urodeles (such as salamanders); in fact, the largest of any vertebrate is that of the marbled lungfish (*Protopterus aethiopicus*) (Roth et al. 1997). The very large genome in urodeles is correlated with paedomorphosis (Joss 2005), which is the retention of juvenile traits into adulthood (think about how an axolotl looks like a giant larval amphibian preserving its external gills into sexual maturity). Similarly, extant lungfish preserve numerous juvenile characteristics which suggest their large genome can also be ascribed to a process called paedomorphosis (Bemis 1984).

The eggs of lungfishes are larger than those of amphibians and have huge amounts of yolk. The lepidosirenid lungfish have spherical eggs, while those of *Neoceratodus* are hemispherical (Kemp 1982). Cleavage patterns and gastrulation are closely similar to what is seen in amphibians (Kemp 1982); however, lungfishes differ in that they do not undergo metamorphosis. They have an inactive thyroid gland and are therefore incapable of producing the hormone required to metamorphose (Joss 2005). Not surprisingly, the cranial neural crest gives rise to large parts of the skull; however, these cells emerge late from the neural tube compared to other vertebrates (Ericsson et al. 2011). In the developing hatchling, the teeth, jaws, (cartilaginous) quadrate and opercular bones develop prior to the braincase and dermal skull. The number and arrangement of skull bones is exactly the same in hatchlings and juveniles and is even retained into adulthood (Kemp 1999).

Lungfish Further Reading

Jørgensen, J. M. & Joss, J. 2011. *The Biology of Lungfishes,* Enfield, USA, Science Publishers.

Miles, R. S. 1977. Dipnoan (lungfish) skulls and the relationships of the group: A study based on new species from the Devonian of Australia. *Zoological Journal of the Linnean Society,* 61, 1–328.

6.7 Finned Stem-Tetrapods

In the next group, we start to see animals with a mixture of both fish and amphibian features (Fig. 6.5). "Tetrapodomorpha" means four-footed-like, and it is in this group where we see the largest changes in the skeleton for moving out of the water and onto land (Clack 2012). There have been volumes of work investigating the transition from fins to limbs in the context of the first land vertebrates (the tetrapods), but for the purpose of this book, we will focus only on the major changes we see in the skull. The term "stem-tetrapods" include tetrapodomorph fishes and all derived taxa on the stem to tetrapods.

Early tetrapodomorphs generally have wide, flat skulls with small eyes but large, deep opercular bones. Most of them possess a choana in addition to a single pair of external nostrils (the choana is a unique "internal nostril" located inside the palate; this is what allows us to breathe when our mouth is closed). The exception is *Kenichthys* from China, one of the earliest known tetrapodomorphs which represents an intermediate stage between other sarcopterygian fishes and tetrapods Chang and Zhu 1993; Zhu and Ahlberg 2004; Janvier 2004). Primitive sarcopterygian fishes like *Youngolepis* have two pairs (front and rear) of external nostrils, while tetrapods have just one nostril and one choana. Instead, the second nostril in *Kenichthys* has "wandered" into an intermediate position separating two of the upper jaw bones, maxilla and premaxilla. The choana represents one of the most vital components in the evolution of the tetrapod respiratory system.

The earliest known tetrapodomorph is *Tungsenia* (Lu et al. 2012), and like *Kenichthys*, it also comes from China. *Tungsenia* (Fig. 6.5f) has a more tetrapod-like lower jaw, whereas the rest of the skull still retains some general sarcopterygian features, such as the compound cheek plate consisting of fused squamosal, quadratojugal and preopercular bones. There is evidence for a structure called *pars tuberalis* in the earliest known stem-tetrapod *Tungsenia.* This is part of the pituitary gland important for sensing photoperiod and regulating circadian rhythm; its presence in *Tungsenia* was uncovered from endocast CT scan data. This structure is not known in fishes and indicates that some important brain modifications for an increasingly terrestrial lifestyle had already occurred by the Early Devonian in the earliest stem-tetrapods (Lu et al. 2012). The earliest stem-tetrapods like *Tungsenia* and *Kenichthys* retained a cover of cosmine on their bones.

The rhizodonts (Andrews and Westoll 1970) are a monophyletic group of stem-tetrapods that reached huge sizes; in fact, *Rhizodus* from

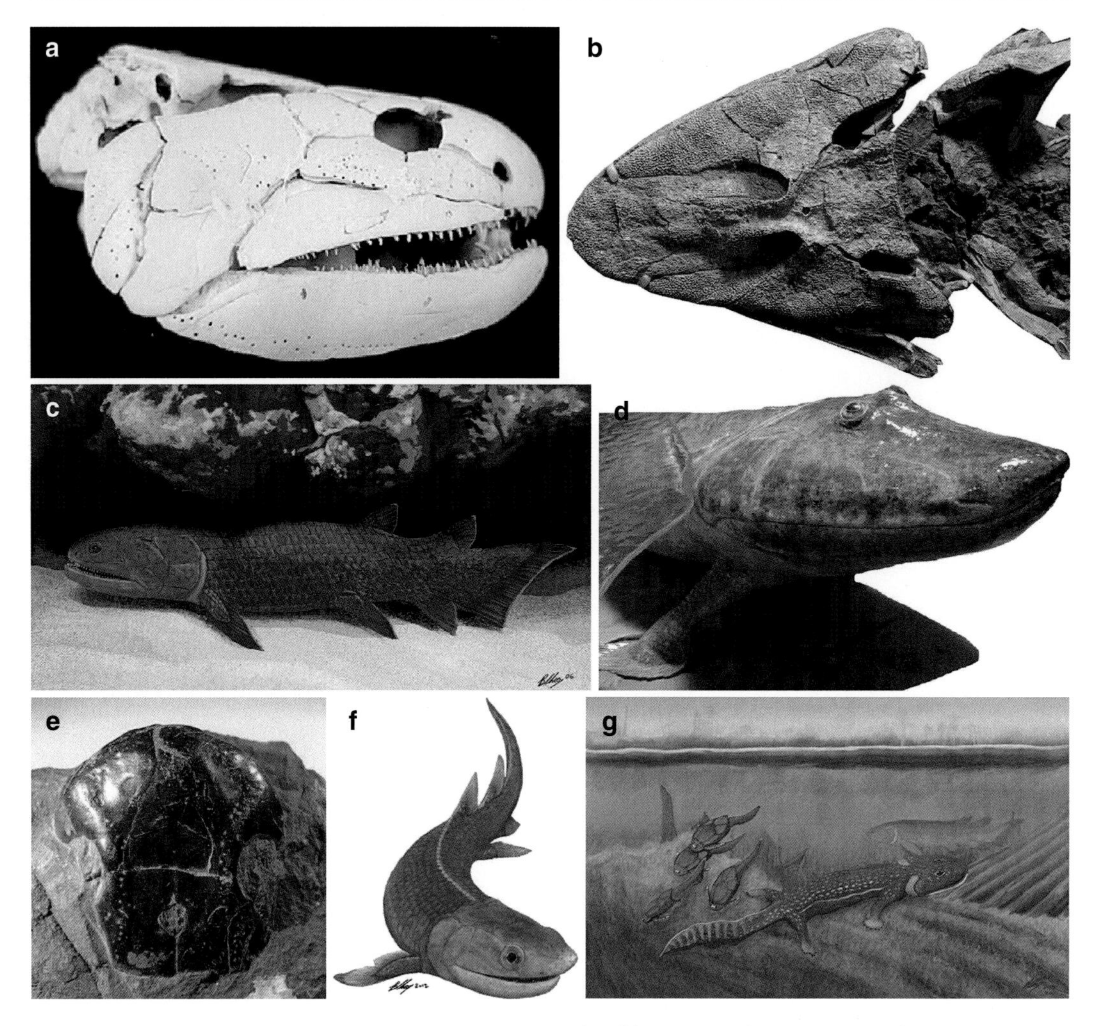

Fig. 6.5 Stem-tetrapods. (**a**) *Gogonasus andrewsae* from the Late Devonian Gogo Formation of Australia, skull in right lateral view (photo courtesy of John Long); (**b**) dorsal view of the skull of a cast of *Tiktaalik roseae* from the Canadian Arctic (photo courtesy of John Long); (**c**) life reconstruction of *Gogonasus andrewsae* (artwork courtesy of Brian Choo); (**d**) life reconstruction model of *Tiktaalik roseae* (photo courtesy of John Long); (**e**) *Osteolepis macrolepidotus* skull in dorsal view (photo courtesy of John Long); (**f**) *Tungsenia paradoxa* (reconstruction artwork courtesy of Brian Choo); (**g**) life reconstruction of *Acanthostega gunnari* (artwork courtesy of Brian Choo)

Scotland (Jeffery 2003) was probably the largest bony fish during the Palaeozoic Era reaching up to 7 metres in length! They had uniquely specialised, stiff pectoral fins and had lost cosmine from their skeleton. They possessed two rows of teeth (on the dentary and coronoid) and large fangs and had a fearsome lower jaw that protruded out from underneath the upper jaw (Long 2011). Some of the earliest rhizodonts are known from Antarctica with *Aztecia*, *Barameda* and *Goologongia* from Australia and *Strepsodus* from Scotland, and even some immature juvenile specimens have been discovered (*Sauripterus*) in North America (Davis et al. 2001). Rhizodonts didn't disappear at the end of the Devonian like many other fishes, instead they held on throughout the Carboniferous and into the beginning of the Permian Period (Long 2011).

The next grade, a paraphyletic group (historically termed "osteolepiforms" or "bony scales"), were widespread around the globe during the Middle and Late Devonian, but only a few lineages survived into the Carboniferous and beyond (Ahlberg and Johanson 1998). Unlike the rhizodonts, most "osteolepiforms" still had cosmine (Fig. 6.5e) on their bones (although this was later lost in some taxa such as *Glyptopomus*), usually seven bones in their cheek, and of course a choana. Two of the best-known stem-tetrapods of this grade are *Gogonasus* from Australia and the tristichopterid *Eusthenopteron* from Quebec. Swedish palaeontologists meticulously serially ground a skull of *Eusthenopteron* over a quarter of a century and reconstructed a wax model of the internal and external anatomy in a similar manner to the way scientists CT scan and then digitally reconstruct specimens today (Jarvik 1942). Consequently, *Eusthenopteron* is still one of the best-known fossil fishes from the Palaeozoic. Tristichopterids are considered advanced "osteolepiforms"; a number of them have a reduced cover of cosmine and have lost the extratemporal bone from the skull (Long 2011). Some other tristichopterids include the globally widespread *Eusthenodon* (Jarvik 1952; Clément 2002), *Tristichopterus* from Scotland and the Northern Isles (Andrews and Westoll 1970), and a number of taxa from Australia including *Mandageria* (Johanson and Ahlberg 1997), *Cabonnichthys* (Ahlberg and Johanson 1997) and *Edenopteron* (Young et al. 2013), as well as *Notorhizodon* from Antarctica (Young et al. 1992) and *Bruehnopteron* from the USA (Schultze and Reed 2012).

Gogonasus (Fig. 6.5a, c) was discovered only 30 years ago, but due to its exceptional 3D preservation, it has contributed much to discussions about stem-tetrapod evolution (Long 1985; Long et al. 2006). Like all tetrapodomorphs, *Gogonasus* has a single external nostril, but it also has some large openings in the skull on top of the head called spiracular openings. These openings were very significant as they are used to take in air for the animal to breathe (rather than obtain all its oxygen through its gills in water) (Holland and Long 2009), and these structures eventually evolve into part of the middle ear found in land vertebrates.

The next group, the elpistostegalids, contains well-known forms like *Tiktaalik* from the Canadian Arctic, *Elpistostege* from Quebec, and *Panderichthys* from Latvia. Their skulls are broad and flat, and they have large heads with their eyes and brow ridge situated on top—superficially like crocodiles. These must have been effective hunters sneaking up stealthily on prey in the waters they inhabited. In *Panderichthys* (Vorobyeva 1980; Ahlberg et al. 1996) and its kin, the intracranial joint has fused, and their snout protrudes further forward than their lower jaws. In elpistostegalids, the hyomandibula is isolated from the palatoquadrate showing the beginning of the transition towards a tetrapod-like middle ear bone (Brazeau and Ahlberg 2006). Elpistostegalids gain a third pair of median skull roof bones (frontals), the snout becomes even longer and they have large spiracular slits on top of the skull for taking in air (Ahlberg and Clack 2006). *Tiktaalik* has even lost the opercular bones (Daeschler et al. 2006). Furthermore, the shoulder bones of a fish are attached to their skull, while those in elpistostegalids do not connect allowing the animal to move its head around with its neck (Fig. 6.5b, d). Indeed, these animals must have developed some of the very first neck muscles in vertebrates. *Elpistostege* was poorly known until recently, but ongoing work led by Richard Cloutier at the University of Quebec on a new complete specimen of this fish will ensure that this will soon be one of the better-known elpistostegalid fishes. It is in this group that we see some of the most important changes in the skeleton enabling fishes that swim to later be able to walk around on land using their powerful, large pectoral and pelvic fins skeletons (Shubin et al. 2004, 2006, 2014). Yet despite all these adaptations in their skeletons, it is unlikely that stem-tetrapods were yet terrestrial.

Tetrapodomorphs Further Reading

Ahlberg, P. E. & Clack, J. A. 2006. A firm step from water to land. *Nature,* 440**,** 747–749.

Clack, J. A. 2012. *Gaining Ground: the Origin and Evolution of Tetrapods,* USA, Indiana University Press.

Shubin, N. 2008. *Your inner fish: a journey into the 3.5-billion-year history of the human body*, Vintage.

6.8 Tetrapods

And the final group we come to in this chapter on sarcopterygians are the digitate stem-tetrapods, or "fish with limbs and digits" (Ahlberg 1995). The first truly terrestrial amphibians did not appear until the Carboniferous Period; but these stem-tetrapods fall outside of the crown groups (Lissamphibia and Amniota) so we still include them in this sarcopterygian fish chapter. It is important to remember that although they were tetrapods, these animals were still living in an aquatic environment (Clack 2012). Many of the features of this group are considered "preconditions" for the eventual transition from water to land. Several of these features relate to the limbs and shoulder and pelvic girdles, but there are also changes in the skull. Three of the most important changes are the formation of the neck through separation of the skull from the pectoral girdle (as discussed for *Tiktaalik*), reduction of the lateral line system and the early evolution of the inner ear.

In fishes, the hyomandibula bone is one of the connections between the jaws and the braincase (the other being a ligament attaching to the palatoquadrate). The hyomandibula originally served to suspend the jaws and steady the jaw joint, but this same bone later became the stapes of the inner ear—a bone involved in airborne hearing in tetrapods (Clack 2002a). Jaw suspension is instead taken over by the articular and quadrate bones. Other inner ear bones also derived from jaw components like the incus and malleus that appeared even later still in mammals.

Alongside the evolution of the stapes and ability for airborne hearing, we see a concurrent reduction in the lateral line sensory system (Long 2011). Fishes rely heavily on their lateral line system to detect movement and vibration in the water around them. It is very important in orientation, predatory and schooling behaviours (Kardong 2006). Fishes have their lateral line enclosed in canals, whereas early tetrapods have large pores or open grooves, and this system is lost entirely in later terrestrial vertebrates.

Two of the best-known early digitate stem-tetrapods are *Ichthyostega* ("fish roof") and *Acanthostega* ("spiny roof"), both known from the Late Devonian of Greenland (Save-Soderbergh 1932; Clack 2002b, 1988; Blom 2005). The intracranial joint has been lost, and like earlier finned stem-tetrapods such as *Tiktaalik*, the eyes are situated close to one another on top of the broad, flat head. The eyes tripled in size, at the same time they shifted to the top of the head in stem-tetrapods, affording these creatures a millionfold increase in visual acuity as they begun to look through air rather than water (Maciver et al. 2017). These animals combined an amphibian-like head and limbs with fish-like tail and gills (Fig. 6.5g). Both *Ichthyostega* and *Acanthostega* are said to have labyrinthodont dentition ("maze-toothed"), owing to the pattern of infolding dentine and enamel on their teeth. The pattern of sutures (joints) of the skull roof bones suggest that *Acanthostega* would have been able to use a more "terrestrial type" of prey capture (direct biting), compared to the suction feeding method of fishes (Markey and Marshall 2007). The spiracular notch increased in width and volume probably resulting in increased air breathing capacity at the same time as reduced reliance on the gills.

There are currently 15 described genera of Devonian tetrapods (Olive et al. 2016; Gess and Ahlberg 2018), but the first truly terrestrial tetrapod known to science is an amphibian called *Pederpes* from the Carboniferous of Scotland (Clack and Finney 2005). You can read more about the evolution and diversity of amphibians in the following chapter.

Tetrapods Further Reading

Clack, J. A. 2012. Gaining Ground: the Origin and Evolution of Tetrapods, USA, Indiana University Press.

Janvier, P. 1996. Early Vertebrates, New York, Oxford University Press.

6.9 Summary

Sarcopterygians were one of the most widespread, diverse and successful groups early in vertebrate history. Today there are only eight extant sarcopterygian *fish* species, but the other sarcopterygians, the tetrapods, dominate the terrestrial and aerial realms (as well as those that returned to the water such as whales). The skulls of primitive sarcopterygian fish are characterised by being divided into two halves (the ethmosphenoid and otoccipital) and were covered in the hard tissue called cosmine. The hinged braincase was lost in many groups such as the lungfish and tetrapods, and today only remains in the coelacanth. Similarly, cosmine was lost in numerous lineages and is not known in any extant animal, meaning its exact purpose remains somewhat of a mystery to modern science. The living sarcopterygian fish show a number of their own specialisations, such as the unique double jaw joint and feeding mechanism in *Latimeria* or the distinctive skull roof and hypermineralised tooth plates of lungfishes. It was in the sarcopterygians that we saw the origin of the internal nostril, the choana, as well as the presence of enamel over dentine in their teeth, traits that we can still see in our bodies today. One of the greatest steps in evolution, the first step from water to land, was enabled by these and other changes in the skeletons of sarcopterygian fishes.

Acknowledgments Firstly I wish to thank John Long, Brian Choo and Min Zhu for their generous supply of images included in this chapter and Jing Lu for her helpful comments on the text of an earlier draft. I would also like to thank the editors, Janine Ziermann, Raul Diaz Jr, and Rui Diogo for the invitation to contribute this chapter. Two helpful reviews from John Long and Sébastian Olive improved the quality of the text.

Glossary

Acanthodian/s A group of stem-chondrichthyans known as "spiny sharks" with bony spines preceding all their fins

Actinistia A group of sarcopterygian fish commonly known as coelacanths

Actinopterygian/s One of the major groups of bony fishes, comprising the vast majority of extant fish species, also known as "ray-finned fishes"

Adductor mandibulae/mandibular/muscle The major jaw muscles

Aestivate To spend a long period in torpor during hot or dry conditions (similar to hibernation)

Amniota A group of vertebrates that can lay eggs with an amnion (a membrane that covers the embryo), this group includes all reptiles, birds and mammals

Ampullae of Lorenzini Special electroreceptive sense organs (jelly-filled pores) found commonly in cartilaginous fish (e.g., sharks, rays and chimaeras)

Articular bone A bone that is part of the lower jaw in most vertebrates

Autostyly/autostylic jaw A type of jaw suspension whereby the upper jaw is connected directly to the cranium

Basicranial muscle Large muscles that span the intracranial joint and attach underneath the braincase that is activated during feeding

Braincase Also known as the "neurocranium", this is the inner part of the skull that encases the brain

Branchial Meaning related to the gills

Carboniferous A geological period that occurred 299–359 million years ago

Cartilaginous A type of connective tissue found in the body, it can be mineralized and form part of the skeleton, but it is not as hard as bone, but far stiffer than muscle

Cenozoic The geological era that occurred from 23 million years ago to the present

Choana A unique internal nostril found in tetrapodomorphs which opens from the nasal sac into the roof of the mouth

Chondrichthyans A group of fish with their skeletons made primarily of cartilage rather than bone (e.g., sharks, rays and chimaeras)

Class A taxonomic rank in biological classification (e.g., Mammalia)

Cleavage Division of the cells in the early embryo

Convergent Independent evolution of similar features in different lineages

Coracomandibularis A muscle found in coelacanths that elevates the palatoquadrate

Coronoid A bone forming part of the lower jaw

Cosmine A hard tissue present on the scales and dermal bones in primitive sarcopterygians that is composed of a mixture of enamel and dentine

Cranial A subdivision of the skull (that together with the mandible comprises the skull)

Cranial ribs Paired structures that attach to the base of the braincase in some lungfishes, distinct from the pleural ribs

Cretaceous A geological period that occurred 66–145 million years ago

Cristae Bony struts supporting the skull roof in primitive lungfishes, allowing large spaces for attachment of jaw muscles

Dendrodont A special type of dentition found in porolepiforms whereby their fangs show a infolding of enamel and dentine

Dentary The main dermal bone of the lower jaw

Denticulated A type of dentition seen in some early lungfishes whereby their buccal cavity is covered with a shagreen of small denticles (e.g., *Griphognathus*)

Dentine A calcified tissue of the body, one of the major components of teeth

Dentine-plated A type of dentition seen in some early lungfishes whereby large areas within their buccal cavity are covered in very hard but usually smooth hypermineralised dentine patches

Dentition Arrangement or condition of teeth

Depressor mandibulae A muscle used to open the jaw (in lungfishes)

Dermal skull The skull roof, or roofing bones of the skull derived from dermal bone

Devonian A geological period that occurred 359–419 million years ago

Dipnoi A group of sarcopterygian fish commonly known as lungfishes

Durophagy Feeding by crushing prey, usually hard-shelled animals

Electroreception/electrosensory The biological ability to perceive electrical stimuli

Elpistostegalids A group of advanced stem-tetrapods such as *Tiktaalik*

Enamel The hardest type of calcified tissue of the body, one of the major components of teeth

Endocast Internal cast of a hollow object, such as the cavity inside the skull

Epaxials Dorsal trunk body muscles

Ethmosphenoid The front portion of the skull

Extant Living or recent, not extinct

Frontals Paired median skull roof bones at the front of the skull in tetrapod-related taxa

Ganoine A bony tissue homologous to enamel that is found in actinopterygians

Gastrulation An early phase of embryonic development

Geniocoracoideus A muscle opening jaw found in the lungfish *Neoceratodus*

Geniothoracicus A muscle opening jaw found in the lepidosirenid lungfishes

Genome The full genetic material of an organism

Gnathostome Jawed vertebrates

Gogo Formation A geological formation and famous Devonian fossil site in North Western Australia

Homologies/homologous/homology Shared ancestry between structures in different taxa

Hyoid A bone derived from the second gill arch in fish, used during feeding and respiration

Hyomandibula One of the jaw attachment bones in fishes that become incorporated into the inner ear of tetrapods

Hypaxials Ventral trunk body muscles

Hypermineralised Highly mineralized

Hypobranchial Meaning located below the gills

In vivo Within a living organism

Inner ear The portion of the ear located within the skull

Interhyoideus A muscle used to help raise the jaw and hyoid in lungfishes

Intermandibularis A muscle used to help raise the jaw and hyoid in lungfishes

Intracranial joint The joint separating two divisions of the skull in early sarcopterygians and still present in the coelacanth

Kinetic/kinesis Moveable

Labyrinthodont Teeth with a complex pattern of infolding of dentine and enamel found in some tetrapods

Lateral line A sense organ of fishes running the length of their body to detect vibration and movement in the surrounding water

Lepidosirenid A member of the lepidosirenid family of lungfishes (*Lepidosiren* and *Protopterus*, but not *Neoceratodus*)

Levator hyoideus A muscle used to help raise the jaw and hyoid in lungfishes

Lip retractor Muscles used to control the lips in lepidosirenid lungfishes

Lissamphibia A taxonomic group that includes all modern (crown group) amphibians but excludes stem members

Mandible Lower jaw

Maxilla An upper jaw bone

Mesozoic Era A geological Era that occurred 66–252 million years ago

Metamorphosis A biological process whereby an animal's body undergoes conspicuous change and development after birth

Monophyletic A cladistic term used to characterise a clade of organisms that share derived characters (synapomoprhies)

Neural crest A temporary group of cells that give rise to a diverse range of cells in the body, unique to vertebrates

Onychodont A group of extinct basal sarcopterygian fish related to Coelacanths

Opercular bones A bony flap covering the gills in fishes

Ossified/ossifications Meaning bony/bones

Osteichthyans The "bony fishes" are a superclass that includes all Actinopterygii and Sarcopterygii

Osteolepiformes A paraphyletic assemblage of finned stem-tetrapods from the Devonian Period (e.g., *Eusthenopteron*)

Otoccipital The rear portion of the skull

Ovoviviparous Live-bearing, embryos develop inside eggs that remain within the mother's body

Paedomorphosis The retention of juvenile traits into adulthood, or "juvenilisation", a type of heterochrony

Palaeozoic Era A geological Era that occurred 252–541 million years ago

Palatoquadrate Endoskeletal part of the upper jaw

Parasphenoid A median (often with teeth) bone of the palate in fishes

Parietal shield A set of dermal bones covering the dorsal surface of the anterior half of the skull

Parietals Paired skull bones that enclose the pineal region in fishes and tetrapods

Pectoral girdle Shoulder girdle

Permian The geological period that occurred 252–299 million years ago

Phylogenetic position/phylogeny Inferred evolutionary relationships among taxa based on differences and similarities in their physical or genetic characteristics

Placoderm An extinct group of fishes that were dominant during the Devonian with thick plated "armour" covering their bodies

Pleural Referring to the body, especially around the chest cavity

Porolepiformes A group of extinct sarcopterygian fishes closely related to lungfishes

Postcranial All parts of the skeleton apart from the skull

Postparietals Paired median skull roof bones situated towards the rear

Premaxilla An upper jaw bone situated in front of the maxilla

Preopercular A dermal cheekbone situated in front of the operculum

Preorbital The area of a skull situated in front of the eyes (orbits)

Quadrate A bone in the skull that contributes to the jaw joint

Quadratojugal A cheekbone found in some fishes

Rectus cervicis A muscle used to open the jaw (in lungfishes)

Rhizodonts A monophyletic group of stem-tetrapods that reached huge sizes during the Palaeozoic

Rostral organ A large sensory organ in the snout of coelacanths

Rostral tubuli Network of small, bony tubules throughout the snout in some sarcopterygian fishes

Sarcopterygian/Sarcopterygii One of the major groups of bony fishes that include lungfish, coelacanths and tetrapods, also known as "lobe-finned fishes"

Silurian A geological period that occurred 419–444 million years ago

Sister taxa/group The closest relatives of another taxon/group in a phylogenetic tree

Skull roof The roofing bones of the skull derived from dermal bone

Spiracular openings/slits Openings on the top of the skull in some fishes and tetrapods, thought to be involved in accessory air breathing

Squamosal A cheekbone found in some fishes

Stapes A bone of the inner ear found in tetrapods

Stem group A phylogenetic term meaning members of a total group that are excluded from the crown group

Sutures A type of fibrous joint between bones of the skull

Tandem jaw joint A specialised jaw arrangement found in coelacanths whereby the lower jaw is attached via two joints on each side of the head

Terrestrial Live predominantly on land (rather than in water)

Tetrapodomorph/s Meaning "four-footed-like" and this includes finned and limbed forms that are more closely related to living tetrapods than to living lungfishes

Tetrapods "Four-footed" vertebrates with digit-bearing limbs and all of their descendants

Tooth whorl A distinctive type of dentition found in some sarcopterygian fishes

Tooth-plated Those lungfishes with tooth plates rather than denticulated or dentine-plated dentition

Topologically Pertaining to position or region

Urodeles A group of amphibians that includes salamanders

Vascular Carrying blood within the body, part of the circulatory system

Vertebrates All animals with a backbone (including fishes, amphibians, reptiles, birds and mammals)

References

Ahlberg PE (1991) A re-examination of sarcopterygian interrelationships, with special reference to Porolepiformes. Zool J Linnean Soc 103:241–287

Ahlberg PE (1995) *Elginerpeton pancheni* and the earliest tetrapod clade. Nature 373:420–425

Ahlberg PE, Clack JA (2006) A firm step from water to land. Nature 440:747–749

Ahlberg PE, Johanson Z (1997) Second tristichopterid (Sarcopterygii, Osteolepiformes) from the upper Devonian of Canowindra, New South Wales, Australia, and phylogeny of the Tristichopteridae. J Vertebr Paleontol 17:653–673

Ahlberg PE, Johanson Z (1998) Osteolepiforms and the ancestory of tetrapods. Nature 395:792–794

Ahlberg PE, Clack JA, Luksevics E (1996) Rapid braincase evolution between *Panderichthys* and the earliest tetrapods. Nature 381:61–64

Ahlberg PE, Johanson Z, Daeschler EB (2001) The late Devonian lungfish *Soederberghia* (Sarcopterygii, Dipnoi) from Australia and North America, and its biogeographical implications. J Vertebr Paleontol 21:1–12

Ahlberg PE, Smith MM, Johanson Z (2006) Developmental plasticity and disparity in early dipnoan (lungfish) dentitions. Evol Dev 8:331–349

Amemiya CT, Alfoldi J, Lee AP, Fan SH, Philippe H, Maccallum I, Braasch I, Manousaki T, Schneider I, Rohner N, Organ C, Chalopin D, Smith JJ, Robinson M, Dorrington RA, Gerdol M, Aken B, Biscotti MA, Barucca M, Baurain D, Berlin AM, Blatch GL, Buonocore F, Burmester T, Campbell MS, Canapa A, Cannon JP, Christoffels A, de Moro G, Edkins AL, Fan L, Fausto AM, Feiner N, Forconi M, Gamieldien J, Gnerre S, Gnirke A, Goldstone JV, Haerty W, Hahn ME, Hesse U, Hoffmann S, Johnson J, Karchner SI, Kuraku S, Lara M, Levin JZ, Litman GW, Mauceli E, Miyake T, Mueller MG, Nelson DR, Nitsche A, Olmo E, Ota T, Pallavicini A, Panji S, Picone B, Ponting CP, Prohaska SJ, Przybylski D, Saha NR, Ravi V, Ribeiro FJ, Sauka-Spengler T, Scapigliati G, Searle SMJ, Sharpe T, Simakov O, Stadler PF, Stegeman JJ, Sumiyama K, Tabbaa D, Tafer H, Turner-Maier J, Van Heusden P, White S, Williams L, Yandell M, Brinkmann H, Volff JN, Tabin CJ, Shubin N, Schartl M, Jaffe DB, Postlethwait JH, Venkatesh B, di Palma F, Lander ES, Meyer A, Lindblad-Toh K (2013) The African coelacanth genome provides insights into tetrapod evolution. Nature 496:311–316

Andrews SM, Westoll TS (1970) The postcranial skeleton of Rhipidistian fishes excluding *Eusthenopteron*. Transactions of the Royal Society of Edinburgh 68:489

Andrews SM, Long J, Ahlberg PE, Barwick RE, Campbell KSW (2006) The structure of the sarcopterygian *Onychodus jandemarrai* n.Sp. from Gogo, Western Australia: with a functional interpretation of the skeleton. Trans Roy Soc Edinb Earth Sci 96:197–307

Barwick RE, Campbell KSW, Markkurik E (1997) *Tarachomylax*: a new early Devonian dipnoan from Severnaya Zemlya, and its place in the evolution of the Dipnoi. Geobios 30:45–73

Bemis WE (1984) Paedomorphosis and the evolution of the Dipnoi. Paleobiology 10:293–307

Bemis WE (1986) Feeding systems of living Dipnoi: anatomy and function. Journal of Morphology Supplement 1:249–275

Bemis WE, Northcutt RG (1992) Skin and blood-vessels of the snout of the Australian lungfish, *Neoceratodus forsteri*, and their significance for interpreting the cosmine of Devonian lungfishes. Acta Zool 73:115–139

Betancur-r R, Broughton RE, Wiley EO, Carpenter K, Lopez JA, Li C, Holcroft NI, Arcila D, Sanciango M, Cureton JC, Zhang F, Buser T, Campbell MA, Ballesteros JA, Roa-Varon A, Willis S, Borden WC, Rowley T, Renaue PC, Hough DJ, Lu G, Grande T, Arratia G, Orti G (2013) The tree of life and a new classification of bony fishes. PloS Curr 1:5

Bishop IR, Foxon GE (1968) The mechanism of breathing in the south American lungfish, *Lepidosiren paradoxa*; a radiological study. J Zool 154:263–271
Blom H (2005) Taxonomic revision of the late Devonian tetrapod Ichthyostega from East Greenland. Palaeontology 48:111–134
Brazeau MD, Ahlberg PE (2006) Tetrapod-like middle ear architecture in a Devonian fish. Nature 439:318–321
Campbell KSW, Barwick RE (1982) A new species of the lungfish *Dipnorhynchus* from New South Wales. Palaeontology 25:509–527
Campbell KSW, Barwick RE (1983) Early evolution of dipnoan dentitions and a new genus *Speonesydrion*. Memoirs of the Association of Australasian Palaeontologists 1:17–49
Campbell KSW, Barwick RE (1990) Paleozoic dipnoan phylogeny: functional complexes and evolution without parsimony. Paleobiology 16:143–169
Campbell KSW, Barwick RE (2000) The braincase, mandible and dental structures of the early Devonian lungfish *Dipnorhynchus kurikae* from wee Jasper, New South Wales. Rec Aust Mus 52:103–128
Challands TJ (2015) The cranial endocast of the middle Devonian dipnoan *Dipterus valenciennesi* and a fossilised dipnoan otoconal mass. Papers in Palaeontology 1:289–317
Chang MM (1982) The braincase of *Youngolepis*, a Lower Devonian crossopterygian from Yunnan, south-western China. PhD, University of Stockholm and Section of Palaeozoology, Swedish Museum of Natural History
Chang MM, Yu X (1984) Structure and phylogenetic significance of *Diabolichthys speratus* gen. Et sp. nov., a new dipnoan-like form from the lower Devonian of eastern Yunnan, China. Proc Linnean Soc NSW 107:171–184
Chang MM, Zhu M (1993) A new middle Devonian osteolepid from Qujing, Yunnan. Memoirs of the Association of Australasian Palaeontologists 15:183–198
Choo B, Zhu M, Zhao W, Jia L, Zhu Y (2014) The largest Silurian vertebrate and its palaeoecological implications. Sci Rep 4:5242
Clack JA (1988) New material of the early tetrapod *Acanthostega* from the upper Devonian of East Greenland. Palaeontology 31:699–724
Clack JA (2002a) Patterns and processes in the early evolution of the tetrapod ear. J Neurobiol 53:251–264
Clack JA (2002b) The dermal skull roof of *Acanthostega gunnari*, an early tetrapod from the late Devonian. Trans R Soc Edinb Earth Sci 93:17–33
Clack JA (2012) Gaining ground: the origin and evolution of tetrapods. Indiana University Press, USA
Clack JA, Ahlberg PE (2016) Sarcopterygians: from lobe-finned fishes to the tetrapod stem group. In: Clack JA, Fay RR, Popper AN (eds) Evolution of the vertebrate ear—evidence from the fossil record. Springer International Publishing, Basel
Clack JA, Finney SM (2005) Pederpes finneyae, an articulated tetrapod from the Tournasian of western Scotland. J Syst Palaeontol 2:311–346
Clément G (2002) Large Tristichopteridae (Sarcopterygii, Tetrapodomorpha) from the late Famennian Evieux formation of Belgium. Palaeontology 45:577–593
Clement AM (2012) A new species of long-snouted lungfish from the late Devonian of Australia, and its functional and biogeographical implications. Palaeontology 55:51–71
Clement AM, Ahlberg PE (2014) The first virtual cranial endocast of a lungfish (Sarcopterygii: Dipnoi). PLoS One 9:19
Clement AM, Long JA (2010) Air-breathing adaptation in a marine Devonian lungfish. Biol Lett 6:509–512
Clement AM, Nysjö J, Strand R, Ahlberg PE (2015) Brain—endocast relationship in the Australian lungfish, *Neoceratodus forsteri*, elucidated from tomographic data (Sarcopterygii: Dipnoi). PLoS One 10(10):e0141277
Clement AM, Challands TJ, Long JA, Ahlberg PE (2016a) The cranial endocast of *Dipnorhynchus sussmilchi* (Sarcopterygii: Dipnoi) and the interrelationships of stem-group lungfishes. Peer J 4:e2539
Clement AM, Long JA, Tafforeau P, Ahlberg PE (2016b) New insights into the origins of buccal pump air breathing in Devonian lungfishes. Paleobiology 1:16
Cloutier R (1996a) Dipnoi (Akinetia: Sarcopterygii). In: Schultze HP, Cloutier R (eds) Devonian fishes and plants of Miguasha, Quebec, Canada. Verlag Dr Friedrich Pfeil, Munich
Cloutier R (1996b) The primitive actinistian *Miguashaia bureaui* Schultze (Sarcopterygii). Devonian fishes and plants of Miguasha, Québec, Canada. Verlag Dr. Friedrich Pfeil, München
Cloutier R, Ahlberg PE (1996) Morphology, characters and interrelationships of basal Sarcopterygians. In: Stiassny MLJ, Parenti LR, Johnson GD (eds) Interrelationships of fishes. Academic Press, San Diego
Daeschler EB, Shubin N, Jenkins FAJ (2006) A Devonian tetrapod-like fish and the evolution of the tetrapod body plan. Nature 440:757–763
Davis MC, Shubin NH, Daeschler EB (2001) Immature rhizodonts from the Devonian of North America. Bull Mus Comp Zool 156:171–187
Denison RH (1968) The evolutionary significance of the earliest known lungfish, *Uranolophus*. In: Orvig T (ed) Current problems in lower vertebrate phylogeny, nobel symposium 4. Almqvist and Wiksell, Stockholm
Diogo R, Abdala V, Lonergan N, Wood BA (2008) From fish to modern humans - comparative anatomy, homologies and evolution of the head and neck musculature. J Anat 213:391–424
Dutel H, Maisey JG, Schwimmer DR, Janvier P, Herbin M, Clément G (2012) The Giant cretaceous coelacanth (Actinistia, Sarcopterygii) *Megalocoelacanthus dobiei* Schwimmer, Stewart & Williams, 1994, and its bearing on Latimerioidei interrelationships. PLoS One 7:e49911
Dutel H, Herrel A, Clément G, Herbin M (2013) A reevaluation of the anatomy of the jaw-closing

system in the extant coelacanth *Latimeria chalumnae*. Naturwissenschaften 100(11):1007–1022
Dutel H, Herbin M, Clément G, Herrel A (2015a) Bite force in the extant coelacanth Latimeria: the role of the intracranial joint and the basicranial muscle. Curr Biol 25(9):1228–1233
Dutel H, Herrel A, Clément G, Herbin M (2015b) Redescription of the hyoid apparatus and associated musculature in the extant coelacanth *Latimeria chalumnae*: functional implications for feeding. Anat Rec (Hoboken) 298:579–601
Dzerzhinsky FY (2016) The mystery of the two-unit skull of the Sarcopterygii: a trap for functional morphologists. J Zool 301:1–17
Ericsson R, Joss J, Olsson L (2011) Early head development in the Australian lungfish, *Neoceratodus forsteri*. In: Jørgensen JM, Joss J (eds) The biology of lungfishes. Science Publishers, Enfield
Etheridge JR (1906) The cranial buckler of a dipnoan fish, probably Ganorhynchus, from the Devonian beds of the Murrumbidgee River, New South Wales. Rec Aust Mus 6:129–132
Forey PL (1998) History of the coelacanth fishes. Chapman and Hall, London
Forster-Cooper C (1937) The middle Devonian fish fauna of Achanarras. Transactions of the Royal Society of Edinburgh 59:223–239
Friedman M (2007) *Styloichthys* as the oldest coelacanth: implications for early osteichthyan interrelationships. J Syst Palaeontol 5:289–343
Gess RW, Ahlberg PE (2018) A tetrapod fauna from within the Devonian Antarctic Circle. Science 360:1120–1124
Gross W (1956) Über Crossopterygier und Dipnoer aus dem baltischen Oberdevon im Zusammenhang einer vergleichenden Untersuchung des Porenkanalsystems paläozoischer Agnathen und Fische. Almqvist and Wiksell, Stockholm
Holland T, Long JA (2009) On the phylogenetic position of *Gogonasus andrewsae* long 1985, within the Tetrapodomorpha. Acta Zool 90:285–296
Inoue JG, Miya M, Venkatesh B, Nishida M (2005) The mitochondrial genome of Indonesian coelacanth *Latimeria menadoensis* (Sarcopterygii: Coelacanthiformes) and divergence time estimation between the two coelacanths. Gene 349:227–235
Jaekel O (1927) Der Kopf der Wirbeltiere. Zeitschrift fur die gesarnte Anatornie 27(3):815–974
Janvier P (2004) Wandering nostrils. Nature 432:23–24
Jarvik E (1942) On the structure of the snout of crossopterygians and lower Gnathostomes in general. Zoologiska bidrag från Uppsala 21:235–675
Jarvik E (1952) On the fish-like tail in the Ichthyostegid Stegocephalians: with descriptions of a new Stegocephalian and a new crossopterygian from the upper Devonian of East Greenland. Meddelelser om Gronland 114:1–90
Jeffery JE (2003) Mandibles of rhizodontids: anatomy, function and evolution within the tetrapod stem-group. Trans R Soc Edinb Earth Sci 93:255–276
Jessen H (1975) A new choanate fish, *Powichthys thorsteinssoni* n.G., n.Sp., from the early lower Devonian of the Canadian Arctic archipelago. In: Lehman JP (ed) Problémes actuels de Paléontologie. Evolution des Vertébrés. Colloques Internationaux du Centre, Paris
Jessen H (1980) Lower Devonian Porolepiformes from the Canadian Arctic with special reference to *Powichthys thorsteinssoni* Jensen. Palaeontographica A 167:180–214
Johanson Z, Ahlberg PE (1997) A new tristichopterid (Osteolepiformes: Sarcopterygii) from the Mandagery sandstone (late Devonian, Famennian) near Canowindra, NSW, Australia. Trans R Soc Edinb Earth Sci 88:39–68
Joss JMP (2005) Lungfish evolution and development. Gen Comp Endocrinol 148:285–289
Kardong KV (2006) Vertebrates: comparative anatomy, function, evolution. McGraw Hill, New York
Kemp A (1982) The embryological development of the Queensland lungfish, *Neoceratodus forsteri*, (Krefft). Memoirs of the Queensland Museum 20:553–597
Kemp A (1999) Ontogeny of the skull of the Australian lungfish *Neoceratodus forsteri* (Osteichthyes: Dipnoi). J Zool 248:97–137
Kemp A (2017) Cranial nerves in the Australian lungfish, Neoceratodus forsteri, and in fossil relatives (Osteichthyes: Dipnoi). Tissue Cell 49(1):45–55
Kemp A, Molnar RE (1981) *Neoceratodus forsteri* from the lower cretaceous of New South Wales, Australia. J Paleontol 55:211–217
King B, Hu Y, Long JA (2018) Electroreception in early vertebrates: survey, evidence and new information. Palaeontology 61:325–358
Long JA (1985) A new osteolepid fish from the upper Devonian Gogo formation of Western Australia. Records of the Western Australian Museum 12:361–377
Long JA (1993) Cranial ribs in Devonian lungfishes and the origin of dipnoan air-breathing. Memoirs of the Association of Australasian Palaeontologists 15:199–209
Long JA (1999) A new genus of fossil coelacanth (Osteichthyes: Coelacanthiformes) from the middle Devonian of southeastern Australia. Records of the Western Australian Museum, Supplement 57:37–53
Long JA (2010) New holodontid lungfishes from the late Devonian Gogo formation of Western Australia. In: Elliott DK, Maisey JG, Yu X, Miao D (eds) Fossil fishes and related biota: morphology, phylogeny and paleobiogeography. D. Verlag Pfeil, München
Long JA (2011) The rise of Fishes-500 million years of evolution. University of New South Wales Press, Sydney
Long JA, Trinajstic K (2010) The late Devonian Gogo formation lägerstatten of Western Australia–exceptional vertebrate preservation and diversity. Annual Review of Earth & Planetary Sciences 38:665–680
Long JA, Young GC, Holland T, Senden TJ, Fitzgerald EMG (2006) An exceptional Devonian fish from

Australia sheds light on tetrapod origins. Nature 444:199–202
Lu J, Zhu M (2010) An onychodont fish (Osteichthyes, Sarcopterygii) from the early Devonian of China, and the evolution of the Onychodontiformes. Proc R Soc B 277:293–299
Lu J, Zhu M, Long JA, Zhao W, Senden TJ, Jia LT, Qiao T (2012) The earliest known stem-tetrapod from the lower Devonian of China. Nat Commun 3:1160
Lu J, Zhu M, Ahlberg PE, Qiao T, Zhu Y, Zhao W, Jia LT (2016) A Devonian predatory fish provides insights into the early evolution of modern sarcopterygians. Sci Adv 2:1–8
Maciver MA, Schmitz L, Mugan U, Murphey TD, Mobley CD (2017) A massive increase in visual range preceded the origin of terrestrial vertebrates. PNAS 114(12):E2375–E2384
Markey MJ, Marshall CR (2007) Terrestrial-style feeding in a very early aquatic tetrapod is supported by evidence from experimental analysis of suture morphology. Proc Natl Acad Sci U S A 104:7134–7138
Miles RS (1977) Dipnoan (lungfish) skulls and the relationships of the group: a study based on new species from the Devonian of Australia. Zool J Linnean Soc 61:1–328
Millot J, Anthony J (1958a) Anatomie de *Latimeria chalumnae*, I—Squelette, Muscles, et Formation de Soutiens. CNRS, Paris
Millot J, Anthony J (1958b) Anatomie de *Latimeria chalumnae*, II—systeme nerveux et organes des sens. CNRS, Paris
Millot J, Anthony J, Robineau D (1978) Anatomie de *Latimeria chalumnae*, III. CNRS, Paris
Mondejar-Fernandez J, Clément G (2012) Squamation and scale microstructure evolution in the Porolepiformes (Sarcopterygii, Dipnomorpha) based on *Heimenia ensis* from the Devonian of Spitsbergen. J Vertebr Paleontol 32:267–284
Musick JA, Bruton MN, Balon EK (1991) The biology of *Latimeria chalumnae* and evolution of coelacanths. Springer, Netherlands
Niedźwiedzki G, Szrek P, Narkiewicz K, Narkiewicz M, Ahlberg PE (2010) Tetrapod trackways from the early middle Devonian period of Poland. Nature 463: 43–48
Olive S, Ahlberg PE, Pernegre VN, Poty E, Steurbaut E, Clément G (2016) New discoveries of tetrapods (ichthyostegid-like and whatcheeriid-like) in the Famennian (Late Devonian) localities of Strud and Becco (Belgium). Palaeontology 59(6):827–840
Ørvig T (1961) New finds of acanthodians, arthrodires, crossopterygians, ganoids and dipnoans in the upper middle Devonian calcareous flags (Oberer Plattenkalk) of the Bergisch Gladbach-Paffrath trough. Paläontol Z 35:10–27
Ørvig T (1969) Cosmine and cosmine growth. Lethaia 2:241–260
Qu Q, Haitina T, Zhu M, Ahlberg PE (2015) New genomic and fossil data illuminate the origin of enamel. Nature 526:108–111
Roth G, Nishikawa KC, Wake DB (1997) Genome size, secondary simplification, and the evolution of the brain in salamanders. Brain Behav Evol 50:50–59
Save-Soderbergh G (1932) Preliminary note on Devonian stegocephalians from East Greenland. Medd Grønland 94:1–211
Säve-Söderbergh G (1952) On the skull of *Chirodipterus wildungensis* gross, an upper Devonian dipnoan from Wildungen. Kunglinga Svenska Vetenskapsakademiens Handlingar 4(3):1–29
Schultze HP (1994) Comparison of hypotheses on the relationships of Sarcopterygians. Syst Biol 43:155–173
Schultze HP, Campbell KSW (1986) Characterization of the Dipnoi, a monophyletic group. Journal of Morphology Supplement 1:25–37
Schultze H-P, Reed JW (2012) A tristichopterid sarcopterygian fish from the upper middle Devonian of Nevada. Hist Biol 24:425–440
Shubin NH, Daeschler EB, Coates MI (2004) The early evolution of the tetrapod humerus. Science 304:90–93
Shubin NH, Daeschler EB, Jenkins FAJ (2006) The pectoral fin of *Tiktaalik roseae* and the origin of the tetrapod limb. Nature 440:764–771
Shubin NH, Daeschler EB, Jenkins FAJ (2014) Pelvic girdle and fin of *Tiktaalik roseae*. PNAS 111: 893–899
Stensiö E (1963) The brain and the cranial nerves in fossil, lower craniate vertebrates, Skrifter utgitt av Det Norske Videnskaps-Akademi. Universitetsforlaget, Oslo, pp 1–120
Thomson KS (1977) On the individual history of cosmine and a possible electroreceptive function of the pore-canal system in fossil fishes. In: Mahala Andrews S, Miles RS, Walker AD (eds) Problems in vertebrate evolution: essays presented to prof. T.S. Westoll. Academic Press, London
Vorobyeva EI (1980) Observations on two rhipidistian fishes from the upper Devonian of lode, Latvia. Zool J Linnean Soc 70:191–201
Wang S, Drapala V, Barwick RE, Campbell KSW (1993) The dipnoan species, *Sorbitorhynchus deleaskitus*, from the lower Devonian of Guangxi, China. Philosophical Trans R Soc Lond (Biol) 340:1–24
Watt M, Evans CS, Joss JMP (1999) Use of electroreception during foraging by the Australian lungfish. Anim Behav 58:1039–1045
Wourms JP, Atz JW, Stribling MD (1991) Viviparity and the maternal-embryonic relationship in the coelacanth *Latimeria chalumnae*. Environ Biol Fish 32:225–248
Young GC, Long JA, Ritchie A (1992) Crossopterygian fishes from the Devonian of Antarctica: systematics, relationships and biogeographic significance. Rec Aust Mus 14:1–77
Young B, Dunstone RL, Senden TJ, Young GC (2013) A gigantic Sarcopterygian (Tetrapodomorph lobe-finned fish) from the upper Devonian of Gondwana (Eden, New South Wales, Australia). PLoS One 8:1–25
Yu X (1998) A new porolepiform-like fish, *Psarolepis romeri*, gen. Et sp. nov. (Sarcopterygii, Osteichthyes)

from the lower Devonian of Yunnan, China. J Vertebr Paleontol 18:261–274
Yu X, Zhu M, Zhao W (2010) The origin and diversification of osteichthyans and sarcopterygians: rare Chinese fossil findings advance research on key issues of evolution. Palaeoichthyology 24:71–75
Zardoya R, Meyer A (1997) Molecular phylogenetic information of the identity of the closest living relative(s) of land vertebrates. Naturwissenschaften 84:389–397
Zhu M, Ahlberg PE (2004) The origin of the internal nostril of tetrapods. Nature 432:94–97
Zhu M, Yu X (2002) A primitive fish close to the common ancestor of tetrapods and lungfish. Nature 418:767–770
Zhu M, Yu X, Janvier P (1999) A primitive fossil fish sheds light on the origin of bony fishes. Nature 397:607–610
Zhu M, Yu X, Ahlberg PE (2001) A primitive sarcopterygian fish with an eyestalk. Nature 410:81–84
Zhu M, Zhao WJ, Jia LT, Lu J, Qiao T, Qu Q (2009) The oldest articulated osteichthyan reveals mosaic gnathostome characters. Nature 458:469–474
Zhu M, Yu X, Lu J, Qiao T, Zhao W-J, Jia LT (2012) Earliest known coelacanth skull extends the range of anatomically modern coelacanths to the early Devonian. Nat Commun 3:772
Ziermann JM, Clement AM, Ericsson R, Olsson L (2018) Cephalic muscle development in the Australian lungfish, *Neoceratodus forsteri*. J Morphol 279(4):494–516

Diversity of Heads, Jaws, and Cephalic Muscles in Amphibians

7

Janine M. Ziermann

7.1 Introduction

Living **amphibians** (**Lissamphibia**) comprise three orders: the **caecilians** (**Gymnophiona**), **frogs** (**Anura**), and **salamander** (**Urodela**) (Fig. 7.1; Hillis 1991; Duellman and Trueb 1994; Pyron and Wiens 2011). Caudata for salamanders or Salientia for frogs include not only extant but also extinct species; still, the terms caudate and urodele are often used interchangeably. Most scientists place amphibians as sister group of amniotes (reptiles, turtles, birds, mammals) to form the Tetrapoda (four-limbed vertebrates). With lungfishes (Dipnoi) being the sister group to Tetrapoda, the comparison of morphology and development of amphibians and lungfishes is often used to shed light on the evolution of tetrapod structures like cranial neural crest and mesoderm (Ericsson et al. 2009), heart anatomy (Johansen and Hanson 1968), and other structures that changed, or not, during the transition from water to land. The tooth development is an example of conserved characters (Kundrát et al. 2008). The water to land transition was accompanied by many anatomical and physiological changes; the former will be shortly discussed here.

J. M. Ziermann
Department of Anatomy, Howard University College of Medicine, Washington, DC, USA
e-mail: janine.ziermann@howard.edu

The currently largely accepted view is that frogs are closer related to salamanders than to caecilians (Fig. 7.1; Hillis 1991; Frost et al. 2006). Most amphibians have a biphasic lifestyle with an aquatic larva and a terrestrial adult, which are connected via a phase called **metamorphosis** (Reiss 2002), which includes drastic changes of the cranium and associated muscles from larva to adulthood. However, many exceptions are present, and in particular salamanders evolved multiple times to direct-developing species (see below).

There are currently 7807 amphibian species known (Fig. 7.1; AmphibiaWeb 2018), which are significantly more than the ca. 5400 known mammal species. Amphibians can be found almost everywhere in the world, from swamps to desserts, except in the polar regions of Antarctica and Greenland; they are especially concentrated in the Neotropical countries. Their habitats (obligate and facultative aquatic, boreal, etc.) are as diverse as their lifestyle with a wide range of feeding modes (classified by feeding: filter feeders, scrapers, herbivores, carnivores, etc., or classified by size of prey: macrophage, microphage, megalophage, etc.) (Altig and Johnston 1989; McDiarmid and Altig 1999). They also have a variety of reproductive and developmental modes (ovopar, ovovivipar, vivipara, with metamorphosis, direct developers, neotene). All the differences in lifestyle and feeding modes are obvious in their head anatomy. For example, the miniature,

J. M. Ziermann et al. (eds.), *Heads, Jaws, and Muscles*, Fascinating Life Sciences,
https://doi.org/10.1007/978-3-319-93560-7_7

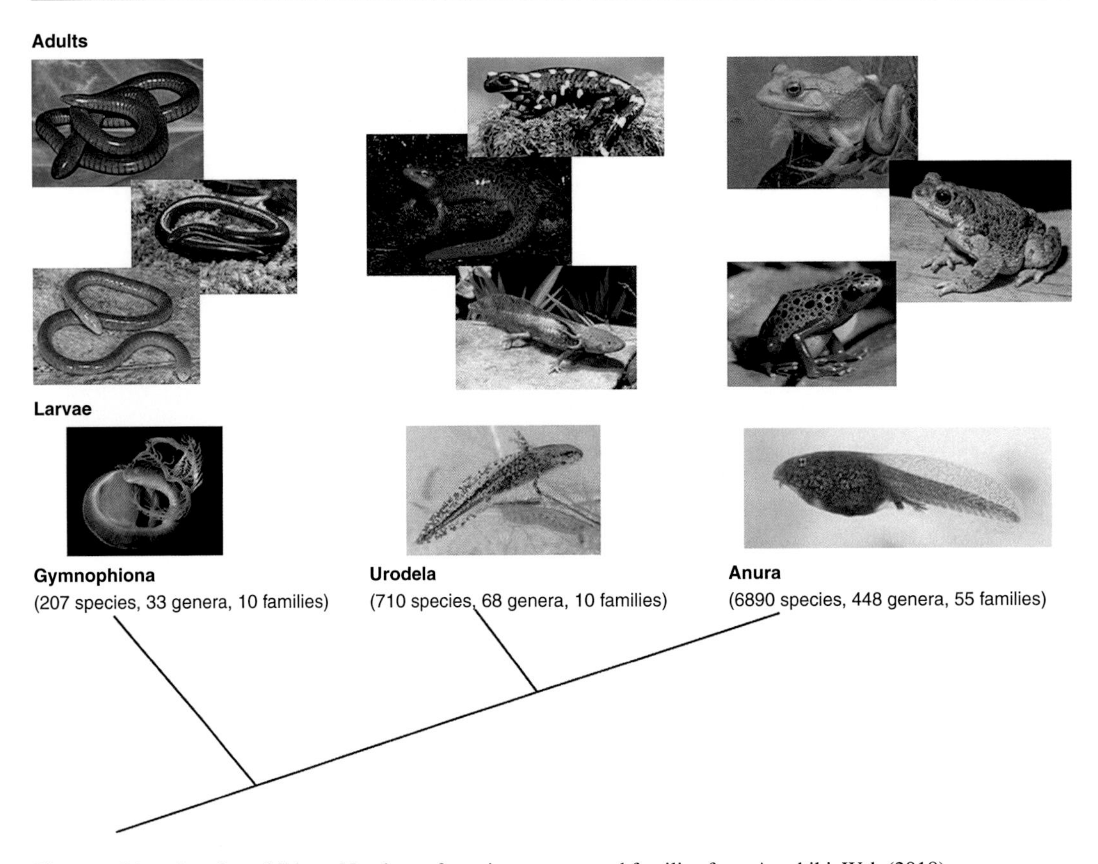

Fig. 7.1 Diversity of amphibians. Numbers of species, genera, and families from AmphibiaWeb (2018)

predatory tadpole of *Hymenochirus boettgeri*, did not only reduce some of the cephalic muscles but also changed the morphology of the jaws in a way that the mouth can form a tubelike structure to enable efficient suction feeding (Sokol 1969; Dean 2003). Not only adults show a variety of skull types and specific adaptations of their cephalic muscles but also their larvae. Caecilians are characterized by the loss of limbs and girdles and the development of an external muscular sheath which is likely related to their fossorial lifestyle; only the members of the family Typhlonectidae are aquatic.

Besides all the diversity, oviparity, free-living aquatic larvae, and a complex life-cycle, which connects the larval with the adult stages via metamorphosis, are considered to be the plesiomorphic (ancestral) condition for lissamphibians (Duellman and Trueb 1994). All aquatic salamander larvae investigated up to date are suction feeders (Deban et al. 2001) as are several larval caecilians (O'Reilly et al. 2002), and there are many similarities between larval caecilians and larval salamanders (see below). It is therefore likely that the ancestral larval amphibians were also suction feeders.

Descriptions of the morphology and/or development of larval and adult crania and cranial muscles are numerous for frogs (e.g., Rocek 1989; Hanken et al. 1992; Reiss 1997; Haas 2001; Cihak et al. 2002; Manzano and Abdala 2003; Vera Candioti 2005; Alcalde and Barg 2006; Ziermann and Olsson 2007; Fabrezi and Lobo 2009; Weisbecker and Mitgutsch 2010; Johnston 2011; Ziermann et al. 2013, 2014a) and salamanders (e.g., Piatt 1935, 1938; Reilly 1987; Bauer 1997; Ericsson et al. 2004; Ericsson and Olsson 2004; Piekarski and Olsson 2007; Dulcey Cala et al. 2009; Ziermann and Diogo 2013; Sefton et al. 2015). However, studies in caecilians mostly concern only the skull morphology and muscle descriptions are rare (e.g., Visser

1963; Taylor 1969; Wake and Hanken 1982; Nussbaum 1983; Wake et al. 1985; Duellman and Trueb 1994; Müller et al. 2005; Müller 2006; Kleinteich and Haas 2007). Even less studies compare members of all three amphibian taxa (Luther 1914; Edgeworth 1935; Iordansky 1996; Haas 2001).

Here, I will focus on general patterns of larval and adult morphology in amphibians and point out changes that are relevant in the water to land transition that helps our current understanding of the evolution of tetrapods.

7.2 Metamorphosis

Several types of **metamorphosis**, which differ in the amount of larval to adult changes, are known in amphibians (Lynn 1961). The "typical" amphibian development starts with external fertilization and passes an embryonic stage in an egg, an aquatic larval stage, followed by metamorphosis to a terrestrial adult. The hatched larva develops gills, two pairs in frogs and three in salamanders and caecilians. The gills are gradually covered by an opercular fold during larval development. The larva starts feeding and depending on their feeding mode shows a variety of specialized mouth and head characters, like horny teeth in frogs for plant scrapers, wide openings for carnivorous larvae, small mouth openings with wide buccal areas for filter feeders. Salamanders have only carnivorous larvae and therefore only wide mouth openings. This feeding larval stage can last from few days to more than a year and hardly any changes occur in the larva during this time (Lynn 1961).

With changes in the thyroid hormone levels, the metamorphosis begins (Tata 2006). Morphological changes can be related to three major functional changes: (1) water to land transition; (2) change in locomotion from swimming to walking, crawling, or hopping; and (3) change of feeding (Reiss 2002). Externally this phase is marked by the appearance of limbs (hind limbs in frogs, forelimbs in salamanders; caecilians have no limbs). The horny teeth, if present, disappear, and the mouth and head form changes (Lynn 1961; Reiss 2002). The tail is resorbed in frogs but persists in salamanders and caecilians. Many other changes occur during this time inside the body: the histology of the skin changes, the intestine shortens, gills are resorbed and lungs take over the breathing, ossification begins, and so forth (Lynn 1961; Reiss 2002). For this chapter, changes of the head and associated structures are the most interesting ones.

Besides this typical metamorphosis, several other forms of development exist in amphibians. Several salamanders of different families (e.g., Proteidae, Ambystomatidae, Sirenidae, Cryptobranchus) are **neotenic**, i.e., they do not undergo full metamorphosis and keep several larval features like external gills, when reaching sexual maturity (Lynn 1961). Some salamanders, like *Ambystoma tigrinum*, even have different developmental modes dependent on their environment, and while some populations undergo full metamorphosis, others are neotenic (Lynn 1961). Other terrestrial salamanders, like *Plethodon*, reduced their time the offspring spends in water to such an extreme that some of them undergo metamorphosis in the egg close to the time of hatching (Lynn 1961). Furthermore, some salamanders (e.g., *Salamandra atra*) are **ovovivipar**, which means that the eggs stay in the oviduct during embryonal and some or all larval stages (Lynn 1961). **Direct development** also occurs in amphibians; however, it was shown that some, like the frog *Eleutherodactylus coqui*, have a **cryptic metamorphosis**, that is, developmental changes regulated by thyroid hormones occur in the egg (Callery and Elinson 2000; Callery et al. 2001; Elinson 2013; Ziermann and Diogo 2014). Caecilians also show a similar variety of developmental modes as salamanders and frogs with aquatic larvae (e.g., *Ichthyophis*), terrestrial development (e.g., *Hypogeophis*), or ovoviviparity (e.g., *Typhlonectes*) (Lynn 1961; Kleinteich and Haas 2007).

A free swimming larval stage that is connected through metamorphosis to a terrestrial adult is considered to be ancestral for amphibians (Duellman and Trueb 1994). As larval and postlarval stages are more similar in caecilians and salamanders than in frogs, the ancestral condition is likely that the changes were not as dramatic as observed today during anuran metamorphosis. Changes of the jaw and cranium during metamorphosis are more dramatic in frogs than in salamanders or caecilians. During metamorphosis, the jaw apparatus morphology in salamanders is relatively stable from larval stages to adulthood (Iordansky 1992). This makes the comparison of muscles between larval and adult salamanders easier than between larval and adult frogs which show dramatic changes of the cranium and associated muscles during metamorphosis (Haas 1996).

Even the last common ancestor (LCA) of anurans had at least a few rearrangements of musculature, and the muscle fiber turnover in all extant frogs was already present (Alley 1989; Nishikawa and Hayashi 1995; Alley and Omerza 1998; Chanoine and Hardy 2003). This process permits the relocation of muscle attachments during metamorphosis (Haas 2001). Only the presence of two distinct cell lineages, which evolved together but with different developmental programs for larval and adult myogenesis, allows those dramatic changes (Alberch et al. 1985; Alberch and Gale 1986). The relocation of muscles during metamorphosis also allows evolutionary changes in larval forms without affecting the adult one and vice versa. The extent of uncoupling larval and adult elements in frogs is unique among amphibians (McDiarmid and Altig 1999).

7.3 Cranium

The skeletal tissues of vertebrates have dual origins, the **neural crest** and mesoderm. The developmental origin of most parts of the cranium from neural crest has been shown through many embryological studies, which included neural crest grafting experiments (e.g., Noden 1983; Creuzet et al. 2005), cell lineage tracing (e.g., Olsson and Hanken 1996), and transgenic techniques (e.g., Rinon et al. 2007). Traditionally the cranium has been divided into **neurocranium**, which is the part that surrounds the brain, and **viscerocranium**, which is the facial skeleton including the jaws (Gegenbaur 1878; Goodrich 1930). Developmental studies have shown that the origin of the neurocranium is mixed, while the viscerocranium is derived from neural crest cells. In the neurocranium, the posterior part derives from so-called parachordal cartilages which are of mesodermal origin, while the otic capsule is of mixed origin (Hirasawa and Kuratani 2015). The cranial base is mostly derived from mesoderm, while its rostral extension called trabeculae derives from neural crest cells (see below; Couly et al. 1993; McBratney-Owen et al. 2008; Wada et al. 2011; Kuratani et al. 2013).

7.3.1 Neural Crest-Derived Cartilages

The most common concept of visceral arches and craniofacial skeletons in vertebrates is the derivation from **neural crest streams** that can be divided into mandibular, hyoid, and branchial arch streams (Fig. 7.2; see below). Conserved cell fates within higher taxa (e.g., tetrapods) can be used to support homology statements based on comparative morphological studies (but see below and Helms et al. 2005). However, transgenic and chimeric experiments revealed that the cranium of frogs challenges this pattern as homologous structures develop from different cell lineages (see mixed origin of parts of the cranium in Fig. 7.3d–f) (Gross and Hanken 2008). That shows the decoupling of embryonic patterns, cell lineages, and adult morphology during skeletal development in frogs (Hirasawa and Kuratani 2015). It was hypothesized that the insertion of larval stages in species that undergo metamorphosis lead to a topographical shift of neural crest-derived chondrogenic cells (cells that develop to cartilages), which will form the adult skeleton (Hirasawa and Kuratani 2015). Grafting experiments between quails and chicks demonstrated the importance of local tissue interaction during the cranial skeleton development. Grafted cranial neural crest cells from quail embryos into mesenchymal populations that make the skull wall in chicken resulted in a neural crest-derived skeletal element that would normally derive from mesoderm (Schneider 1999).

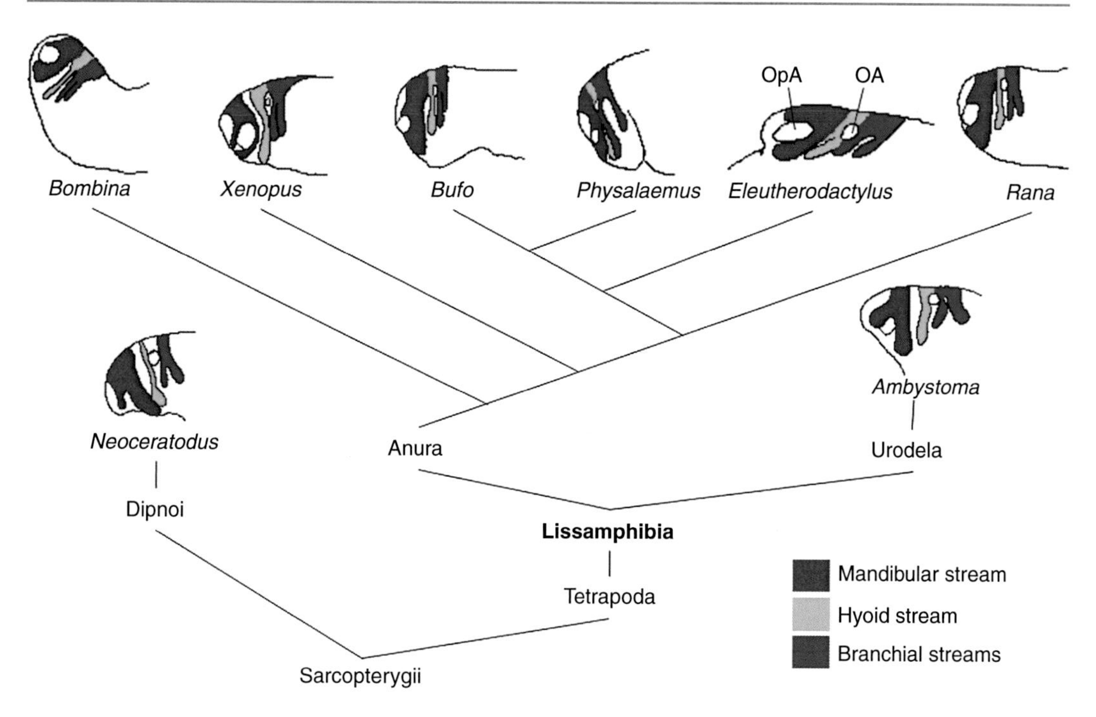

Fig. 7.2 Cranial neural crest stream in frogs (Anura), salamander (Urodela), and lungfish (Dipnoi). Specimens not to scale. Modified from Ericsson et al. (2009), phylogeny based on AmphibiaWeb.com (2018). *OA* otic anlage, *OpA* optic anlage

The importance of local tissue interactions was discussed by Noden (1978), Hall (1980), and others.

Focusing on the craniofacial morphogenesis of the visceral skeleton, conserved developmental patterns are obvious throughout (jawed) vertebrates. The visceral skeleton is organized into an anteroposterior bilateral series of **branchial arches** that are colonized by cranial neural crest cells, which migrate along precise pathways from the mid-hindbrain segments of the neural tube. Each neural crest stream associated with a branchial arch gives rise to different skeletal elements (Fig. 7.3), a process which is orchestrated by a gene regulatory network (Sauka-Spengler and Bronner-Fraser 2008; Reisoli et al. 2010). For example, the distinct differentiation of specific skeletal elements deriving from specific branchial arches is dependent on the expression pattern of genes from the Hox gene family. The first arch (mandibular arch) is the only arch without *Hox* gene expression (e.g., Rijli et al. 1998; Pasqualetti et al. 2000; Trainor and Krumlauf 2001). Additionally, signals from surrounding tissue like endoderm and ectoderm, which are adjacent to the neural crest tissue internally and externally respectively, are crucial for the specification of the mandibular arch (e.g., Couly et al. 2002; Miller et al. 2003; Sato et al. 2008) but also for the proper development of the other arches (Trainor and Krumlauf 2001; Fan et al. 2016).

The patterning and migration of neural crest streams are highly conserved in vertebrates (Falck et al. 2000; Graham and Smith 2001; Olsson et al. 2005; Ericsson et al. 2008); therefore, the labeling of those streams results in similar patterns (Fig. 7.2). The neural crest cells emerge from the dorsal part of the neural tube and form two streams in front of the ear anlage and one behind it. The latter divides later into two streams and will form the skeleton of the gill arches. In larval amphibians, most of the cranial elements derive from neural crest cells (Fig. 7.3b, c) (Olsson and Hanken 1996; Ericsson et al. 2004). The neural crest cells migrate from the dorsal neural tube in three distinct streams called

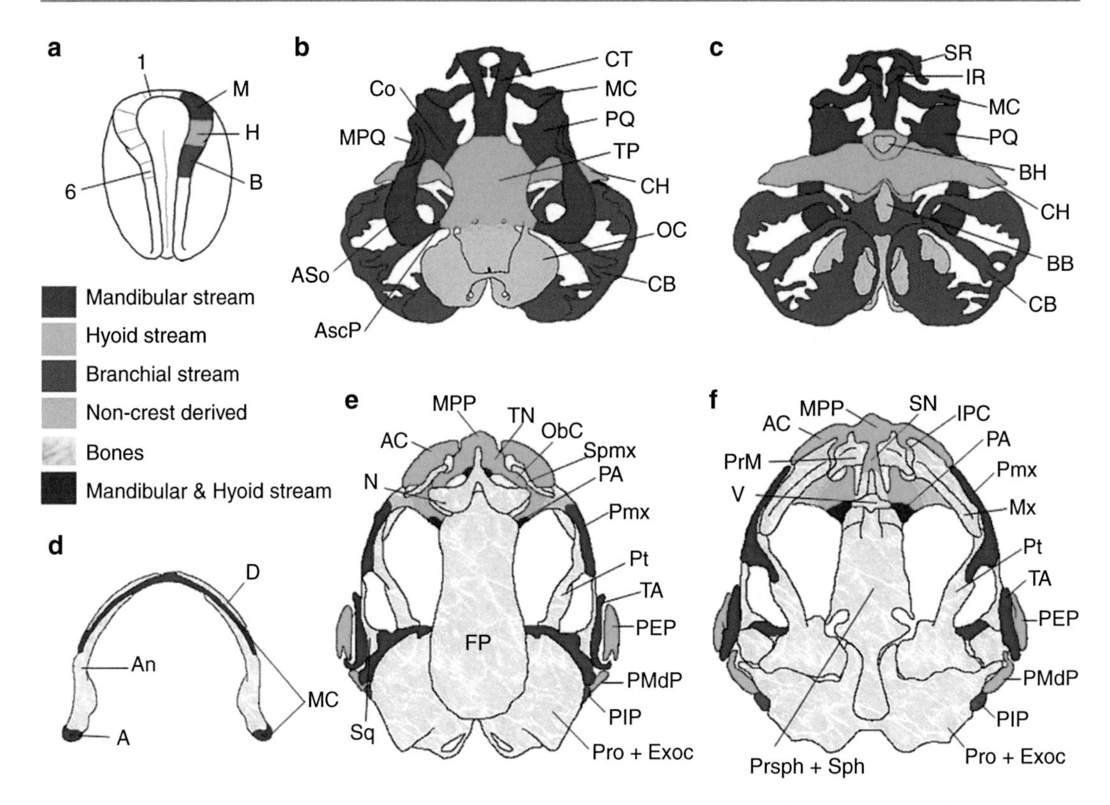

Fig. 7.3 Neural crest contribution to the cranium of frogs. Colors correspond to the contribution of specific neural crest streams. (**a**) Dorsal view of an embryo (stage 14) of *Bombina orientalis*; cranial region to the top; left side shows the six regions of the neural fold used for DiI injections; right side shows the approximate origins of the neural crest cell streams. (**b**) Dorsal and (**c**) ventral view of larval skull (stage 36) of *Bombina orientalis*. (**b**) Note that the trabecular plate (TP), generally derived from the premandibular crest cells (Hirasawa and Kuratani 2015), is mapped on the hyoid crest in *Bombina*. (**d**) Dorsal view of the lower jaw (mandible), (**e**) dorsal and (**f**) ventral view of post-metamorphic (stage 66+ 1 month) *Xenopus laevis* cranium. Note that hyoid crest cells are widely distributed in the sphenethmoidal region of the cranium and the articular end of Meckel's cartilage also contains hyoid cells. **a–c** Recolored from Olsson and Hanken (1996) and Trueb and Hanken (1992); **d–f** recolored from Gross and Hanken (2008). Nomenclature follows previously mentioned publications and McDiarmid and Altig (1999). *A* articular, *AC* alar cartilage, *An* angular, *AscP* ascending process of the palatoquadrate, *Aso* arcus subocularis of the palatoquadrate, *BB* basibranchial, *BH* basihyal, *CB* ceratobranchial, *CH* ceratohyal, *Co* commissura quadratocranialis anterior, *CT* cornual trabecula/cornua trabeculae, *D* dentary, *Exoc* exoccipital bone, *FP* frontoparietal, *IPC* inferior prenasal cartilage, *IR* infrarostral, *MC* Meckel's cartilage, *MPP* median prenasal process of septum nasi, *MPQ* muscular process of the palatoquadrate, *Mx* maxilla, *N* nasal, *ObC* oblique cartilage, *OC* otic capsule, *PA* planum antorbitale, *PEP* pars externa plectri, *PIP* pars interna plectra, *PMdP* pars media plectra, *Pmx* posterior maxillary process, *Pro* prootic bone, *PQ* palatoquadrate, *Prsph* parasphenoid, *PrM* premaxilla, *Pt* pterygoid process, *Sph* sphenoid, *Spmx* septomaxilla, *SN* septum nasi, *SR* suprarostral, *TA* tympanic annulus, *Sq* squamosal, *TP* trabecular plate, *TN* tectum nasi, *V* vomer, "+" fused elements

mandibular, hyoid, and branchial streams (Figs. 7.2 and 7.3a), which colonize the first (mandibular), second (hyoid), and third and fourth (branchial) arches, respectively. There they differentiate into visceral skeletal derivatives (Fig. 7.3b, c): the mandibular stream gives rise to the upper (**palatoquadrate**) and lower (**Meckel's cartilage**) jaw and the **ethmoid** (= **trabecular plate**); the hyoid stream gives rise to the **ceratohyal**; the branchial arch streams give rise to the **ceratobranchialia** (gill cartilages) (Sadaghiani and Thiébaud 1987).

Interestingly, many postmetamorphic cartilages in the adult anuran skull form *de novo* at metamorphosis without larval precursors (Gross and Hanken 2008). Most adult cartilages derive from the neural crest streams of the mandibular and hyoid arches, while the branchial arch streams have only minor contributions (Fig. 7.3e, f). In *Xenopus*, there are four cartilages that derive from at least two cell populations (see boundaries in Fig. 7.3d–f). Such composite elements can also be found in other species and in a variable amount. Several boundaries are cryptic; that is, adjacent neural crest territories do not coincide with anatomical boundaries. Fate-mapping studies revealed that adult structures differ from larval structures in amniotes as is most strikingly seen in the changes of neural crest-derived cartilages in adult frogs where mandibular stream-derived elements differentiate caudal to hyoid stream-derived elements (Fig. 7.3e, f). As mentioned above, this new pattern might be a consequence of the biphasic life history in amphibians, in which the cranial architecture is remodeled during metamorphosis, in particular in frogs (Figs. 7.3 and 7.4) (Sedra and Michael 1957; Gross and Hanken 2008; Slater et al. 2009). Likewise, individual elements of the cranial skeleton in vertebrates are variable in their derivation from embryonic tissues like neural crest or mesoderm.

7.3.2 Metamorphosis and Ossification

In frogs, the larval braincase is open dorsally via the frontoparietal fenestra; the floor is formed by the trabecular plate (ethmoid). The ossification of the braincase starts with the **frontoparietals** (Fig. 7.4i, j), followed by the parasphenoid (Trueb and Hanken 1992; Slater et al. 2009). The paired frontoparietals will form the roof of the braincase (Figs. 7.3e and 7.4l), while the single parasphenoid forms the floor. The latter forms also the otic capsules in non-pipid frogs (NB: *Xenopus* in Figs. 7.3 and 7.4 is a pipid frog). The **otic capsules** are spherical and join the braincase posteriorly (Figs. 7.3b and 7.4i). They start their development as hemispheres that open medially, and then they fuse to the parachordals and attach to the basal plate ventrally, the tectum synoticum dorsally, and the taeniae tecti marginales anterodorsally (McDiarmid and Altig 1999). During metamorphosis, the ossification of the otic capsules and the occipital arch includes the **prootics**, the **exoccipitals** (Figs. 7.3e, f and 7.4j–l), and the stapes, which are the bones of the inner ear. The combination of prootics and exoccipitals forms solid otic capsules in adults. The sphenethmoid appears only late during metamorphosis (Trueb and Hanken 1992).

The second structure to ossify is the **mandible** (lower jaw), followed by the upper jaw (Trueb and Hanken 1992). The anterior end of the larval chondrocranium supports the nasal capsules and larval jaws (McDiarmid and Altig 1999). Anuran tadpoles have supra- and infrarostral cartilages (Fig. 7.3c), which are unique among amphibians. The **suprarostral cartilages** (cartilago labialis superior) articulate with or are fused to the cornua trabeculae and are present in the moveable upper jaw of tadpoles (Fig. 7.3b, c). The suprarostral cartilages are resorbed during metamorphosis, and the **infrarostral cartilages** (cartilago labialis inferior) are incorporated into the Meckel's cartilages. The **cornua trabeculae** extend anteriorly or anterolaterally from the trabecular plate (= ethmoid). Two ligaments join the trabecula and the trabecular plate to the **palatoquadrate**. In salamander larvae, the cornua trabeculae are anterior extensions of the floor of the braincase and therefore in similar position as the ones found in frog larvae. The suprarostral cartilages are absent in salamanders, but it was suggested that the broadened anterior ends of the cornua trabeculae might be homologous to those anuran structures, which were fused to the proximal part of the cornua trabeculae (Haas 2001). This might be related to the change form terminal mouth in salamanders to subterminal mouth in frog larvae.

The lower jaw consists of the infrarostral and Meckel's cartilages in anuran tadpoles (Fig. 7.3c). Meckel's cartilages articulate with the infrarostral cartilages anteriorly and the pars articularis quadrati of the palatoquadrate posteriorly. Bones that develop during metamorphosis and that are associated with the lower jaw are the paired angulars, dentaries, articulars, and mentomeckelians (McDiarmid and Altig 1999). The latter bone

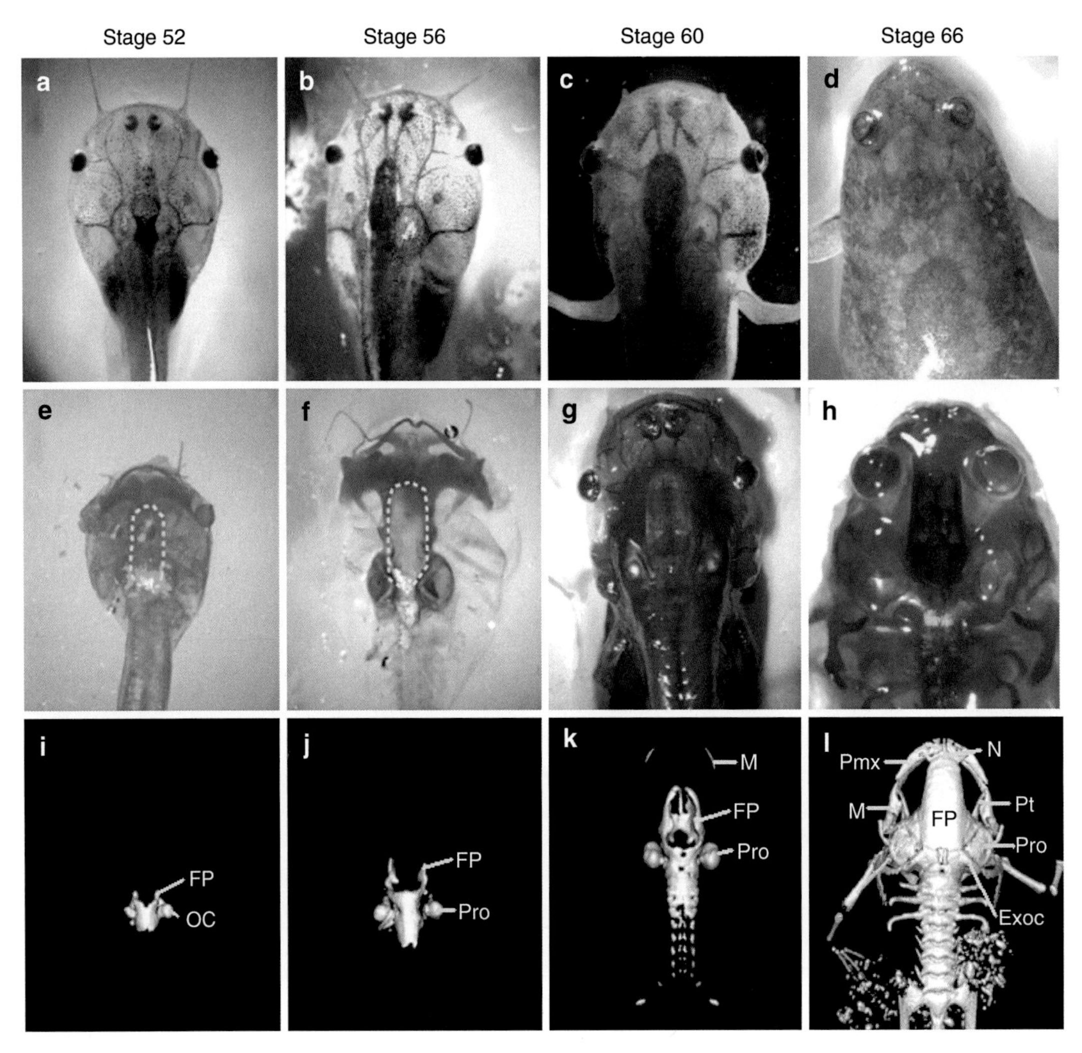

Fig. 7.4 Cranium of *Xenopus laevis* during metamorphosis. (**a–d**) Dorsal view of unstained *Xenopus* in stages 52–66; stage 52 = late larva, stage 66 = postmetamorphic juvenile. (**e–h**) Alizarin red and Alcian blue whole-mount staining of specimens from the same stages as row above; cartilages are blue, bones are red; yellow dashed lines indicate the region of the frontoparietal. (**i–l**) Micro-CT scans delineate ossified portions of cranium. Ossification is complete by stages 64–66. Modified from Slater et al. (2009). *Exoc* exoccipital bone, *FP* frontoparietal, *M* mandible, *N* nasal, *OC* otic capsule, *Pmx* posterior maxillary process, *Pro* prootic bone, *Pt* pterygoid process

ossifies endochondral in the medial ends of the infrarostral cartilages. The posteromedial angular and anterolateral dentary are elongated dermal bones that invest Meckel's cartilages (Fig. 7.3d). The articular ossifies endochondral and is located at the posterior end of Meckel's cartilage.

The larval **palatoquadrate** forms most of the larval upper jaw and is an elongated cartilage connected to the neurocranium by four processes (McDiarmid and Altig 1999): (1, 2) **otic process** and **ascending process** connect the palatoquadrate to the otic capsule (Fig. 7.3b); (3) the **basal process** articulates in most vertebrates with the basitrabecular process, which is a lateral projection of the trabecula (this lateral projection, however, is lost in most frogs during ontogeny, and the basal process fuses or articulates with the neurocranium posterior to the palatine nerve); and (4) the **commissura quadrato-cranialis anterior** (Fig. 7.3b). The larval otic process is destroyed during metamorphosis, but its cells are used to form the adult otic process. The part of the palatoquadrate that underlies the orbit is called the **arcus subocularis** (Fig. 7.3b); the gap between this arch and the neurocranium is called the subocular fenestra. Anteriorly to the fenestra is the body of the palatoquadrate, which is also called the **quadrate**. The most prominent part of the quadrate is the muscular process (Fig. 7.3b). Anteromedial of the quadrate rises the commissura quadrato-cranialis anterior, which extends dorsally to join the palatoquadrate with the floor of the braincase near the trabecular plate (= ethmoid; Fig. 7.3b).

During metamorphosis, most of the processes of the palatoquadrate are resorbed, and it rotates posteriorly. The commissura quadrato-cranialis anterior is only partially destroyed, and its remaining piece forms part of the adult pterygoid process (Figs. 7.3e and 7.4l). As most of the palatoquadrate disappears during metamorphosis, ossifications (maxilla, premaxilla, quadrate, quadratojugal, pterygoid, squamosal) take over some of its functions (Figs. 7.3e, f and 7.4l). The adult quadrate is an endochondral ossification that forms at a similar location as the larval quadrate (McDiarmid and Altig 1999). The arch of premaxilla, maxilla, and quadratojugals (under squamosal) forms in the adult frog the upper jaw (Fig. 7.3e, f). However, several bones associated with the suspensorium develop much later (suspensorium *sensu* Huxley 1858: "The hyoid and the mandibular arches are thus suspended to the skull by a common peduncle, which, to avoid all theoretical suggestion, I will simply term the 'suspensorium'."); the pterygoid is one of the last elements to ossify during metamorphosis in *Xenopus*, and the squamosal and the plectoral apparatus ossify even later until several months post-metamorphosis (Trueb and Hanken 1992).

The fourth structures to ossify are the nasal capsules, nasals, and septomaxillae (Trueb and Hanken 1992). The **nasal capsule** is separated from the orbit by the planum antorbitale (Fig. 7.3e), which is confluent with the tectum nasi rostrally and the orbital cartilage via the sphenethmoid commissure medially. During metamorphosis, the nasal cartilages and the olfactory epithelium become elaborated into a highly folded saclike structure (McDiarmid and Altig 1999). The **nasals** ossify and form the dorsomedial roof of the nasal capsule (Fig. 7.4l), while the **vomers**, which are palatal bones, form the floor. By the end of the metamorphosis, the dorsal-lateral roof of the nasal capsule is formed by the enlarged septomaxilla. The oblique cartilages are located dorsally between the septomaxillae and the tectum nasi (Fig. 7.3e).

The cranium of amphibians is in principle build from the same elements in all taxa; however, huge variations exist among caecilians, salamanders, and frogs (Fig. 7.5). The caecilians possess the most robust cranium with several complex, fused bones, e.g., maxillary-palatine or os basale (Mickoleit 2004). The latter includes the exoccipitalia, otica, and parasphenoid, which are fused to a massive block. This complex cranium corresponds to the burrowing lifestyle of caecilians. Other adaptations are reduced eyes, which often are below the skin and in some species even under bones. The skull in salamanders and anurans is clearly slenderer and less robust than in caecilians (Fig. 7.5). The salamanders are characterized by a gap between the maxilla and the squamosal, the quadratojugal is missing in most species, and frontal and parietal bones are separated (Mickoleit 2004). Frogs have the least robust cranium, which are marked by large fenestrae (Fig. 7.5); the frontal and parietal bones are fused and form together with the sphenethmoid the roof of the skull.

A comparison of neural crest contribution to skull bones revealed that those bones seem to be more evolutionary labile than the cartilages (Hanken and Gross 2005). Furthermore, sometimes even the results of fate maps in the same species differ between labs (Santagati and Rijli 2003; Hanken and Gross 2005; Matsuoka et al. 2005). Amphibians could be helpful to reveal the basic tetrapod pattern of skull development, but due to the extreme changes during metamorphosis and the late ossification of the adult skull, cell lineage tracing studies are challenging and rare for frogs and salamanders. Still, some characters seem to be evolutionary conserved (see

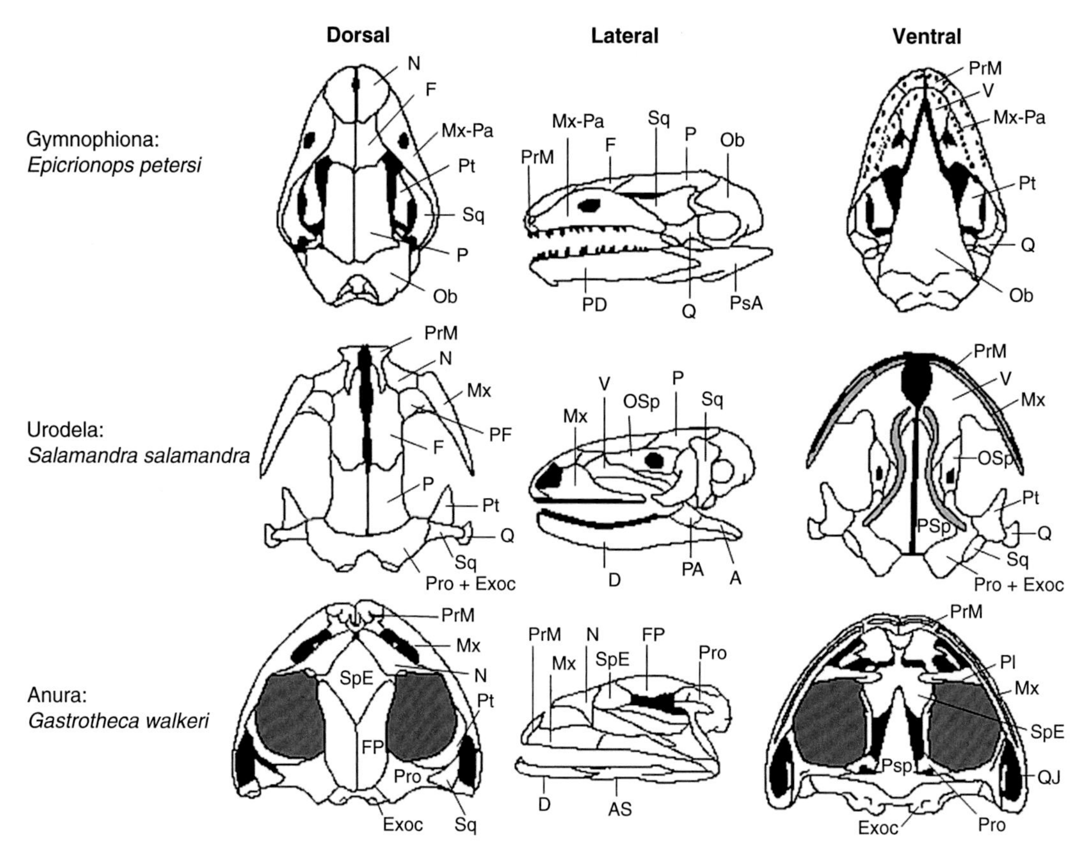

Fig. 7.5 Crania of extant amphibians. Modified from Duellman and Trueb (1994). *A* articular, *AS* angulosplenial, *D* dentary, *Exoc* exoccipital, *F* frontal, *FP* frontoparietal, *Mx* maxillary, *Mx-Pa* maxillary-palatine, *N* nasal, *Ob* os basale, *Osp* orbitosphenoid, *P* parietal, *PA* prearticular, *PD* pseudodentary, *PF* prefrontal, *Pl* palatine, *PrM* premaxillary, *Pro* prootic, *PsA* pseudoangular, *Psp* parasphenoid, *Pt* pterygoid, *Q* quadrate, *QJ* quadratojugal, *SpE* sphenethmoid, *Sq* squamosal, *V* vomer

above), while others seem to be more flexible. One example of the differences in vertebrates is the development of the bony skull roof (cranial vault). As detailed by Hanken and Gross (2005), the skull roof consists mainly of the parietal and frontal bones, which can be separated or fused (Fig. 7.5). However, the development of those bones differs between vertebrates. In the mouse, the frontal but not the parietal derives from neural crest (Jiang et al. 2002). In chicken, both bones are neural crest derived (Couly et al. 1993), but earlier studies claim that only the rostral portion of the frontal is neural crest derived, while the rest is mesodermal (Noden 1978). The frontoparietal bone in frogs derives completely from neural crest cells (Gross and Hanken 2005).

A more recent study of the skull development comparing chicken, axolotl, and *Xenopus* shows again different results (Piekarski et al. 2014). From this study, it seems that the frontal bone in chicken is derived from mandibular neural crest cells (anterior part) and mesoderm (posterior part) and the parietal is also of mesodermal origin. The axolotl has a fully mandibular neural crest cell-derived frontal bone and a mesodermal parietal bone. In *Xenopus*, both bones are fused to the frontoparietal bone, and neural crest cells from mandibular, hyoid, and branchial streams contribute in an anterior to posterior fashion to the development of the bone. Other skull elements also differ between axolotl and *Xenopus*. It is clear that further studies and the inclusion of a wider spectrum of animals is needed to unravel the development of the skull roof.

7.4 Muscles

Amphibians play an important role in the study of the evolutionary changes from fishes to tetrapods. There are extensive studies on the comparative morphology between amphibians and other vertebrates (Lubosch 1914; Edgeworth 1935; Iordansky 1996; Haas 2001, 2003). However, drastic changes during metamorphosis in salamanders and frogs make it often difficult to homologize structures, in particular muscles (Duellman and Trueb 1994). Many muscles present early during larval stages disappear during development and specifically during metamorphosis. The observation from muscle development from early embryological stages through metamorphosis until adulthood is the best way to establish homology hypotheses between taxa that differ from each other (e.g., Haas 2001). While doing this, several **morphogenetic gradients** during early muscle development were observed in amphibians (Ziermann 2008; Ericsson et al. 2009; Ziermann and Diogo 2013, 2014), cartilaginous fishes (Ziermann et al. 2014b), zebrafishes (Schilling and Kimmel 1997), lungfishes (Ziermann et al. 2017b), and quails (McClearn and Noden 1988). With only few exceptions, those gradients include the development of cranial muscles from anterior to posterior (nose to shoulder girdle), lateral to medial (side to middle), and origin to insertion (attachments, with insertion being the attachment on the moving structure upon contraction of the muscle).

Neural crest cells are not only important for the development of the cranium, but they also pattern head structures derived from other tissues like mesoderm that gives rise to cranial muscles (for a recent review, see Ziermann et al. 2018). Neural crest-derived connective tissue of a given branchial muscle develops from the same migratory crest stream as the skeletal attachment sides of the muscle (e.g., Olsson et al. 2001), which derives from the mesoderm from the same branchial arch (e.g., Sadaghiani and Thiébaud 1987; Schilling and Kimmel 1997; but see Graham and Smith 2001). However, exceptions and "cryptic" segmental boundaries, where one skeletal structure derives from more than one source or branchial stream, exist (see above and Fig. 7.3d–f).

The terminology of larval musculature often differs from that of adult morphology as the musculature gets extensively remodeled during metamorphosis (Haas 2001; Ziermann and Diogo 2014). The head muscles can be divided into several groups based on developmental and comparative anatomical studies: (1) extraocular muscles, (2) branchiomeric muscles, and (3) hypobranchial muscles. The branchiomeric muscle can be subdivided into muscles deriving from the mesodermal core of the (first) mandibular arch, (second) hyoid arch, or (third to sixth) caudal branchial arches. The branchiomeric muscles usually attach to cartilages or bones that derive from neural crest streams of the same branchial arch. Furthermore, besides using attachments, the innervation via specific cranial nerves is often used to establish homologies (e.g., Schlosser and Roth 1995). Rarely, exceptions to the expected innervation pattern can occur (e.g., Bauer 1997), but mostly there is a tight correlation between the nerve and the muscle that allows to assess from which developmental anlage a muscle develops (Luther 1914; Edgeworth 1935).

In the following, I focus on few species that represent the different amphibian groups: *Ichthyophis kohtaoensis* (caecilian), *Ambystoma mexicanum* (axolotl, Mexican axolotl, salamander), and *Xenopus laevis* (African clawed toad/frog). Based on several comparative studies, it can be inferred that the last common ancestor (LCA) of larval amphibians had the following muscles: mandibular arch muscles (intermandibularis, at least four adductor mandibulae: A2, A2-PVM, A3′, A3″; Diogo and Abdala 2010), hyoid arch muscles (interhyoideus anterior and posterior, several larval muscles that give rise to the depressor mandibulae in adults), branchial arch muscles (true branchial arch muscles: subarcuales recti and obliqui; subarcuales obliqui II and III; levatores arcuum branchialium I, II, III, and IV; and transversus ventralis IV; other branchial arch muscles: protractor pectoralis; and laryngeal muscles: dilatator laryngis and constrictor laryngis), and hypobranchial muscles (sternohyoideus, geniohyoideus) (Kleinteich and Haas 2007; Diogo and Abdala 2010). Commonly used alternative terminologies for muscles are placed in brackets, e.g., adductor mandibulae (levator mandibulae).

7.4.1 Mandibular Arch Muscles

The **mandibular arch musculature** is innervated by the mandibular branch of the trigeminal nerve (fifth cranial nerve, CNV3; but see below levator bulbi) and includes mainly a group of jaw-closing muscles, the **adductor mandibulae** (add. mand.) or **levatores mandibulae** (lev. mand.), and the intermandibular muscles (Luther 1914; Edgeworth 1935; Iordansky 1992, 1996; Haas 2001). The latter term (lev. mand.) is commonly used by herpetologist for adult and larval amphibians, while comparative morphologists tend to use the former term (add. mand.) to simplify comparisons with other vertebrates (Luther 1914; Haas 2001; Diogo and Abdala 2010; Porro and Richards 2017) or use for larval muscles lev. mand. and for adult ones add. mand. (Ziermann and Diogo 2014).

The ventral **intermandibularis posterior** spans between the cartilago Meckeli (larvae) and dentale (adults) and lowers the floor of the buccal cavity (Fig. 7.6). Frog and salamander larvae also possess a ventral **intermandibularis anterior** muscle (**submentalis** *sensu*, e.g., Iordansky 1992), which is absent in caecilians and spans between the infrarostral cartilages (Fig. 7.6f). In salamanders, the intermandibularis anterior originates also from Meckel's cartilage and inserts posteroventral onto the jaw symphysis. This muscle forms late in larval live in frogs and salamanders or even during metamorphosis (Haas 2001). The anuran **mandibulolabialis** muscle is likely derived from the intermandibularis anterior (Diogo and Abdala 2010) and originates from Meckel's cartilage and inserts onto the lower lip in frog larvae (Fig. 7.6f) but is absent in *Xenopus* and *Ascaphus*. It may present as a muscle with inferior (onto lower lip) and superior (onto upper lip) portions (e.g., *Bombina*, *Pelobates*); the latter part is originating from the lower lip in *Bombina* Haas 2001).

Four of the add. mand. muscles seem to be homologous throughout amphibian larvae: add. mand. A2 (lev. mand. externus), add. mand. A2-PVM (lev. mand. articularis), add. mand. A3′ (lev. mand. longus), and add. mand. A3″ (lev. mand. internus) (Haas 2001).

Laterally in frog larvae, there are a variable number of add. mand. muscles usually extending between the palatoquadrate or its processus and the cartilago Meckeli (Fig. 7.6e). Exceptions to this pattern exist: the add. mand. A3′ profundus (lev. mand. longus profundus, Ziermann and Diogo 2014) and the add. mand. A2 profundus (lev. mand. externus profundus) insert onto the alar cartilages of the upper jaw (Haas 2001; Ziermann and Olsson 2007). The add. mand. A2 (lev. mand. externus) muscle develops late during larval development or even only during metamorphosis in some species (Haas 2001; Ziermann and Olsson 2007). The anuran add. mand. A3″ (levator mandibulae internus, Ziermann and Diogo 2014) has in larvae a far posterior origin from the ascending process of the palatoquadrate and from the cupula anterior of the otic capsule. Besides the three long muscles (add. mand. A3″, A3′ superficialis, A3′ profundus = lev. mand. internus, longus superficialis, longus profundus), up to four smaller muscles originate from the palatoquadrate (lev. mand. externus superficialis = add. mand. A2 superficialis, lev. mand. externus profundus = A2 profundus, lev. mand. articularis = A2-PVM, lev. mand. lateralis = A2 lateralis). The least number (3) of add. mand. muscles has the basal species *Ascaphus truei* (Haas 2001).

In salamander larvae, there are usually four add. mand. muscles observed (Fig. 7.6c, d; lev. mand. longus = lev. mand. anterior = add. mand. A3′, lev. mand. externus = add. mand. A2, lev. mand. internus = add. mand. A3″, lev. mand. articularis = add. mand. A2-PVM). The add.

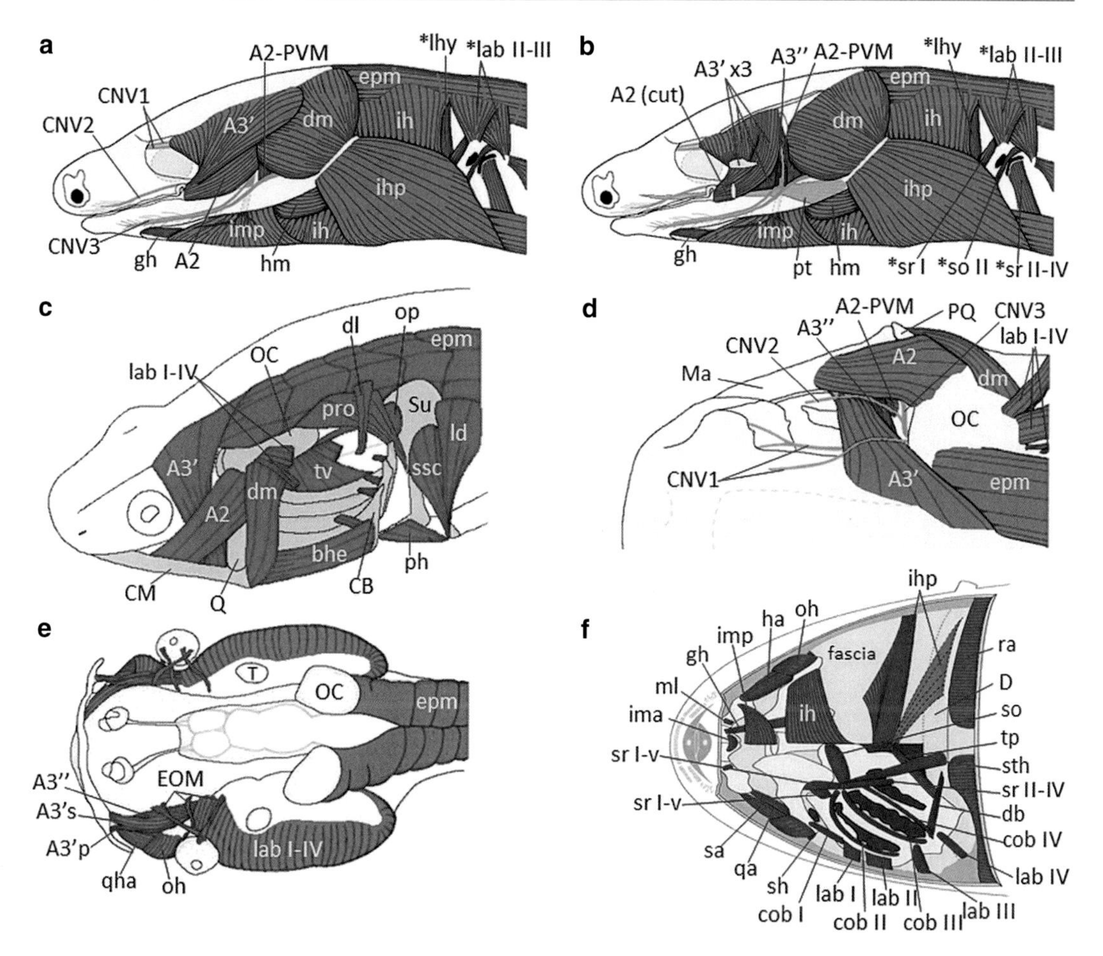

Fig. 7.6 Cranial muscles in larval representatives of amphibians. (**a**, **b**) Left lateral views of *Ichthyophis kohtaoensis*, stage 37; modified from Haas (2001); the pterygoideus (pt in **b**) is hidden as it attaches medially onto the lower jaw and vomer; * indicates muscles not labeled in the original figure; those muscles were labeled after comparison with figures in Kleinteich and Haas (2007). (**c**) Dorsolateral view of *Ambystoma mexicanum*; modified from Piekarski and Olsson (2007). (**d**) Dorsal view of *Ambystoma mexicanum*; modified from Olsson (2007). (**e**) Dorsal view of *Xenopus laevis*, stage 20; modified from Ziermann and Olsson (2007). (**f**) Ventral view of a standardized tadpole; modified from McDiarmid and Altig (1999). Specimens are not to scale. *A2* adductor mandibulae (add. mand.) A2 (levator mandibulae externus); *A2-PVM* A2 postero-ventro-mesial (lev. mand. articularis); *A3′* add. mand. A3′ (lev. mand. longus); *A3′ ×3* three layers of A3′; *A3′p/s* A3′ profundus/superficialis; *A3″* add. mand. A3 (lev. mand. internus); *bhe* branchiohyoideus externus; *CB* ceratobranchial; *CM* cartilago Meckeli; *CNV1* ramus ophthalmicus of trigeminal nerve; *CNV2* ramus maxillaris of trigeminal nerve; *CNV3* ramus mandibularis of trigeminal nerve; *cob I, II, III, IV* constrictor branchialis I, II, III, IV; *D* diaphragm; *db* diaphragmatic branchialis; *dl* dilatator laryngis; *dm* depressor mandibulae; *EOM* extraocular muscles; *epm* epaxial trunk musculature; *gh* geniohyoideus; *ha* hyoangularis; *hm* hyomandibularis; *ih* interhyoideus (anterior); *ihp* interhyoideus posterior; *ima* intermandibularis anterior; *imp* intermandibularis posterior; *lab I–IV* levator arcuum branchialium I–IV; *ld* latissimus dorsi; *lhy* levator hyoideus; *Ma* mandible; *ml* mandibulolabialis; *OC* otic capsule; *oh* orbitohyoideus; *op* opercularis; *ph* procoraco humeralis; *pro* protractor pectoralis; *pt* pterygoideus; *qa* quadratoangularis; *qha* quadratohyoangularis; *Q* quadrate; *ra* rectus abdominis; *sa* suspensorioangularis; *sh* suspensoriohyoangularis; *so* subarcualis obliquus; *sr I, II–IV* subarcualis rectus I, II–IV; *sr I-d/v* subarcualis rectus I dorsal/ventral head; *ssc* suprascapularis; *Su* suprascapular; *sth* sternohyoideus; *T* thymus gland; *tp* tympanopharyngeus; *tv* transversus ventralis

mand. A3′ is the longest and originates high on the cranium, and its origin can be as far posterior as the parietal and occipital bones. The add. mand. A3″ originates mainly from the frontal and parietal. The add. mand. A2 and A2-PVM originate from the palatoquadrate and sometimes also the otic capsule. All those muscles insert onto the lower jaw.

The caecilian larvae studied so far are similar to the caudate larvae (Haas 2001). The add. mand. A2 (lev. mand. externus) is the most lateral muscle, originating from the palatoquadrate (Fig. 7.6a). Medial to it is the add mand. A2-PVM (lev. mand. articularis) that originates from the palatoquadrate and squamosal. The largest area of origin has the add. mand. A3′ (lev. mand. longus) with fibers originating between the otic and orbital region and also from the palatoquadrate (Haas 2001). The most medial muscle, the add. mand. A3″ (lev. mand. internus), originates from the lateral wall of the braincase (taenia tecti marginalis). Interestingly, the adult caecilians seem to have no post-metamorphic successor of the add. mand. A2 (Luther 1914; Iordansky 1996), which could be either lost during metamorphosis or merged with the add. mand. A3′ (Kleinteich and Haas 2007).

Caecilian larvae also have the **pterygoideus** muscle (from vomer to medial side of the lower jaw, close to articulation; Fig. 7.6b) and the **levator quadrati** (from junction between taenia tecti marginalis and pila metoptica to processus pterygoideus palatoquadrati) (Haas 2001; Kleinteich and Haas 2007). The pterygoideus moves the pterygoid caudally upon contraction. It is said to be present only in caecilians (Haas 2001) or in some caecilians as well as a few salamanders and seems to correspond to the **pterygomandibularis** (add. mand. A3″or A3′ + A3″, pterygoideus) of reptiles (Kleinteich and Haas 2007; Diogo and Abdala 2010). The levator quadrati is the most medial muscle of the jaw levators and rotates the quadrate; it is unique in caecilian larvae (Luther 1914; Iordansky 1996; Haas 2001; Kleinteich and Haas 2007).

During **metamorphosis,** the larval adductores mandibulae degenerate and are replaced by new, differentiating myoblasts. Those myoblasts are usually found between the old larval fibers. As old and new muscles can be seen simultaneously during mid-metamorphosis, a relationship between larval and adult muscles can be established (Haas 2001). In species where two larval heads of the add. mand. A2 (lev. mand. externus) are present, it was suggested that both correspond to the single add. mand. A2 in adults (Haas 2001). The postmetamorphic adductores mandibulae can be grouped into two groups depending on their insertion onto the mandible (lower jaw). The medial group comprises the add. mand. A3″, A3′, and A2-PVM (lev. mand. internus, longus, and articularis), while the lateral group includes the add. mand. A2 and A2 lateralis (lev. mand. externus and lateralis). The add. mand. lateralis (lev. mand. lateralis) is in some species, like *Xenopus*, indistinguishable from the add. mand. A2 (lev. mand. externus). The adult **pseudotemporalis** likely developed during metamorphosis by fusion of add. mand. A3′ and A3″ (Diogo and Abdala 2010; Ziermann and Diogo 2014). The muscle also has often two portions as recognized by different fiber orientations (Iordansky 1992; Haas 2001; Ziermann and Diogo 2013). In the direct-developing frog *Eleutherodactylus coqui*, four add. mand. muscles can be distinguished in adults: add. mand. A2, A3″, A3′ + A2-PVM, and A2 lateralis (Ziermann and Diogo 2014). The A3′ + A2-PVM developed through the fusion of the A3′ and A2-PVM, but in other adult frogs like *Rana* or *Xenopus*, all add. mand. can be distinguished separately.

The larvae of *Ascaphus truei,* which is one of the most basal extant frogs (Ford and Cannatella 1993; Frost et al. 2006), share several features with salamanders and caecilians. Those features are likely plesiomorphic, i.e., they were present in the last common ancestors (LCAs) of amphibians: small number of jaw muscles, absence of the mandibulolabialis (absent in caecilians, salamanders, and basal frog larvae like *Ascaphus* and *Xenopus*), and origin of the add. mand. muscles from the skull roof and otic capsule (Haas 2001). In other anuran larvae, the add. mand. muscles originate from the palatoquadrate and the otic capsule. *Ascaphus* and salamanders also have no direct muscle attachments to the upper jaw elements, which are exclusively cartilaginous in frog larvae but ossify during metamorphosis. Therefore, the attachments to the suprarostral cartilages likely evolved only in anurans but were likely not present in the LCA of anuran larvae (Haas 2001). The add. mand. muscles of all amphibians seem to be homologous based on comparative anatomical and developmental studies (Pusey 1943; Haas 2001). However, variable attachments, developmental patterns, and confusing terminology between species and between larval and adult specimens still lead to confusions about the homology of specific add. mand. muscles. The position of the muscles to their innervating nerves (CNV3 = mandibular branch of trigeminal nerve; Fig. 7.6a, b, d) was used to homologize the muscles (e.g., Lubosch 1914; Carroll and Holmes 1980), but it was shown that the nerve can have variable relations to the same muscle (Haas 1996, 2001; Haas and Richards 1998).

Finally, there is the **levator bulbi** muscle, which, unlike all other mandibular muscles, is innervated by the maxillary division of the trigeminal nerve (CNV2). It is only present in frogs and salamanders; it originates in the orbit and attaches to the eye bulb. The contraction of this muscle moves the eye deeper into the skull and supports the swallowing process. The levator bulbi is likely derived from the dorsal mandibular anlage (Iordansky 1996; Diogo and Abdala 2010). However, it might also partially derive from the pseudotemporalis profundus (lev. mand. internus; add. mand. A3″) as the position and fiber orientation are similar in both muscles in adult axolotls (Ziermann and Diogo 2014). This would support Edgeworth's (1935) idea that the levator bulbi is a derivative of the adductor mandibulae. However, the innervation of the levator bulbi makes the latter hypothesis less likely, as it is reported to be innervated by CNV1 and CNIV or CNV2 (Francis 1934; Ziermann and Diogo 2013), while the adductor mandibulae muscles are innervated by CNV3. On the other hand, the levator arcus palatini muscles in fishes derive also from the dorsal mandibular anlage and are innervated by CNV2 and/or CNV3 (Schilling and Kimmel 1997). Developmental studies (gene expression patterns, cell lineage tracing) will be necessary to really determine the origin of the levator bulbi muscle.

7.4.2 Hyoid Arch Muscles

The **hyoid arch musculature** is innervated by the facial nerve (seventh cranial nerve, CNVII) and includes the ventral **interhyoideus** and the lateral jaw-opening muscles that form in adults the **depressor mandibulae** (dep. mand.) in all amphibians (Fig. 7.6). The interhyoideus can often be divided into an anterior and a posterior portion (Fig. 7.6a, b, f; Piatt 1938; Nussbaum 1983; Carroll 2007). The **interhyoideus anterior** originates from the ceratohyal and inserts into a median raphe with its contralateral muscle. The intermandibularis posterior and interhyoideus anterior form often a continuous muscle sheet stretching along the ventral side of the head in larval and adult amphibians. The **interhyoideus posterior** develops from a separate anlage (Ericsson and Olsson 2004) and attaches to the ceratobranchial I and the opercular fold in larvae but shifts their attachments to the mandible in adults as the ceratobranchial degenerates during metamorphosis (Francis 1934; Bauer 1992). Neotene salamander retains the larval condition. Adult caecilians have a unique dual jaw-closing mechanism where the interhyoideus posterior acts together with the adductor mandibulae musculature (Francis 1934; Bauer 1992). The interhyoideus posterior in frogs and salamanders acts as a ventral constrictor of the hyobranchium.

The adult dep. mand. develops from several larval muscles in caecilians, frogs, and salamanders (Drüner 1901; Edgeworth 1935; Duellman and Trueb 1994; Bauer 1997; Haas 2003). The larval frog muscles **orbitohyoideus** and **suspensoriohyoideus** (all attach to the ceratohyal; Fig. 7.6e, f) correspond likely to the dep. mand. posterior in adult frogs, and the larval **hyoangularis**, **suspensorioangularis**, and **quadratoangularis** (= angularis group; all attach to the cartilago Meckeli; Fig. 7.6e, f) correspond to the dep. mand. anterior in adults (Edgeworth 1935; Hanken et al. 1997; Pasqualetti et al. 2000; Haas

2003; Diogo and Abdala 2010). In salamanders, the **ceratomandibularis** (branchiomandibularis *sensu* Edgeworth 1935, subhyoideus *sensu* Duellman and Trueb 1994), **branchiohyoideus** (branchiohyoideus externus *sensu* Edgeworth 1935, subhyoideus *sensu* Carroll 2007), and **levator hyoideus** are present (Fig. 7.6c; Diogo and Abdala 2010; Ziermann and Diogo 2013). The levator hyoideus becomes during development completely integrated into the adult depressor mandibulae in axolotl (Ziermann and Diogo 2013). During this process, it changes its attachment from the ceratohyal (larvae) to the mandible (adult). Also during the development of the non-neotenic species *Lissotriton helveticus* and *Ichthyosaura alpestris*, some fibers of the depressor mandibulae change its fiber orientation so that those become part of the larval branchiohyoideus (Ziermann 2008). A hyomandibularis muscle was also described to be present in larval and paedomorphic salamanders (Edgeworth 1935), but the attachments correspond to the ceratomandibularis described by others (Edgeworth 1935). The ceratomandibularis and branchiohyoideus seem to develop from the same anlage (Piatt 1938; Bauer 1997). Both muscles seem to fuse to the depressor mandibulae in fully metamorphosed salamanders (Edgeworth 1935; Bauer 1997).

A levator hyoidei muscle that acts on the ceratohyal was described for larval *Ichthyophis* (Fig. 7.6a, b) and several larval salamander species and was homologized with the dep. mand. posterior (Edgeworth 1935; Bauer 1997; Kleinteich and Haas 2007). The levator hyoideus is either included into the dep. mand. in adults or disappears during metamorphosis (Edgeworth 1935). The caecilian muscle **hyomandibularis** corresponds likely to the ceratomandibularis or the ceratomandibularis + branchiohyoideus of salamanders; it is present in larval specimens and perhaps also in adults (Edgeworth 1935; Kleinteich and Haas 2007). In adult caecilians, the dep. mand. develops from the merging of the larval dep. mand., levator hyoidei, and hyomandibularis (Edgeworth 1935). However, as pointed out by Kleinteich and Haas (2007), there is still some controversy which muscles correspond to which in the dep. mand. complex. Bauer (1997) described three larval dep. mand. muscles: dep. mand. anterior and posterior and ceratomandibularis. In salamander larvae, the dep. mand. anterior inserts onto the lower jaw (Piatt 1938; Fox 1959; Bauer 1997) and is homologous to the same muscle in larval *Ichthyophis kohtaoensis* (Kleinteich and Haas 2007).

Furthermore, there seem to be still some problems when trying to identify the homology of the branchiohyoideus and depressores branchiales muscles (e.g., constrictores branchiales; see below) (Haas 1996; Cannatella 1999). The branchiohyoideus changes during development both origin (from palatoquadrate to ceratobranchial I) and insertion (from Meckel's cartilage to ceratohyal), which increases the confusion about its homology. A hypothesis based mainly on muscle attachments is that the branchiohyoideus of salamanders is homologous to the constrictor branchialis I of frogs (Ziermann 2008). The four constrictores branchiales in anurans are innervated by the glossopharyngeal (CNIX) and vagus (CNX) nerves (Schlosser and Roth 1995). However, the branchiohyoideus is innervated by the facial nerve (CNVII) what would contradict a homology with the constrictor branchialis I (Diogo and Abdala 2010). Lightoller (1939) suggested that the branchiohyoideus could be homologous to the depressor mandibulae. Ziermann and Diogo (2013) described the development of the branchiohyoideus from the ventromedial fibers of the depressor mandibulae in the axolotl, which fits the observations in *Lissotriton helveticus* and *Ichthyosaura alpestris* (Ziermann 2008). However, as the hyoid muscles in salamanders appear almost simultaneously (Edgeworth 1935; Ericsson and Olsson 2004), it is still hard to analyze if the branchiohyoideus derives from the dorsomedial or ventral hyoid musculature. The dorsomedial origin is supported by Piatt (1935) and Ziermann and Diogo (2013).

7.4.3 Branchial Arch Muscles

The caudal **branchial arch muscles** are innervated by cranial nerves IX (glossopharyngeal nerve), X (vagus nerve), or XI (accessory spinal nerve). They are similar in all amphibians and are associated with gill movements and swallowing (Fig. 7.6; branchial and laryngeal muscles; pharyngeal muscles only evolve in mammals; see Chap. 3: Fig. 3.3). The true branchial arch muscles *sensu stricto* (Diogo and Abdala 2010) include in larval amphibians the **levatores arcuum branchialium, constrictores branchiales, transversi ventrales**, and/or **subarcuales recti**, among others (Edgeworth 1935; Kesteven 1942–45). Most branchial arch muscles degenerate during the climax of metamorphosis (Edgeworth 1935); muscles that persist are mentioned in the text below.

The **levatores arcuum branchialium** I, II, III, and IV cover the larval branchial basket dorsolateral and can be present as a flat sheet as in *Xenopus* (Fig. 7.6e; Ziermann and Olsson 2007) or, more commonly, as distinctive muscles as in *Ichthyophis*, *Ambystoma*, or *Rana* (Fig. 7.6a–d, f; Kleinteich and Haas 2007; Ziermann and Diogo 2013, 2014). In frog tadpoles and salamander larvae, they originate from the posterior side of the otic capsule and extend inferiorly to attach latero-inferiorly at the ceratobranchialia; contraction enlarges the branchial cavity which leads to a posterior streaming of water. In caecilian larvae, they all originate from the dorsal trunk fascia and extend ventrally. Edgeworth (1935) stated that the levator arcuum branchialium I is absent in caecilians. Nevertheless, in the larval *Ichthyophis*, the muscle clearly exists but does not reach the ceratobranchial I as in other amphibians but inserts onto soft tissue close to the pharynx (Wilkinson and Nussbaum 1997; Kleinteich and Haas 2007). Note that Wilkinson and Nussbaum (1997) called this muscle cephalodorsosubpharyngeus. The levator arcuum branchialium I was also described as inserting onto the lateral tip of fused ceratobranchials III and IV in the caecilian family Typhlonectidae and onto the pharyngeal wall in Caeciliidae. The levatores arcuum branchialium II, III, and IV insert onto ceratobranchials II, III, and III and IV, respectively (Kleinteich and Haas 2007). In adult amphibians, the levatores arcuum branchialium form the **petrohyoidei** (Ziermann and Diogo 2014); *Xenopus* has only one petrohyoideus (number IV), while other amphibians have three to four petrohyoidei (Duellman and Trueb 1994; Ziermann and Diogo 2014). Interestingly, in the direct-developing frog *Eleutherodactylus coqui*, the petrohyoidei develop without an earlier development of levatores arcuum branchialium (Ziermann and Diogo 2014).

The subarcuales recti and obliqui are present in all amphibians (Edgeworth 1935; Haas 2003). The **subarcualis rectus** I (ceratohyoideus internus *sensu* Drüner 1901) originates from the ceratobranchial I and inserts onto the ceratohyal and/or basihyal (Fig. 7.6b, f). The subarcualis rectus II–IV extend between ceratobranchial IV and ceratobranchial II (Fig. 7.6b, f); however, variations might exist where the muscle only extends between ceratobranchials III and I (Kleinteich and Haas 2007). In *Ichthyophis*, the subarcualis rectus I persists into adulthood (Kleinteich and Haas 2007). In anuran larvae, several variations can be found for the subarcualis recti muscles (Haas 2003).

The subarcuales obliqi II and III are present in larval amphibians (Edgeworth 1935; Pusey 1943; Haas 2003; Kleinteich and Haas 2007). The **subarcualis obliquus** II originates from the ceratobranchial II and inserts onto ceratobranchial I and basibranchial II in *Ichthyophis* larvae (Fig. 7.6b). The subarcualis obliquus III in *Ichthyophis* originates from the ceratobranchial III and inserts onto the ceratobranchial II, and some fibers merge with the contralateral muscle (Kleinteich and Haas 2007). The subarcualis obliquus II does not persist to the postmetamorphic stages in caecilians; however, the subarcualis obliquus III was described in adults (Edgeworth 1935; Wilkinson and Nussbaum 1997; Kleinteich and Haas 2007). It should be noted that the subarcuales recti terminology was in the past often used to name muscles that extend between the ceratobranchials independent of their actual position on the hyobranchial apparatus (e.g., Edgeworth 1935; Wilkinson and Nussbaum 1997). However, more recently only the lateral muscles are termed subarcualis recti, while the ventral ones are termed subarcuales obliqui (Kleinteich and Haas 2007; Ziermann and Olsson 2007).

The **constrictores branchiales** extend between two adjacent ceratobranchialia from the lateral end of the posterior one to the medial end of the anterior one (Fig. 7.6f). The **transversi ventrales** connect the ventromedial parts of the same ceratobranchialia from both sides of the branchial basket (Fig. 7.6c, f). The transversus ventralis IV in the caecilian *Ichthyophis*, for example, originates from the latero-caudal ends of ceratobranchial IV and inserts into a median raphe with its contralateral muscle, and some fibers reach over the fascia of the rectus abdominis (Kleinteich and Haas 2007). The constrictores branchiales, subarcuales recti, and transversus ventralis never develop in the direct-developing species *Eleutherodactylus coqui* (Schlosser and Roth 1997; Ziermann and Diogo 2014) and degenerate during metamorphosis in biphasic frogs (Edgeworth 1935; Ziermann and Diogo 2014).

Another true branchial muscle, but not *sensu stricto*, is the **protractor pectoralis**, which originates high on the cranium (otic capsule and/or fascia above the epaxial musculature) and inserts onto the scapula (Fig. 7.6c). The **interscapularis** is uniquely present in adult anurans and probably derived from the protractor pectoralis (Diogo and Abdala 2010).

The protractor pectoralis is present in frogs and salamanders but absent in caecilians that is likely related to the loss of the pectoral girdle (Edgeworth 1935; Carroll 2007). The protractor pectoralis is often termed **cucullaris**; however, the "true" cucullaris gives rise to the protractor pectoralis and levatores arcuum branchialium in the last common ancestor (LCA) of osteichthyans (Ziermann et al. 2014b). The evolution of the cucullaris was broadly discussed over the last decades (see Chap. 3; e.g., Kusakabe and Kuratani 2005; Noden and Francis-West 2006; Theis et al. 2010; Ericsson et al. 2013, Ziermann et al. 2014b, 2017c; Sefton et al. 2016; Naumann et al. 2017). The protractor pectoralis connects the head with the shoulder girdle.

Cell lineage tracing studies in the axolotl indicated that the protractor pectoralis and the dilatator laryngis develop also with contributions from somitic material (Piekarski and Olsson 2007), as was previously shown in quail-chick chimera. For both muscles, those findings were surprising as both should derive from the branchiomeric mesoderm (Edgeworth 1935; Piatt 1938). Studies in transgenic mice also indicate the contribution of somitic cells to the trapezius (homologous to part of protractor pectoralis) (Matsuoka et al. 2005). However, it should be pointed out that the somitic contribution to muscles does not mean that the muscles are not head muscles or more specifically branchial muscles. The somitic contribution was also shown in one specimen to the interhyoideus posterior and in two specimens to the levatores arcuum branchialium (Piekarski and Olsson 2007), and no one would argue those are trunk muscles. Furthermore, the contribution of presomitic mesoderm to the protractor pectoralis cannot be excluded due to the difficulty of marking and tracing the first somite only (Piekarski and Olsson 2007). Importantly, it was shown that the protractor pectoralis derives ontogenetically from the anlage of the levator arcuum branchialium IV (Ziermann and Diogo 2013, 2014). Evolutionary, as mentioned above, the cucullaris gives rise to the protractor pectoralis and levatores arcuum branchialium in the LCA of osteichthyans (Ziermann and Diogo 2013; Ziermann et al. 2014b, 2017c). The somitic contribution to those muscles seems therefore to be a derived character in tetrapods (Piekarski and Olsson 2007) but does not exclude the contribution of branchiomeric mesoderm. This is also supported by the expression of *Tbx1* in the trapezius (protractor pectoralis derived), which is a marker for branchiomeric muscles, whereas *Pax3* is expressed in somite-derived muscles tissue (limb muscles, hypobranchial muscles, etc.), but not in the trapezius (Theis et al. 2010; Sambasivan et al. 2011).

The laryngeal muscles are grouped into "other branchial muscles" (Diogo and Abdala 2010). The **dilatator laryngis** (Fig. 7.6c) and **constrictor laryngis** are present in all amphibians. The latter can be missing in some taxa (Duellman and Trueb 1994; Johnston 2011). The constrictor laryngis from both sides enclose the larynx. In frogs and caecilians, the constrictor laryngis can be divided into several portions that are then called constrictor laryngis ventralis and dorsalis (Kleinteich and Haas 2007; Diogo and Abdala 2010). Both parts arise from the ceratobranchial IV in *Ichthyophis*, the ventralis part from the lateral edge, and the dorsal portion from the medial edge (Kleinteich and Haas 2007). Close to the arytenoid cartilage, the fibers of both parts merge again and then attach to the arytenoid cartilage and to the trachea. The dilatator laryngis originates from the ceratobranchial IV and inserts onto the arytenoid cartilage in caecilian larvae (Kleinteich and Haas 2007).

The **laryngeus** seems only to be present in salamanders (Piatt 1938), but one of the three constrictor laryngis muscles identified by Edgeworth (1935) might correspond to the laryngeus as both muscles derive ontogenetically from the same anlage. As the protractor pectoralis the laryngeal muscles too appear to have somitic contribution in tetrapods (Noden 1983; Couly et al. 1993; Huang et al. 1999; Piekarski and Olsson 2007). As discussed above for the other muscles, also the laryngeal muscles are evolutionary and developmentally branchiomeric muscles.

7.4.4 Hypobranchial Muscles

The **hypobranchial muscles** derive from ventrolateral processes of occipital somites forming the hypoglossal chord and move ventrally and into the head region (Edgeworth 1935; Birchmeier and Brohmann 2000). They are innervated by the hypoglossal nerve (CNXII) and 1–3 cervical spinal nerves and include the infrahyoid strap muscles (e.g., sternohyoideus) and tongue muscles (Edgeworth 1935). For example, cell lineage tracing studies in the Mexican axolotl showed that the **geniohyoideus** (coracomandibularis) and **sternohyoideus** (rectus cervicis) derive from somite 2 and somites 2 and 3, respectively (Piekarski and Olsson 2007). Both muscles are invariably present in amphibians, which is the plesiomorphic and larval condition. The geniohyoideus originates from the hypobranchial plate and inserts onto the infrarostral cartilage (frog larvae) or the cartilago Meckeli (salamander larvae) and originates from the fascia of the sternohyoideus and inserts onto the pseudodentary (caecilian larvae) (Kleinteich and Haas 2007). The sternohyoideus stretches between the medial part of the diaphragm (anterior continuation of the rectus abdominis) or the sternum (adults) and the medial part of the ceratobranchial II (Fig. 7.6f) (McDiarmid and Altig 1999), or basibranchial II (Kleinteich and Haas 2007), or hyoid (adult) (Ziermann and Diogo 2014). The **omohyoideus** is only present in frogs and salamanders but might be deeply blended with the sternohyoideus in some species. It originates in most specimens from the scapula and inserts onto the hyoid in adult specimens but was described in *Xenopus* as originating from the sternum (Ziermann and Diogo 2014; Porro and Richards 2017).

With the transition from water to land, a new group of hypobranchial muscles evolved: the tongue muscles; they develop during metamorphosis and are present in most adult amphibians (Ziermann and Diogo 2014). As *Xenopus* lacks a tongue, it also has no tongue muscles. The tongue muscles include the **genioglossus** and **hyoglossus**, but the latter is absent in caecilians (Edgeworth 1935). The former muscle is absent in some salamanders, such as *Siren*. The **interradialis** is a muscle only found in salamanders and derives ontogenetically from the genioglossus (Piatt 1938).

Several studies revealed that, dependent on the species investigated, hypobranchial muscles derive from a variable number of somites. In birds, somites 2–6 contribute equally to the development of hypobranchial muscles (Couly et al. 1993; Huang et al. 1999). However, this might not be the case in species other than chicken, as several studies of hypobranchial muscles in amphibians convey in different results regarding which somites contribute to the muscles (Platt 1898; Edgeworth 1935; Piatt 1938; Piekarski and Olsson 2007). Piekarski and Olsson (2007) conclude that their results, at least for the development of the sternohyoideus, are consistent with previous results in quail-chick chimeras (Huang et al. 1999), where it was demonstrated that a single somite contributes equally to the hypobranchial musculature. This is explained by the formation of the hypoglossal cord in which the myogenic cells of different somites mix completely. Huang et al. (1999) state that this mixing is the reason why tongue muscles are derived from myogenic cells from both sides of the embryo, but Piekarski and Olsson (2007) found only unilateral contribution to the marked muscles.

7.4.5 Extraocular Muscles

Extraocular muscles (eye muscles; Fig. 7.6e) appear to form a developmental module that can evolve relatively independently from other cranial muscles in terms of their developmental timing (Ericsson et al. 2009; Ziermann et al. 2014a). However, cell lineage tracing and clonal studies in mice have shown that the mesoderm that gives rise to the eye muscles is actually closely related to the mandibular arch mesoderm (Lescroart et al. 2015). Most vertebrates have six eye muscles that are innervated by the cranial nerves III, IV, and VI. Usually, there are four rectus muscles that originate from an area around the optic nerve (CNII) and two obliquus muscles that originate from the anterior corner of the orbit. There are also the retractor bulbi and levator bulbi in adult anurans. The latter is actually a mandibular arch muscle and is innervated by the maxillary division of the trigeminal nerve (CNV2; see above).

7.5 Evo-Devo

In the neotenic salamander axolotl, only few minor changes are observed during development (Ziermann and Diogo 2013). This is to be expected as the lifestyle of the aquatic larvae and aquatic adult do not differ much (Lauder and Reilly 1988). Neotenic species retain larval characters like levatores arcuum branchialium and external gills, but they also miss some adult characters as tongue and tongue muscles. Most muscles simply increase their mass and change slightly their orientation while the adult cranium forms (Lauder and Reilly 1988). The adductor mandibulae muscles are diagonally directed in feeding larvae, while they are more vertical in feeding adults due to the backward elongation of the lower jaw during ossification (Ziermann and Diogo 2013). Another interesting observation is that in neotenic salamanders, several muscles become indistinct because of fusion or reduction during development (Ziermann and Diogo 2013). Similar data were found during the studies of humans (Diogo and Wood 2012) and frogs with different developmental modes (Ziermann and Diogo 2014). This contradicts a commonly accepted view that development is toward the differentiation of muscles.

Comparing the appearance of muscles during the ontogeny, some scientist showed that the pattern parallels the muscles' appearance in phylogeny with only a few exceptions (Diogo et al. 2008; Ziermann and Diogo 2013). Through analyses of development and morphology of many vertebrate taxa, it is possible to reconstruct the evolutionary history of those muscles (Ziermann et al. 2014b). For example, the jaw-moving adductores mandibulae appear first in the last common ancestor (LCA) of gnathostomes (jawed vertebrates) and later divide into many subunits in the LCA of osteichthyans (bony fishes including tetrapods) (Ziermann et al. 2014b). The same is observable during ontogeny: there is one adductor mandibulae anlage before this anlage divides into several subunits (Ziermann and Diogo 2013). An exception to this parallelism of ontogeny and phylogeny is the development of the dilatator laryngis in the Mexican axolotl (*Ambystoma mexicanum*), where the muscle develops early (before the intermandibularis – a mandibular arch muscle) compared to the late evolutionary appearance (after many mandibular, hyoid, and branchial arch muscles). Tongue muscles evolve later than other hypobranchial muscles and derive from the coracomandibularis muscle that is present in sarcopterygian fishes (lungfishes and *Latimeria*). Similarly, the tongue muscles in frogs also derive only late in or even after metamorphosis and seemingly from another hypobranchial muscle (genioglossus; see above).

During the past decade, it was shown that the development of the head and heart musculature differs from the myogenesis of trunk muscles and both, head and heart development, are tightly connected (Noden and Francis-West 2006; Lescroart et al. 2010, 2015; Sambasivan et al. 2011), which led to the definition of the cardiopharyngeal field (Diogo et al. 2015). Simplified this field is a mesodermal progenitor that gives rise to branchiomeric (head) and cardiac (heart) musculature. Specific gene expression patterns define the differentiation of specific branchial arch muscles and cardiac regions. Those regulatory factors include cardiogenic (*Isl1, Nkx2.5)* and branchiomeric [*Tbx1*, *MyoR* (Msc), *Capsulin* (*Tcf21*), *Pitx2*] factors (Bothe and Dietrich 2006). *Tbx1*, for example, plays a crucial role in the extension of the arterial pole of the heart but also in the activation of branchiomeric myogenesis (Castellanos et al. 2014; Rana et al. 2014). *Isl1* is also expressed in branchiomeric and cardiac progenitor cells, but not in hypobranchial or extraocular muscle precursors (Nathan et al. 2008). Another example is *Pitx2* that specifies mandibular arch mesoderm but not hyoid arch mesoderm in mice (Shih et al. 2007). In contrast to cranial muscle, the formation of trunk muscle is *Pax3*-dependent (Tajbakhsh et al. 1997). Yet, it was shown that *Pax3* is also expressed in tongue muscles (Harel et al. 2009).

It would be exceeding the purpose of this chapter to explain all the regulatory genes. However, what is important is that all those mechanisms are evolutionary conserved and therefore also regulate the head (and heart) development in amphibians. As amphibians, in particular frogs and salamanders, are already good established as laboratory species, studies of gene regulatory networks or of mutations in specific genes can shed light onto gene defects and cardiopharyngeal syndromes in humans (Ziermann et al. 2017a).

7.6 Heterochrony

Changes in the adult and larval morphology result from changes in developmental processes, which are the product of evolutionary changes of developmental mechanisms that control those developmental processes. **Allometric changes** cause differences in the size of body parts due to an extended or reduced developmental period. The skull shape changes during development in the *Leptodactylus fuscus* group (Anura: Leptodactylidae) revealed, for example, that allometric, non-heterochronic changes explain the differences observed between species (Ponssa and Candioti 2012).

The spatial pattern of tissue formation (heterotopy) or its timing can be altered. A change in the timing of developmental events is called **sequence heterochrony**, which can be important for morphological evolution (e.g., Gould 1977; Smith 2002; Bininda-Emonds et al. 2003). The patterning of cranial neural crest and cranial mesoderm cells are conserved within amphibians as is their developmental anatomy. Yet, anuran larvae show a huge variety of cranial morphology and feeding modes, ranging from the basal herbivorous and microphagous feeding mode as of the filter feeder *Xenopus* larvae to herbivorous and microphagous algae scrapers like *Rana* or *Bufo*, carnivorous suction feeders with a tube-shape mouth as *Hymenochirus* that is closely related to *Xenopus*, or megalophagous species like *Lepidobatrachus* (Ruibal and Thomas 1988; Deban and Olson 2002; Dean 2003; Ziermann and Olsson 2007; Fabrezi and Lobo 2009; Ziermann et al. 2013). The specialized morphology of the tiny *Hymenochirus* larva was shown to be related to an elongation of the skeletal structures, as Meckel's cartilage, and hyoid and hypobranchial muscles in comparison to the *Xenopus* larva development (Ericsson et al. 2009). Additionally, branchial arch muscles are reduced in the tiny predator, while the *Xenopus* larva has a well-developed large branchial basked covered by musculature (Fig. 7.6e).

The evolution of carnivorous larvae in terminal taxa was correlated with **heterochronic** changes in patterning of head and skeletal muscle development (Ericsson et al. 2009). However, an examination of a wide range of anurans with different feeding modes and their cranial muscle development revealed no obvious correlation between feeding mode and heterochronic changes in basal branches (i.e., family level or higher) of phylogeny as the majority of changes occurred in terminal branches (i.e., on species level; Ziermann et al. 2014a).

The comparison of the ossification patterns in a population of *Xenopus* larvae with previously published studies showed that the timing but not the sequence of ossification is variable (Trueb and Hanken 1992). But, it should be noted that differences in source (wild-catch vs. lab-reared specimens), rearing (temperature, population density, etc.), and analytical methods (cleared-and-stained vs. serial sectioned specimens) can result in different developmental descriptions within the same species (Trueb and Hanken 1992). Different staging methods also make comparisons more complicated (length vs. age vs. external character assessment), in particular as specimens develop at different speed and have different sizes in particular stages, both dependent on extrinsic and intrinsic factors. Yet, the assessment of ossification sequences in *Xenopus,* which is a phylogenetic basal frog, revealed that the sequence is more similar to the pelobatid *Spea*, which is also relatively basal, than to neobatrachian species (e.g., *Bufo*, *Rana*), which are phylogenetically derived frogs (Trueb and Hanken 1992). The comparison between the ossification patterns shows variability between all species (Trueb and Hanken 1992), indicating a variation in developmental sequence on terminal branches rather than on nonterminal ones (i.e., not species or even genus level), as was observed for the cranial muscle developmental sequences (Ziermann et al. 2014a).

In contrast to the cranium and cranial muscle studies mentioned above, the analysis of the development of 21 embryonic and larval characters in toads (Anura: Bufonidae) revealed changes that can be best explained by phylogeny (e.g., embryos with kyphotic body curvature, Type C adhesive glands, and a tiny third pair of gills occur in early divergent taxa of bufonids), while other changes seem to be correlated with the reproductive mode (e.g., embryos developing in phytotelmata, i.e., small water-filled cavities in terrestrial plants, hatch late and have an accelerated hind limb development as compared to embryos developing in streams or ponds) (Vera Candioti et al. 2016).

In the direct-developing frog *Eleutherodactylus coqui*, the developmental sequence of cranial muscles also differs from that of other frogs in that the hypobranchial muscles develop earlier than in other species (Ziermann and Diogo 2014). Callery and Elinson (2000) also suggested that heterochronic shifts during development cause the boundary between embryonic and metamorphic phase to blur. Furthermore, the limb bud (Hanken et al. 2001) and spinal cord (Schlosser 2003) development seems to be accelerated in *E. coqui*. However, cranial nerve orientation (Schlosser and Roth 1997) and changes in cranial muscles, cartilages, and bones are similar to changes that occur in other frogs during metamorphosis (Hanken et al. 1992; Ziermann and Diogo 2014). Most differences between *E. coqui* and other frogs can be related to a very fast developmental mode, and when *E. coqui* hatches, most cranial structures resemble those of adult frogs. Studies of other direct-developing species, like *Oreobates barituensis* (Anura: Strabomantidae), revealed that morphological changes are similar to those observed in other Neotropical direct-developing species, including the absence of several embryonic and larval characters (e.g., external gills) (Goldberg et al. 2012). Heterochronic changes, like early developing limbs, seem to be a common trait in direct-developing species.

7.7 Conclusion and Further Reading

Amphibians are well studied, but many open questions remain that will provide exciting research opportunities for a new generation of herpetologists, comparative anatomists, and developmental and evolutionary biologists. Recent advances in genetic manipulation

(CRISPR, TALENs, etc.) and the easy access of developmental stages in amphibians make them also an ideal model organism to study developmental processes as well as pathology. This is already a helpful tool to understand our own evolution and pathology and might in future reveal even further mechanism to shed light onto complex human syndromes.

From the extensive literature list provided here, I would like to point out some books that are an amazing source to study amphibian morphology and ecology. Duellman and Trueb (1994) cover the biology and diversity of amphibians, while McDiarmid and Altig (1999) focus on amphibian tadpoles. Both cover not only head structures but also postcranial elements, ecology, and many other topics.

Acknowledgment The helpful review from Virginia Abdala improved the quality of the text.

References

Alberch P, Gale EA (1986) Pathways of cytodifferentiation during the metamorphosis of the epibranchial cartilage in the salamander *Eurycea bislineata*. Dev Biol 117:233–244

Alberch P, Lewbart G, Gale EA (1985) The fate of larval chondrocytes during the metamorphosis of the epibranchial in the salamander, *Eurycea bislineata*. Development 88:71–83

Alcalde L, Barg M (2006) Chondrocranium and cranial muscle morphology in *Lysapsus* and *Pseudis* tadpoles (Anura: Hylidae: Hylinae). Acta Zool 87:91–100

Alley KE (1989) Myofiber turnover is used to retrofit frog jaw muscles during metamorphosis. Am J Anat 184:1–12

Alley KE, Omerza FF (1998) Neuromuscular remodeling and myofiber turnover in *Rana pipiens*' jaw muscles. Cell Tissues Organs 164:46–58

Altig R, Johnston GF (1989) Guilds of anuran larvae: relationships among developmental modes, morphologies, and habitats. Herpetol Monogr 3:81–109

AmphibiaWeb (2018) University of California, Berkeley. https://amphibiaweb.org. Accessed 14 Feb 2018

Bauer WJ (1992) A contribution to the morphology of the m. interhyoideus posterior (VII) of urodele Amphibia. Zool Jb Anat 122:129–139

Bauer WJ (1997) A contribution to the morphology of visceral jaw-opening muscles of urodeles (Amphibia: Caudata). J Morphol 233:77–97

Bemis WE, Schwenk K, Wake M (1983) Morphology and function of the feeding apparatus in *Dermophis mexicanus* (Amphibia: Gymnophiona). Zool J Linnean Soc 77:75–96

Bininda-Emonds ORP, Jeffery JE, Richardson MK (2003) Is sequence heterochrony an important evolutionary mechanism in mammals? J Mamm Evol 10:335–361

Birchmeier C, Brohmann H (2000) Genes that control the development of migrating muscle precursor cells. Curr Opin Cell Biol 12:725–730

Bothe I, Dietrich S (2006) The molecular setup of the avian head mesoderm and its implication for craniofacial myogenesis. Dev Dyn 235:2845–2860

Callery EM, Elinson RP (2000) Thyroid hormone-dependent metamorphosis in a direct developing frog. PNAS 97:2615–2620

Callery EM, Fang H, Elinson RP (2001) Frogs without polliwogs: evolution of anuran direct development. BioEssays 23:233–241

Cannatella DC (1999) 4. Architecture: cranial and axial musculoskeleton. In: McDiarmid RW, Altig R (eds) Tadpoles—the biology of anuran larvae, vol 1. The University of Chicago Press, Chicago, pp 52–81

Carroll RL (2007) The Palaeozoic ancestry of salamanders, frogs and caecilians. Zool J Linnean Soc 150:1–140

Carroll RL, Holmes R (1980) The skull and jaw musculature as guides to the ancestry of salamanders. Zool J Linnean Soc 68:1–40

Castellanos R, Xie Q, Zheng D, Cvekl A, Morrow BE (2014) Mammalian TBX1 preferentially binds and regulates downstream targets via a tandem T-site repeat. PLoS One 9:e95151

Chanoine C, Hardy S (2003) *Xenopus* muscle development: from primary to secondary myogenesis. Dev Dyn 226:12–23

Cihak R, Kralovec K, Rocek Z (2002) Developmental origin of the frontoparietal bone in *Bombina variegata* (Anura: Discoglossidae). J Morphol 255:122–129

Couly GF, Coltey PM, Douarin NML (1993) The triple origin of skull in higher vertebrates: a study in quail-chick chimeras. Development 117:409–429

Couly G, Creuzet S, Bennaceur S, Vincent C, Douarin NML (2002) Interactions between Hox-negative cephalic neural crest cells and the foregut endoderm in patterning the facial skeleton in the vertebrate head. Development 129:1061–1073

Creuzet S, Couly G, Douarin NML (2005) Patterning the neural crest derivatives during development of the vertebrate head: insights from avian studies. J Anat 207:447–459

Dean MN (2003) Suction feeding in the Pipid frog, *Hymenochirus boettgeri*: kinematic and behavioral considerations. Copeia 4:879–886

Deban SM, Olson WM (2002) Suction feeding by a tiny predatory tadpole. Nature 420:41–42

Deban SM, O'Reilly JC, Nishikawa KC (2001) The evolution of the motor control of feeding in amphibians. Am Zool 41:1280–1298

Diogo R, Abdala V (2010) Muscles of vertebrates—comparative anatomy, evolution, homologies and development. CRC Press, Enfield

Diogo R, Wood B (2012) Violation of Dollo's law: evidence of muscle reversions in primate phylogeny and their implications for the understanding of the ontogeny, evolution, and anatomical variations of modern humans. Evolution 66:3267–3276

Diogo R, Hinits Y, Hughes SM (2008) Development of mandibular, hyoid and hypobranchial muscles in the zebrafish: homologies and evolution of these muscles within bony fishes and tetrapods. BMC Dev Biol 8:1–22

Diogo R, Kelly RG, Christiaen L, Levine M, Ziermann JM, Molnar JL, Noden DM, Tzahor E (2015) A new heart for a new head in vertebrate cardiopharyngeal evolution. Nature 520:466–473

Drüner L (1901) Zungenbein-, Kiemenbogen- und Kehlkopf-Skelet, -Muskeln und Nerven von Siredon (Larvenform). Studien zur Anatomie der Zungenbein-, Kiemenbogen-und Kehlkopfmusculatur der Urodelen. 2. T. Z. J Anat 468–593

Duellman WE, Trueb L (1994) Biology of Amphibians. Johns Hopkins University Press, Baltimore

Dulcey Cala CJ, Tarazona OA, Ramìrez-Pinilla MP (2009) The morphology and post-hatching development of the skull of *Bolitoglossa nicefori* (Caudata: Plethodontidae): developmental implications of recapitulation and repatterning. Zoology 112:227–239

Edgeworth FH (1935) The cranial muscles of vertebrates. Cambridge at the University Press, London

Elinson RP (2013) Metamorphosis in a frog that does not have a tadpole. In: Shi Y-B (ed) Current topics in developmental biology, vol 103. Academic Press, Burlington, pp 259–276

Ericsson R, Olsson L (2004) Patterns of spatial and temporal visceral arch muscle development in the Mexican Axolotl (*Ambystoma mexicanum*). J Morphol 261:131–140

Ericsson R, Cerny R, Falck P, Olsson L (2004) The role of cranial neural crest cells in visceral arch muscle positioning and morphogenesis in the Mexican axolotl, *Ambystoma mexicanum*. Dev Dyn 231:237–247

Ericsson R, Joss J, Olsson L (2008) The fate of cranial neural crest cells in the Australian lungfish (*Neoceratodus forsteri*). J Exp Zool B Mol Dev Evol 310:345–354

Ericsson R, Ziermann JM, Piekarski N, Schubert G, Joss J, Olsson L (2009) Cell fate and timing in the evolution of neural crest and mesoderm development in the head region of amphibians and lungfishes. Acta Zool 90:264–272

Ericsson R, Knight R, Johanson Z (2013) Evolution and development of the vertebrate neck. J Anat 222:67–78

Fabrezi M, Lobo F (2009) Hyoid skeleton, its related muscles, and morphological novelties in the frog *Lepidobatrachus* (Anura, Ceratophryidae). Anat Rec 292:1700–1712

Falck P, Joss J, Olsson L (2000) Cranial neural crest cell migration in the Australian lungfish, *Neoceratodus forsteri*. Evol Dev 2:179–185

Fan X, Loebel DA, Bildsoe H, Wilkie EE, Qin J, Wang J, Tam PP (2016) Tissue interactions, cell signaling and transcriptional control in the cranial mesoderm during craniofacial development. AIMS Genet 3(1):74–98

Ford LS, Cannatella DC (1993) The major clades of frogs. Herpetol Monogr 7:94–117

Fox H (1959) A study of the development of the head and pharynx of the larval Urodele *Hynobius* and its bearing on the evolution of the vertebrate head. Philos Trans R Soc Lond 242:151–204

Francis ETB (1934) IV The muscles. In: The anatomy of the salamander. Oxford University Press, London, pp 48–75

Frost DR, Grant T, Faivovich J, Bain RH, Haas A, Haddad CFB, de Sá RO, Channing A, Wilkinson M, Donnellan SC, Raxworthy CJ, Campbell JA, Blotto BL, Moler P, Drewes RC, Nussbaum RA, Lynch JD, Green DM, Wheeler WC (2006) The amphibian tree of life. Bull Am Mus Nat Hist 297:370–371

Gegenbaur C (1878) Elements of comparative anatomy. Macmillan, London

Goldberg J, Candioti FV, Akmentins MS (2012) Direct-developing frogs: ontogeny of *Oreobates barituensis* (Anura: Terrarana) and the development of a novel trait. Amphibia-Reptilia 33:239–250

Goodrich ES (1930) Studies on the structure and development of vertebrates. Dover, London

Gould SJ (1977) Ontogeny and phylogeny. Harvard University Press, Cambridge, MA

Graham A, Smith A (2001) Patterning the pharyngeal arches. BioEssays 23:54–61

Gross JB, Hanken J (2005) Cranial neural crest contributes to the bony skull vault in adult *Xenopus laevis*: insights from cell labeling studies. J Exp Zool B Mol Dev Evol 304:1–8

Gross JB, Hanken J (2008). Segmentation of the vertebrate skull: neural-crest derivation of adult cartilages in the clawed frog, *Xenopus laevis*. Annual meeting of the Society for Integrative and Comparative Biology, p 1–16

Haas A (1996) Das larvale Cranium von *Gastrotheca riobambae* und seine Metamorphose (Amphibia, Anura, Hylidae), vol 36. Verhandlungen des naturwissenschaftlichen Vereins, Hamburg, pp 33–162

Haas A (2001) Mandibular arch musculature of anuran tadpoles; with comments on homologies of amphibian jaw muscles. J Morphol 247:1–33

Haas A (2003) Phylogeny of frogs as inferred from primarily larval characters (Amphibia: Anura). Cladistics 19:23–89

Haas A, Richards SJ (1998) Correlations of cranial morphology, ecology, and evolution in Australian suctorial tadpoles of the Genera *Litoria* and *Nyctimystes* (Amphibia: Anura: Hylidae: Pelodryadinae). J Morph 238:109–141

Hall BK (1980) Tissue interactions and the initiation of osteogenesis and chondrogenesis in the neural crest-derived mandibular skeleton of the embryonic mouse as seen in isolated murine tissues and in recombinations of murine and avian tissues. Development 58:251–264

Hanken J, Gross JB (2005) Evolution of cranial development and the role of neural crest: insights from amphibians. J Anat 207:437–446

Hanken J, Klymkowsky MW, Summers CH, Seufert DW, Ingebrigsten N (1992) Cranial ontogeny in the direct-

developing frog, *Eleutherodactylus coqui* (Anura: Leptodactylidae), analysed using whole-mount immunohistochemistry. J Morphol 211:95–118
Hanken J, Klymkowsky MW, Alley KE, Jennings DH (1997) Jaw muscle development as evidence for embryonic repatterning in direct-developing frogs. Proc R Soc Lond B 264:1349–1354
Hanken J, Carl TF, Richardson MK, Olsson L, Schlosser G, Osabutey CK, Klymkowsky MW (2001) Limb development in a "nonmodel" vertebrate, the direct-developing frog *Eleutherodactylus coqui*. J Exp Zool B Mol Dev Evol 291:375–388
Harel I, Nathan E, Tirosh-Finkel L, Zigdon H, Guimaraes-Camboa N, Evans SM, Tzahor E (2009) Distinct origins and genetic programs of head muscle satellite cells. Dev Cell 16:822–832
Helms JA, Codero D, Tapadia MD (2005) New insights into craniofacial morphogenesis. Development 132:851–861
Hillis DM (1991) The phylogeny of amphibians: current knowledge and the role of cytogenetics. In: Amphibian cytogenetics and evolution. Academic Press, San Diego, pp 7–31
Hirasawa T, Kuratani S (2015) Evolution of the vertebrate skeleton: morphology, embryology, and development. Zool Lett 1:2
Huang R, Zhi Q, Izpisua-Belmonte J-C, Christ B, Patel K (1999) Origin and development of the avian tongue muscles. Anat Embryol 200:137–152
Huxley TH (1858) On the theory of the vertebrate skull. The Croonian lecture. Proc. Roy. Soc., London
Iordansky NN (1992) Jaw muscles of the Urodela and Anura: some features of development, functions, and homology. Zool Jb Anat 122:225–232
Iordansky NN (1996) Evolution of the musculature of the jaw apparatus in the Amphibia. Advances in Amphibian Research in the Former Soviet Union 1:3–26
Jiang X, Iseki S, Maxson RE, Sucov HM, Morriss-Kay GM (2002) Tissue origins and interactions in the mammalian skull vault. Dev Biol 241:106–116
Johansen K, Hanson D (1968) Functional anatomy of the hearts of lungfishes and amphibians. Am Zool 8:191–210
Johnston P (2011) Cranial muscles of the anurans *Leiopelma hochstetteri* and *Ascaphus truei* and the homologies of the mandibular adductors in Lissamphibia and other gnathostomes. J Morphol 272:1492–1512
Kesteven HL (1942–45) The evolution of the skull and the cephalic muscles: a comparative study of their development and adult morphology. Part I. The fishes. Australian Museum Memoir 8:1–63
Kleinteich T, Haas A (2007) Cranial musculature in the larva of the caecilian, *Ichthyophis kohtaoensis* (Lissamphibia: Gymnophiona). J Morph 268:74–88
Kundrát M, Joss JM, Smith MM (2008) Fate mapping in embryos of *Neoceratodus forsteri* reveals cranial neural crest participation in tooth development is conserved from lungfish to tetrapods. Evol Dev 10:531–536
Kuratani S, Adachi N, Wada N, Oisi Y, Sugahara F (2013) Developmental and evolutionary significance of the mandibular arch and prechordal/premandibular cranium in vertebrates: revising the heterotopy scenario of gnathostome jaw evolution. J Anat 222:41–55
Kusakabe R, Kuratani S (2005) Evolution and developmental patterning of the vertebrate skeletal muscles: perspectives from the lamprey. Dev Dyn 234:824–834
Lauder GV, Reilly SM (1988) Functional design of the feeding mechanism in salamanders: causal bases of ontogenetic changes in function. J Exp Biol 134:219–233
Lescroart F, Kelly RG, Le Garrec J-F, Nicolas J-F, Meilhac SM, Buckingham M (2010) Clonal analysis reveals common lineage relationships between head muscles and second heart field derivatives in the mouse embryo. Development 137:3269–3279
Lescroart F, Hamou W, Francou A, Théveniau-Ruissy M, Kelly RG, Buckingham M (2015) Clonal analysis reveals a common origin between nonsomite-derived neck muscles and heart myocardium. Proc Natl Acad Sci 112:1446–1451
Lightoller G (1939) Probable homologues. A study of the comparative anatomy of the mandibular and hyoid arches and their musculature. Part I. Comparative myology. Trans Zool Soc London 24:349–402
Lubosch W (1914) Vergleichende Anatomie der Kaumuskeln der Wirbeltiere, in fünf Teilen. Erster Teil: Die Kaumuskeln der Amphibien. Jen Z Naturwissenschaften 53:51–188
Luther A (1914) Über die vom N. trigeminus versorgte Muskulatur der Amphibien mit einem vergleichenden Ausblick über den Adductor mandibulae der Gnathostomen, und einem Beitrag zum Verständnis der Organisation der Anurenlarven. Acta Societatis Scientiarum Fennicæ 7:1–151
Lynn WG (1961) Types of amphibian metamorphosis. Am Zool 1:151–161
Manzano A, Abdala V (2003) The depressor mandibulae muscle in Anura. Alytes 20:93–131
Matsuoka T, Ahlberg PE, Kessaris N, Iannarelli P, Dennehy U, Richardson WD, McMahon AP, Koentges G (2005) Neural crest origins of the neck and shoulder. Nature 436:347–355
McBratney-Owen B, Iseki S, Bamforth S, Olsen B, Morriss-Kay G (2008) Development and tissue origins of the mammalian cranial base. Dev Biol 322:121–132
McClearn D, Noden DM (1988) Ontogeny of architectural complexity in embryonic quail visceral arch muscles. Am J Anat 183:277–293
McDiarmid RW, Altig R (1999) Tadpoles: the biology of anuran larvae. University of Chicago Press, Chicago, IL
Mickoleit G (2004) Phylogenetische Systematik der Wirbeltiere. Verlag Dr. Friedrich Pfeil, München
Miller CT, Yelon D, Stainier DYR, Kimmel CB (2003) Two *endothelin 1* effectors, *hand2* and *bapx1*, pattern ventral pharyngeal cartilage and the jaw joint. Development 130:1353–1365
Müller H (2006) Ontogeny of the skull, lower jaw, and hyobranchial skeleton of *Hypogeophis rostratus*

(Amphibia: Gymnophiona: Caeciliidae) revisited. J Morphol 267:968–986

Müller H, Oommen OV, Bartsch P (2005) Skeletal development of the direct-developing caecilian *Gegeneophis ramaswamii* (Amphibia: Gymnophiona: Caeciliidae). Zoomorphology 124:171–188

Nathan E, Monovich A, Tirosh-Finkel L, Harrelson Z, Rousso T, Rinon A, Harel I, Evans SM, Tzahor E (2008) The contribution of Islet1-expressing splanchnic mesoderm cells to distinct branchiomeric muscles reveals significant heterogeneity in head muscle development. Development 135:647–657

Naumann B, Warth P, Olsson L, Konstantinidis P (2017) The development of the cucullaris muscle and the branchial musculature in the Longnose Gar (*Lepisosteus osseus*, Lepisosteiformes, Actinopterygii) and its implications for the evolution and development of the head/trunk interface in vertebrates. Evol Dev 19(6):263–276

Nishikawa A, Hayashi H (1995) Spatial, temporal and hormonal regulation of programmed muscle cell death during metamorphosis of the frog *Xenopus laevis*. Differentiation 59:207–214

Noden DM (1978) The control of avian cephalic neural crest cytodifferentiation: I. Skeletal and connective tissues. Dev Biol 67:296–312

Noden DM (1983) The role of the neural crest in patterning of avian cranial skeletal, connective, and muscle tissues. Dev Biol 96:144–165

Noden DM, Francis-West P (2006) The differentiation and morphogenesis of craniofacial muscles. Dev Dyn 235:1194–1218

Nussbaum RA (1983) The evolution of a unique dual jaw-closing mechanism in caecilians (Amphibia: Gymnophiona) and its bearing on caecilian ancestry. J Zool 199:545–554

O'Reilly JC, Deban SM, Nishikawa KC (2002) Derived life history characteristics constrain the evolution of aquatic feeding behavior in adult amphibians. In: Aerts P, D'Août K, Herrel A, Van Damme R (eds) Topics in functional and ecological vertebrate morphology. Shaker, Maastricht, pp 153–190

Olsson L, Hanken J (1996) Cranial neural-crest migration and chondrogenic fate in the oriental fire-bellied toad *Bombina orientalis*: defining the ancestral pattern of head development in anuran amphibians. J Morphol 229:105–120

Olsson L, Falck P, Lopez K, Cobb J, Hanken J (2001) Cranial neural crest cells contribute to connective tissue in cranial muscles in the anuran amphibian, *Bombina orientalis*. Dev Biol 237:354–367

Olsson L, Ericsson R, Cerny R (2005) Vertebrate head development: segmentation, novelties, and homology. Theory Biosci 124:145–163

Pasqualetti M, Ori M, Nardi I, Rijli FM (2000) Ectopic *Hoxa2* induction after neural crest migration results in homeosis of jaw elements in *Xenopus*. Development 127:5367–5378

Piatt J (1935) A comparative study of the hyobranchial apparatus and throat musculature in the Plethodontidae. J Morphol 57:213–251

Piatt J (1938) Morphogenesis of the cranial muscles of *Ambystoma punctatum*. J Morphol 63:531–587

Piekarski N, Olsson L (2007) Muscular derivatives of the cranial most somites revealed by long-term fate mapping in the Mexican axolotl (*Ambystoma mexicanum*). Evol Dev 9:566–578

Piekarski N, Gross JB, Hanken J (2014) Evolutionary innovation and conservation in the embryonic derivation of the vertebrate skull. Nat Commun 5:5661 9pp

Platt JB (1898) The development of the cartilaginous skull and of the branchial and hypoglossal musculature in *Necturus*. Morphologisches Jahrbuch 25:377–463

Ponssa ML, Candioti MFV (2012) Patterns of skull development in anurans: size and shape relationship during postmetamorphic cranial ontogeny in five species of the *Leptodactylus fuscus* Group (Anura: Leptodactylidae). Zoomorphology 131:349–362

Porro LB, Richards CT (2017) Digital dissection of the model organism *Xenopus laevis* using contrast-enhanced computed tomography. J Anat 231:169–191

Pusey HK (1943) On the head of the liopelmid frog, *Ascaphus truei*. I. The chondrocranium, jaws, arches, and muscles of a partly-grown larva. Quart J Micr Sci 84:105–195

Pyron RA, Wiens JJ (2011) A large-scale phylogeny of Amphibia including over 2800 species, and a revised classification of extant frogs, salamanders, and caecilians. Mol Phylogenet Evol 61:543–583

Rana MS, Théveniau-Ruissy M, De Bono C, Mesbah K, Francou A, Rammah M, Domínguez JN, Roux M, Laforest B, Anderson RH, Mohun T, Zaffran S, Christoffels VM, Kelly RG (2014) Tbx1 coordinates addition of posterior second heart field progenitor cells to the arterial and venous poles of the heart. Circ Res 115:790–799

Reilly SM (1987) Ontogeny of the Hyobranchial apparatus in the salamanders *Ambystoma talpoideum* (Ambystomatidae) and *Notophthalmus viridescens* (Salamandridae): the ecological morphology of two neotenic strategies. J Morphol 191:205–214

Reisoli E, De Lucchini S, Nardi I, Ori M (2010) Serotonin 2B receptor signaling is required for craniofacial morphogenesis and jaw joint formation in *Xenopus*. Development 137:2927–2937

Reiss JO (1997) Early development of chondrocranium in the tailed frog *Ascaphus truei* (Amphibia: Anura): implications for anuran palatoquadrate homologies. J Morphol 231:63–100

Reiss JO (2002) The phylogeny of amphibian metamorphosis. Zoology 105:85–96

Rijli FM, Gavalas A, Chambon P (1998) Segmentation and specification in the branchial region of the head: the role of the *Hox* selector genes. Int J Dev Biol 42:393–401

Rinon A, Lazar S, Marshall H, Büchmann-Møller S, Neufeld A, Elhanany-Tamir H, Taketo MM, Sommer L, Krumlauf R, Tzahor E (2007) Cranial neural crest cells regulate head muscle patterning and differentiation during vertebrate embryogenesis. Development 134:3065–3075

Rocek Z (1989) Developmental patterns of the ethmoidal region of the anuran skull. In: Fortschritte der Zoologie/progress in zoology, Splechtna and Hilgers, vol 35. Gustav Fischer, Stuttgart, pp 412–415

Ruibal R, Thomas E (1988) The obligate carnivorous larvae of the frog, *Lepidobatrachus laevis* (Leptodactylidae). Copeia 1988(3):591–604

Sadaghiani B, Thiébaud CH (1987) Neural crest development in the *Xenopus laevis* embryo, studied by interspecific transplantation and scanning electron microscopy. Dev Biol 124:91–110

Sambasivan R, Kuratani S, Tajbakhsh S (2011) An eye on the head: the development and evolution of craniofacial muscles. Development 138:2401–2415

Santagati F, Rijli FM (2003) Cranial neural crest and the building of the vertebrate head. Nat Rev Neurosci 4:806–820

Sato T, Kurihara Y, Asai R, Kawamura Y, Tonami K, Uchijima Y, Heude E, Ekker M, Levi G, Kurihara H (2008) An endothelin-1 switch specifies maxillomandibular identity. Proc Natl Acad Sci U S A 105:18806–18811

Sauka-Spengler T, Bronner-Fraser M (2008) A gene regulatory network orchestrates neural crest formation. Nat Rev Mol Cell Biol 9:557

Schilling TF, Kimmel CB (1997) Musculoskeletal patterning in the pharyngeal segments of the zebrafish embryo. Development 124:2945–2960

Schlosser G (2003) Mosaic evolution of neural development in anurans: acceleration of spinal cord development in the direct developing frog *Eleutherodactylus coqui*. Anat Embryol 206:215–227

Schlosser G, Roth G (1995) Nerves in tadpoles of *Discoglossus pictus*: distribution of cranial and rostral spinal nerves in tadpoles of the frog *Discoglossus pictus* (Discoglossidae). J Morph 226:189–212

Schlosser G, Roth G (1997) Evolution of nerve development in Frogs II: modified development of the peripheral nervous system in the direct-developing frog *Eleutherodactylus coqui* (Leptodactylidae). Brain Behav Evol 50:94–128

Schneider RA (1999) Neural crest can form cartilages normally derived from mesoderm during development of the avian head skeleton. Dev Biol 208:441–455

Sedra SN, Michael IM (1957) The development of the skull, visceral arches, larynx and visceral muscles of the South African clawed toad, *Xenopus laevis* (Daudin) during the process of metamorphosis (from Stage 55 to Stage 66). Amsterdam, Verhandelingen der Koninklijke Nederlandse Akademie van Wetenschappen, AFD. Natuurkunde

Sefton EM, Piekarski N, Hanken J (2015) Dual embryonic origin and patterning of the pharyngeal skeleton in the axolotl (*Ambystoma mexicanum*). Evol Dev 17:175–184

Sefton EM, Bhullar B-AS, Mohaddes Z, Hanken J (2016) Evolution of the head-trunk interface in tetrapod vertebrates. elife 5:e09972

Shih HP, Gross MK, Kioussi C (2007) Cranial muscle defects of Pitx2 mutants result from specification defects in the first branchial arch. Proc Natl Acad Sci 104:5907–5912

Slater BJ, Liu KJ, Kwan MD, Quarto N, Longaker MT (2009) Cranial osteogenesis and suture morphology in *Xenopus laevis*: a unique model system for studying craniofacial development. PLoS One 4:e3914

Smith KK (2002) Sequence heterochrony and the evolution of development. J Morphol 252:82–97

Sokol OM (1969) Feeding in the pipid frog *Hymenochirus boettgeri* (Tornier). Herpetologica 25:9–24

Tajbakhsh S, Rocancourt D, Cossu G, Buckingham M (1997) Redefining the genetic hierarchies controlling skeletal myogenesis: *Pax-3* and *Myf-5* act upstream of *MyoD*. Cell 89:127–138

Tata JR (2006) Amphibian metamorphosis as a model for the developmental actions of thyroid hormone. Mol Cell Endocrinol 246:10–20

Taylor EH (1969) Skulls of gymnophiona and their significance in the taxonomy of the group. University of Kansas Publications, Lawrence, Kan

Theis S, Patel K, Valasek P, Otto A, Pu Q, Harel I, Tzahor E, Tajbakhsh S, Christ B, Huang R (2010) The occipital lateral plate mesoderm is a novel source for vertebrate neck musculature. Development 137:2961–2971

Trainor PA, Krumlauf R (2001) *Hox* genes, neural crest cells and branchial arch patterning. Curr Opin Cell Biol 13:698–705

Trueb L, Hanken J (1992) Skeletal Development in *Xenopus laevis* (Anura: Pipidae). J Morphol 214:1–41

Vera Candioti MF (2005) Morphology and feeding in tadpoles of *Ceratophrys cranwelli* (Anura: Leptodactylidae). Acta Zool 86:1–11

Vera Candioti F, Grosso J, Haad B, Pereyra MO, Bornschein MR, Borteiro C, Costa P, Kolenc F, Pie MR, Proaño B (2016) Structural and heterochronic variations during the early ontogeny in toads (Anura: Bufonidae). Herpetol Monogr 30:79–118

Visser MHC (1963) The cranial morphology of *Ichthyophis glutinosus* (Linné) and *Ichthyophis monochrous* (Bleeker). Ann Univ Stellenbosch A 38:67–102

Wada N, Nohno T, Kuratani S (2011) Dual origins of the prechordal cranium in the chicken embryo. Dev Biol 356:529–540

Wake MH, Hanken J (1982) Development of the skull of *Dermophis mexicanus* (Amphibia: Gymnophiona), with comments on skull kinesis and amphibian relationships. J Morphol 173:203–223

Wake MH, Exbrayat J-M, Delsol M (1985) The development of the chondrocranium of *Typhlonectes compressicaudus* (Gymnophiona), with comparison to other species. J Herpetol 19:68–77

Weisbecker V, Mitgutsch C (2010) A large-scale survey of heterochrony in anuran cranial ossification patterns. J Zool Syst Evol Res 48:332–347

Wilkinson M, Nussbaum RA (1997) Comparative morphology and evolution of the lungless caecilian *Atretochoana eiselti* (Taylor) (Amphibia: Gymnophiona: Typhlonectidae). Biol J Linn Soc 62:39–109

Ziermann JM (2008) Evolutionäre Entwicklung larvaler Cranialmuskulatur der Anura und der Einfluss von Sequenzheterochronien. Dr. PhD, Friedrich Schiller University Jena

Ziermann JM, Diogo R (2013) Cranial muscle development in the model organism *Ambystoma mexicanum*: implications for tetrapod and vertebrate comparative and evolutionary morphology and notes on ontogeny and phylogeny. Anat Rec 296:1031–1048

Ziermann JM, Diogo R (2014) Cranial muscle development in frogs with different developmental modes: direct development vs. biphasic development. J Morphol 275:398–413

Ziermann JM, Olsson L (2007) Patterns of spatial and temporal cranial muscle development in the African clawed frog, *Xenopus laevis* (Anura: Pipidae). J Morphol 268:791–804

Ziermann JM, Infante C, Hanken J, Olsson L (2013) Morphology of the cranial skeleton and musculature in the obligate carnivorous tadpole of *Lepidobatrachus laevis* (Anura: Ceratophryidae). Acta Zool 94:101–112

Ziermann JM, Mitgutsch C, Olsson L (2014a) Analyzing developmental sequences with Parsimov—a case study of cranial muscle development in anuran larvae. J Exp Zool B Mol Dev Evol 322B:584–604

Ziermann JM, Miyashita T, Diogo R (2014b) Cephalic muscles of cyclostomes (hagfishes and lampreys) and Chondrichthyes (sharks, rays and holocephalans): comparative anatomy and early evolution of the vertebrate head muscles. Zool J Linnean Soc 172:771–802

Ziermann JM, Fahimuddin F, Forrester A, Singh S (2017a) The cardiopharyngeal field in the light of evolutionary medicine—implications for human syndromes. J Hum Anat 1:10 https://medwinpublishers.com/JHUA/JHUA16000110.pdf

Ziermann JM, Clement AM, Ericsson R, Olsson L (2017b) Cephalic muscle development in the Australian lungfish, *Neoceratodus forsteri*. J Morphol 279:494. https://doi.org/10.1002/jmor.20784

Ziermann JM, Freitas R, Diogo R (2017c) Muscle development in the shark *Scyliorhinus canicula*: implications for the evolution of the gnathostome head and paired appendage musculature. Front Zool 14:1–17. https://doi.org/10.1186/s12983-12017-10216-y.

Ziermann JM, Diogo R, Noden DM (2018) Neural crest and the patterning of vertebrate craniofacial muscles. Genesis J Genet Dev 56:e23097

8 Evolution, Diversity, and Development of the Craniocervical System in Turtles with Special Reference to Jaw Musculature

Gabriel S. Ferreira and Ingmar Werneburg

8.1 Origin of Turtles and Their Cranial Anatomy

The phylogenetic position of turtles among amniotes has been a highly debated issue for the last 150 years (Rieppel 2008; Joyce 2015) and much of this controversy is related to the greatly modified body plan of these reptiles, especially in their postcranium (Scheyer et al. 2013). The body of turtles is entirely encapsulated inside a bony shell, and, consequently, most anatomical parts, such as their limbs, girdles, and their respiratory system, were greatly modified with the emergence of this structure (Nagashima et al. 2012). Several lines of investigation, including paleontological (Li et al. 2008; Lyson et al. 2010, 2013, 2014, 2016; Lyson and Joyce 2012; Schoch and Sues 2015) and developmental studies (Burke 1989; Clark et al. 2001; Gilbert et al. 2001, 2007, 2008; Loredo et al. 2001; Cebra-Thomas et al. 2005; Nagashima et al. 2007, 2009, 2013, 2015; Kuratani et al. 2011; Rieppel 2013; Cordero and Quinteros 2015), contributed to this debate, and now we have a comprehensive scenario for the origin of the shell from a phylogenetic and ontogenetic perspectives (Nagashima et al. 2012; Joyce 2015; Ferreira 2016; Rice et al. 2016).

Nevertheless, the phylogenetic origin of turtles remains somewhat controversial (Fig. 8.1). Traditional classifications of amniotes (fully land-adapted tetrapods with a cleidoic egg) have considered the **temporal region** of the skull as the most important character for defining large group interrelationships (e.g., Osborn 1903; Williston 1917; Gregory 1946; Olson 1947; Romer 1956). Of special importance was the number of openings, the so-called **fenestrae**, in the temporal skull region. Using this feature, Osborn (1903) divided amniotes (and some non-amniotes) in two main lineages: the '**Synapsida**', with one or no **fenestra**, and the '**Diapsida**', with two **fenestrae** in the temporal region (Rieppel 2000). Later, Williston (1917) modified Olson's (1947) definition of '**Synapsida**' by classifying the amniotes without temporal openings as another group that he called '**Anapsida**', a group

G. S. Ferreira (✉)
Senckenberg Center for Human Evolution and Palaeoenvironment, Eberhard Karls Universität, Tübingen, Germany

Fachbereich Geowissenschaften, Eberhard-Karls-Universität, Tübingen, Germany

Laboratório de Paleontologia de Ribeirão Preto, FFCLRP, Universidade de São Paulo, Ribeirão Preto, SP, Brazil
e-mail: gsferreira@usp.br

I. Werneburg (✉)
Senckenberg Center for Human Evolution and Palaeoenvironment, Eberhard Karls Universität, Tübingen, Germany

Fachbereich Geowissenschaften, Eberhard-Karls-Universität, Tübingen, Germany

Museum für Naturkunde, Leibniz-Institut für Evolutions- und Biodiversitätsforschung, Humboldt-Universität zu Berlin, Berlin, Germany
e-mail: ingmar.werneburg@senckenberg.de

J. M. Ziermann et al. (eds.), *Heads, Jaws, and Muscles*, Fascinating Life Sciences,
https://doi.org/10.1007/978-3-319-93560-7_8

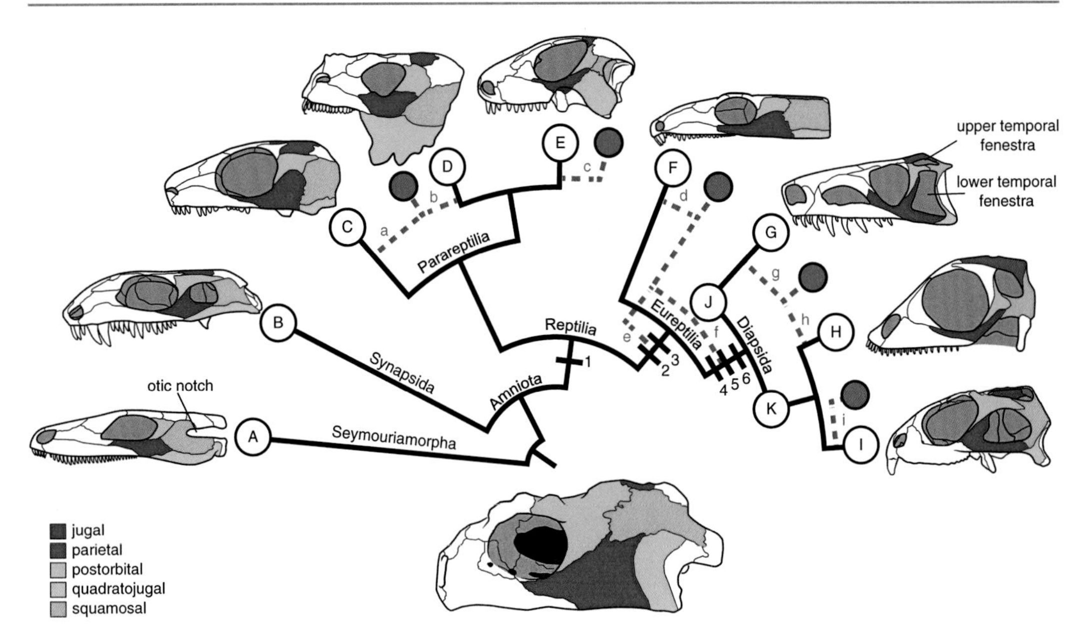

Fig. 8.1 Phylogenetic tree of Reptiliomorpha based on Laurin and Reisz (1995), with the skulls of representative taxa of each lineage plotted in lateral view with highlighted temporal region bones. **A**, *Seymouria sanjuanensis*; **B**, *Eothyris parkeyi*; **C**, *Milleretta rubidgei*; **D**, *Scutosaurus karpinskii*; **E**, *Procolophon trigoniceps*; **F**, *Captorhinus aguti*; **G**, *Euparkeria capensis*; **H**, *Palatodonta bleekeri*; **I**, *Sphenodon punctatus*; **J**, Archosauromorpha; **K**, Lepidosauromorpha. The skull on the bottom depicts the stem-turtle *Proganochelys quenstedti* and the blue circles and dotted lines point to previously proposed relationships of turtles: **a**, Millerettidae (e.g., Lyson et al. 2010); **b**, Pareiasauria (e.g., Lee 1993); **c**, Procolophonidae (e.g., Laurin and Reisz 1995); **d**, Captorhinidae (e.g., Gaffney and Meylan 1988); **e**, stem-Diapsida (e.g., Werneburg and Sánchez-Villagra 2009); **f**, Diapsida (e.g., Neenan et al. 2013); **g**, Archosauria (e.g., Wang et al. 2013); **h**, Sauropterygia (e.g., Rieppel and Reisz 1999); **i**, Lepidosauria (e.g., Müller 2003). The numbers represent selected possible synapomorphies for the respective clades (based on Laurin and Reisz 1995 and Müller 2003): 1, large posttemporal fenestra; 2, supratemporal bone small or absent; 3, long interpterygoid vacuity; 4, upper temporal fenestra; 5, lower temporal fenestra; 6, loss of lower temporal bar

from which fenestrated reptiles presumably arose. In these early schemes of classification, the turtle skull was considered "primitively" **anapsid** (Baur 1889; Cope 1896; Hay 1905; Watson 1914), and the group was classified either **synapsid** (Osborn 1903) or **anapsid** (Williston 1917). Later and more comprehensive classifications considering both extant and extinct taxa (Gregory 1946; Olson 1947; Romer 1956) included not only turtles but also the extinct diadectomorphs (stem amniotes), capthorhinids (stem diapsids), and placodonts (stem lepidosaurs, although they have a large **upper temporal fenestra**) into '**Anapsida**'; together with pareiasaurs and procolophonids, these groups were united as 'Cotylosauria.' Today, the cotylosaurs are considered a paraphyletic assemblage of early amniotes and non-amniotes (e.g., seymouriamorphs; Laurin 2002), but those traditional classifications were highly influential on later phylogenetic studies, which grouped turtles along with different **anapsid** reptiles such as capthorhinids (Gaffney 1980), pareiasaurs (Lee 1995, 1997), and procolophonids (Laurin and Reisz 1995).

Despite this influential view that considered turtles as "primitive" amniotes, survivors of an extinct lineage of reptiles (the '**Parareptilia**' of Gauthier et al. 1988 or '**Anapsida**' of Laurin and Reisz 1995), several alternative early studies raised doubts about those affinities. Based on postcranial characters, Goodrich (1916) suggested close affinities between turtles and other living reptiles within **Sauropsida** (= crown

Reptilia, including birds). Goodrich (1930), Boulenger (1918), and Broom (1924) also emphasized that the arrangement of the bones in the **temporal region** of turtles is not comparable to that of other **anapsid** reptiles. They concluded that this morphology in turtles has been secondarily acquired (see also Müller 2003). Ontogenetic studies (de Beer 1937; Hofsten 1941) of jaw adductor muscle anatomy (Lakjer 1926) revealed several features shared by turtles and other extant **diapsids** (Rieppel 2000). Some phylogenetic studies based on morphological characters also retrieved turtles closely related to **diapsids**, either to Lepidosauromorpha [a lineage that comprises tuatara, lizards, snakes, and the extinct marine reptiles, the sauropterygians (deBraga and Rieppel 1997; Rieppel and Reisz 1999)], or to Archosauromorpha [a lineage that includes crocodiles and birds, among others (Løvtrup 1977, 1985; Gardiner 1993)]. Molecular-based phylogenetic reconstructions usually result in a closer relationship to one of the diapsid clades as well, more commonly among the archosaurs (Zardoya and Meyer 1998, 2001; Hedges and Poling 1999; Mannen and Li 1999; Hedges 2012; Wang et al. 2013).

Most recent studies that examine fossils of extinct amniotes also support the view of turtles as **diapsids**. In one of the oldest turtles, *Proganochelys quenstedti* (Gaffney 1990), marginal teeth were absent, the preorbital region was short, and the **temporal region** was completely closed. The proto-turtle *Odontochelys semitestacea* (Li et al. 2008), although not defined as a member of Testudinata (*sensu* Joyce et al. 2004) due to the lack of a complete turtle carapace, is closer to this lineage than to other reptiles and greatly contributed to the debate of the origin of the turtle shell (Nagashima et al. 2009). Although retrieved together with turtles deeply nested within **Diapsida** (Li et al. 2008), the **anapsid** skull of *Odontochelys semitestacea* did not help to clarify how the transition from a diapsid to an **anapsid** skull could have happened. However, reinterpretations of the skull of *Eunotosaurus africanus* (Bever et al. 2015), a reptile known since the nineteenth century (Seeley 1892) and recently considered part of the turtle stem lineage due to morphological similarities of their postcranial skeleton (Lyson et al. 2010), as well as the recent descriptions of *Pappochelys rosinae* (Schoch and Sues 2015) and *Eorhynchochelys sinensis* (Li et al. 2018), provided some scenarios for this transition. However, the positioning of those taxa along the stem lineage of turtles and the morphological interpretation of their temporal skull region remain open to debate, and some phylogenetic analyses that include *E. africanus* and turtles have retrieved a possible parareptilian affinity for this clade as well (e.g., Lyson et al. 2010; Lee 2013). As such, the turtle skull continues to be a controversial and very important morphological structure to understand not only the relationship of the different turtle lineages but also their origin among amniotes.

In a recent study, the original dataset of Laurin and Reisz (1995) was expanded by adding new information on the parareptilian clade Mesosauria and updating other information. The analysis resulted in a paraphyletic assemblage of parareptiles with mesosaurs being the sister taxon to eureptilians with the remainder of parareptiles, including turtles (as sister to pareiasaurs), nested inside diapsids (Laurin and Piñeiro 2017; but see MacDougall et al. 2018). This result could resolve the obstacle why turtles have shown both parareptilian as well as diapsid affinities in previous studies; however, further fossils need to be included to this analysis in the future to strengthen this promising hypothesis.

8.2 Cranial Diversity of Extant and Fossil Taxa

Compared to those of early reptilian lineages, the skulls of turtles are highly modified, which makes it difficult to trace their morphological origin. Most of the features directly affect the morphology of the **adductor chamber** but are not limited to this region of the skull. For example, one of the most unusual characters is the absence of teeth and the presence of keratinous rhamphothecae (horny beaks) over the upper and **lower jaws**, similar to that of birds (Romer 1956; Li et al. 2018). The Late Triassic (ca. 215 Ma, Ma = mega-annum, million years; million years ago) stem-turtle *Proganochelys quenstedti* presumably already had rhamphothecae and, although palatal teeth were still present [as in some Jurassic (201–145 Ma) turtles such as *Kayentachelys aprix* and *Sichuanchelys palatodentata*: Gaffney et al. 1987; Joyce et al. 2016], marginal teeth were lost. The also Late Triassic (ca. 220 Ma) proto-turtle *Odontochelys semitestacea* had marginal and palatal teeth (Li et al. 2008), so the rhamphotheca should have been present in the common ancestor between *P. quenstedti* and all other turtles. The recent discovery of *Eorhynchochelys sinensis* finally proofed that ramphothecae were developed at the dawn of turtle evolution (Li et al. 2018).

From a superficial perspective, the skull of extant turtles is greatly expanded posterior to the orbits (**temporal region**) and greatly shortened anterior to them (preorbital region) (Romer 1956). A shortened preorbital region of the skull is also seen in *P. quenstedti*, but, as in other stem-turtles such as *Australochelys africanus* (Gaffney and Kitching 1994) and *Palaeochersis talampayensis* (Rougier et al. 1995), the **temporal region** is not elongated as in crown turtles (Gaffney 1990). *O. semitestacea*, on the other hand, possessed a more elongated skull, both in the temporal and in the preorbital regions (Li et al. 2008). In all these proto- and stem-turtles, the dermal roof is completely closed above the **adductor chamber**, without **fenestrae** or any deep **emargination**, resulting in an **anapsid** morphotype (Werneburg 2012). The bones forming this dermatocranial covering are the jugal, quadratojugal, postorbital, squamosal, and parietal (Fig. 8.1). A largely reduced supratemporal is putatively identifiable in *O. semitestacea*, *P. quenstedti*, and *Pa. talampayensis* (Gaffney 1990; Rougier et al. 1995; Li et al. 2008), but absent in *Eorhynchochelys sinensis* (Li et al. 2018). In fact, reduction of dermal bones in the skull of turtles is recurrent: the supratemporal is lost in all other turtles; there is no sign of lacrimal, tabular, or postparietal bones; the postfrontal is fused with the postorbital; and the nasals are reduced or lost in many groups (Romer 1956). Several of those bones are commonly found in the skull of **parareptiles,** which, in addition to the different shape of the jugal and quadratojugal of *P. quenstedti* (i.e., an elongated jugal and a short but high quadratojugal), led some authors to propose that the **anapsid** condition of the turtle skull is actually a secondary derivation (e.g., Goodrich 1930; Müller 2003). That means that in the ancestral lineage of turtles, the plesiomorphic present **temporal fenestrae** were closed and their absence is not evidence of a closer relationship to other **anapsid** reptiles. The potential closure of temporal **fenestrae** in turtle evolution would not be very surprising because increasing evidence suggests that changes in this region, the appearance and disappearance of **fenestrae**, frequently occurred among **Reptilia**, including **parareptiles** (Müller 2003; Tsuji and Müllcr 2009). Additionally, the three fossil taxa putatively assigned to the turtles' stem lineage prior to the divergence of *O. semitestacea* and Testudinata, *Eorhynchochelys sinensis* (ca. 230 Ma), *Pappochelys rosinae* (ca. 240 Ma) and *Eunotosaurus africanus* (ca. 260 Ma) (Bever et al. 2015; Schoch and Sues 2015), suggest that those still had one or two **temporal fenestrae** (with the lower one opened ventrally) resembling the condition in crown-diapsid reptiles (Bever et al. 2015, 2016; Schoch and Sues 2015, 2016; Li et al. 2018). If the relationship of those taxa is further confirmed, then the closure of the **fenestrae** in the turtle lineage would have occurred simultaneously with the first steps in the acquisition of the shell in the common ancestor of *Eorhynchochelys sinensis*, *O. semitestacea*, and Testudinata.

Although the condition among proto-turtles older than *E. sinensis* remains unknown, it is certain that in the stem lineage to the crown clade **Testudines**, the plesiomorphic state is an **anapsid** skull. Most taxa diverging prior to the origin of **Testudines** had a closed **temporal region**, surrounding the **adductor chamber** by bone (Werneburg 2012). This condition, however, was modified several times in different turtle lineages, and most extant taxa exhibit reduction in bones that arch above the **adductor chamber** (Zdansky 1923; Romer 1956; Rieppel 1993; Werneburg 2012). The type of dermatocranial reduction of turtles is, nevertheless, different from that found in other living reptiles that exhibit **fenestrae**. The

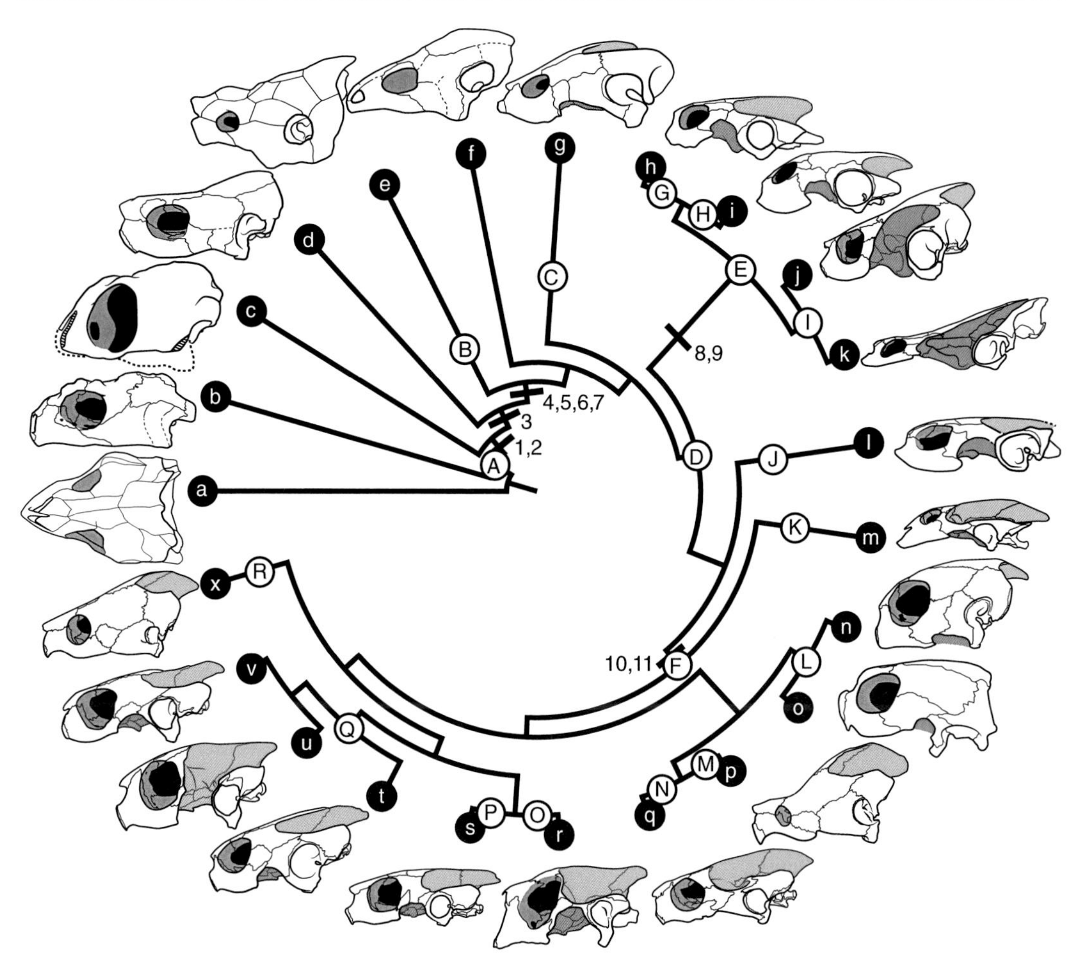

Fig. 8.2 Diversity of turtle skull (phylogenetic hypothesis based on Joyce et al. 2016) with posterodorsal (orange) and anteroventral (blue) emarginations highlighted. Uppercase letters indicate the large groups **A**, Meiolaniformes; **B**, Baenidae; **C**, Pelomedusidae; **D**, Podocnemididae; **E**, Chelidae; **F**, Sinemydidae; **G**, Trionychia; **H**, Chelonioidea; **I**, Chelydridae; **J**, Kinosternidae; **K**, Testudinidae; **L**, Geoemydidae; **M**, Emydidae; **N**, Platysternidae. Lowercase letters indicate the following taxa: **a**, *Odontochelys semitestacea* (in dorsal view); **b**, *Proganochelys quenstedti*; **c**, *Australochelys africanus*; **d**, *Kayentachelys aprix*; **e**, *Meiolania platyceps*; **f**, *Kallokibotion bajazidi*; **g**, *Plesiobaena antiqua*; **h**, *Pelusios sinuatus*; **i**, *Podocnemis expansa*; **j**, *Emydura macquarii*; **k**, *Chelodina expansa*; **l**, *Sinemys gamera*; **m**, *Lissemys punctata*; **n**, *Chelonia mydas*; **o**, *Dermochelys coriacea*; **p**, *Macrochelys temminckii*; **q**, *Kinosternon subrubrum*; **r**, *Testudo graeca*; **s**, *Cuora trifasciata*; **t**, *Pseudemys concinna*; **u**, *Terrapene ornata*; **v**, *Emys orbicularis*; **x**, *Platysternon megacephalum*. The numbers represent selected synapomorphies for the respective clades (based on Sterli and de la Fuente 2010; Rabi et al. 2013; and Werneburg et al. 2015a): 1, opisthotic tightly sutured to squamosal; 2, basipterygoid process sutured; 3, interpterygoid vacuity partially or completely closed; 4, processus inferior parietalis closing foramen nervi trigemini; 5, crista supraoccipitalis posteriorly developed; 6, posterodorsal emargination developed; 7, processus trochlearis otici; 8, processus trochlearis pterygoidei and pleurodiran trochlear mechanism; 9, pleurodiran neck retraction; 10, cryptodiran trochlear mechanism; 11, cryptodiran neck retraction mode. Pictures from different sources (see Werneburg 2012 for details)

temporal skull reduction in turtles is named **emargination** (Fig. 8.2), which presents marginal excavations either at the ventrolateral border of the skull or at the dorsal margin of the posttemporal fenestra (Romer 1956; Kilias 1957; Rieppel 1993; Werneburg 2012). The former is known as anteroventral (*sensu* Werneburg 2013b) or cheek **emargination** and proceeds by usually reducing the jugal and quadratojugal bones. The latter is known as posterodorsal (*sensu* Werneburg 2013b), occiput, or temporal **emargination**, shown as reductions primarily of the parietal and squamosal. The extent of these **emarginations** varies greatly among different turtles (Fig. 8.2), but in their general constitution, they can be used to characterize the main turtle clades (Werneburg 2012).

Most cryptodires possess a well-developed posterodorsal **emargination** (Fig. 8.2), with the exception of the big-headed turtle *Platysternon megacephalum* and of sea turtles (Chelonioidea), in which it is only shallow or almost not present (Romer 1956; Werneburg 2012). Likewise, the anteroventral **emargination** is absent or very shallow in several taxa, such as *Pl. megacephalum*, sea turtles, and snapping turtles (Chelydridae), but can be moderately to well developed in other cryptodires, such as in *Terrapene ornata*, in which it is confluent with the posterodorsal **emargination** (Zdansky 1923; Werneburg 2012). In pleurodires, the degree of **emargination** is also variable (Fig. 8.2). In chelids, there is a large anteroventral **emargination**, sometimes merged with the shallow posterodorsal excavation, as seen in *Chelodina* (Romer 1956; Kilias 1957; Gaffney 1979). In pelomedusids and podocnemidids, there is only a shallow anteroventral **emargination**, but the former also shows a well-developed posterodorsal one similar to that found in several cryptodires, while in podocnemidids there is a larger dermatocranial coverage with shallow posterodorsal **emargination** (Romer 1956; Werneburg 2012). Among extinct turtle lineages (Fig. 8.2), xinjianchelyids and sinemydids usually possess deep posterodorsal and moderate anteroventral **emarginations** (Rabi et al. 2014; Zhou and Rabi 2015), and pleurosternids and baenids have moderate posterodorsal and moderate to well-developed anteroventral **emarginations** (Gaffney 1975; Joyce and Lyson 2015). However, these general descriptions may apply only to the most common members in each of these clades, and the development of the **emarginations** in individual taxa may be highly variable (Zangerl 1948; Gaffney 1979; Werneburg 2012).

Numerous factors have been raised to explain the repeated evolution of dermatocranial bone reductions (see Werneburg 2012), either forming **fenestrae** or **emarginations**, including phylogenetic constraints (particularly in cases when a whole clade possess the same pattern, such as **synapsids** or trionychids; Kilias 1957), reducing skull weight (Frolich 1997), skull dimensions (Tarsitano et al. 2001), diet (Versluys 1919), ear anatomy, **jaw** muscle bending mechanism (in the case of turtles; Karl 1997), plasticity of bones influenced by internal forces on the skull (Kilias 1957; Frazzetta 1968; Tarsitano et al. 2001), and environmental pressures (Gaupp 1895; Nick 1912; Zdansky 1923). The most common of those, however, are the anatomy and function of the **jaw adductor musculature** (Gregory and Adams 1915; Zdansky 1923; Rieppel 1993; Werneburg 2012). In this case, the contraction of the **jaw musculature** would pressure the bones and bony bars in the **temporal region** resulting in the modification of this area (discussed by Werneburg 2012). More recently, Werneburg (2015) showed that the **neck**-bending mechanisms are strongly correlated to type and degree of temporal skull reduction in turtles. It is more likely, however, that no single factor causes the reduction of the dermatocranial bones but actually that several of those in conjunction influence the shaping of the **temporal region** in turtles and other amniotes (Werneburg 2012).

Reductions of the dermatocranial coverage of the **adductor chamber** may have become possible after reinforcement of the attachment of the braincase to the palate and the ear capsule (i.e., palatoquadrate-related structures) in the stem lineage of turtles (Gaffney 1990; Eßwein 1992; Sterli and de la Fuente 2010; Werneburg 2012; Werneburg and Maier 2018). *P. quenstedti, O. semitestacea and E. sinensis* possess robust basipterygoid processes that articulate with the pterygoid ventrolaterally (Rabi et al. 2013; Li et al. 2018), which results in a less rigid, possibly kinetic, basicranial articulation as found in stem amniotes and stem tetrapods (Gaffney 1979; Rabi et al. 2013). In addition, the parietal of *P. quenstedti* does not develop a descending process anterior to the **trigeminal nerve** foramen that would connect the pterygoid ventrally (as it does in crown turtles), and the opisthotic is not strongly sutured to the quadrate (Gaffney 1990; Joyce et al. 2016). As such, the dermatocranial temporal bone coverage may have been the only structure giving mechanical support to the quadrate while developing stronger **bite** forces in stem-turtles (Werneburg 2012). A suture between the basisphenoid and the pterygoid is already seen in *A. africanus* and *Pa. talampayensis* (Rabi et al. 2013), providing evidence for the trend to increasingly strengthen the contact between the braincase and palate by the basipterygoid articulation (Fig. 8.2). These changes preceded the closure of the interpterygoid vacuities realized by an extension of the contact between the pterygoids and an anterior extension of the basisphenoid (Sterli and de la Fuente 2010), the development of the descending process of the parietal, and the sutural contact between the opisthotic and the quadrate (Eßwein 1992; Werneburg and Maier 2018). All those features made the skull more rigid in crown turtles, releasing the **temporal region** from supporting the **jaw** articulation (quadrate-articular) during biting movements and allowing the development of temporal and cheek **emarginations** (Romer 1956; Werneburg 2012).

Closure of the interpterygoid vacuity and the reinforcement of the basipterygoid articulation had consequences on the **carotid** circulation in turtles as well (Sterli et al. 2010; Müller et al. 2011; Rabi et al. 2013). The internal **carotid** artery in most amniotes bifurcates into cerebral and palatal branches before the former pierces the basisphenoid and enters the pituitary fossa, whereas the latter continues anterior and ventral to the braincase. In squamates and **parareptiles** (similar to birds and some sauropterygians), however, the internal **carotid** enters the braincase, and only afterward do the cerebral and palatal branches separate from each other (Müller et al. 2011). Among turtles, two general patterns can be identified: (1) in stem-turtles, such as *P. quenstedti* (Gaffney 1990), the cerebral branch separates from the palatal branch before entering the skull, and (2) in all crown turtles, it is ventrally floored by bone and bifurcates inside the skull (Sterli and de la Fuente 2010; Sterli et al. 2010). Several variations of those two basic morphotypes exist in turtles (Sterli and de la Fuente 2010; Sterli et al. 2010; Müller et al. 2011; Rabi et al. 2013), but the increasing ossification of the basisphenoid-pterygoid articulation and the posterior projection of the pterygoid, which closes the interpterygoid vacuities, enclose the **carotid** branches inside the skull (Sterli et al. 2010; Rabi et al. 2013); this is also seen in therapsids, sauropterygians, and crocodyliformes, which also evolved more rigid skulls (Romer 1956; Rabi et al. 2013).

The stronger attachment of the quadrate to the braincase and to the palate is also thought to be related to the **trochlear mechanism**, a structure specific to turtles (Schumacher 1954a, b, 1954/55, 1956; Werneburg and Maier 2018) (Fig. 8.3). All crown turtles possess an enlarged otic chamber, with the quadrate forming a wall that separates the middle ear into two distinct portions: the cavum tympani (laterally) and the cavum acustico-jugulare (medially) (Gaffney 1979). This separation is not seen in the stem-turtles *P. quenstedti* or *A. africanus* (Gaffney 1990; Gaffney and Kitching 1995), but in *Kayentachelys aprix* it is already well formed (Sterli and Joyce 2007). Enlargement of the quadrate, which becomes cup-shaped and forms the otic chamber, fills a large portion of the **adductor chamber** and separates it into an upper and lower temporal fossa. This condition imposes an obstacle to the course of the **jaw musculature** from its origin on the skull roof to its insertion on the coronoid process of the **lower jaw** (Fig. 8.3), which, as a consequence, is redirected around the otic chamber by a pulley system named **trochlear mechanism** (Schumacher 1956, 1973; Gaffney 1975, 1979; Joyce 2007). The pressure thought to be exerted by the trochlea during jaw movements can only be accommodated because of the more rigid skull of crown turtles.

Although all crown turtles developed the same solution to this problem of limited space, two different mechanisms are found in **Cryptodira** and **Pleurodira** (Fig. 8.3), each enabling a similar pulley system. The former developed its trochlea on the anterodorsal aspect of the otic chamber itself (called processus trochlearis otici) and the quadrate and prootic may form a protuberant or roughened surface (Schumacher 1954b; Gaffney 1975). Where the coronar aponeurosis of the external **jaw adductors** (see below) contacts this

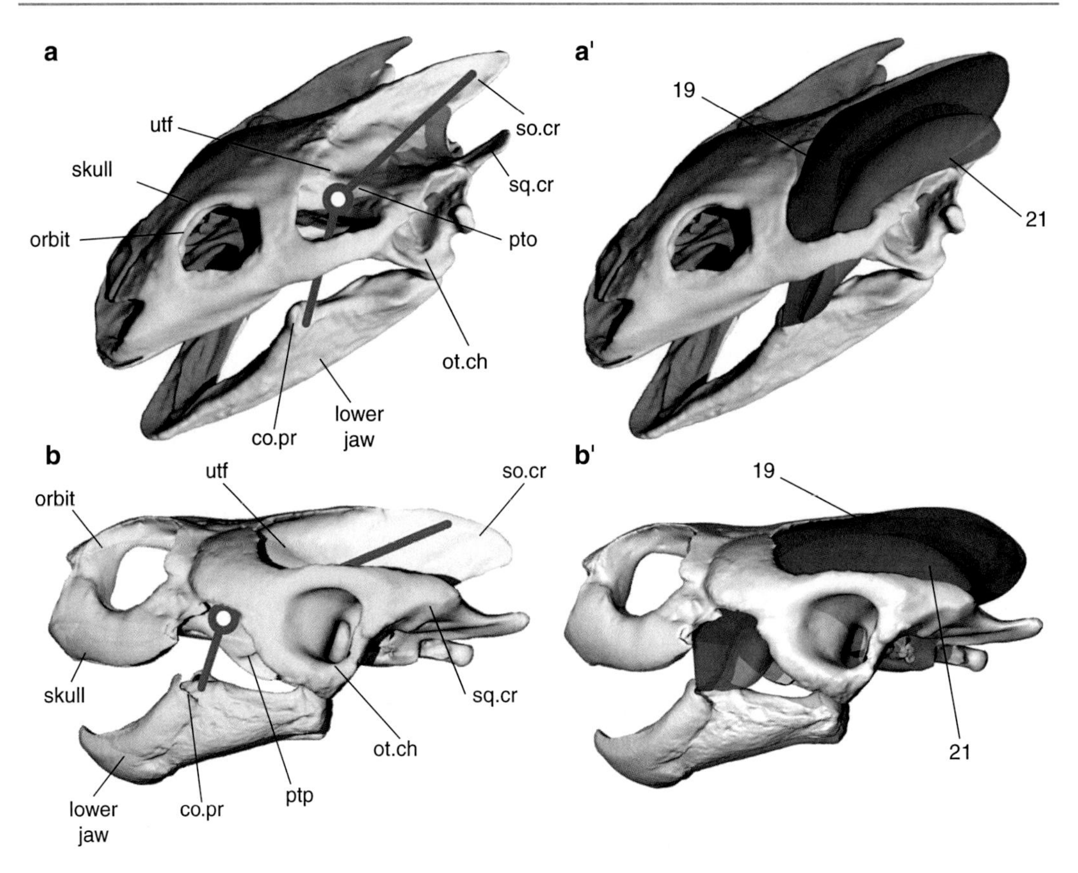

Fig. 8.3 The trochlear mechanism in (**a, a′**) the cryptodire *Pelodiscus sinensis* and in (**b, b′**) the pleurodire *Pelomedusa subrufa*. The external jaw muscle portions partes profundus (19) and superficialis (21) originate, respectively, on the supraoccipital (so.cr) and squamosal (sq.cr) crests in the upper temporal fossa (utf) and insert to the coronoid process (co.pr) of the lower jaw. On this course (simplified in **a** and **b** by the purple line), they bend (circle) around the expanded otic chamber (ot.ch), turning their fibers almost vertically in the lower temporal fossa. This bending is realized by the processus trochlearis otici (pto) in Cryptodira and by the processus trochlearis pterygoidei (ptp) on Pleurodira. The simplified view (**a, b**) with just one line neglects the complexity indicated by the more realistic 3D reconstruction of the whole muscle mass (**a′**, **b′**)

bony process, a sesamoidal cartilago transiliens (Schumacher 1954a, 1956) or a bony os transiliens in *Gopherus polyphemus* (Ray 1959; Bramble 1974) is developed. In this true articulation, cryptodires develop a gliding joint, surrounded by a capsule that involves the cartilage and a bony process (Schumacher 1973; Gaffney 1979). On the other hand, pleurodires possess a trochlea anterior to the otic chamber, around an enlarged flange of the pterygoid called processus trochlearis pterygoidei (Schumacher 1973; Gaffney 1975). The coronar aponeurosis also develops a transiliens cartilage in the contact to this process of the pterygoid in pleurodires, but there is no joint capsule around them. Alternatively, a fold of the oral mucosa (the ductus angularis oris) is enlarged in pleurodires (Fuchs 1931) and forms a pocket that extends between the processus trochlearis pterygoidei and the transiliens cartilage and provides a lubricated surface over which the structure glides (Schumacher 1973; Gaffney 1979). These differences led Gaffney (1975) to conclude that these **trochlear mechanisms** are nonhomologous structures that arose independently in cryptodires and pleurodires. However, the analysis of new fossils, as well as the growing support for a long stem lineage to **Testudines**, including several taxa previously considered cryptodires (e.g., Joyce 2007; Sterli et al. 2010; Rabi et al. 2013; Joyce et al. 2016), suggest that a cryptodiran-like **trochlear mechanism** is the plesiomorphic condition for the crown turtles (Joyce and Sterli 2012). Considering this, Joyce (2007) proposed a scenario for the evolution of the pleurodiran **trochlear mechanism** through a transfer of function from the processus trochlearis otici to the processus trochlearis pterygoidei, with a hypothetical ancestor possessing both types of trochleas (Joyce 2007; Joyce and Sterli 2012). However this has been criticized elsewhere (e.g., Gaffney and Jenkins 2010; Werneburg and Maier 2018) because it lacks mechanical and paleontological support, and there is no known extinct taxon that could represent this intermediate condition. Furthermore, preliminary observations suggest that the **trochlear mechanism**, at least in pleurodires, may be related to other features of the skull, such as larger origin sites for the pterygoid muscles, aside from redirecting the external muscles around the otic chamber.

Regardless of the origin of the **trochlear mechanism** in turtles, its appearance represented a new possibility for enabling more powerful muscle function in an **adductor chamber**, which was limited in its volume by the expanded otic chamber. With the pulley system, the **external adductor muscles** became delimited into two parts, one behind the trochlea with horizontally oriented fibers closer to the muscle's origin on the skull and another in front of the trochlea with vertically oriented fibers closer to the muscle's insertion to the **lower jaw** (Schumacher 1973; Gaffney 1975; Joyce 2007; Werneburg 2013a). In several taxa, the supraoccipital and the squamosal develop elongated projections posteriorly in the latter part of the **adductor chamber** (Sterli and de la Fuente 2010; Werneburg 2012). These posterior crests increase the area for fiber attachment and, by subsequent elongation of muscle fibers, increase the generation of muscle power of the **external adductors**. It is important to note that these modifications in the turtle skull (stronger attachment of the braincase to the palate and the skull roof, development of **temporal emarginations**, extension of supraoccipital and squamosal crests, as well as the origin of **trochlear mechanisms**) occur after the Jurassic (ca. 200 Ma). Additionally, these changes occurred in the clade containing the last common ancestor of Meiolaniformes and **Testudines** (Sterli and de la Fuente 2010; Rabi et al. 2013), highlighting the correlation between a more rigid skull and stronger and more robust jaw muscles in turtles (Werneburg and Maier 2018).

8.3 Importance of the Turtle Neck

To understand the diversity of the turtle skull, it is particularly important to consider postcranial characteristics as well. Obviously, a highly integrated functional chain exists among the carapace, **neck**, and skull (Werneburg 2015). The functional origin of the turtle shell has been controversial. If turtles evolved from marine ancestors, heavy abdominal ribs (gastralia) could have permitted controlling buoyancy (Schoch and Sues 2015), resulting in the primary emergence of a plastron as seen in *Odontochelys semitestacea* (Li et al. 2008; but see Reisz and Head 2008; Scheyer et al. 2013). The carapace might have evolved as a defensive structure (Romer 1956), but recently, Lyson et al. (2016) hypothesized that turtles might have had fossorial ancestors in which a strong ossification of the whole body would have developed to withstand external pressure of the soil when digging; curiously, however, other fossorial vertebrates reduce ossifications and form a slender body (Gauthier et al. 2012). Whatsoever the origin of the shell was, the emergence of a stiffened bony armor influenced a great set of anatomical features, including the ventilatory system (Lambertz et al. 2010; Lyson et al. 2014) and the whole locomotory apparatus (Walker 1973; Joyce et al. 2013b).

In addition to the limbs and girdles, the cervical region also had to correspond with such a comprehensive stiffening of the turtle body. Crown turtles evolved a great flexibility of their neck, which, as compensation, enables fast and elaborated nutrition strategies (Herrel et al. 2008). As such, **neck mobility** and **feeding** behavior appears to be strongly connected (Natchev et al. 2015). In some forms, particularly in chelids, the **neck** may be longer than the shell, and many taxa, including trionychids and kinosternids, are able to stretch their heads completely over the carapace for defense or hunting purposes (pers. obs. IW). Related to such great mobility of the **neck**, the **cervical vertebrae** of extant turtles are highly modified compared to other reptiles, including stem-turtles (Williams 1950; Werneburg et al. 2015b).

A unique feature among all vertebrates is the ability of turtles to retract their **neck** and head inside the shell. Each of the two major extant turtle groups evolved a highly specialized mode of retraction, but both fold their **neck** in an S-shaped manner (Werneburg et al. 2015b). In a horizontal plane, pleurodires lay their head and their **neck** below the anterior edge of the shell. Cryptodires, in contrast, retract their **neck**s in a vertical plane and withdraw the **neck** in between the shoulder girdles (Herrel et al. 2008). Cervical ribs, which were still present in stem-turtles, are reduced in extant taxa throughout ontogeny, enhancing mobility and facilitating **neck** retraction into the shell and between the shoulder girdles (Werneburg et al. 2013). *Proganochelys quenstedti* and *Meiolania platyceps*, like most other stem-turtles, have relatively compact **cervical vertebrae** and short **neck**s, and as such, they were hypothesized to have performed only limited mobility (Gaffney 1985, 1990; Jannel 2015). However, Werneburg et al. (2015a) have shown using radiographs, CT scan images, and morphometrics that also stem-turtles like *P. quenstedti* might have even been able to laterally tuck their **neck** below the anterior edge of the carapace similarly to pleurodires but in a simpler manner. Protective osteoderms on the dorsal surface of the **neck** of this stem-turtle support the hypothesis that defense might have been a major pressure for the evolution of **neck retraction**. It must be noted, however, that stem-turtles were mostly terrestrial herbivores as the reduction of marginal teeth and presence of keratinized rhamphotheca clearly indicate. The movements that later allowed **neck retraction** might have evolved initially in relation to this **feeding** behavior, enabling those turtles to pull down plants while maintaining the cumbersome body steady (compare to extant land tortoises, in which, however, the limbs also "still" support; Natchev et al. 2015). Recently, Anquetin et al. (2017) hypothesized that in the early evolution of crown turtles, a specialized foraging strategy underwater (suction **feeding**) might have been related to increased **neck mobility** in general and might have enforced the origin of the cryptodiran **neck retraction**. In both of these scenarios, the protective function of retracting the neck and head inside the shell would be an exaptation of the high mobility already present (Anquetin et al. 2017).

Specialized **neck** muscles, linked to **neck retraction**, have evolved in turtles, and, as for some skull features (see above), a more rigid skull may be related to their appearance. The major **neck** retractor, m. retrahens collique (Werneburg 2011), broadly attaches to the basicranium, in which an immobile basicranial articulation supports stronger forces. Dorsal **neck** muscles attach to the temporal skull region and are in close topographic relation to the external **jaw musculature**. One of the major challenges in future turtle research is to establish homology of turtle **neck** musculature with that of other amniotes (Gasc 1981; Werneburg 2011), as several reorganizations must have occurred in relation to the novel movement abilities of turtles (Werneburg et al. 2013).

Werneburg (2015) tested the influence of **neck mobility** for shaping the temporal skull region in turtles and suggested that ventral flexing of the **neck** and the cryptodiran mode of retraction significantly influenced the size of the posterodorsal **emargination**. As Werneburg (2013a) highlighted, jaw muscle attachments are highly dependent on bone arrangement, and the indirect influence of **neck mobility** also for jaw muscle anatomy may have been underestimated. Werneburg (2015) has also shown that the expansion of posterodorsal and anteroventral **emarginations** are significantly and highly correlated to each other. When one **emargination** expands (e.g., influenced by **neck mobility**), the other shows a correlated change in size. The broad tendinous insertion of dorsal **neck** muscles to the posterodorsal region of the skull enables a better force distribution when moving the **neck** under force. Expanded supraoccipital and squamosal crests also provide broader origin/insertion sites for **neck** musculature. These observations led Werneburg (2015) to formulate a hypothesis for the origin of the **anapsid** skull in turtles (Fig. 8.4).

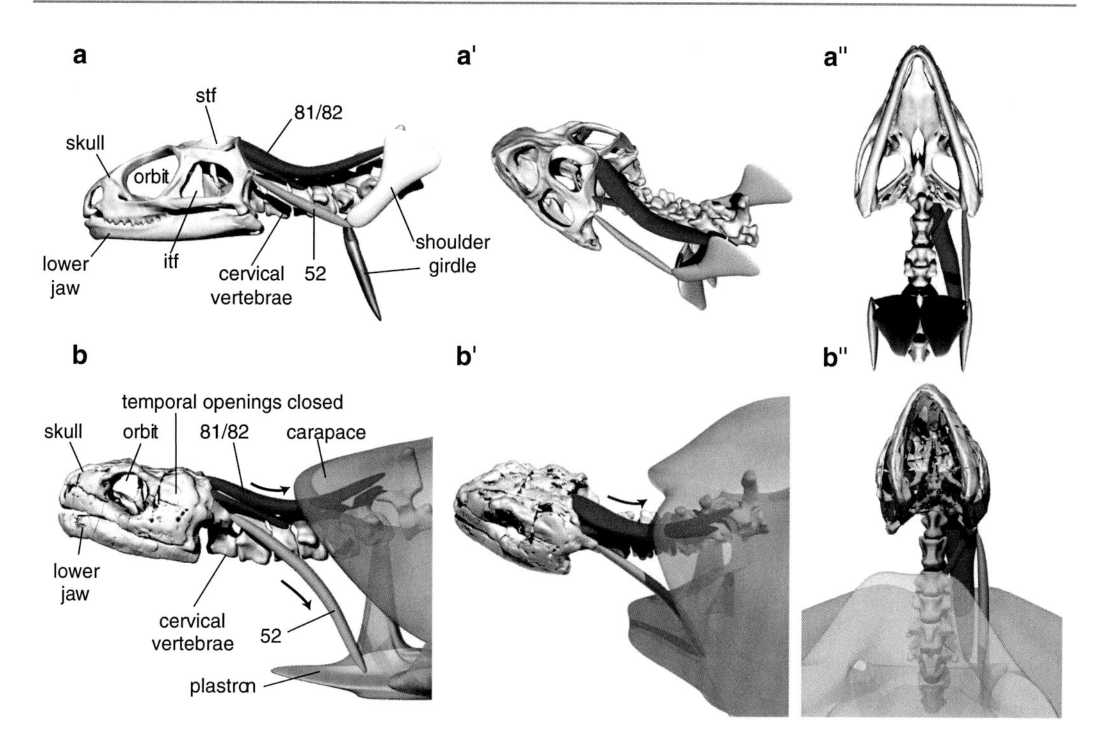

Fig. 8.4 Scenario for the origin of the anapsid skull and temporal emarginations in turtles as proposed by Werneburg (2015). The 3D models were built using CAD software Rhinoceros 3D (McNeel and Associates 2003). Further information on the models can be found in Werneburg et al. (2015a, b). All models are shown in left lateral (left column), oblique dorsolateral (middle column), and ventral (right column) view. (**a**) In the ancestral diapsid condition (visualized by *Sphenodon punctatus*), the selected neck muscles contact the shoulder girdle posteriorly and stabilize the head anteriorly. (**b**) In stem-turtles (exemplified by *Proganochelys quenstedti*), the trapezius (81/82 of Werneburg 2011) and sternocleidomastoideus (52 = m. plastrosquamosus in turtles) muscles lost contact to the shoulder girdle (see Lyson et al. 2013) and posteriorly attach to the carapace and plastron, respectively. Stem-turtles were already able to simply retract their head and neck inside the shell (Werneburg et al. 2015a, b). For that, large tension forces of the trapezius and sternocleidomastoideus muscles acted on the temporal skull region. As a response to withstand those forces and to maintain skull integrity, the infratemporal (ift) and supratemporal fenestrae (stf) were closed in the potential diapsid ancestor of Testudinata. (**c**) Cryptodirans (exemplified by *Graptemys pseudogeographica*) retract their neck in a vertical plane inside the shell. For that, strong dorsal neck musculature (82, a cryptodiran derivative of m. trapezius, = m. carapacocervicocapitis medialis pars capitis) acts on the temporal skull region. To withstand those neck forces, which largely increased compared to those of stem-turtles, marginal posterodorsal emarginations (pd.em) evolved providing broader insertion sites and better distributing neck tension forces. (**d**) In pleurodirans (exemplified by the chelid *Phrynops hilarii*), the pleurodiran derivative of m. trapezius (81 = m. carapacocervicocapitis medialis pars capitis) inserts to the base of the skull. As such, neck muscles do not have a comparable influence on the temporal region as in cryptodires. In pleurodires, several neck muscles enable large lateral neck movement (exemplified by 57, m. collosquamosus) and might influence the shaping of the skull. The origin of the anteroventral emargination (av.em)—in pleurodires and cryptodires alike—is not fully understood (see Werneburg 2015). However, the extent of both anteroventral and posterodorsal emarginations appears to influence each other enabling—associated to particular skull dimensions—a stable, bridge-like construction. The reduction of the dermal armor in the temporal region is certainly associated with a number of intrinsic and extrinsic factors (Werneburg 2012), which need to be identified and quantified in the future. Homology, function, and diversity of turtle neck musculature are hardly understood and require comprehensive research programs in the future. As such, the hypothesis on neck muscle influence for shaping the turtle skull (Werneburg 2015) has to be understood only as a first attempt to incorporate this type of data. Models modified after Werneburg et al. (2015a, b)

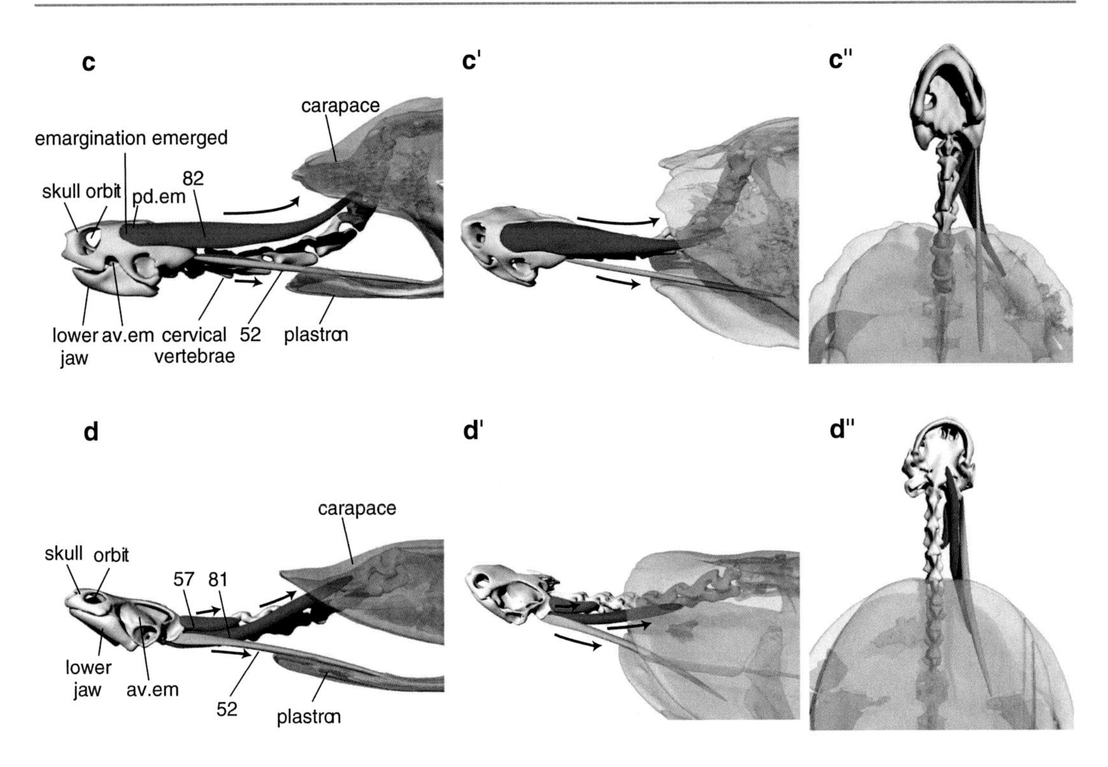

Fig. 8.4 (continued)

Following this hypothesis, the stepwise emergence of the turtle shell (Li et al. 2008; Schoch and Sues 2015) was highly correlated with increased mobility of the turtle **neck**. The forces related to **neck** movement—via strong dorsal and ventral **neck** muscles—greatly influenced the shaping of the turtle skull. If the turtle ancestor actually had a diapsid morphology of the skull (Fig. 8.4), temporal **fenestrae** could have closed in response to the increased pulling force of the **neck** muscles, resulting in an **anapsid** condition. The skulls of *Pappochelys rosinae* (Schoch and Sues 2015), with its small dorsal **fenestra**, and the skull of *Eunotosaurus africanus* (Bever et al. 2015), in which the ventral and perhaps the dorsal **fenestrae** are present, could illustrate this stepwise closure of the skull opening. Later, among crown turtles (**Testudines**), **neck**s became increasingly elongated (Williams 1950; Werneburg et al. 2015b), thereby enabling complex **neck retraction**s. These further stressed the turtle skull, resulting in the formation of the posterodorsal **emargination**, which distributed **neck** forces evenly in the skull. The anteroventral excavation developed as a counterpart to the posterodorsal one, enabling the integrity of the temporal skull region in a bridge-like construction. It is worth mentioning that the dorsal **neck** muscles in many pleurodires (which are related to horizontal **neck retraction**) insert near to the ear capsule, more laterally to the back of the skull than in cryptodires and other sauropsids. This means a less powerful force distribution to the dermal coverage of the **adductor chamber** and a less excavated posterodorsal **emargination** in many forms. In most pleurodires, the anteroventral **emargination** is more prominent than in cryptodires. This might be related to the relatively flattened skull of many pleurodires, such as Chelidae.

8.4 General Morphology of the Cranial Musculature in Turtles

Several authors in the last two centuries dissected and described cranial musculature in turtles. Some focused on specific taxa (Bojanus 1819;

Ogushi 1913a), whereas others applied a comparative approach to understand the general structure and diversity of this part of the turtle body (Hoffmann 1890; Lakjer 1926; Poglayen-Neuwall 1953; Schumacher 1954a, b, 1954/55, 1956, 1973; Kilias 1957). However, those who took a comparative approach concluded that although turtles show a variety of diets and behaviors related to their diverse habitats (Ernst and Barbour 1992), the cranial muscles, and especially the jaw muscles, are highly conserved (e.g., Iordansky 1996). Although this is accurate in a general view and the observed variation is usually related to relative sizes of muscles and tendinous structures, some portions or entire muscles may be present or absent in different taxa (Werneburg 2011), resulting in more profound differences.

Muscles associated with the skull of turtles follow the general pattern of innervation by cranial nerves in gnathostomes (Edgeworth 1935; Diogo et al. 2008). Cranial nerves III–XII are responsible for the movements of the muscles in the testudine head (Werneburg 2011). Nn. oculomotorius (III), trochlearis (IV), and abducens (VI) innervate the muscles related to eye and eyelid movement, whereas the jaw depressor, superficial **neck** musculature, and some muscles related to the ear capsule are innervated by n. facialis (VII). The muscles related to the branchiovisceral region and to the larynx-related musculature are innervated by the glossopharyngeus (IX), vagus (X), and spinal accessorius (XI) nerves, whereas n. hypoglossus (XII) innervates the musculature related to the hyoid apparatus, including the tongue. Some of those posterior cranial nerves also innervate the **neck** musculature, although this region is mainly innervated by spinal nerves (Werneburg 2011). Finally, the most prominent muscle group in the turtle head, the **jaw adductor musculature**, is innervated by the **trigeminal nerve** (V) and represents the most-studied muscular complex in turtles (and in reptiles in general; Lakjer 1926; Edgeworth 1935; Schumacher 1973; Diogo and Abdala 2010; Werneburg 2013a). This is mainly due to its direct relation to dietary preferences and **feeding** mechanisms (Schumacher 1973) and, putatively, to their relation to dermatocranial bone reductions (**fenestrae** and **emarginations**; see Werneburg 2012, 2013b).

In addition to the external eye muscles found in all tetrapods (two mm. obliqui, four mm. recti, and perhaps m. retractor bulbi), several others were described in reptiles and birds (**Sauropsida**) (Underwood 1970; Løvtrup 1985; Werneburg 2011). Their presence and variation among different sauropsid groups is in most cases related to the presence, extent, and mobility of the greatly developed "third eyelid," the membrana nictitans (Werneburg 2011). Of those, the m. pyramidalis (which is innervated by the n. abducens [VI]) (Edgeworth 1935; Werneburg 2011) has its origin on the medial surface of the eye bulbus and inserts via one tendon to the membrana nictitans and via a second tendon to the lower eyelid (Edgeworth 1935; Werneburg 2011). This muscle is of special interest because it is only found in crocodiles and turtles, representing a potential shared character (synapomorphy) of a possible clade containing those taxa (Thomson 1932; Underwood 1970; Schumacher 1972; Løvtrup 1985; Rieppel 2000, 2004; Eger 2006). However, extrinsic musculature of the eye is not well documented in most reptilian taxa, and greater taxon sampling and ontogenetic studies will be useful to identify homologies among these muscular units (Werneburg 2011; see also there for basic muscle terminology). A m. levator bulbi, innervated by the trigeminal nerve (V), is only found in rare cases and is reduced during ontogeny (see Werneburg 2011 for discussion). In trionychian turtles, which have a moveable nose, particular nose muscles, also innervated by n. trigeminus (V), are present.

Three main units are generally recognized in the **jaw adductor musculature** (the external, internal, and posterior muscles) and are established according to their relation to the **n. trigeminus** (V) branches (Luther 1914; Lakjer 1926; Schumacher 1973; Werneburg 2011). Among those, the **external adductors** are the strongest and most prominent in turtles, as in squamates (Rieppel 1980, 1984) and contrary to crocodiles (Schumacher 1973), and may be subdivided at least in three portions (Lakjer 1926): pars profundus, originating on the lateral wall of the braincase and on the supraoccipital crest; pars superficialis, lateral to the previous and originating on the lateral wall of the skull and squamosal crest; and pars medialis, more anteriorly located than the other portions, originating mainly on the quadrate surface anteriorly on the otic chamber (Werneburg 2011). These portions may not be clearly distinguishable in some turtles, and some variation of the relative size and shape of parts may occur (e.g., the pars profundus is slightly reduced in *Dermochelys coriacea*; Lakjer 1926; Schumacher 1972; Poglayen-Neuwall 1966). The general pattern is that the three portions fuse together anteriorly and insert to a large and strong tendon that attaches to the coronar process of the **lower jaw**, the so-called coronar aponeurosis (also called bodenaponeurosis or central tendon; Schumacher 1973; Rieppel 1990; Werneburg 2011). This tendon is important, not only because of its main function in transferring the main contraction forces of these muscles to execute the adductor movements in the **lower jaw** (Iordansky 1996), but also because in its ventral aspect the cartilago / os transiliens of the **trochlear mechanism** develops (Schumacher 1973). The position of this structure varies with the different trochlear processes, being on the dorsal and anterior surfaces of the prootic and quadrate in cryptodires and on the lateral pterygoid process (the processus trochlearis pterygoidei; Gaffney 1979) in pleurodires (Schumacher 1973; Werneburg 2011). It develops as a sesamoid cartilage (or bone in *Gopherus polyphemus*; Ray 1959; Bramble 1974), meaning it likely arises as a result of mechanical stress (Ray 1959; Bramble 1974; Iordansky 1994) across this structure.

An additional jaw adductor muscle, the m. zygomaticomandibularis, may be found in the clade Trionychia, which includes soft-shelled turtles (Trionychidae) and the pig-nosed turtle (*Carettochelys insculpta*) (Ogushi 1913a; George and Shah 1955; Dalrymple 1975; Werneburg 2011; Werneburg et al. unpublished data for *C. insculpta*). Analogous to the masseter muscle of mammals, this muscle originates ventrally and laterally on the jugal and the quadratojugal on the "zygomatic bar" and inserts laterally to the **lower jaw**, near the insertion of the **external adductors** (Werneburg 2011, 2013a, b). Based on its position relative to the other **external adductors**, some authors have described this unit as part of the pars superficialis of the external musculature (Lakjer 1926; Poglayen-Neuwall 1953). Indeed, this muscle in trionychids is comparable to the postorbital head of the pars superficialis found in some turtles with a stronger postorbital/**temporal region**, such as in snapping turtles (Chelonioidea) (Rieppel 1990; Werneburg 2011; Jones et al. 2012). In the chelonioid *Caretta caretta*, this muscle head can be almost completely separated from the rest of the pars superficialis (Jones et al. 2012), becoming very similar to the topology of m. zygomaticomandibularis in trionychids. However, whether these results of convergent evolution actually represent homologous structures remains unresolved, and comparative anatomical and developmental studies should be conducted to test this hypothesis (Werneburg 2011).

The internal and posterior adductors form a fan-shaped arrangement of muscles spanning in the lower temporal fossa of turtles, below the **external adductor** layer (Schumacher 1973). The **internal adductors**, located anteriorly to the posterior adductor in the fan, may be subdivided into two main portions, the partes pseudotemporalis and pterygoideus. The latter is the anterior-most portion and originates on the dorsal, lateral, and ventral surfaces of the pterygoid bone, reaching the palatine near the orbit cavity and the parietal on the medial wall of the temporal fossa (Werneburg 2011). The fibers of pars pterygoideus insert on the posteromedial surface of the **lower jaw** near the jaw joint (Iordansky 2010) either directly or by the subarticular (internal) tendon (Schumacher 1973) and/or the pterygoidal aponeuroses (Schumacher 1973; Werneburg 2011).

The pars pseudotemporalis is the central portion of the muscle series (Lakjer 1926; Schumacher 1973; Rieppel 1990) and originates mainly on the descending process of the parietal bone. It inserts directly or via the subarticular aponeurosis on the medial surface of the **lower jaw,** anteriorly to the insertion of pars pterygoideus (Werneburg 2011). A third and smaller part of the **internal adductors**, the pars intramandibularis, can be found in several turtles. It originates from a tendon that connects it to the pars pseudotemporalis, the so-called *zwischensehne* (Poglayen-Neuwall 1953; Schumacher 1973; Iordansky 1994, 1996), and in this case the latter does not attach to the **lower jaw**, but to the *zwischensehne*. The pars intramandibularis inserts laterally to the Meckel's cartilage, inside the fossa primordialis of the **lower jaw** (Werneburg 2011).

The pars pseudotemporalis may be closely associated to the posterior adductor in some turtles (Werneburg 2011), which led some authors to consider it as a portion of m. adductor mandibulae posterior (Schumacher 1954a, 1954/55; Hacker 1954). However, the innervation patterns of both structures, and their development, are completely different (Poglayen-Neuwall 1953, 1954, 1966; Werneburg 2011), which suggests they are not homologous. Finally, the posterior adductor originates mainly on the anterior surface of the quadrate medially to m. adductor mandibulae externus pars medialis and inserts directly, with its own tendon, or via the subarticular aponeurosis, on the posteromedial surface of the **lower jaw**, near the insertion sites of the **internal adductors** (Werneburg 2011).

8.5 Development of Jaw Musculature

The adductor musculature in turtles starts to develop as a single homogeneous cell aggregation (Fig. 8.5) that surrounds the mandibular (V_3) branch of the **trigeminal nerve**. Although there is no sign of compartmentalization in this aggregate, the portion lateral to the nerve branch will differentiate into the **external adductor** and the one medial to it into the internal and posterior adductors (Rieppel 1990; Tvarožková 2006). At early stages this cluster of cells is restricted anteroposteriorly, the medial portion of which extends anteriorly along the pterygoid process of the palatoquadrate cartilage (i.e., known only for the cryptodire *Chelydra serpentina*), posteriorly to the quadrate cartilage, dorsally to a level below the Gasserian ganglion (where the trigeminal foramen will be formed), and posteroventrally to Meckel's cartilage (Rieppel 1990; Tvarožková 2006). Contrary to squamates (e.g., *Podarcis*; Rieppel 1987), the portion that will differentiate into the **external adductors** begins to extend posterodorsally before the posterodorsal extension of the prospective **internal adductors** (Rieppel 1990).

Subsequently (at stage 15; Rieppel 1990), both rudiments (internal and external) begin to differentiate and become compartmentalized (Fig. 8.5). For the **external adductors**, the differentiation of the coronar aponeurosis (Lakjer 1926; Schumacher 1973) subdivides their anlage into medial (pars profundus) and lateral (partes superficialis and medialis) portions. These rudiments continue to extend posterodorsally across the paroccipital process of the chondrocranium until they finally reach their origin sites, when the parietal, postorbital, and jugal ossifications are already expanded (Rieppel 1990; Tvarožková 2006). The medial portion, the prospective pars profundus, follows the posterior elongation of the supraoccipital crest, attaching to it. The lateral portion also extends posteriorly and, although some fibers become attached to the quadrate and others continue their posterior elongation, the superficial and medial parts will become compartmentalized only later during ontogeny (Rieppel 1990).

The **internal adductors**, distributed between the mandibular branch of the **trigeminal nerve** (laterally) and the pterygoid process of the palatoquadrate (medially) (as is typical for all reptiles; Lakjer 1926; Edgeworth 1935), start to become compartmentalized by the development of an anterior ventromedial projection, which will become the anlage of the partes pterygoidei (Rieppel 1990). This portion grows posteroventrally to the **lower jaw** and then anteriorly to the dorsal and ventral surfaces of the developing pterygoid (Tvarožková 2006). Another projection extends dorsally along the lateral flange of the parietal and represents the prospective pars pseudotemporalis. In contrast to squamates (Rieppel 1987), both parts do not become fully separated and share a horizontal tendinous structure (the anlage of the subarticular aponeurosis (Werneburg 2011) that divides the partes pterygoidei ventrally and the pars pseudotemporalis dorsally (Rieppel 1990)). However, distally they become well separated,

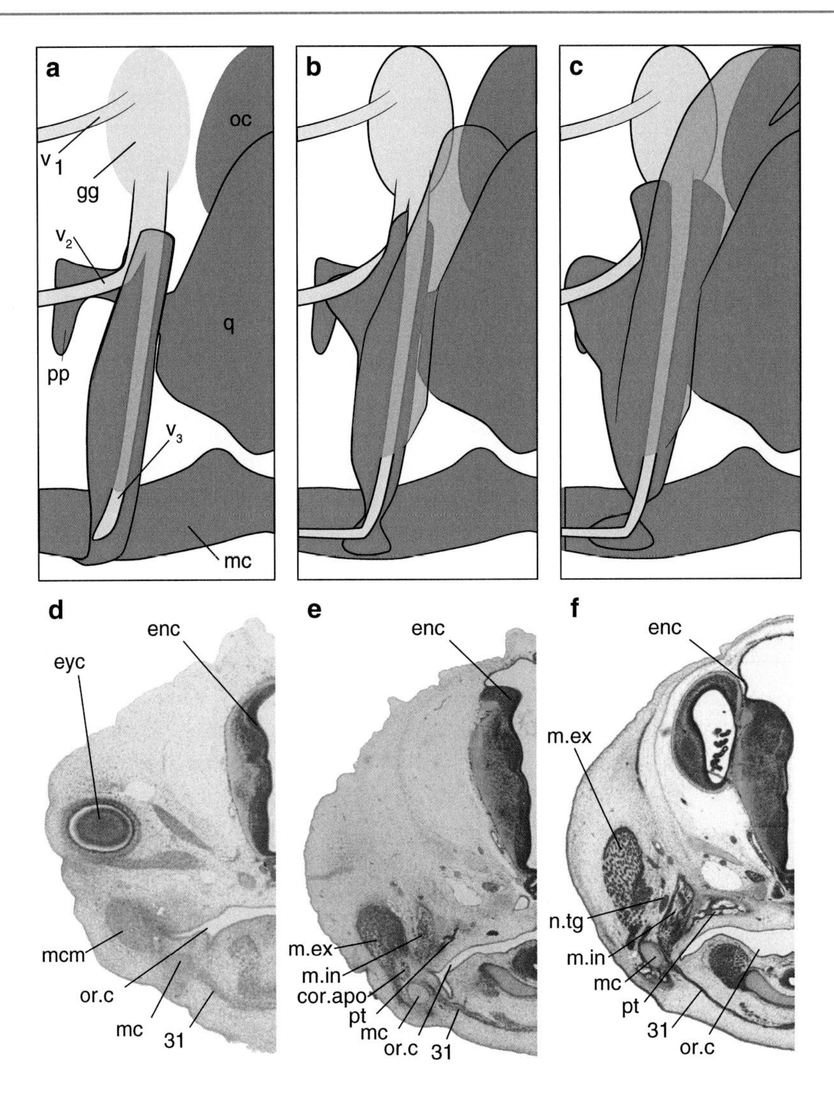

Fig. 8.5 Schematic drawings (**a–c**) representing three stages of development of the muscles in *Chelydra serpentina*, based on Rieppel (1990), and histological slices with Azan-staining after Haidenhain (Mulisch and Welsch 2015) of *Emydura subglobosa* embryos (**d–f**) in different stages (**d**, Y15; **e**, Y17; **f**, Y18; Wolfgang Maier collection Tübingen; Y = Yntema staging system, Yntema 1968). Development starts from an indistinguishable muscle cell mass (mcm) around the mandibular branch (V_3) of the trigeminal nerve anlage (n.tg). It extends from near the Gasserian ganglion (gg) dorsally and to the Meckel's cartilage (mc) ventrally (**a**, **d**). The cell aggregate progressively differentiates (**b**, **e**, **f**) into two portions lateral and medial to the V_3, which will become the external (m.ex) and internal (m.in) muscles, respectively. The latter also becomes progressively projected anteriorly and dorsally, which will differentiate into the pterygoid portions and the pseudotemporalis/posterior muscle anlage (**c**). Note how the initial cell mass is connected to the ventrally located intermandibularis muscle (31) in *C. serpentina* (**a**) and how they become distinct latter during development (**b**, **c**). **Additional abbreviations:** *cor.apo* coronar aponeurosis, *enc* encephalon, *eyc* eye capsule, *oc* otic capsule, *or.c* oral cavity, *pp* palatal process, *pt* pterygoid, *q* quadrate cartilage, *V1, V2* ophthalmic and maxillary branches of trigeminal nerve, respectively

with the pterygoid portions elongating anteromedially and the pseudotemporalis dorsally (Fig. 8.5). The former also extend posteroventrally to attach to the **lower jaw**, near the jaw joint, and the pseudotemporalis elongates ventrally to reach the Meckel's cartilage (by the intramandibularis muscle in taxa that have it; Rieppel 1990).

Two muscles, the posterior adductor and the intramandibularis, feature noteworthy ontogenies in turtles relative to other reptiles. The former originates not from its own rudiment but rather from the m. internus anlage in **Testudines**, after the compartmentalization of the pterygoideus rudiment (Rieppel 1990; Tvarožková 2006). The anlage of the internal muscles posterior to the pterygoideus rudiment, in its dorsal expansion, begins to surround the exit of the maxillary and mandibular branches from the Gasserian ganglion until it is finally pierced by those branches. The portion of this rudiment anterior to the trigeminal mandibular branch corresponds to the pseudotemporalis anlage, whereas the posterior adductor develops from the posterodorsal portion behind the mandibular branch (Rieppel 1990). This corresponds to the topological criteria proposed by Lakjer (1926) to identify the adductor muscle portions. In contrast, in lepidosaurs (Rieppel 1987) the posterior adductor differentiates from the external anlage and becomes topologically equivalent to the posterior muscle of turtles. While most authors consider those as posterior adductors based on its adult topology, origin, and insertion sites (Lakjer 1926; Poglayen-Neuwall 1953; Schumacher 1973; Werneburg 2011), from a developmental perspective, it seems that they are analogous, not homologous (Rieppel 1990).

The intramandibularis, in contrast, starts its development in *Chelydra serpentina* (Rieppel 1990; Tvarožková 2006) in the ventral part of the same homogeneous cell aggregation (Fig. 8.5), deep to the mandibular branch of the **trigeminal nerve**, in continuity to another bunch of cells ventromedially to the Meckel's cartilage. The latter corresponds to the intermandibularis rudiment that will expand between the two rami of the **lower jaws** (Rieppel 1990). It becomes gradually separated from the intramandibularis anlage, which attaches dorsally to the Meckel's cartilage and remains continuous to the dorsal pseudotemporalis rudiment. This close association to the pseudotemporalis and, earlier, to the intermandibularis is also found in crocodiles (Schumacher 1973; Rieppel 1990) but differs from the development of the intramandibularis in lizards. In the latter group, this muscle develops closer to the Meckel's cartilage as an anterior extension of the posterior adductor, rather than the **internal adductor** anlage (Rieppel 1987). As for the posterior adductors, the intramandibularis of turtles and squamates seem to be nonhomologous structures from a developmental point of view, and evidence suggests that this portion is also not related to the crocodilian intramandibularis, since only a few taxa nested deep and separated from each other within **Cryptodira** develop it (Werneburg 2011, 2013a).

8.6 Functional Anatomy of Jaw Muscles and Feeding

The movements related to **feeding** in turtles are generally executed by a set of motions by the jaws, **neck**, and forelimbs (Bramble 1974; Dalrymple 1977; Iordansky 1987, 1996) Some variation has been observed among turtles **feeding** in terrestrial environments or under water, but the generalized behavior includes the jaws closing to hold the prey and the forelimbs and **neck** moving to tear it in smaller pieces that can be swallowed (Iordansky 1996). The **feeding** behavior of extant aquatic turtles involves movements of the head toward the prey and a suction **feeding** mechanism (Schumacher 1973; Lemell et al. 2002), followed by the closure of the jaw holding the prey (Natchev et al. 2015). Some aquatic turtles feature extremely well-developed suction mechanisms, swallowing the food item without grabbing it with their jaws (e.g., *Chelus fimbriatus* and *Apalone spinifera*; Lemell et al. 2002; Anderson 2009), but most turtles use only a weak suction flow and, hence, holding the prey with the closure of jaws is an important part of the **feeding** behavior (Natchev et al. 2015).

Although stem-turtles most probably occupied terrestrial habitats, the ancestral testudine was certainly aquatic (Joyce 2015), and the mode of **feeding** seen in extant terrestrial turtles (Testudinoidea) evolved independently several times from aquatic ancestors (Summers et al. 1998; Natchev et al. 2009; Anquetin et al. 2017). Most testudinoids (i.e., Emydidae and Geoemydidae) use their jaws to grab food items (known as "jaw prehension") on land or in water (Bels et al. 1997, 2008; Summers et al. 1998; Heiss et al. 2008; Natchev et al. 2009; Stayton

2011). However, the exclusive terrestrial tortoises (Testudinidae) developed a different way to grab food items in which they first touch the food with their tongue ("tongue prehension") and then bring the food item to the mouth (Wocheslânder et al. 1999; Bels et al. 2008). Natchev et al. (2015) proposed a four-stage scenario in which this terrestrial **feeding** behavior might have originated from an aquatic ancestor. First, amphibious (but predominantly aquatic) turtles with this feeding mechanism might have explored terrestrial environments and taken food items with their jaws but would have to drag the food into the water to swallow it (as seen in some emydids; Weisgram 1985; Stayton 2011). In the second step, these turtles might have still been able to use hydrodynamic mechanisms to swallow food underwater but might also use their tongue to swallow food items, allowing complete intake of food on land (a behavior that has been documented for the geoemydid *Cuora*; Heiss et al. 2008; Natchev et al. 2009, 2010). The ability to swallow underwater would have been lost during the third stage (seen in *Manouria emys*; Natchev et al. 2015). Finally, tortoises started to use their tongues to grasp food items on land (Weisgram 1985; Wocheslânder et al. 1999; Bels et al. 2008). How stem-turtles fed is still a controversial issue, but the apparent completely terrestrial behavior (Joyce et al. 2004; Scheyer and Sander 2007; Joyce 2015; Anquetin et al. 2017) and the presence of palatal teeth (Gaffney et al. 1987; Gaffney 1990; Joyce et al. 2016) suggest they held and processed food items with their jaws (Matsumoto and Evans 2017).

The closure mechanisms of the **lower jaw** are extremely important for **feeding** behavior in turtles, and **bite** force or speed of closure may vary, depending on diet. The main force component of adduction of the **lower jaw** is generated by the large **external adductors** (Schumacher 1973; Iordansky 1996). Although originating posteriorly (mainly on the walls of the temporal fossa, supraoccipital and squamosal crests) and running anteriorly in a horizontal plane, this large muscle mass is redirected by the **trochlear mechanism** and inserts almost vertically on the coronoid process, providing an adduction as well as a retraction component to the **lower jaw** (Iordansky 1996). This force vector is compensated by the internal pterygoid muscles which, originating anteriorly mostly on the pterygoid and inserting near to the jaw joint, produces a protraction component (Schumacher 1973; Iordansky 1996, 2010). Finally, the internal pseudotemporalis and the posterior adductor run almost entirely vertically relative to their insertion point and generate greater adductive forces; this results in a strong adductor vector during **lower jaw** closure (Schumacher 1973; Iordansky 1996). It is important to highlight that these force vectors were hypothesized simply based on the topological position of the different muscular units in some turtle taxa (e.g., Iordansky 1996). A comparative approach using biomechanical models to infer direction and strength of muscle vectors, considering more taxa with different muscle arrangements, would provide a more detailed view of the functions developed by the adductor musculature in turtles (Ferreira et al. 2018).

The size and shape of the adductor muscles are greatly affected by the shape of the skull, which can expand or limit the relative areas of origin and insertion of those muscles. **Bite-force** (Herrel et al. 2002) and skull shape morphometric analyses of durophagous (i.e., eaters of hard food items) turtles (Claude et al. 2004) found that higher skulls tend to produce more powerful **bites**. At the same time, aquatic turtles **feeding** on fast and elusive preys usually rely on a powerful suction mechanism, produced mostly by an increase in buccal volume associated with movements of the hyoid apparatus (Lemell et al. 2002, 2010) and a rapid lowering of the floor of the buccal cavity. Increasing any dimension of the skull can generate larger buccal volumes, but increasing height can compromise the ability to withdraw the head inside the shell. Considering this, Herrel et al. (2002) proposed that specialized suction **feeding** turtles, such as *Chelus fimbriatus* (Herrel et al. 2002) and *Apalone spinifera* (Pritchard 1984) have relatively flattened skulls with expanded posterior and lateral regions, permitting those turtles to still maintain their **neck retraction** mechanism but compromising their **bite** performance. Nevertheless, some taxa, such as *Phrynops geoffroanus* and *Pelusios castaneus*, seem to combine suction **feeding** with a strong **bite** force (Lemell and Weisgram 1997; Herrel et al. 2002)even with relatively flat skulls. Thus, it seems likely that, aside from general skull proportions, different factors (e.g., different fiber types, differences between relative size of the internal and external muscles) also affect the force and speed of contraction of the jaw adductors in turtles.

8.7 Evolution of Cranial Musculature

Approximately 40 of the 356 known species of extant turtles (Rhodin et al. 2017) have been dissected and had their **jaw musculature** described, including at least one representative of each main lineage of **Testudines** (Werneburg 2011). These data allowed for a comprehensive study that considers the evolution of this region of the turtle head, such as that of Werneburg (2013a). In that study, the variation of the jaw muscles observed in several taxa was included into a taxon-character matrix to investigate the general trends during the evolution of turtles and ancestral state reconstructions for jaw muscle anatomy in the last common ancestor of **Pleurodira**, **Cryptodira**, and **Testudines**. For the present chapter, we created 3D-models (see details on Figures captions) for the skulls and jaw muscles of a cryptodiran and a pleurodiran turtle, *Pelodiscus sinensis* (Fig. 8.6) and *Pelomedusa subrufa* (Fig. 8.7), respectively, to compare topology, origin, and insertion of their jaw muscles to the reconstructed ancestral conditions of **Cryptodira** and **Pleurodira**. Using the predicted states for the Testudines crown-node, we also plotted the **jaw musculature** on a 3D model of the skull of *Proganochelys quenstedti* (Fig. 8.8) (scan data from Werneburg et al. 2015a).

As suggested by Werneburg (2013a), modeling reconstructed states for **Testudines** onto a stem-turtle, such as *P. quenstedti*, should be interpreted tentatively because those states are inferred for a different node on the turtle tree and we usually cannot directly access soft tissue data in extinct taxa. However, this represents a first step toward a better understanding of the **adductor chamber** in stem- and ancestral turtles (*P. quenstedti*, **Cryptodira,** and **Pleurodira** ancestors) and may be useful to infer general trends that occurred during the evolution of these lineages. Careful analyses of bone surfaces in the **adductor chamber** and on the **lower jaw** (e.g., Araújo and Polcyn 2013; Witzmann and Werneburg 2017) and internal bone structures, such as Sharpey's fibers (e.g., Scheyer and Sander 2007), to identify attachment sites could provide direct evidence about the arrangement of the **jaw musculature** in stem-turtles. That being said, we describe below first the **jaw musculature** in cryptodires and pleurodires, considering the changes between the ancestral condition and that of the chosen extant taxa, and then in *P. quenstedti*, commenting on the general transformations of the **jaw musculature** from stem to crown turtles.

8.7.1 Jaw Muscles in Cryptodira

The extant cryptodire chosen for jaw muscle visualization, *P. sinensis* (Fig. 8.6), belongs to Trionychia, a clade that forms the sister taxon to all other cryptodires based on molecular data (Shaffer 2009; Guillon et al. 2012; Crawford et al. 2015). In many regards, Trionychia show plesiomorphic skull conditions; however, a list of unique characters also exists (Vitek and Joyce 2015). Their skulls possess long supraoccipital and squamosal crests and a broad **adductor chamber** that affect not only the condition of the **external adductors** but also that of the internal and posterior adductors. In *P. sinensis*, one of the best-described trionychid species regarding its jaw muscles (Ogushi 1911, 1913a, b, 1914), the

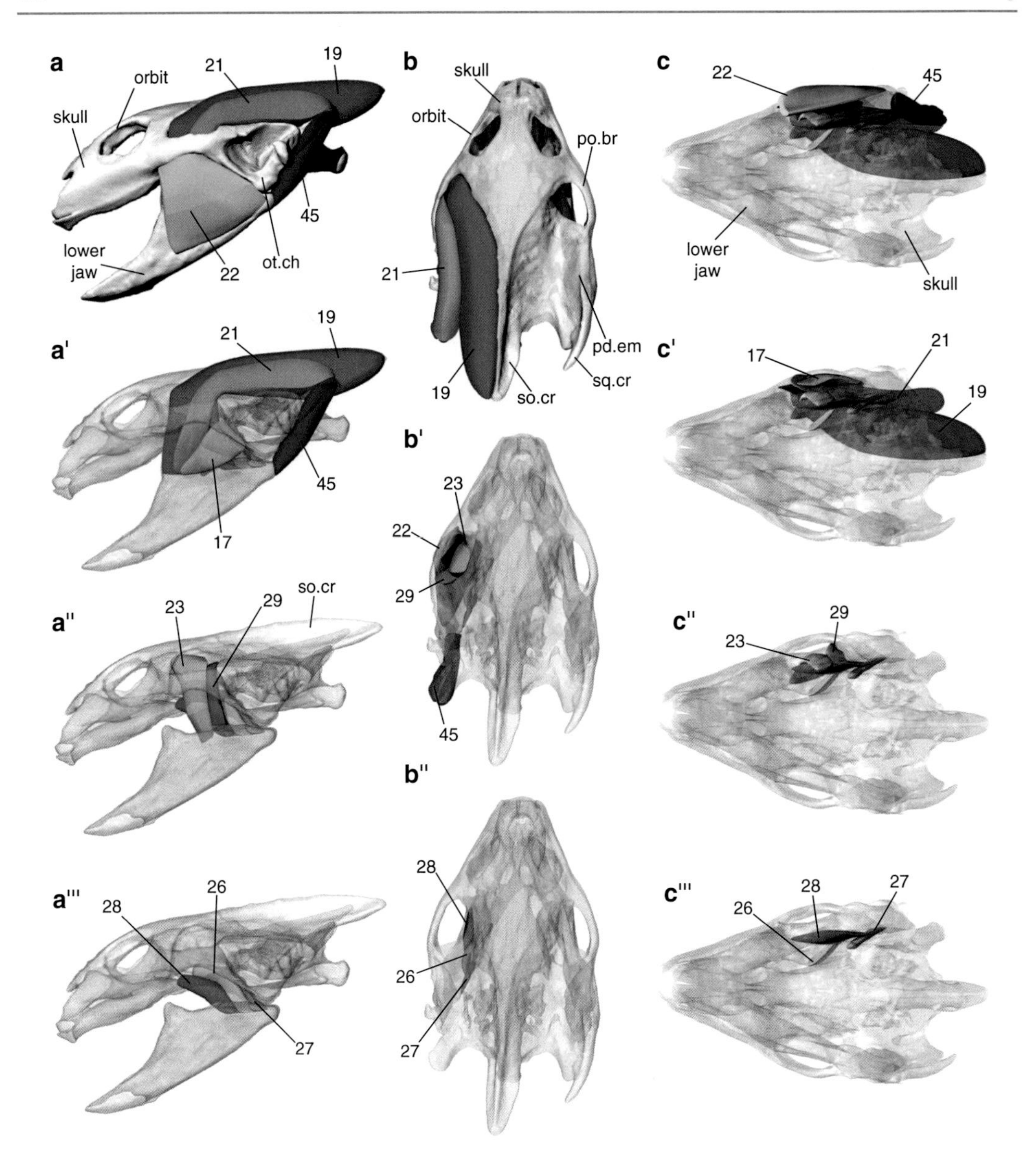

Fig. 8.6 Digital three-dimensional reconstruction of the jaw adductor and depressor musculature of the cryptodire *Pelodiscus sinensis* based on Ogushi (1913a, b), Schumacher (1954a, b), and Werneburg (2011, 2013a) made with CAD software Rhinoceros 3D (Robert McNeel & Associates 2003). Skull with solid (**a**, **b**) and transparent (**a′**–**a″′**, **b′**–**b″**, **c′**–**c′″**) textures in left lateral view (left column), dorsal (middle column), and ventral (right column) view. (**a**) All jaw adductor and depressor muscles plotted. (**a′**) M. zygomaticomandibularis (22; numbers following the proposal of Werneburg 2011) removed. (**a″**) 22, all external adductor (m. add. mandibulae ext.) portions, and m. depressor mandibulae (man.) (45) removed. (**a′″**) 22, all ext. add. portions, 45, m. add. man. internus (int.) pars pseudotemporalis (23), and m. add. man. posterior (29) removed. (**b**) Dorsal view with all adductor and depressor muscles plotted. (**b′**) m. add. man. ext. pars profundus (19) and superficialis (21) removed. (**b″**) All external adductor portions, 22, 23, 29, and 45 removed. (**c**) Ventral view with all adductor and depressor muscles plotted. (**c′**) 22 and 45 removed. (**c″**) 22, 45, and all external adductor portions removed. (**c′″**) 22, 23, 29, 45, and all external adductor portions removed. **Additional abbreviations:** 17, m. add. man. ext. pars medialis; 26, m. add. man. int. pars pterygoideus dorsalis; 27, m. add. man. int. pars pterygoideus posterior; 28, m. add. man. int. pars pterygoideus ventralis; ot.ch, otic chamber; pd.em, posterodorsal emargination; po.br, postorbital bridge; so.cr, supraoccipital crest; sq.cr, squamosal crest

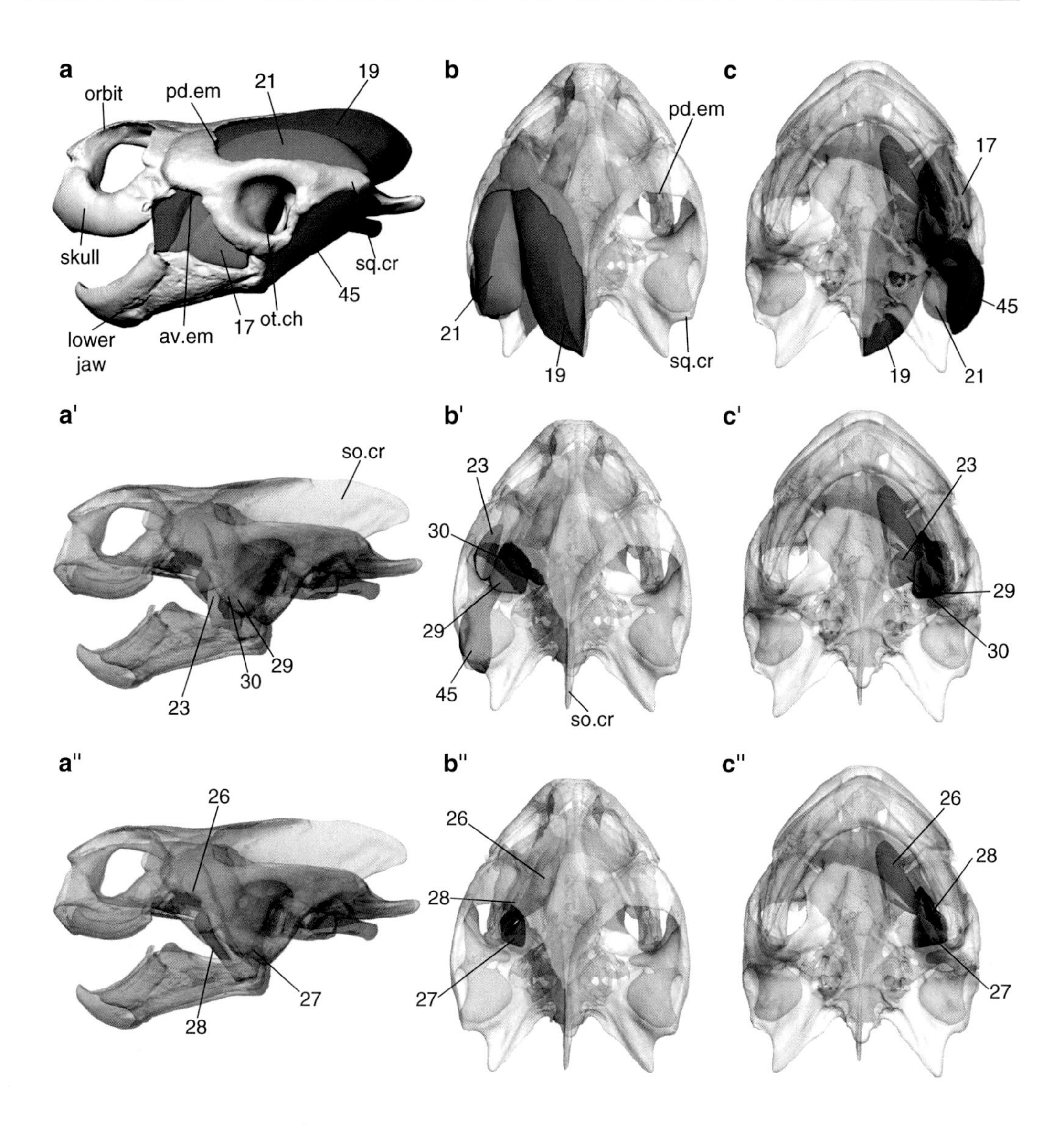

Fig. 8.7 Digital three-dimensional reconstruction of the jaw adductor and depressor musculature of *Pelomedusa subrufa* based on Schumacher (1954b, 1954/55) and Werneburg (2011, 2013a) made with CAD software Rhinoceros 3D (Robert McNeel & Associates 2003). Skull with solid (**a**) and transparent (**a′**–**a″**, **b′**–**b″**, **c′**–**c″**) textures in left lateral (left column), dorsal (middle column), and ventral (right column) views. (**a**, **b**, **c**) Skulls with all jaw adductor and depressor muscles plotted. (**a′**, **b′**, **c′**) All external adductor (m. add. mandibulae ext.) portions and m. depressor mandibulae (man.) (45; numbers following the proposal of Werneburg 2011) removed (45 plotted in **b′**). (**a″**, **b″**, **c″**) All external adductor muscles, 45, m. add. man. internus (int.) pars pseudotemporalis (23) and m. add. man. posterior (29) removed. **Additional abbreviations:** 17, m. add. man. ext. pars medialis; 19, m. add. man. ext. pars profundus; 21, m. add. man. ext. pars superficialis; 26, m. add. man. int. pars pterygoideus dorsalis; 27, m. add. man. int. pars pterygoideus posterior; 28, m. add. man. int. pars pterygoideus ventralis; 30, m. add. man. posterior pars rostralis; av.em, anteroventral emargination; ot.ch, otic chamber; pd.em, posterodorsal emargination; so.cr, supraoccipital crest; sq.cr, squamosal crest

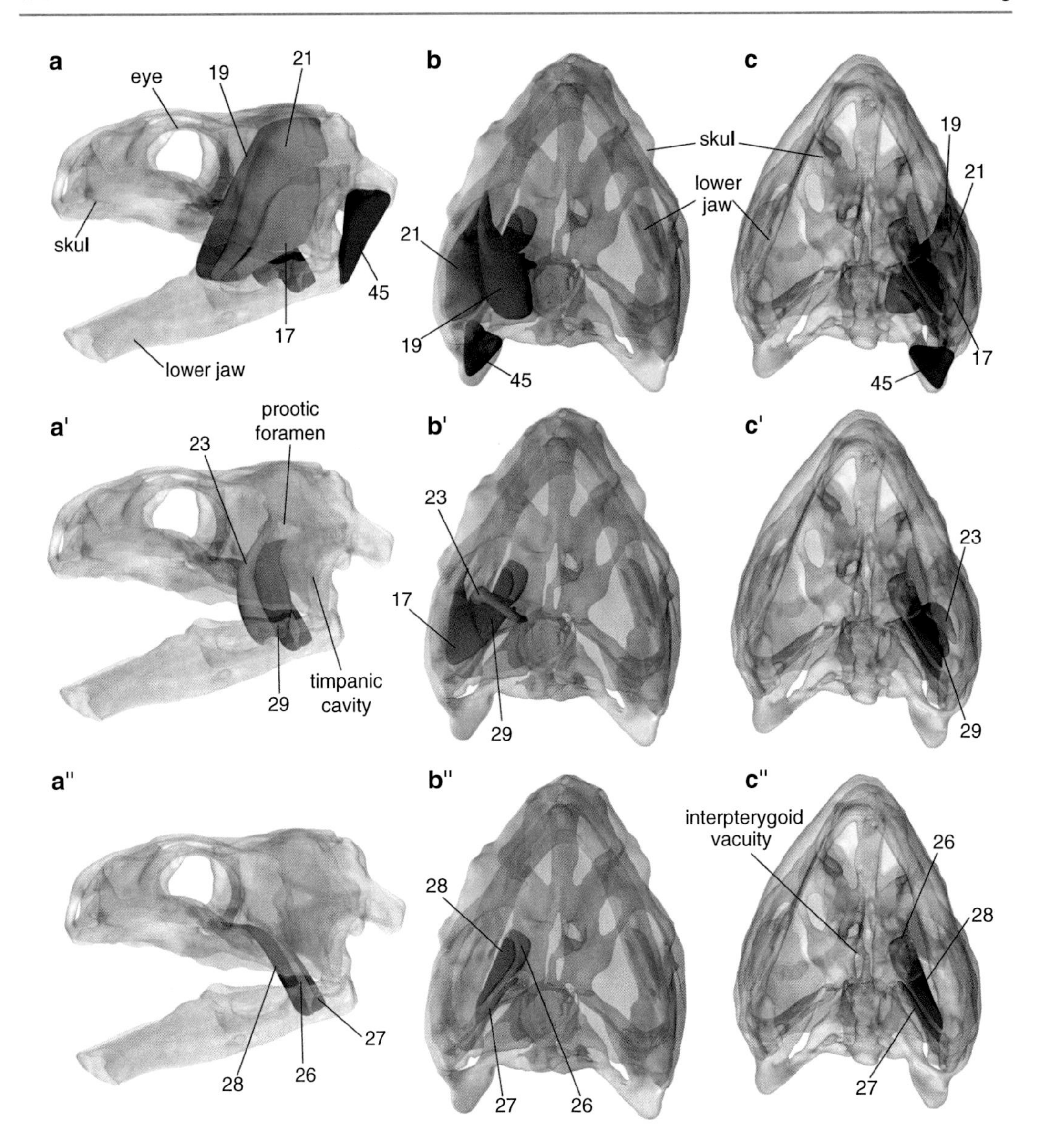

Fig. 8.8 Digital three-dimensional reconstruction of the jaw adductor and depressor musculature of *Proganochelys quenstedti* made with CAD software Rhinoceros 3D (Robert McNeel & Associates 2003). Skull with transparent textures in left lateral (left column), dorsal (middle column), and ventral (right column) views. (**a**, **b**, **c**) Skulls with all jaw adductor and depressor muscles plotted. (**a′**, **b′**, **c′**) All external adductor (m. add. mandibulae ext.) portions and m. depressor mandibulae (man.) (45; numbers following the proposal of Werneburg 2011) removed (note: m. add. man. externus pars medialis in **b′**). (**a″**, **b″**, **c″**) All external adductor muscles, 45, m. add. man. internus (int.) pars pseudotemporalis (23), and m. add. man. posterior (29) removed. **Additional abbreviations:** 26, m. add. man. int. pars pterygoideus dorsalis; 27, m. add. man. int. pars pterygoideus posterior; 17, m. add. man. externus (ext.) pars medialis; 19, m. add. man. ext. pars profundus; 21, m. add. man. ext. pars superficialis; 28, m. add. man. int. pars pterygoideus ventralis

partes profundus and superficialis of the **external adductor** have extended their origins far posteriorly due to its elongated crests (Fig. 8.6b) but also to other bones when compared to the reconstructed ancestral cryptodiran condition. The pars profundus extends its origin to the postorbital, opisthotic, and prootic beyond the ancestral cryptodiran origins on the parietal and supraoccipital, and the pars superficialis in *P. sinensis* originates on the quadratojugal, jugal, and opisthotic (Ogushi 1913a; Lakjer 1926; George and Shah 1955), in addition to the squamosal and quadrate, as in the ancestral cryptodire (Werneburg 2013a). Furthermore, the origin of both portions are strengthened by several tendinous sheets, extending from the coronar aponeurosis to the supraoccipital and squamosal crests of trionychids (Schumacher 1956; Werneburg 2011). The pars medialis (Fig. 8.6a′), which lies ventral to those parts and originates more anteriorly than the other portions of the **external adductors** (Werneburg 2013a), remained attached to the anterior surfaces of the quadrate and squamosal in the ancestral cryptodire, but in *P. sinensis*, it extends its origin posteroventrally near to the jaw joint on the quadrate, below the tympanic cavity (Ogushi 1913a; Werneburg 2011). All three portions fuse distally during their course to the insertion sites on the dorsal and lateral aspects of the **lower jaw** in all cryptodires (Werneburg 2011).

The three portions overlay a strong tendinous bundle, the coronar aponeurosis, which inserts on the coronoid process on the **lower jaw** of all turtles (Werneburg 2011). Although all cryptodires possess a trochlear articulation on the external surface of the otic chamber (Schumacher 1973; Gaffney 1975), there is variation among the different groups on the components of this mechanism (i.e., the surface of the bones, the size and shape of the tendon structure, and the type of gliding joint or surface developed between them). In *P. sinensis*, the quadrate and prootic form an anterodorsal projection onto the lower temporal fossa, pushing the trochlear articulation to a position just above the high coronoid process of the **lower jaw** where the coronar aponeurosis inserts. As in the ancestral cryptodiran condition, several fibers surrounding the coronar aponeurosis also insert directly to the dorsal and lateral surfaces of the **lower jaw**. In the ancestral cryptodire, the direct fibers were restricted to the coronoid bone, but in *P. sinensis*, they expand anteriorly and posteriorly to the surangular and dentary (Ogushi 1913a; Werneburg 2013a). The m. zygomaticomandibularis (Fig. 8.6a), found only in Trionychia and not in the ancestral cryptodire (Werneburg 2011), originates with two heads lateral to the **external adductors** in *P. sinensis* (Ogushi 1913a). The anterior head originates anteriorly on the medial and ventral surfaces of the jugal and quadratojugal on the postorbital bridge (Werneburg 2013a), and the posterior head runs laterally to it, attaching to the ventral and lateral aspects of this bone bridge but also on the temporal fascia and the anterolateral surface of the quadrate. The heads run ventrad and fuse on their way to insert broadly, partly via a tendon but mostly directly into a lateral depression of the dentary and surangular (Ogushi 1913a; Werneburg 2011, 2013a).

The arrangement of the **internal adductors** (Fig. 8.6a″, a‴) of *P. sinensis* is influenced by the broad horizontal plate of the pterygoid that extends posteriorly to the quadrate and by the anterodorsal projection of the prootic and quadrate described above. The fibers of the pterygoid muscle, as in the ancestral cryptodire, run posteriorly from their origin on the pterygoid and palatine, mostly above the horizontal plate of the pterygoid (Werneburg 2011). The enlarged palate of trionychids supports broader pterygoid muscles, in contrast to that found in other cryptodires, such as the marine turtle *Lepidochelys kempii* (Jones et al. 2012) and the snapping turtle *Chelydra serpentina* (Rieppel 1990). In *P. sinensis*, these are more robust, and the dorsalis portion reaches the descending process of the parietal (Fig. 8.6b″). The pars pseudotemporalis also expands its origin anteriorly to reach the posteroventral face of the frontal (Ogushi 1913a; Werneburg 2013a), differing from the ancestral cryptodiran condition, in which this muscle portion originated only on the descending process of the parietal. As in all turtles, the pars pseudotemporalis is very closely related to the posterior adductor (Rieppel 1990), which originates posteriorly to the former (Fig. 8.6a″) on the quadrate and prootic in the reconstructed ancestral cryptodire and in *P. sinensis*.

All the **internal adductors** insert to the medial aspect of the posterior half of the **lower jaw** (Fig. 8.6c″), in the region between the coronoid process and the retroarticular process near the jaw joint, mainly via tendinous structures (Schumacher 1973; Werneburg 2011). The insertions in *P. sinensis* and in the ancestral cryptodire do not differ much. The pars pseudotemporalis runs ventrad and inserts entirely to the subarticular aponeurosis, which inserts on the prearticular, surangular, and articular and to which some fibers of the pterygoid portions and the posterior adductor also insert. Most fibers of the posterior muscle, however, insert directly to the medial surfaces of the articular and prearticular. The pterygoid muscles insert on a broader area (Fig. 8.6a″), ranging from the coronoid to the articular on the retroarticular process (pars pterygoideus posterior) via the subarticular aponeurosis and direct fibers. Although seemingly absent in trionychids (Werneburg 2013a), in several other cryptodires and in the ancestral cryptodiran condition, the pterygoid muscles develop their own tendon (pterygoid aponeurosis = lamina anterior of the subarticular aponeurosis of Schumacher 1973), to which several fibers insert (Werneburg 2011, 2013a). Lastly, the m. depressor mandibulae has almost the same condition in *P. sinensis* and in the ancestral cryptodire, originating on the ventral and lateral surfaces of the squamosal, running ventrad, and inserting to the posterior and ventral faces of the articular on the retroarticular process (Fig. 8.6a′) via the retroarticular tendon (Werneburg 2013a). In *P. sinensis*, however, it originates with two heads that fuse distally, and may also insert on the surangular (Ogushi 1913a).

8.7.2 Jaw Muscles in Pleurodira

Among the three major side-necked turtle lineages, Chelidae, Podocnemididae, and Pelomedusidae (Gaffney and Meylan 1988), the skull of the latter is relatively similar (in many general aspects) to that of some cryptodire turtles (Fig. 8.7). It is dorsoventrally flattened, but not as much as in some chelids (e.g., *Chelodina oblonga*; Gaffney 1977). It has large posterodorsal and shallow-to-moderate anteroventral **emarginations**, and its supraoccipital and squamosal crests are posteriorly elongated (but not as extreme as in trionychids). As such, the skull of pelomedusids may be morphologically closer to that of the pleurodiran ancestors than the skulls of Chelidae and Podocnemididae. Indeed, by comparing the condition of *Pelomedusa subrufa* (Schumacher 1954b) to that of the reconstructed ancestral states for the **jaw musculature** of the **Pleurodira** node, we find only minor variations (Werneburg 2013a).

The **external adductors** in the ancestral pleurodires originated (Werneburg 2013a) on the parietal and supraoccipital, extending posteriorly to cover the crista supraoccipitalis (pars profundus), on the medial faces of the quadrate and squamosal, spanning to the lateral wall of the upper temporal fossa (pars superficialis), and to the anterior face of the quadrate (pars medialis). Their fibers fuse together soon after their points of origin and insert to the coronar aponeurosis, which runs anteroventrad underneath the muscle layers and bends over the processus trochlearis pterygoidei, where it develops a transiliens cartilage and forms a gliding surface on the dorsal aspect of the pterygoid. Most fibers insert via the coronar aponeurosis on the dorsal and lateral aspects of the coronoid process, but some fibers of the partes profundus and superficialis insert directly to the coronoid, whereas some fibers of the pars medialis pass directly to the dorsal and lateral surfaces of the surangular, posteriorly to the other portions. In *P. subrufa* (Fig. 8.7a, b), the only modifications from this morphology are the anterior expansion of the superficialis origin also to the quadratojugal, postorbital, and parietal and its insertion more posteriorly, with some fibers also reaching the surangular (Schumacher 1954b; Iordansky 1996).

Similarly, the internal and posterior adductors maintain, with some small changes, the inferred ancestral condition for pleurodires in *P. subrufa*.

The pars pterygoideus dorsalis originated anteriorly on the dorsal faces of the pterygoid and palatine, entering the tunnellike structure formed by the pterygoid ventrally and the postorbital and parietal dorsally (= sulcus palatino-pterygoideus of Gaffney et al. 2006) and inserting also on its medial wall (Fig. 8.7a″) on the descending process of the parietal (Schumacher 1954b, 1954/55). The pars pterygoideus ventralis originated on the dorsal and lateral aspects of the pterygoid, bordered laterally by the processus trochlearis pterygoidei. The pars pterygoideus posterior originated on the dorsal and ventral surfaces of the pterygoid, posteriorly to the origins of the partes pterygoideus ventralis and dorsalis (Fig. 8.7b″). In *P. subrufa*, the pars pterygoideus posterior expands its origins to the ventral aspect of the skull (Fig. 8.7c″), a trend that is observed to a greater extent in podocnemidids. In *Podocnemis expansa*, for example, the pars pterygoideus posterior inserts inside a cavity (cavum pterygoidei *sensu* Gaffney et al. 2006) formed by the pterygoid ventrally and the basisphenoid, quadrate, and prootic dorsally (Schumacher 1954a, 1973; Gaffney 1979).

As mentioned above, the pars pseudotemporalis and the posterior adductor are integrated among pleurodires, as in cryptodires (Werneburg 2011), and are positioned anteriorly and posteriorly, respectively, to the mandibular branch (V_3) of the **trigeminal nerve** (Lakjer 1926; Schumacher 1973; Werneburg 2011). The former originated, in the reconstructed pleurodire ancestor, on the descending process of the parietal, ran ventrad, and inserted on the subarticular aponeurosis (together with some fibers of the partes pterygoidei), which inserted on the medial aspect of the prearticular (Werneburg 2013a). In *P. subrufa* (Fig. 8.7a′) the origin of the pars pseudotemporalis extends to the prootic, above the **foramen nervi trigemini** (Schumacher 1954b, 1954/55; Iordansky 1996). The posterior adductor originates broadly on the anterior surfaces of the quadrate and prootic in the ancestral pleurodire. It is more restricted laterally in *P. subrufa* (Fig. 8.7b′), reaching only the anterolateral surface of the prootic (Werneburg 2013a). In *P. subrufa* and the ancestral pleurodire, the posterior adductor fibers run ventrad and insert to the medial face of the surangular on the **lower jaw**. Similarly to the ancestral condition of cryptodires, the m. depressor mandibulae of the ancestral pleurodire originated only with one head but develops a second head (Fig. 8.7a) in *P. subrufa* (Schumacher 1954b, 1954/55; Iordansky 1996). On the other hand, the depressor mandibulae origin site is broader in all pleurodires, spanning from the ventral and lateral surfaces of the squamosal to the opisthotic in the ancestral condition and in *P. subrufa* also on the medial face of the quadrate. In the latter, it runs ventrad and inserts to the posterior surface of the articular, but in the inferred pleurodire ancestor, it inserted more broadly, via a retroarticular tendon to the retroarticular process of the **lower jaw**.

8.7.3 Jaw Muscles in *Proganochelys quenstedti*

Although the skull of *Proganochelys quenstedti* may seem similar to that of some crown turtles, especially to those with a full dermal bone covering such as sea turtles, there are several important differences that most likely made the **jaw musculature** of that taxon distinct from the crown turtle general pattern (Fig. 8.8). *P. quenstedti* lacked supraoccipital and squamosal crests (Gaffney 1990), which, together with the smaller otic capsule than that of crown turtles, suggests a more vertical orientation of the **external adductors** inside the closed **adductor chamber**. As inferred from the reconstructed pattern of the **Testudines** ancestor, the pars profundus originated on the dorsal plate of the parietal and on the lateral aspect of the supraoccipital, occupying approximately half of the upper temporal fossa (Fig. 8.8b). The pars superficialis likely was attached, based on our reconstruction, to the medial faces of the quadrate and the squamosal (certainly, it also expanded to the medial face of the quadratojugal, jugal, and postorbital as in extant sea turtles, but this we could not reconstruct with the available data), mostly on the lateral braincase wall, and possibly some fibers to the roof of the **adductor chamber**. This arrangement is very similar to that of

diapsids, such as *Sphenodon punctatus* (Jones et al. 2009) and *Alligator mississippiensis* (Holliday and Witmer 2007), except for the pars medialis. The inferred position for this muscle in the ancestral **Testudines** resembles that of other turtles, and in *P. quenstedti* it should also be similar, based on the arrangement of the surrounding structures. The site of origin for the pars medialis likely was much more restricted than that of the other portions of the **external adductors**, only on the anterior surface of the quadrate (Fig. 8.8a) and extending slightly onto the lateral wall of the **adductor chamber**. This is more similar to the condition of *A. mississippiensis* (Holliday and Witmer 2007), in which it is also restricted to the anterior face of the quadrate, than to that of *S. punctatus*, in which it has a broad origin on the posttemporal bar, near to the site of insertion of the pars profundus (Jones et al. 2009).

The **external adductor** fibers in *P. quenstedti* likely ran ventrad to insert on the dorsal and lateral aspects of the **lower jaw** (Fig. 8.8a), on the low coronoid process, the dentary, and the surangular. Although all extant turtles possess a coronar aponeurosis to which most fibers of the **external adductors** attach, we cannot be certain if it was present in *P. quenstedti*, because the conditions in the outgroup representatives are diverse. The coronar aponeurosis of lepidosaurs (Lakjer 1926; synonym: basal aponeurosis, Jones et al. 2009; *bodenaponeurosis* see Werneburg 2013b) develops in a different way than that of turtles, namely, between the internal and **external adductor** anlagen (Rieppel 1987, 1990), and is shared only by the profundus and medialis portions of the **external adductors** (Jones et al. 2009). Among archosaurs, a similar structure could not be identified in birds and, although crocodiles possess a tendon shared by all **external adductors**, as in turtles (a potential synapomorphy), it is significantly folded (due to the suturing of the quadrate to the braincase; Holliday and Witmer 2007) and shared also by the posterior adductor (Iordansky 1994; Werneburg 2013b). It is important to note that the coronar aponeurosis would not form a transiliens cartilage in *P. quenstedti*. The almost vertical arrangement of the **external adductors** in the **adductor chamber** of *P. quenstedti*, confirmed by our reconstructions (Fig. 8.8a), suggests that it was not affected by the otic capsule in its course to the **lower jaw**, so there was no mechanical stress on the coronar aponeurosis to develop a transiliens cartilage (also present on the internal muscles of crocodiles and possibly temnospondyls; Tsai and Holliday 2011; Witzmann and Werneburg 2017) as found in crown turtles.

The internal and posterior adductors of *P. quenstedti* were certainly more restricted than in crown turtles (Fig. 8.8a′), especially by two features of its skull. First, this stem-turtle lacked a descending process of the parietal, which contributes to the secondary lateral braincase wall in extant turtles (Gaffney 1990; Eßwein 1992; Werneburg and Maier 2018), and from which most of the pars pseudotemporalis and some fibers of the pars pterygoideus dorsalis originate (Werneburg 2011). Secondly, *P. quenstedti* possessed very large eyes relative to the size of the rest of the skull (Gaffney 1990, Fig. 41, p. 51; Fig. 8.8), which most likely limited the space available for the anterior expansion of the pterygoid muscle portions. This also results in a much more restricted **adductor chamber** in comparison to crown turtles and, hence, the **internal adductor** should have been smaller as well. The pterygoid muscle portions should have been more integrated than in other turtles, given their topology: they all originated on the dorsal surface of the pterygoid (Fig. 8.8b″). The dorsalis portion was more anteriorly located, extending on a dorsal depression, just posterior to the border of the foramen palatinum posterius (Gaffney 1990). This portion was followed by the ventralis, which would have originated on the lateral border of the pterygoid as well, and by the posterior portion, originating also on a small dorsal projection near the suture with the quadrate. The ancestral state reconstruction does not support a ventral origin of any portion of the pterygoid muscle for the ancestral **Testudines** (Werneburg 2013a). Given the presence of teeth on the ventral surface of the pterygoid bone in *P. quenstedti* (Werneburg 2013a), this was likely the same state in this taxon and other stem-turtles (compare to Witzmann and Werneburg 2017). Thus, the expansion of the pterygoid origin to the ventral aspect of this bone should have happened at least twice in **Testudines**, once within **Cryptodira**, in Americhelydia (*sensu* Joyce et al. 2013a), and once within **Pleurodira**, among Pelomedusoides (Werneburg 2013a).

In *P. quenstedti*, the pars pseudotemporalis likely originated anterior to the **foramen nervi trigemini** (Fig. 8.8a′) as in all extant turtles (Werneburg 2011), on the small, thin process of the parietal that closed the prootic foramen anteriorly and contacted the basisphenoid ventrally, near the processus clinoideus (Gaffney 1990). It may have been continuous with the origin of the posterior adductor, which attached to the anterior faces of the prootic and quadrate, medial to the origin of the pars medialis of the **external adductor** (Fig. 8.8b′). If so, this would not differ much from the condition seen in several extant turtles in which the pseudotemporalis and posterior are highly integrated (Werneburg 2011). In crown turtles, the pseudotemporalis and the posterior are only recognized as separate unities by the passage of the mandibular branch (V_3) of the **trigeminal nerve** that pierces this otherwise continuous muscle mass (Lakjer 1926; Rieppel 1990). The main difference would be that the pars pseudotemporalis in **Testudines** extends more anteriorly than that of *P. quenstedti*, following the anteroventral expansion of the descending process of the parietal, which approaches the palate in the former. In the ancestral **Testudines**, the posterior adductor originated on the prootic and quadrate, probably reaching the pars pseudotemporalis. However in *P. quenstedti*, it is not clear if this condition was present, given the open **foramen nervi trigemini** that forms a groove that extends to the prootic foramen in this taxon (Gaffney 1990). It is possible that in this stem-turtle, the origins of the posterior adductor and that of the pars pseudotemporalis were separated by that space between the prootic and **trigeminal nerve** foramina. A well separated pars pseudotemporalis and a posterior adductor is displayed by archosaurs (Holliday and Witmer 2007) and lepidosaurs (Jones et al. 2009), and, if this was also the case in *P. quenstedti*, a greater integration between those muscles would have been a feature acquired during the evolution of turtles. This had the possible advantage of them acting as one united powerful muscle vector during jaw closure.

The insertions of the internal and posterior adductors in the ancestral **Testudines** and *P. quenstedti* roughly corresponded to those of crown turtles. The pars pseudotemporalis inserted via the subarticular aponeurosis on the medial and dorsal surfaces of the prearticular, inside the Meckel's fossa. A pars intramandibularis, as found in some extant turtles (Werneburg 2011), was most likely absent. Some fibers of the partes pterygoidei shared the subarticular aponeurosis, but a pterygoidal aponeurosis was most likely present in the ancestral **Testudines** (Werneburg 2013a) and some fibers would also insert directly on the medial surfaces of the prearticular and articular. The posterior adductor inserted dorsal to the pterygoid portions insertion sites and posterior to the pars pseudotemporalis insertion (Fig. 8.8a′), also on the prearticular and articular, closer to the jaw joint. Finally, the m. depressor mandibulae originated with only one head on the ventral and lateral aspects of the squamosal in the ancestor of **Testudines** and likely also on the lateral portions of the opisthotic in *P. quenstedti*. The depressor mandibulae ran ventrad (as in all turtles) to insert posteriorly on the retroarticular process of the articular bone (Fig. 8.8a), possibly via the retroarticular aponeurosis and direct fibers.

Extant turtles do not show a m. levator pterygoidei, a muscle, which connects the primary braincase wall with the pterygoid and enables movement of the palate in reptiles with kinetic skulls. A rudimentary anlage of this muscle, however, was found in the emydid cryptodire *Emys orbicularis* (Fuchs 1915). A kinetic basicranial articulation was likely present in *P. quenstedti* (Gaffney 1990) and other stem-turtles (Rabi et al. 2013), which is further supported by the recapitulation of this embryonic structure in an extant species.

The ancestor of **Testudines** possessed more robust external adductors relative to the internal and posterior adductors, as in all other turtles (Iordansky 1996; Schumacher 1973; Werneburg 2011), and likely the same condition was found in

P. quenstedti. Lepidosaurs show the same relation (Jones et al. 2009), whereas in archosaurs the opposite is observed (Holliday and Witmer 2007). Regarding the portions of the **external adductors**, all turtles (including *P. quenstedti*) show a greater integration between the partes medialis and superficialis (Rieppel 1990; Werneburg 2011), contrasting to the condition observed in lepidosaurs, in which the partes medialis and profundus are more closely related (Jones et al. 2009; Daza et al. 2011; Diogo and Abdala 2010). The strong coronar aponeurosis of turtles is developed between the **external adductor** layers, as in crocodiles, contrasting to the condition found in lepidosaurs, in which it develops between the internal and external rudiments (Rieppel 1987, 1990; Holliday and Witmer 2007; Jones et al. 2009). More integrated pseudotemporalis and posterior adductors, on the other hand, are features exclusive to turtles, whereas in archosaurs and lepidosaurs, these parts are more extensively separated (Holliday and Witmer 2007; Jones et al. 2009). In addition to other features reported by previous studies supporting relationships of turtles either with archosaurs [i.e., external eye pyramidalis muscle (Werneburg 2011) and closely related internal and posterior adductors and inter-/intramandibularis (Rieppel 1990; Werneburg 2011)] or with lepidosaurs [i.e., compartmentalization of m. intramandibularis from the internus anlage (Rieppel 1990; Werneburg 2011)], available data seem to present a mosaic of archosaurian and lepidosaurian features in the turtle head musculature. This may suggest that these are actually symplesiomorphies retained alternatively in those groups and in turtles and, as such, do not support closer relationships with either archosaurs or lepidosaurs.

The homology of specific parts of the **jaw musculature** (i.e., the different portions of the external, internal, and posterior adductors) between different extant sauropsid groups (archosaurs, lepidosaurs, and turtles) is still poorly understood (Holliday and Witmer 2007; Werneburg 2011). This is largely due to the fact that we can only directly assess the morphology of the muscles of extant reptilian taxa, which diverged from each other and started to diversify as early as in the Permian, approximately 300–250 million years ago (Mulcahy et al. 2012). Hence, using ancestral state estimates in conjunction with detailed anatomical analysis of fossils that aim to reconstruct the **jaw musculature** in extinct taxa will provide real potential to help resolving these issues. The known skull material of *P. quenstedti* does not have well-preserved surfaces in this region, so it is difficult to infer sites of muscle attachment with sufficient detail to musculature reconstruction (see Witzmann and Werneburg 2017). Nevertheless, further analyses of additional stem-turtles can generate more data on those potential attachment sites expanding our knowledge of the early evolution of the **jaw musculature** in turtles.

8.8 Concluding Remarks

Although turtles have attracted the attention of researchers for a long time, important aspects of their evolution are still debated. New anatomical, paleontological, and developmental data have the potential to increase our understanding of the nature of certain derived morphological structures, such as the turtle shell (Li et al. 2008; Nagashima et al. 2012). Additionally, considerable controversy still surrounds the phylogenetic origin and the interrelationship of turtles among reptiles. In this context, the turtle skull presents one of the most promising structures to resolve those open questions.

One of the open questions is the multiple origins of turtle temporal **emarginations** and dermatocranial bone reductions in amniotes, in general. For many years (Zdansky 1923; Romer 1956; Kilias 1957), the bulging of the jaw adductor muscles were thought to be primarily responsible for those reductions, but current interpretations suggest that other factors can be related to them. In the case of turtles, the modifications induced by the origin of the shell and alternate mechanisms of **neck retraction** seem to play prominent roles in the emergence of **emarginations** (Werneburg 2015; Werneburg et al. 2015a, b). However, variation among taxa with similar **neck retraction** mechanisms (such as chelids and pelomedusids) suggests other factors may also contribute to shape their morphology. Additionally, one can ask if the factors shaping the temporal **emarginations** in turtles also affect the repeated origin of temporal openings in other amniotes. New studies of functional morphology (e.g., finite element analyses), allied to comparative developmental analyses focusing on those structures, may help to clarify their role and evolution in the amniote head.

Developmental studies with larger samples can also be useful to assess some putative synapomorphies of turtles with other amniotes. For example, m. pyramidalis of the eye is found only in turtles, crocodiles, and birds and is used as morphological evidence supporting the prevailing results of molecular phylogenetics that place turtles closer to archosaurs (Rieppel 1990; Eger 2006). Although morphologically they seem to be homologous structures, their ontogenetic origin has never been studied in a more comprehensive reptile framework. At the same time, the pattern of differentiation of the posterior adductor and intermandibularis muscles in turtles are more similar to those of crocodiles than to that of squamates (Rieppel 1990). Further support or rejection of these primary statements of homology may be obtained via studies of the ontogeny of these structures in turtles. To date, muscle development has been studied in detail only in *Chelydra serpentina* (Rieppel 1990). However, though Tvarožková (2006) studied other taxa, she had access to only a few embryos at later stages of development, and although Edgeworth (1935) studied three other turtle species, this was done only superficially.

Several other evolutionary questions remain open. For example, the relation of the m. zygomaticomandibularis in Trionychia to the other adductors and to structures in other turtles, such as the postorbital head of the pars superficialis in chelydrids and chelonioids (Werneburg 2011; Jones et al. 2012), can be an important issue, revealing new synapomorphies that can contribute to the debated interrelationships of cryptodiran clades (Sterli 2010). Also, functional anatomical analyses of this muscle may also contribute to broader questions, since the repeated evolution of comparable structures in other non-related taxa, such as parrots and mammals (Schulman 1906; Tokita 2004), suggests specific functional roles for this muscle unit perhaps related to particular **feeding** preferences. The **trochlear mechanism** found in the **adductor chamber** of turtles is another debated evolutionary question. The employment of functional models (e.g., finite element analyses) to further explore its role during the contraction of the **external adductors** would be worthy of attention, especially considering the different morphotypes found among cryptodires and pleurodires. Their origin is relevant to understand whether or not stem-turtles had a **trochlear mechanism** and, if so, how was its anatomy presented. Was it similar to that of cryptodiran taxa only in its position (on the outer surface of the otic chamber) or in other aspects (e.g., bone projections and synovial capsule like the cryptodiran trochlea)? Or, how did the pleurodiran mechanism evolve? Was it independently acquired or derived from a cryptodiran-like mechanism of stem-turtles? And if the latter is more likely, did the pleurodiran ancestors possess an intermediate stage with two trochleas (as hypothesized by Joyce 2007)? Those questions can be approached comparatively from the perspective of developmental data (comparing the ontogeny of those structures in pleurodires and cryptodires), functional and structural anatomy of extant species (comparing the mechanisms in different taxa to see how much variation exists), and paleontology (to evaluate the condition in extinct turtles), i.e., in sum, a 'holistic organism approach' (Maier and Werneburg 2014).

Finally, our approach to reconstruct the **jaw musculature** in *Proganochelys quenstedti* by combining 3D models with ancestral state reconstructions (Werneburg 2013a) is an important step toward the understanding of the evolution of jaw muscles in turtles and amniotes in general. This approach has been already used elsewhere with other groups, such as temnospondyls (Lautenschlager et al. 2016; Witzmann and Werneburg 2017) and sauropterygians (Araújo and Polcyn 2013; Foffa et al. 2014) to understand the evolution of some morphological structures and the biomechanics of their skulls and should be more applied when considering the questions highlighted above.

Acknowledgments Janine Ziermann, Raul Diaz Jr., and Rui Diogo are thanked for the invitation to write this chapter. We would like to thank Juliane Hinz (Tübingen) for help with the 3D models and Wolfgang Maier for access to the histological sections. We also thank two anonymous reviewers for their suggestions. GSF was supported by FAPESP (Fundação de Amparo à Pesquisa do Estado de São Paulo) grants 2016/03934-2 and 2014/2539-5. IW was supported by SNF advanced postdoc mobility grant P300PA_164720.

References

Anderson NJ (2009) Biomechanics of feeding and neck motion in the Softshell turtle, *Apalone spinifera*, Rafinesque [doctor of arts thesis in the Department of Biology, Idaho state university]. ProQuest, Ann Arbor

Anquetin J, Tong H, Claude J (2017) A Jurassic stem pleurodire sheds light on the functional origin of neck retraction in turtles. Sci Rep 7:42376. https://doi.org/10.1038/srep42376 http://www.nature.com/articles/srep42376#supplementary-information

Araújo R, Polcyn MJ (2013) A biomechanical analysis of the skull and adductor chamber muscles in the Late Cretaceous plesiosaur *Libonectes*. Palaeontologica Electronica 16(2:10A):25

Baur G (1889) On the morphology of the vertebrate skull. J Morphol 3:471–474

de Beer GR (1937) The development of the vertebrate skull. The University of Chicago Press, Chicago

Bels VL, Davenport J, Delheusy V (1997) Kinematic analysis of the feeding behavior in the box turtle *Terrapene carolina* (L.), (Reptilia: Emydidae). J Exp Zool 277:198–212

Bels V, Baussart S, Davenport J, Shorten M, O'Riordan RM, Renous S, Davenport JL (2008) Functional evolution of feeding behavior in turtles. In: Wyneken J, Godfrey MH, Bels V (eds) Biology of turtles. CRC Press, Boca Raton, pp 187–212

Bever GS, Lyson TR, Field DJ, Bhullar B-AS (2015) Evolutionary origin of the turtle skull. Nature 525:239–242

Bever GS, Lyson TR, Field DJ, Bhullar B-AS (2016) The amniote temporal roof and the diapsid origin of the turtle skull. Zoology 119:471–473. https://doi.org/10.1016/j.zool.2016.04.005

Bojanus LH (1819) Anatome testudinis europaeae. Isis 11:1766–1769 1762 plates

Boulenger GA (1918) Sur la place des cheloniens dans la classification, vol 167. Comptes Tendues A l'Academie Des Sciences, Paris, pp 614–618

deBraga M, Rieppel O (1997) Reptile phylogeny and the interrelationships of turtles. Zool J Linnean Soc 120:281–354

Bramble DM (1974) Occurrence and significance of the os transiliens in gopher tortoises. Copeia:102–109

Broom R (1924) On the classification of the reptiles. Bull Am Mus Nat Hist 51:39–65

Burke AC (1989) Development of the turtle carapace: implications for the evolution of a novel Bauplan. J Morphol 199:363–378

Cebra-Thomas J, Tan F, Sistla S, Estes E, Bender G, Kim C, Riccio P, Gilbert SF (2005) How the turtle forms its shell: a paracrine hypothesis of carapace formation. J Exp Zool B Mol Dev Evol 304B:558–569

Clark K, Bender G, Murray BP, Panfilio K, Cook S, Davis R, Murnen K, Tuan RS, Gilbert SF (2001) Evidence for the neural crest origin of turtle plastron bones. Genesis 31:111–117

Claude J, Pritchard P, Tong H, Paradis E, Auffray JC (2004) Ecological correlates and evolutionary divergence in the skull of turtles: a geometric morphometric assessment. Syst Biol 53:933–948

Cope E (1896) The ancestry of the Testudinata. Am Nat 30:398–400

Cordero GA, Quinteros K (2015) Skeletal remodelling suggests the turtle's shell is not an evolutionary straitjacket. Biol Lett 11:20150022. https://doi.org/10.1098/rsbl.2015.0022

Crawford NG, Parham JF, Sellas AB, Faircloth BC, Glenn TC, Papenfuss TJ, Henderson JB, Hanson MH, Simison WB (2015) A phylogenomic analysis of turtles. Mol Phylogenet Evol 83:250–257

Dalrymple GH (1975) Variation in the cranial feeding mechanism of turtles of the genus *Trionyx* Geoffroy. PhD thesis, University of Toronto, Toronto

Dalrymple GH (1977) Intraspecific variation in the cranial feeding mechanism of turtles of the genus *Trionyx* (Reptilia, Testudines, Trionychidae). J Herpetol 11:255–285

Daza JD, Diogo R, Johnston P, Abdala V (2011) Jaw adductor muscles across lepidosaurs: a reappraisal. Anat Rec 294:1765–1782. https://doi.org/10.1002/ar.21467

Diogo R, Abdala V (2010) Muscles of vertebrates. CRC Press/Science Publishers, Boca Bacon, New York; Oxon/Enfield

Diogo R, Abdala V, Lonergan N, Wood BA (2008) From fish to modern humans–comparative anatomy, homologies and evolution of the head and neck musculature. J Anat 213:391–424

Edgeworth FH (1935) The cranial muscles of vertebrates. Cambridge University Press, London

Eger SC (2006) Morphologische und phylogenetische Untersuchungen an der Nickhautmuskulatur bei Sauropsiden (unter besonderer Berücksichtigung der Chelonia). Universität Tübingen, Tübingen

Ernst CH, Barbour RW (1992) Turtles of the world. Smithsonian Institution Scholarly Press, Washington, DC

Eßwein SE (1992) Zur phylogenetischen und ontogenetischen Entwicklung des akinetischen Craniums der Schildkröten. Natürliche Konstruktionen-Mitteilungen des SFB 230 7 (Proceedings of the II. International Symposium of the Sonderforschungsbereich 230, Stuttgart, 1.-4.10.1991):51–55

Ferreira GS (2016) Abordagens convergentes, novidades evolutivas e a origem da carapaça das tartarugas. Revista da Biologia 16:6

Ferreira GS, Lautenschlager S, Langer MC, Evers SW, Rabi M, Werneburg I (2018) Biomechanical analyses suggest relation between neck-retraction and the trochlear mechanism in extant turtles. In: Turtle evolution symposium. Scidinge Hall, Tübingen, pp 38–40

Foffa D, Cuff AR, Sassoon J, Rayfield EJ, Mavrogordato MN, Benton MJ (2014) Functional anatomy and feeding biomechanics of a giant Upper Jurassic pliosaur (Reptilia: Sauropterygia) from Weymouth Bay, Dorset, UK. J Anat 225:209–219. https://doi.org/10.1111/joa.12200

Frazzetta TH (1968) Adaptive problems and possibilities in the temporal fenestration of tetrapod skulls. J Morphol 125:145–157

Frolich LM (1997) The role of the skin in the origin of amniotes: permeability barrier, protective covering and mechanical support. In: Sumida SS, Martin KLM (eds) Amniote origins. Completing the transition to land. Academic Press, San Diego

Fuchs H (1915) Über den Bau und die Entwicklung des Schädels der *Chelone imbricata*. Ein Beitrag zur Entwicklungsgeschichte und vergleichenden Anatomie des Wirbeltierschädels. Erster Teil: Das Primordialskelett des Neurocraniums und des Kieferbogens. In: Voeltzkow A (ed) Reise in Ostafrika in den Jahren 1903–1905, Wissenschaftliche Ergebnisse, vol 5. Schweizerbart, Stuttgart, pp 1–325

Fuchs H (1931) Von dem Ductus angularis oris der Arrauschildkröte (*Podocnemis expansa*). (Ein neues Organ?). Nachrichten von der Gesellschaft der Wissenschaften zu Göttingen. Mathematisch-Physikalische Klasse:131–147

Gaffney ES (1975) A phylogeny and classification of the higher categories of turtles. Bull Am Mus Nat Hist 155:387–436

Gaffney ES (1977) The side-necked turtle family Chelidae: a theory of relationships using shared derived characters. Am Mus Novit 2620:1–28

Gaffney ES (1979) Comparative cranial morphology of recent and fossil turtles. Bull Am Mus Nat Hist 164:67–376

Gaffney ES (1980) Phylogenetic relationships of the major groups of amniotes. In: Panchen AL (ed) The terrestrial environment and the origin of land vertebrates, Systematic Association, vol 15. Academic Press, London

Gaffney ES (1985) The cervical and caudal vertebrae of the cryptodiran turtle, *Meiolania platyceps*, from the Pleistocene of Lord Howe Island. Australia American Museum Novitates 2805:1–29

Gaffney ES (1990) The comparative osteology of the Triassic turtle *Proganochelys*. Bull Am Mus Nat Hist 194:1–263

Gaffney ES, Jenkins FA Jr (2010) The cranial morphology of *Kayentachelys*, an early Jurassic cryptodire, and the early history of turtles. Acta Zool 91:335–368

Gaffney ES, Kitching JW (1994) The most ancient African turtle. Nature 369:55–58

Gaffney ES, Kitching JW (1995) The morphology and relationships of *Australochelys*, and early Jurassic turtle from South Africa. Am Musem Novitates 3130:29

Gaffney ES, Meylan PA (1988) A phylogeny of turtles. In: Benton MJ (ed) The phylogeny and classification of the Tetrapods. Volume 1: Amphibians, reptiles, birds, vol 35A. Clarendon Press, Oxford, pp 157–219

Gaffney ES, Hutchison JH, Jenkins AF, Meeker LJ (1987) Modern turtle origins: the oldest known cryptodire. Science 237:289–291

Gaffney ES, Tong H, Meylan PA (2006) Evolution of the side-necked turtles: the families Bothremydidae, Euraxemydidae, and Araripemydidae. Bull Am Mus Nat Hist 300:700

Gardiner BG (1993) Haematothermia: warm-blooded amniotes. Cladistics 9:369–395

Gasc JP (1981) Axial musculature. In: Gans C, Parsons TS, Parsons TS (eds) Biology of the Reptilia, (Morphology F). morphology D, vol 11. Academic Press, London, pp 355–435

Gaupp E (1895) Zur vergleichenden Anatomie der Schläfengegend am knöchernen Wirbeltierschädel. Morphologische Arbeiten 4:77–131

Gauthier J, Kluge AG, Rowe T (1988) Amniote phylogeny and the importance of fossils. Cladistics 4:105–209

Gauthier JA, Kearney M, Anderson Maisano J, Rieppel O, Behlke ADB (2012) Assembling the Squamate Tree of Life: Perspectives from the Phenotype and the Fossil Record. Bulletin of the Peabody Museum of Natural History 53(1):3–308

George JC, Shah RV (1955) The myology of the head and the neck of the common Indian pond turtle, *Lissemys punctata granosa* Schoepff. J Anim Morphol Physiol 1:1–12

Gilbert SF, Loredo GA, Brukman A, Burke AC (2001) Morphogenesis of the turtle shell: the development of a novel structure in tetrapod evolution. Evol Dev 3:47–58

Gilbert SF, Bender G, Betters E, Yin M, Cebra-Thomas JA (2007) The contribution of neural crest cells to the nuchal bone and plastron of the turtle shell. Integr Comp Biol 47:401–408. https://doi.org/10.1093/icb/icm020

Gilbert SF, Cebra-Thomas JA, Burke AC (2008) How the turtle gets its shell. In: Wyneken J, Godfrey MH, Bels V (eds) Biology of turtles. CRC Press, Boca Raton, pp 1–16

Goodrich ES (1916) On the classification of the Reptilia. Proc R Soc London Ser B, Containing Papers of a Biological Character 89:261–276

Goodrich ES (1930) Studies on the structure and development of vertebrates. Macmillan and Co, London

Gregory WK (1946) Pareiasaurs versus placodonts as near ancestors to the turtles. Bull Am Mus Nat Hist 86:277–326

Gregory WK, Adams LA (1915) The temporal fossæ of vertebrates in relation to the jaw muscles. Science 41:763–765

Guillon JM, Guéry L, Hulin V, Girondot M (2012) A large phylogeny of turtles (Testudines) using molecular data. Contrib Zool 81:147–158

Hacker G (1954) Über Kiefermuskulatur und Mundfascien bei *Testudo graeca*. PhD thesis, Ernst-Moritz-Arndt-Universität, Greifswald

Hay OP (1905) On the group of fossil turtles known as the Amphichelydia; with remarks on the origin and relationships of the suborders, superfamilies, and families of Testudines. Bull Am Mus Nat Hist 21:137–175

Hedges SB (2012) Amniote phylogeny and the position of turtles. BMC Biol 10:64

Hedges SB, Poling LL (1999) A molecular phylogeny of reptiles. Science 283:998–1001

Heiss E, Plenk H, Weisgram J (2008) Microanatomy of the palatal mucosa of the semiaquatic Malayan box turtle, *Cuora amboinensis*, and functional implications. Anat Rec 291:10

Herrel A, O'Reilly JC, Richmond AM (2002) Evolution of bite performance in turtles. J Evol Biol 15:1083–1094

Herrel A, Van Damme J, Aerts P (2008) Cervical anatomy and function in turtles. In: Wyneken J, Godfrey MH, Bels V (eds) Biology of turtles. CRC Press, Boca Raton, pp 163–185

Hoffmann CK (1890) Reptilien. 1. Schildkröten, 6(3). Dr. H.G. Bronn's Klassen und Ordnungen des Thier-Reichs, wissenschaftlich dargestellt in Wort und Bild. C.F. Winter'sche Verlagshandlung, Leipzig

Hofsten N (1941) On the phylogeny of the Reptilia. Zool Bidrag Fran Uppsala 20:501–521

Holliday CM, Witmer LM (2007) Archosaur adductor chamber evolution: integration of musculoskeletal and topological criteria in jaw muscle homology. J Morphol 268:457–484. https://doi.org/10.1002/Jmor.10524

Iordansky NN (1987) Morphological and functional features of mandibular apparatus in turtles (Reptilia, Chelonia) and the problem of their origin [in Russian] (English abstract), МОРФО-ФУНКЦИОНАЛЬНЫЕ ОСОБЕННОСТИ ЧЕЛЮСТНОГО АППАРАТА ЧЕРЕПАХ (REPTILIA, CHELONIA) И ПРОБЛЕМА ИХ ПРОИСХОЖДЕНИЯ. Zoologichesky Zhurnal 66:1716–1729

Iordansky NN (1994) Tendons of jaw muscles in Amphibia and Reptilia: homology and evolution. Russ J Herpetol 1:13–20

Iordansky NN (1996) Jaw musculature of turtles: structure, functions, and evolutionary conservatism. Russ J Herpetol 3:49–57

Iordansky NN (2010) Pterygoideus muscles and other jaw adductors in amphibians and reptiles Biol Bull 37:905–914. [English version of Russian original text]

Jannel A (2015) Neck mobility, grazing habits, and intraspecific combat behaviour in the Giant Pleistocene horned turtle *Meiolania platyceps*. Uppsala Universitet

Jones MEH, Curtis N, O'Higgins P, Fagan M, Evans SE (2009) The head and neck muscles accociated with feeding on *Sphenodon* (Reptilia: Lepidosauria: Rynchocephalia). Palaeontologia Electronica 12(7A):56 http://palaeo-electronica.org/2009_2002/2179/index.html

Jones MEH, Werneburg I, Curtis N, Penrose R, O'Higgins P, Fagan MJ, Evans SE (2012) The head and neck anatomy of sea turtles (Cryptodira: Chelonioidea) and skull shape in Testudines. PLoS One 7:e47852. https://doi.org/10.1371/journal.pone.0047852

Joyce WG (2007) Phylogenetic relationships of Mesozoic turtles. Bull Peabody Mus Nat Hist 48:3–102

Joyce WG (2015) The origin of turtles: a paleontological perspective. J Exp Zool B Mol Dev Evol 324:181–193

Joyce WG, Lyson TR (2015) A review of the fossil record of turtles of the clade Baenidae. Bull Peabody Mus Nat Hist 56:147–183. https://doi.org/10.3374/014.056.0203

Joyce WG, Sterli J (2012) Congruence, non-homology, and the phylogeny of basal turtles. Acta Zool 93:149–159. https://doi.org/10.1111/j.1463-6395.2010.00491.x

Joyce WG, Parham JF, Gauthier JA (2004) Developing a protocol for the conversion of rank-based taxon names to phylogenetically defined clade names, as exemplified by turtles. J Paleontol 78:989–1013

Joyce WG, Werneburg I, Lyson TR (2013a) The hooked element in the pes of turtles (Testudines): a global approach to exploring homology. J Anat 223:421–441

Joyce WG, Parham JF, Lyson TR, Warnock RCM, Donoghue PCJ (2013b) A divergence dating analysis of turtles using fossil calibrations: an example of best practice. J Paleontol 87:612–634

Joyce WG, Rabi M, Clark JM, Xu X (2016) A toothed turtle from the late Jurassic of China and the global biogeographic history of turtles. BMC Evol Biol 16:236. https://doi.org/10.1186/s12862-016-0762-5

Karl HV (1997) Zur Taxonomie und Morphologie einiger tertiärer Weichschildkröten unter besonderer Berücksichtigung von Trionychinae Zentraleuropas (Testudines: Trionychidae). PhD thesis, Universität Salzburg, Salzburg

Kilias R (1957) Die funktionell-anatomische und systematische Bedeutung der Schläfenreduktion bei Schildkröten. Mitteilungen aus dem Zoologischen Museum in Berlin 33:307–354

Kuratani S, Kuraku S, Nagashima H (2011) Evolutionary developmental perspective for the origin of turtles: the folding theory for the shell based on the developmental nature of the carapacial ridge. Evol Dev 13:1–14

Lakjer T (1926) Studien über die Trigeminus-versorgte Kaumuskulatur der Sauropsiden. C.A. Reitsel Buchhandlung, Copenhagen

Lambertz M, Böhme W, Perry SF (2010) The anatomy of the respiratory system in *Platysternon megacephalum* Gray, 1831 (Testudines: Cryptodira) and related species, and its phylogenetic implications. Comp Biochem Physiol A Mol Integr Physiol 156:7. https://doi.org/10.1016/j.cbpa.2009.12.016

Laurin M (2002) Tetrapod phylogeny, amphibian origins, and the definition of the name Tetrapoda. Syst Biol 51:6

Laurin M, Piñeiro GH (2017) A reassessment of the taxonomic position of mesosaurs, and a surprising phylogeny of early amniotes. Front Earth Sci 5:88

Laurin M, Reisz RR (1995) A reevaluation of early amniote phylogeny. Zool J Linnean Soc 113:165–223

Lautenschlager S, Witzmann F, Werneburg I (2016) Palate anatomy and morphofunctional aspects of interpterygoid vacuities in temnospondyl cranial evolution. Sci Nat 103:79. https://doi.org/10.1007/s00114-016-1402-z

Lee MSY (1993) The origin of the turtle body plan: bridging a famous morphological gap. Science 261:1716–1720

Lee MSY (1995) Historical burden in systematics and the interrelationships of 'parareptiles'. Biol Rev 70:459–547

Lee MSY (1997) Pareiasaur phylogeny and the origin of turtles. Zool J Linnean Soc 120:197–280

Lee MSY (2013) Turtle origins: insights from phylogenetic retrofitting and molecular scaffolds. J Evol Biol 26:2729–2738

Lemell C, Weisgram J (1997) Feeding patterns of *Pelusios castaneus* (Chelonia: Pleurodira). Neth J Zool 47:429–441

Lemell P, Lemell C, Snelderwaard P, Gumpenberger M, Wocheslånder R, Weisgram J (2002) Feeding patterns of *Chelus fimbriatus* (Pleurodira: Chelidae). J Exp Biol 205:1495–1506

Lemell P, Beisser CJ, Gumpenberger M, Snelderwaard P, Gemel R, Weisgram J (2010) The feeding apparatus of *Chelus fimbriatus* (Pleurodira; Chelidae)–adaptation perfected? Amphibia-Reptilia 31:97–107

Li C, Wu XC, Rieppel O, Wang LT, Zhao LJ (2008) An ancestral turtle from the Late Triassic of southwestern China. Nature 456:497–501

Li C, Fraser NC, Rieppel O, Wu X-C (2018) A Triassic stem turtle with an edentulous beak. Nature 560:476–479 https://doi.org/10.1038/s41586-018-0419-1

Loredo GA, Brukman A, Harris MP, Kagle D, Leclair EE, Gutman R, Denney E, Henkelman E, Murray BP, Fallon JF, Tuan RS, Gilbert SF (2001) Development of an evolutionarily novel structure: fibroblast growth factor expression in the carapacial ridge of turtle embryos. J Exp Zool Mol Dev Evol 291:274–281

Løvtrup S (1977) The phylogeny of Vertebrata. John Wiley, London

Løvtrup S (1985) On the classification of the taxon Tetrapoda. Syst Zool 34:463–470

Luther A (1914) Über die vom N. trigeminus versorgte Muskulatur der Amphibien mit einem vergleichenden Ausblick über den Adductor mandibulae der Gnathostomen, und einem Beitrag zum Verständnis der Organisation der Anurenlarven. Acta Societatis Scientiarum Fenniciae 44:1–151

Lyson TR, Joyce WG (2012) Evolution of the turtle bauplan: the topological relationship of the scapula relative to the ribcage. Biol Lett 8:1028–1031

Lyson T, Bever GS, Bhullar BAS, Joyce WG, Gauthier JA (2010) Transitional fossils and the origin of turtles. Biol Lett 6:830–833. https://doi.org/10.1098/rsbl.2010.0371

Lyson TR, Bever GS, Scheyer TM, Hsiang AY, Gauthier JA (2013) Evolutionary origin of the turtle shell. Curr Biol 23:1–7

Lyson TR, Schachner ER, Botha-Brink J, Scheyer TM, Lambertz M, Bever GS, Rubidge BS, Queiroz K (2014) Origin of the unique ventilatory apparatus of turtles. Nat Commun 5:1–11

Lyson TR, Rubidge Bruce S, Scheyer TM, de Queiroz K, Schachner Emma R, Smith Roger MH, Botha-Brink J, Bever GS (2016) Fossorial origin of the turtle shell. Curr Biol 26:1887–1894. https://doi.org/10.1016/j.cub.2016.05.020

MacDougall MJ, Modesto SP, Brocklehurst N, Verrière A, Reisz RR, Fröbisch J (2018) Response: a reassessment of the taxonomic position of mesosaurs, and a surprising phylogeny of early amniotes. Front Earth Sci 6:99

Maier W, Werneburg I (2014) Schlüsselereignisse der organismischen Makroevolution. Scidinge Hall, Tübingen

Mannen H, Li SSL (1999) Molecular evidence for a clade of turtles. Mol Phylogenet Evol 13:144–148

Matsumoto R, Evans SE (2017) The palatal dentition of tetrapods and its functional significance. J Anat 230:47–65. https://doi.org/10.1111/joa.12534

Mulcahy DG, Noonan BP, Moss T, Townsend TM, Reeder TW, Sites JW Jr, Wiens JJ (2012) Estimating divergence dates and evaluating dating methods using phylogenomic and mitochondrial data in squamate reptiles. Mol Phylogenet Evol 65:974–991. https://doi.org/10.1016/j.ympev.2012.08.018

Mulisch M, Welsch U (2015) Romeis-Mikroskopische Technik. Springer, Berlin

Müller J (2003) Early loss and multiple return of the lower temporal arcade in diapsid reptiles. Naturwissenschaften 90:473–476

Müller J, Sterli J, Anquentin J (2011) Carotid circulation in amniotes and its implications for turtle relationships. Neues Jahrbuch der Gelogie und Paläontologie, Abhandlungen 261:289–297

Nagashima H, Kuraku S, Uchida K, Ohya YK, Narita Y, Kuratani S (2007) On the carapacial ridge in turtle embryos: its developmental origin, function and the chelonian body plan. Development 134:2219–2226

Nagashima H, Sugahara F, Takeshi M, Ericsson R, Kawashima-Ohya Y, Narita Y, Kuratani S (2009) Evolution of the turtle body plan by the folding and creation of new muscle connections. Science 325:193–196

Nagashima H, Kuraku S, Uchida K, Kawashima-Ohya Y, Narita Y, Kuratani S (2012) Body plan of turtles: an anatomical, developmental and evolutionary perspective. Anat Sci Int 87:1–13

Nagashima H, Kuraku S, Uchida K, Kawashima-Ohya Y, Narita Y, Kuratani S (2013) Origin of the turtle body plan: the folding theory to illustrate turtle-specific developmental repatterning. In: Morphology and evolution of turtles. Springer, Dordrecht, pp 37–50

Nagashima H, Sugahara F, Takechi M, Sato N, Kuratani S (2015) On the homology of the shoulder girdle in turtles. J Exp Zool B Mol Dev Evol 324(3):244–254

Natchev N, Heiss E, Lemell P, Stratev D, Weisgram J (2009) Analysis of prey capture and food transport kinematics in two Asian box turtles, *Cuora amboinensis* and *Cuora flavomarginata* (Chelonia, Geoemydidae), with emphasis on terrestrial feeding patterns. Zoology 112:113–127

Natchev N, Lemell P, Heiss E, Beisser C, Weisgram J (2010) Aquatic feeding in a terrestrial turtle: a

functional-morphological study of the feeding apparatus in the Indochinese box turtle *Cuora galbinifrons* (Testudines, Geoemydidae). Zoomorphology 129:111–119
Natchev N, Tzankov N, Werneburg I, Heiss E (2015) Feeding behaviour in a 'basal' tortoise provides insights on the transitional feeding mode at the dawn of modern land turtle evolution. PeerJ 3:e1172. https://doi.org/10.7717/peerj.1172
Neenan JM, Klein N, Scheyer TM (2013) European origin of placodont marine reptiles and the evolution of crushing dentition in Placodontia. Nat Commun 4:1621. https://doi.org/10.1038/ncomms2633
Nick L (1912) Das Kopfskelett von *Dermochelys coriacea* L. Zoologische Jahrbücher. Abteilung für Anatomie und Ontogenie der Tiere 33:1–238
Ogushi K (1911) Anatomische Studien an der japanischen dreikralligen Lippenschildkröte (*Trionyx japanicus*). I. Mitteilung. Morphologisches Jahrbuch 43:1–106
Ogushi K (1913a) Anatomische Studien an der japanischen dreikralligen Lippenschildkröte (*Trionyx japanicus*). II. Mitteilung: Muskel- und peripheres Nervensystem. Morphologisches Jahrbuch 46:299–562
Ogushi K (1913b) Zur Anatomie der Hirnnerven und des Kopfsympathicus von *Trionyx japonicus* nebst einigen kritischen Bemerkungen. Morphologisches Jahrbuch 45:441–480
Ogushi K (1914) Der Kehlkopf von *Trionyx japonicus*. Anat Anz 45:481–503
Olson EC (1947) The family Diadectidae and its nearing on the classification of turtles. Fieldiana Geology 11:1–53
Osborn HF (1903) On the primary division of the Reptilia into two sub-classes, Synapsida and Diapsida. Science 17:275–276
Poglayen-Neuwall I (1953) Untersuchungen der Kiefermuskulatur und deren Innervation bei Schildkröten. Acta Zool 34:241–292
Poglayen-Neuwall I (1954) Die Kiefermuskulatur der Eidechsen und ihre Innervation. Z Wiss Zool 158:79–132
Poglayen-Neuwall I (1966) Bemerkungen zur Morphologie und Innervation der Trigeminusmuskulatur von *Chelus fimbriatus* (Schneider). Zoologische Beiträge 12:43–65
Pritchard PCH (1984) Piscivory in turtles, and evolution of long-necked Chelidae. In: Ferguson MWJ (ed) The structure, development and evolution of Reptiles. A Festschrift in honour of Professor A.d'A. Bellairs on the occasion of his retirement, Symposia of the Zoological Society of London, vol 52. Academic Press, London
Rabi M, Zhou C-F, Wings O, Ge S, Joyce WG (2013) A new xinjiangchelyid turtle from the middle Jurassic of Xinjiang, China and the evolution of the basipterygoid process in Mesozoic turtles. BMC Evol Biol 13:1–28
Rabi M, Sukanov VB, Egorova VN, Danilov I, Joyce WG (2014) Osteology, relationships, and ecology of *Annemys* (Testudines, Eucryptodira) from the Late Jurassic of Shar Teg, Mongolia and phylogenetic definitions for Xinjiangchelyidae, Sinemydidae, and Macrobaenidae. J Vertebr Paleontol 34:327–352
Ray CE (1959) A sesamoid bone in the jaw musculature of *Gopherus polyphemus* (Reptilia: Testudininae). Anat Anz 107:85–91
Reisz RR, Head JJ (2008) Turtle origins out to sea. Nature 456:450–451
Rhodin AGJ, Iverson JB, Bour R, Fritz U, Georges A, Shaffer HB, van Dijk PPJ (2017) Turtles of the world: annotated checklist and atlas of taxonomy, synonymy, distribution, and conservation status. In: AGJ R, Iverson JB, van Dijk PP, Saumure RA, Buhlmann KA, Pritchard PCH, Mittermeier RA (eds) Conservation biology of freshwater turtles and tortoises: a compilation project of the IUCN/SSC tortoise and freshwater turtle specialist group, vol 7, 8th edn. Chelonian Research Monographs, Lunenburg, pp 1–292
Rice R, Kallonen A, Cebra-Thomas J, Gilbert SF (2016) Development of the turtle plastron, the order-defining skeletal structure. Proc Natl Acad Sci U S A 113:6. https://doi.org/10.1073/pnas.1600958113
Rieppel O (1980) The trigeminal jaw adductor musculature of *Tupinambis*, with comments on the phylogenetic relationship of the Teiidae (Reptilia, Lacertilia). Zool J Linnean Soc 69:1–29
Rieppel O (1984) The structure of the skull and jaw adductor musculature in the Gekkota, with comments on the phylogenetic-relationships of the Xantusiidae (Reptilia, Lacertilia). Zool J Linnean Soc 82:291–318
Rieppel O (1987) The development of the trigeminal jaw adductor musculature and associated skull elements in the lizard *Podarcis sicula*. J Zool 212:131–150
Rieppel O (1990) The structure and development of the jaw adductor musculature in the turtle *Chelydra serpentina*. Zool J Linnean Soc 98:27–62
Rieppel O (1993) Patterns of diversity in the reptilian skull. In: Hanken J, Hall BK (eds) The skull, patterns of structural and systematic diversity, vol 2. University of Chicago Press, Chicago, pp 344–389
Rieppel O (2000) Turtles as diapsid reptiles. Zool Scr 29:199–212
Rieppel O (2004) Kontroversen innerhalb der Tetrapoda—die Stellung der Schildkröten (Testudines). Sitzungsberichte der Gesellschaft Naturforschender Freunde zu Berlin 43:201–221
Rieppel O (2008) The relationships of turtles within amniotes. In: Wyneken J, Godfrey MH, Bels V (eds) Biology of turtles. CRC Press, Boca Raton, pp 345–353
Rieppel O (2013) The evolution of the turtle shell. In: Gardner J, Brinkman D, Holroyd P (eds) Vertebrate paleobiology and paleoanthropology series, Morphology and evolution of turtles. Springer, Dordrecht, pp 51–61
Rieppel O, Reisz RR (1999) The origin and early evolution of turtles. Annu Rev Ecol Syst 30:1–22
Robert McNeel & Associates (2003) Rhinoceros 3D. Version 3.0 SR3, November. Barcelona
Romer AS (1956) Osteology of the reptiles. The University of Chicago Press, Chicago

Rougier GW, De La Fuente MS, Arcucci AB (1995) Late Triassic turtles from South America. Science 268:855–858
Scheyer TM, Sander PM (2007) Shell bone histology indicates terrestrial palaeoecology of basal turtles. Proc R Soc B Biol Sci 274:1885–1893. https://doi.org/10.1098/rspb.2007.0499
Scheyer TM, Werneburg I, Mitgutsch C, Delfino M, Sánchez-Villagra MR (2013) Three ways to tackle the turtle: integrating fossils, comparative embryology and microanatomy. In: Gardner J, Brinkman D, Holroyd P (eds) Vertebrate paleobiology and paleoanthropology series. Springer, Dordrecht, pp 63–70
Schoch RR, Sues HD (2015) A middle Triassic stem-turtle and the evolution of the turtle body plan. Nature 523:584–587. https://doi.org/10.1038/nature14472
Schoch RR, Sues HD (2016) The diapsid origin of turtles. Zoology 119:3. https://doi.org/10.1016/j.zool.2016.01.004
Schulman H (1906) Vergleichende Untersuchungen über die Trigeminus-Muskulatur der Monotremen, sowie die dabei in Betracht kommenden Nerven und Knochen. In: Jenaische Denkschriften, Zoologische Forschungsreisen in Australien und dem Malayischen Archipel. Mit Unterstützung des Herrn Dr. Paul von Ritter ausgeführt in den Jahren 1891–1893 (III. 2. Teil), vol 2, vol 6. G. Fischer, Jena, pp 297–400
Schumacher GH (1954/55) Beiträge zur Kiefermuskulatur der Schildkröten: II. Mitteilung. Bau des M. adductor mandibularis unter spezieller Berücksichtigung der Fascien des Kopfes bei *Platysternon megacephalum, Emys orbicularis, Testudo graeca, Pelomedusa subrufa, Clemmys caspica riculata, Graptemys geographica, Hardella thurrjii, Macrochelys temminckii, Emydura krefftii, Hydromedusa tectifera, Chelodina longicollis, Trionyx punctatus, Amyda sinensis* und *Dogania subplana*. Wissenschaftliche Zeitschrift der Ernst Moritz Arndt-Universität Greifswald—Mathematisch-naturwissenschaftliche Reihe 4:501–518
Schumacher GH (1954a) Beiträge zur Kiefermuskulatur der Schildkröten. I. Mitteilung. Bau des M. adductor mandibularis unter spezieller Berücksichtigung des M. pterygoideus bei *Chelone, Podocnemis, Sternothaerus* und *Testudo elephantopus*. PhD thesis, Ernst-Moritz-Arndt-Universität, Greifswald
Schumacher GH (1954b) Beiträge zur Kiefermuskulatur der Schildkröten: III. Mitteilung. Bau des M. Adductor mandibularis bei *Macrochelys temminckii, Platysternon megacephalum, Clemmys caspica rivulata, Emys orbicularis, Graptemys geographica, Hardella thurjii, Testudo graeca, Amyda sinensis, Dogania subplana, Trionyx punctatus, Pelomedusa subrufa, Chelodina longicollis, Hydromedusa tectifera* und *Emydura krefftii*. Wissenschaftliche Zeitschrift der Ernst Moritz Arndt-Universität Greifswald—Mathematisch-naturwissenschaftliche Reihe 4:559–588
Schumacher GH (1956) Morphologische Studie zum Gleitmechanismus des M. adductor mandibulae externus bei Schildkröten. Anat Anz 103:1–12
Schumacher GH (1972) Die Kopf- und Halsregion der Lederschildkröte *Dermochelys coriacea* (LINNAEUS 1766)—Anatomische Untersuchungen im Vergleich zu anderen rezenten Schildkröten—Mit 7 Figuren im Text und 31 Tafeln, Abhandlungen der Akademie der Wissenschaften der DDR, vol 2. Akademie, Berlin
Schumacher GH (1973) The head muscles and hyolaryngeal skeleton of turtles and crocodilians. In: Gans C, Parsons TS (eds) Biology of the Reptilia, morphology D, vol 4. Academic Press, London, pp 101–199
Seeley HG (1892) On a new reptile from Welte Vreden (Beaufort West) *Eunotosaurus africanus* (Seeley). Quat J Geol Soc 48:3
Shaffer HB (2009) Turtles (Testudines). In: Hedges SB, Kumar S (eds) The TimeTree of life. Oxford University Press, New York, pp 398–401
Stayton CT (2011) Terrestrial feeding in aquatic turtles: environment-dependent feeding behavior modulation and the evolution of terrestrial feeding in Emydidae. J Exp Biol 214:4083–4091
Sterli J (2010) Phylogenetic relationships among extinct and extant turtles: the position of Pleurodira and the effects of the fossils on rooting crown-group turtles. Contrib Zool 79:93–106
Sterli J, de la Fuente M (2010) Anatomy of *Condorchelys antiqua* STERLI, 2008, and the origin of the modern jaw closure mechanism in turtles. J Vertebr Paleontol 30:351–366
Sterli J, Joyce WG (2007) The cranial anatomy of the early Jurassic turtle *Kayentachelys aprix*. Acta Palaeontol Pol 52:675–694
Sterli J, Müller J, Anquetin J, Hilger A (2010) The parapasisphenoid complex in Mesozoic turtles and the evolution of the testudinate basicranium. Can J Earth Sci 47:1337–1346
Summers AP, Darouian KF, Richmond AM, Brainerd EL (1998) Kinematics of aquatic and terrestrial prey capture in *Terrapene carolina*, with implications for the evolution of feeding in cryptodire turtles. J Exp Zool 281:280–287
Tarsitano SF, Oelofsen B, Frey E, Riess J (2001) The origin of temporal fenestra. S Afr J Sci 97:334–336
Thomson JT (1932) The anatomy of the tortoise. Sci Proc R Dublin Soc New Series 20:359–461 324 plates
Tokita M (2004) Morphogenesis of parrot jaw muscles: understanding the development of an evolutionary novelty. J Morphol 259:69–81. https://doi.org/10.1002/Jmor.10172
Tsai HP, Holliday CM (2011) Ontogeny of the *Alligator* cartilago transiliens and its significance for sauropsid jaw muscle evolution. PLoS One 6:e24935
Tsuji LA, Müller J (2009) Assembling the history of the Parareptilia: phylogeny, diversification, and a new definition of the clade. Fossil Record 12:71–81
Tvarožková B (2006) Development of the temporal emargination in turtles and the temporal fenestration in crocodilians: the origin of an anapsid-like chelonian skull. Masters thesis, Charles University in Prague, Prague

Underwood G (1970) The eye. In: Gans C, Parsons TS (eds) Biology of the Reptilia, morphology D, vol 2. Academic Press, London, pp 1–97

Versluys J (1919) Über die Phylogenie der Schläfengruben und Jochbogen bei den Reptilia. Sitzungsberichte der Heidelberger Akademie der Wissenschaften, Mathematisch-naturwissenschaftliche Klasse, Abteilung B, Biologische Wissenschaften 13:1–29

Vitek NS, Joyce W (2015) A review of the fossil record of new world turtles of the clade pan-Trionychidae. Bull Peabody Mus Nat Hist 56:185–244

Walker WF Jr (1973) The locomotor apparatus of Testudinines. In: Gans C, Parsons TS (eds) Biology of the Reptilia, morphology D, vol 4. Academic Press, London, pp 1–100

Wang Z, Pascual-Anaya J, Zadissa A, Li W, Niimura Y, Huang Z, Li C, White S, Xiong Z, Fang D, Wang B, Ming Y, Chen Y, Zheng Y, Kuraku S, Pignatelli M, Herrero J, Beal K, Nozawa M, Li Q, Wang J, Zhang H, Yu L, Shigenobu S, Wang J, Liu J, Flicek P, Searle S, Wang J, Kuratani S, Yin Y, Aken B, Zhang G, Irie N (2013) The draft genomes of soft-shell turtle and green sea turtle yield insights into the development and evolution of the turtle-specific body plan. Nat Genet 45:701–706. https://doi.org/10.1038/ng.2615

Watson DMS (1914) *Eunotosaurus africanus* Seeley, and the ancestry of the Chelonia. Proc Zool Soc London 11:1011–1020

Weisgram J (1985) Feeding mechanisms of *Claudius angustatus* COPE 1865. In: Dunker HR, Fleischer G (eds) Fortschritte der Zoologie: Functional morphology in vertebrates, vol 30. Gustav Fischer, Stuttgart, pp 256–260

Werneburg I (2011) The cranial musculature in turtles. Palaeontol Electron 14:15a 99 pages

Werneburg I (2012) Temporal bone arrangements in turtles: an overview. J Exp Zool B Mol Dev Evol 318:235–249

Werneburg I (2013a) Jaw musculature during the dawn of turtle evolution. Org Divers Evol 13:225–254

Werneburg I (2013b) The tendinous framework in the temporal skull region of turtles and considerations about its morphological implications in amniotes: a review. Zool Sci 31:141–153

Werneburg I (2015) Neck motion in turtles and its relation to the shape of the temporal skull region. Comptes Rendus Palevol 14:527–548

Werneburg I, Maier W (2018) Considerations on the development of the akinetic skull in pleurodire and cryptodire turtles. In: Turtle Evolution Symposium. Scidinge Hall, Tübingen, pp 90–91

Werneburg I, Sánchez-Villagra MR (2009) Timing of organogenesis support basal position of turtles in the amniote tree of life. BMC Evol Biol 9:1–9. https://doi.org/10.1186/1471-2148-9-82

Werneburg I, Maier W, Joyce WG (2013) Embryonic remnants of intercentra and cervical ribs in turtles. Biol Open 2:1103–1107

Werneburg I, Hinz JK, Gumpenberger M, Volpato V, Natchev N, Joyce WG (2015a) Modeling neck mobility in fossil turtles. J Exp Zool B Mol Dev Evol 324:230–243

Werneburg I, Wilson LAB, Parr WCH, Joyce WG (2015b) Evolution of neck vertebral shape and neck retraction at the transition to modern turtles: an integrated geometric morphometric approach. Syst Biol 64:187–204

Williams EE (1950) Variation and selection in the cervical central articulations of living turtles. Bull Am Mus Nat Hist 94:509–561

Williston SW (1917) The phylogeny and classification of reptiles. J Geol 25:411–421

Witzmann F, Werneburg I (2017) The palatal interpterygoid vacuities of temnospondyls and the implications for the associated eye- and jaw musculature. Anat Rec 300:1240–1269. https://doi.org/10.1002/ar.23582

Wochesländer R, Hilgers H, Weisgram J (1999) Feeding mechanism of *Testudo hermanni boettgeri* (Chelonia, Cryptodira). Neth J Zool 49:1–13

Yntema CL (1968) A series of stages in the embryonic development of *Chelydra serpentina*. J Morphol 125:219–251

Zangerl R (1948) The methods of comparative anatomy and its contribution to the study of evolution. Evolution 2:351–374

Zardoya R, Meyer A (1998) Complete mitochondrial genome suggests diapsid affinities of turtles. Proc Natl Acad Sci U S A 95:14226–14231

Zardoya R, Meyer A (2001) The evolutionary position of turtles revised. Naturwissenschaften 88:193–200

Zdansky O (1923) Über die Temporalregion des Schildkrötenschädels. Bull Geol Inst Univ Upsala 19:89–114

Zhou C-F, Rabi M (2015) A sinemydid turtle from the Jehol Biota provides insights into the basal divergence of crown turtles. Sci Rep 5:16299. https://doi.org/10.1038/srep16299

An Integrative View of Lepidosaur Cranial Anatomy, Development, and Diversification

9

Raul E. Diaz Jr and Paul A. Trainor

9.1 Introduction to Lineages

9.1.1 Rhynchocephalia

At first glance, the New Zealand tuatara (*Sphenodon punctatus*) appears to be a "lizard," a member of the reptilian assemblage known as the squamates (lizards, snakes, and amphisbaenians). While superficially quite similar, the tuatara (actually belonging to the lineage known as **Rhynchocephalia**) and squamates last shared a common ancestor approximately 250 million years ago (MYA) in the Triassic (Fraser and Benton 1989; Heckert et al. 2008; Jones et al. 2013). The tuatara was formally considered to *not* be a lizard by Gunther (1867) and was placed into the order Rhynchocephalia, a rank "equivalent" to other lineages like Squamata, Chelonia, Crocodilia, and Aves (Pough et al. 2004; Kardong 2012). The Rhynchocephalia is thought to persist as a single species based on recent genetic data (Hay et al. 2010). However, fossil evidence suggests that Rhynchocephalia was composed of at least 40 other recognized taxa, with recently discovered specimens showing that the New Zealand tuatara belonged to a lineage more morphologically and ecologically diverse than previously recognized and with a more global distribution (Evans and Jones 2010; Apesteguía and Jones 2012; Herrera-Flores et al. 2017). In contrast, for living squamate reptiles, which are an equally ancient lineage, many new extant species are reported on a yearly basis, with the Lepidosaur species number already surpassing 10,300 (Gauthier et al. 2012; Pyron et al. 2013; Reeder et al. 2015; Uetz et al. 2018; with monumental contributions by L. Lee Grismer and colleagues). Recently, it was shown that the oldest known squamate fossil, a dentary bone originating during ***Tikiguania*** from the Late Triassic Tiki Formation of Central India (Hutchinson et al. 2012), may have actually been a younger specimen that fossilized in older exposed beds in Asia. Romer (1956) coined the clade name Lepidosauria to encompass the non-archosaurian (crocodilians and birds) amniote lineage with two temporal openings (fenestrations; diapsids), thus including Rhyncocephalia and Squamata as each other's closest living relatives, a relationship with continued molecular and morphological support (see also Chaps. 8 and 10).

R. E. Diaz Jr (✉)
Department of Biological Sciences, Southeastern Louisiana University, Hammond, LA, USA

Natural History Museum of Los Angeles County, Los Angeles, CA, USA
e-mail: raul.diaz@selu.edu

P. A. Trainor
Stowers Institute for Medical Research, Kansas City, MO, USA

Department of Anatomy and Cell Biology, University of Kansas Medical Center, Kansas City, KS, USA

J. M. Ziermann et al. (eds.), *Heads, Jaws, and Muscles*, Fascinating Life Sciences,
https://doi.org/10.1007/978-3-319-93560-7_9

As the only living member of **Rhynchocephalia**, the tuatara has been considered to be a "living fossil," a title presented to taxa in which fossils were discovered prior to the extant taxon. While this title holds for *Sphenodon punctatus*, a broader look at the other members of this lineage presents an underappreciated diversity across morphospace and ecological space (Herrera-Flores et al. 2017). Living *Sphenodon* are cold adapted, have slow reproductive and growth rates, and are omnivorous. They also have a unique mode of jaw movement, with anteroposterior displacement of the lower jaw within an akinetic cranium where mandibular teeth move between parallel maxillary and palatal tooth rows, "**propalinal movement**," unique among amniotes (Fig. 9.1; Cree 2014). While thought to have the primitive condition of a complete **lower temporal bar** (jugal-quadratojugal contact below lower temporal fenestra), it is considered to be secondarily complete as earlier diverged members of Rhynchocephalia present an incomplete lower temporal bar (Whiteside 1986; Apesteguia and Novas 2003; Evans 2003; Apesteguía and Jones 2012; Herrera-Flores et al. 2017). Diets were diverse spanning carnivory, omnivory, and herbivory, and even included a durophagous diet of crustaceans or mollusks (Herrera-Flores et al. 2017).

Fig. 9.1 Osteo- and chondrocranium of the tuatara (*Sphenodon punctatus*; Rhynchocephalia). (**a–c**) Drawn from adult in digimorph.org (YPM9194) in lateral, dorsal, and ventral view, respectively. Muscle attachments in (**a**) added based on Haas (1973), Wu (2003), and Jones et al. (2009). The lateral chondrocranium (**d**) is from Werner (1962) but in Bellairs and Kamal (1981). Labels are defined below, with this list to be the reference for subsequent figure labels. 2 Optic nerve, CNII, *amem* M. adductor mandibulae externus medialis, *amem 3b* M. adductor mandibulae externus medialis 3b, *ames* M. adductor mandibulae externus superficialis, *amp* M. adductor mandibulae posterior, *ang* angular, *ap* anterior process of extracolumella, *ar* articular, *bf* basicranial fenestra, *bh* basihyal, *bptp* basipterygoid process, *bo* basioccipital, *bsp* basisphenoid, *ca* columella auris, *cbI* ceratobranchial I, *cbII* ceratobranchial II, *cer man* M. cervicomandibularis, *ch* ceratohyal, *cl delt* M. claviculodeltoideus, *co* coronoid, *co coll* M. constrictor colli, *com* compound bone, *cp* crista parotica, *cr* crista sellaris, *cu* cupola, *d* dentary, *dep mnd* M. depressor mandibulae, *dpo*: days post-oviposition, *ec* ectochoanal cartilage, *ect* ectopterygoid, *egp* entoglossal process, *eh* epihyal, *els* endolymphatic sac, *epax* epaxial, *hypax* hypaxial, *escml–3* M. episternocleidomastoid branch 1–3, *epi* epipterygoid, *ex* exoccipital, *f* frontal, *fm* fenestra metoptica, *fn* fenestra narina, *fpl* footplate of columella, *fpr* fenestra prootica, *fs* fenestra in interorbital septum, *gg* M. genioglossus, *gen hy lat* M. geniohyoideus lateralis, *gen hy med* M. geniohyoideus medialis, *gl* glottis, *ih* interhyoideus, *intm* M. intermandibularis, *intmd ant* M. intermandibularis anterior, *intmd post* M. intermandibularis posterior, *int hyo post* M. intermandibularis hyoideus posterior, *is* interorbital septum, *j* jugal, *l* lacrimal, *lao* M. levator anguli oris, *lat dors* M. latissimus dorsi, *lev scap* M. levator scapulae, *mc* Meckel's cartilage, *mdn hy* M. mandibulohyoideus, *mgh* M. geniohyoideus, *mghp* M. geniohyoideus profundus, *mx* maxilla, *om hy* M. omohyoideus, *n* notochord, *na* nasal, *nc* nasal cartilage, *ncm* M. neurocostomandibularis, *ns* nasal septum, *oa* occipital arch, *oc* Otic capsule, *onf* orbitonasal fissure, *oo* otooccipital = opisthotic + exoccipital, *op* opisthotic, *os* orbitosphenoid, *p* parietal, *pa* pila antotica, *pacc* pila accessoria, *pai* processus alaris inferior, *pal* palatine, *pasu* processus alaris superior, *pat* processus anterior tecti, *pd* dorsal process of columella, *pf* pituitary fossa, *pla* planum antorbitale, *pm* pila metoptica, *pma* anterior maxillary process, *pmp* posterior maxillary process, *pmx* premaxilla, *pnc* paranasal cartilage, *po* postorbital, *pocc p* paroccipital process, *postf* postfrontal, *postorb* postorbital, *ppr* pterygoid process of pterygoquadrate, *preart* prearticular, *pref* prefrontal, *pro* prootic, *prsph* parasphenoid, *ps* planum supraseptale, *pst* M. pseutodemporalis, *ptc* parietotectal cartilage, *pt p of q* pterygoid process of the quadrate, *pt* pterygoid, *ptg* M. pterygoideus, *q* quadrate, *qj* quadratojugal, *rect abd* M. rectus abdominis, *ret quad* M. retractor quadrati, *s* supratemporal, *sc* sphenethmoid commissure, *sc delt* M. scapulodeltoideus, *scl* scleral ossicles, *semi, sp cap* M. semispinalis capitis, *si* subiculum infundibula, *so* supraoccipital, *som* somites, *sp cap* M. spinalis capitis, *spl* splenial, *spmx* septomaxilla, *sq* Squamosal, *st* stapes, *st hy, sup* M. sternohyoideus superior, *st hy pr* M. sternohyoideus profundus, *sur* surangular, *tma* taenia marginalis, *tm* taenia medialis, *tr* trabecula, *tra* trachea, *trap* M. trapezius, *ts* tectum synoticum, *tymp* tympanic membrane, *v* vomer

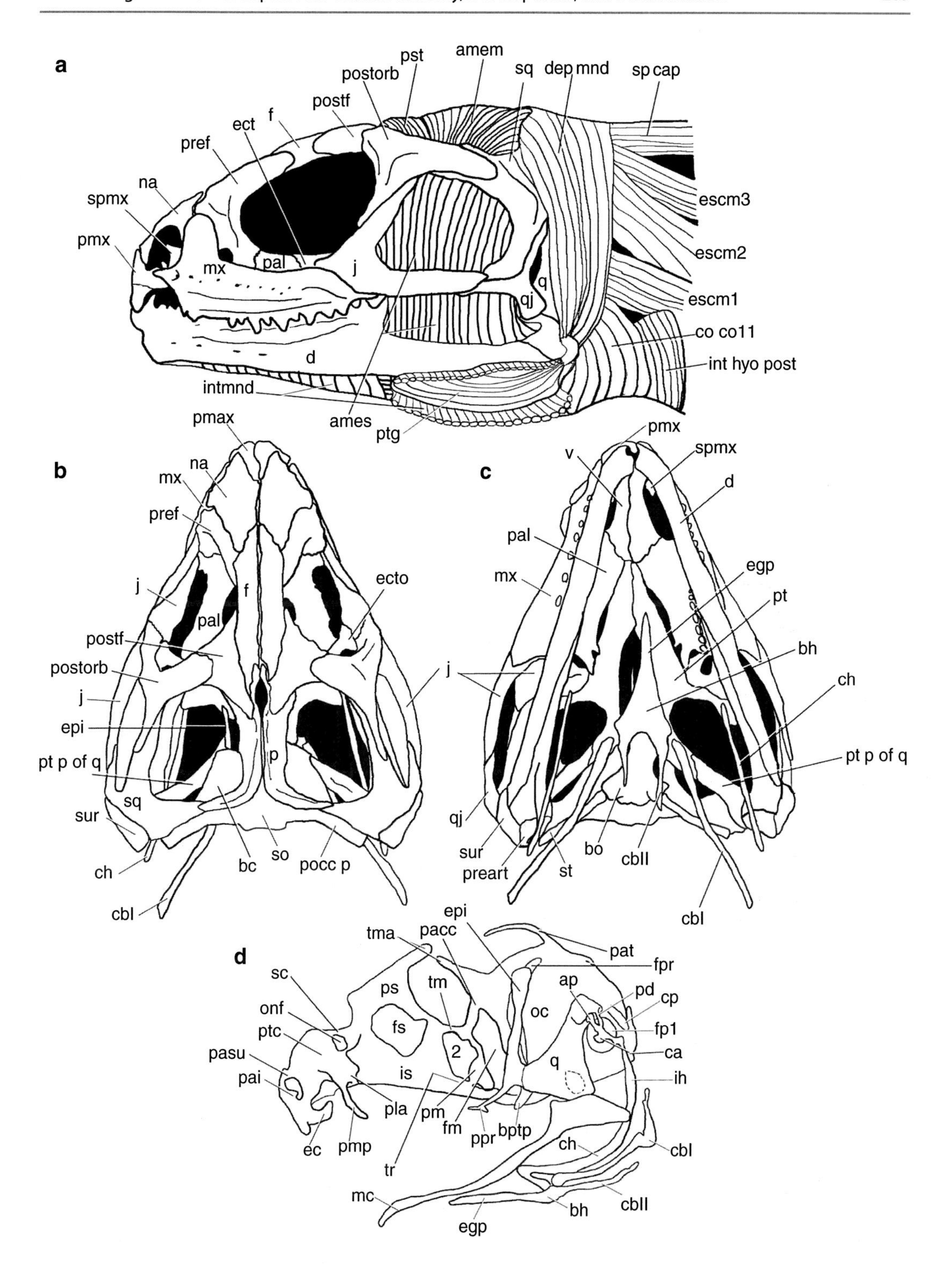
a
pst
amem
postorb
sq
dep mnd
sp cap
postf
ect
f
pref
na
spmx
pmx
mx
pal
j
q
qj
escm3
escm2
escm1
co co11
int hyo post
d
intmnd
ames
ptg
pmax
b
na
mx
pref
j
f
pal
ecto
postf
postorb
j
epi
p
pt p of q
sq
sur
ch
bc
so
pocc p
cbl
c
pmx
spmx
v
d
pal
mx
egp
pt
bh
j
ch
pt p of q
qj
sur
preart
st
bo
cbII
cbl
d
epi
tma
pacc
pat
fpr
sc
ps
tm
ap
pd
cp
onf
fs
oc
fp1
ptc
2
ca
pasu
q
is
ih
pai
pla
pm
fm
ppr
bptp
ec
pmp
ch
cbl
tr
mc
bh
cbII
egp

9.1.2 Squamate Reptiles (Lizards and Limbless Snakes and Amphisbaenians)

Within the squamate lineage, there are approximately 32 lizard, 6 amphisbaenian, and 23 snake lineages recognized at the rank level of family (Pyron et al. 2013). The phylogenetic relationships at higher levels, are however contentious due to differences arising from using morphological (Gauthier et al. 2012) and molecular (Pyron et al. 2013) data. Recent efforts have been made to merge these two data types which also allows for the inclusion of fossils to be framed phylogenetically alongside extant taxa (Reeder et al. 2015). Among extant tetrapods, squamates present some of the most dramatic modifications to the body such as lateral rib expansion for gliding in the genus *Draco*, calvarial bone-derived horns in the genus *Phrynosoma*, or keratin-derived horns in the genus *Trioceros*. They also include some of the smallest amniotes (*Brookesia* and *Sphaerodactylus*) while also showing large body size in various snake lineages such as *Python* and *Eunectes* and the large varanid the Komodo Dragon. Diets are insectivorous, primarily herbivorous, and carnivorous, while taxa also present teeth made for crushing mollusks such as in the family Teiidae. Limb reduction has occurred multiple times in squamates (Greer 1991; Shapiro 2002; Brandley et al. 2008). Serpentes (snakes) is a lineage of specialized limb reduced lizards ranging from loss of the anterior limb skeleton and retainment of vestigial hind limbs to complete loss of both (Woltering 2012). Snakes have also undergone an expansion of trunk vertebrae, together with a reduction in cranial bones to increase movement between cranial bones (kinesis), in contrast to the **akinetic** skull of rhynchocephalians. In addition, macrostomatan snakes utilize a loosely connected cranium to eat meals several times their body diameter.

9.2 Formation of the Cranium and its Tissues

The cranium is the most complex anatomical structure of the vertebrate skeleton. The embryonic **ectoderm** gives rise to the central nervous system which patterns surrounding embryonic tissues during the formation of the head; neural crest cells also arise from the ectoderm where neural and non-neural (**epidermis**) epithelia contact (generally during the process of neurulation). **Neural crest cells**, a vertebrate innovation and transient cell population, differentiate, delaminate, and migrate to target sites along the body and contribute to various organs [and are considered to be the fourth germ layer (Hall 2008; Le Douarin and Dupin 2014)]. The **endoderm** (a germ layer) will form the lining of the respiratory and digestive tract) signals to pattern the dorsal cranium as well as the pharyngeal arches (the segmented domains of the embryonic face and neck). Between the ectoderm and endoderm, cranial neural crest cells migrate and proliferate to differentiate into a plethora of tissues, such as the connective tissue of cranial muscles, lining of blood vessels, the dermis, and skeletal tissue of the anterior portion of the cranium. **Mesoderm** contributes muscle cells and is the source for the cranioskeletal tissue of the posterior cranium (Couly et al. 1993; Chai et al. 2000; Jiang et al. 2002; McBratney-Owen et al. 2008; Bhatt et al. 2013; Trainor 2014). Muscles connecting the head to the shoulder girdle are in a transitional zone with debate as to whether these muscles are of "jaw" (branchial arch) muscle type with respect to their origin from embryonic branchial arches or are derived from the trunk somitic mesoderm. Those muscles are also highly modified across major groups (Sambasivan et al. 2011; Diogo et al. 2015; Michailovici et al. 2015; Sefton et al. 2016).

While several classic publications have been produced that describe lepidosaur cranial musculature (Edgeworth 1935; Langebartel 1968; Haas 1973; Diogo and Abdala 2010), the chondrocranium and bony skull architecture (De Beer 1937; Bellairs and Kamal 1981; Estes et al. 1988; Gans et al. 2008), and to some extent the nervous system (Gans et al. 1979), few works include embryonic development which has been useful in resolving issues of **homology** by providing a better understanding of the origin and development of anatomical features (such as wrist and ankle bones; Diaz and Trainor 2015). Even fewer (if any) publications exist that describe all these tissue types for a single taxon. As one can expect in a lineage of more than 10,000 recognized species, lepidosaur cranial anatomy is diverse with respect to presence, absence, and fusions of bones, cartilages, muscles, and even neural tissue (see references above). Herein we focus on four species of lepidosaurs that belong to three different squamates families, are morphologically different, and emphasize a different suite of sensory systems that maybe drive their cranial anatomical differences. For *Sphenodon punctatus*, the only extant rhynchocephalian, general anatomy was obtained from published literature as embryos were not available for a parallel study. Displaying a "typical" lizard body plan is the desert grassland whiptail lizard (Teiidae: *Aspidoscelis uniparens*), which is diurnal, utilizes lingual "olfaction," and has a well-developed snout for an insectivorous diet and prey apprehension while also having a well-developed **tympanic membrane**. The veiled chameleon (Chamaeleonidae: *Chamaeleo calyptratus*) exhibits a modified calvarial roof as the bones grow away from the primary axis of the head. It has a short snout and reduced olfaction (*sensu* Gisi 1907 *in* Starck 1979), captures prey with a ballistic tongue, has highly specialized vision but a relatively akinetic skull, and does not have a tympanic membrane which is associated with a reduction in hearing (Wever 1968; though they state that the temporal fascia serves as a sound receiving membrane). Representing serpentes is the African house snake, *Boaedon (Lamprophis) fuliginosus* (Lamprophiidae), with a typical bone reduced cranium that is also highly **kinetic**, a heavy dependence on olfaction, potentially reduced dependence on vision [within the framework that snakes may have "rebuilt" their eyes from a nearly blind **fossorial** ancestor (Yi and Norell 2015)], and lack a tympanic membrane. The former three species were bred in-house and present differences in chondrocranial and bone architecture, superficial jaw, throat, cranial, and neck musculature, and cranial nerves. **Antibody** (MF20 and Tuj1) and alcian/alizarin staining followed published methods (see Diaz and Trainor 2015; Diaz et al. 2017). Embryos were collected in triplicate at stages spanning the entire length of development, but only a subset of images are used in this chapter. This provides a snapshot of development for the various taxa and highlights similarities and major differences.

9.3 Cranioskeleton of Sphenodon punctatus

While the adult lepidosaur skull is heavily ossified (as in most tetrapods), during embryogenesis the first skeletal scaffold to appear are parts of the the **neurocranium** and the **viscerocranium**. The former begins as paired sets of ventral and lateral cartilaginous bars and struts providing support, protection, and serving as a scaffold for sensory and neural tissue of the central nervous system. The viscerocranium is derived from the pharyngeal arches and contributes to skeletal elements of the face and neck. The third portion of the skull, the **dermatocranium**, is composed of the flat bones that directly ossify on the more superficial portions of the skull. The cartilaginous **chondrocranium** primarily undergoes endochondral ossification where the cartilage is replaced by bone tissue, although some regions remain as cartilage (see Chap. 7 for more details on cranium development in vertebrates). Thus, the cranium is composed of three different skeletal complexes that play a role in defining a species-specific skull. With of **microcomputed tomography** (microCT) being used to study skull formation, one must realize that soft tissues of the cranium differentiate prior to ossification during the stages of chondrocranial differentiation, and, thus, to understand cranioskeletal development, one must look earlier and use a different array of tools. In agreement with Bellairs and Kamal's (1981) statement "such techniques supplement, they do not supplant, the more laborious procedures of serial sectioning and reconstruction," though contrast-enhanced microCT methods are now a necessary component to visualize soft tissues (Metscher 2009; Gignac and Kley 2014; Gignac et al. 2016; Molnar et al. 2017).

Basic components of the chondrocranium can be seen in Fig. 9.1d with the schematic of a 54 mm total length *Sphenodon punctatus* (redrawn from Werner 1962 *in* Bellairs and Kamal 1981). The anterior cranial base is composed of a pair of long trabecular cartilages (tr) medial to the forming optic cups that later fuse medially and dorsally develop the interorbital septum (ios). Anteriorly, the nasal cartilages fuse medially to form the internasal septum (is) medially and posteriorly fuse with the trabecular cartilages. Obliquely and dorsal to the ios is the planum supraseptale (ps) which continues posterodorsally to meet with the trabeculae marginalis (tma; when present) that form an arc along the dorsolateral roof of the forming neurocranium to connect with the processus accessorius (pacc) dorsal to the pila metotica (pm) and ultimately with the dorsolateral surface of the otic capsule (oc). This arc generally prefigures the initial arcs of intramembranous ossification of the frontal (f) and parietal (p) bones, with the ps underlying the nasals (na) and f bones. The trabeculae lie anterior to where the **Rathke's pouch** (anterior pituitary gland) will form, with this crucial gland in vertebrates marking the posterior trabecular cartilages but also the posterior domain of cranial neural crest cell-derived tissue (Couly et al. 1993; McBratney-Owen et al. 2008; Trainor 2014). Posterior to Rathke's pouch [which lies within the pituitary fenestra (pf)], two parachordal cartilages (pa) are present lateral to the notochord (n) whose anteriormost extent is posterior to the pituitary gland superficial to the basicranial fenestra (bf). Transverse between the pairs of tr and parachordal cartilages forms the crista sellaris (cs) cartilage. Posterior to the cs will form the basisphenoid bone, and anterior ventral to the trabeculae will form the parasphenoid. Both parachordal cartilages expand laterally and together form the basal plate (bpl) which **synchondroses** with the overlying otic capsules (oc) and form struts called commissures, which also form spaces through which cranial nerves can pass through. Posterior to the **otic capsules** [past a space called the metotic fissure (mf)] is the occipital arch (oa) that will ultimately give rise to the basioccipital (bo), exoccipital (exo), and the supraoccipital (so) cartilages that surround the foramen magnum. The taenia medialis (tm) connects the ios to the pm at a deeper level. The nasal cartilages produce anterolateral and posterior projections that form struts against the medial wall of the more superficial intramembranous skeleton. The viscerocranial cartilage derived from the first arch gives rise to the anteromedial epipterygoid (epi) and the posterolateral quadrate (q), of which the latter will form the articulating bone of the upper jaw with the lower jaw. The mandibular portion of the first embryonic pharyngeal arch (mandibular arch) produces the long **Meckel's cartilage** (mc). The second pharyngeal arch (hyoid arch) gives rise to elements of the hyoid skeleton, which, as can be seen in this figure (Fig. 9.1d), highlight the interconnected nature of the columella auris (ca) of the middle ear with the ceratohyal (ch) cartilages of the hyoid through an intermediate interhyal (ih) which is differentially lost across reptilian taxa. Three pairs of cartilages extend mediolaterally away from the central basihyoid (bh) cartilage, with the more lateral being the ceratohyal, medially is the ceratobranchial I (cbI), and along the ventral midline attached to the bh is the cbII. Anterior to the bh is the entoglossal process (egp) of the hyoid which supports the tongue internally. The hypohyal (hh) which bridges the ch to the bh remains cartilaginous in adults and is thus not visible in this illustration (based on a microCT scan of an adult; YPM9194 from Digimorph.org).

Figure 9.1a–c present the cranioskeleton of an adult *Sphenodon punctatus* in lateral, dorsal, and ventral views, respectively. Unique to rhynchocephalians is the dentition of an **acrodont**, but which fuses along the mandibular surface obscuring a distinction between the tooth and surrounding bone (generally misinterpreted as not being real dentition but rather serrations of the intramembranous skeleton). As mentioned previously, being a **diapsid** reptile, *Sphenodon punctatus* exhibits a complete lower temporal bar through a junction formed between the jugal and posterior quadratojugal (secondary contact as earlier diverged members of this lineage have an incomplete lower temporal bar).

Squamates are "modified" diapsids. This latter term refers to amniotes that have a **lower temporal fenestra** (an opening or space between bones) and a **supratemporal fenestra**, like Rhynchocephalia, but are modified in that the lower bar of the lateral fenestra is lost in some lineages. The lateral bar of the upper fenestra is also absent in some species, as in geckos, beaded lizards, and snakes, what leaves the temporal region devoid of skeletal tissue. In the tuatara, unlike in squamates, the quadrate bone presents with a medial pterygoid process that locks the quadrate with the pterygoid bone and does not allow for **streptostyly** (free movement of the ventral end of the quadrate bone around a dorsal articulation with the squamosal—or supratemporal in snakes). Thus, the tuatara has a very akinetic skull and which is also enclosed relative to that of squamates (Figs. 9.2, 9.3 and 9.4). The palatal series of *Sphenodon,* as in most reptiles, consists of an anterior vomer (v), a palatine (pal), and the pterygoid (pt), with a septomaxilla (spmx) present along the posteroventral surface of the nasal capsule cartilages and dorsal to the vomer.

The skull of *Sphenodon* is triangular in shape, with the broad base being the area between the angles of the jaws. This is due to both a broad pair of bpl of the chondrocranium and enlarged otic capsules. The initial intramembranous slivers of the frontal bones appear to follow the continuous arc of the ps through the anterior component of the tma up to the pacc, with the parietals commencing around the pacc and continuing posterior to the insertion point of the tma onto the otic capsule (as in squamates, Figs. 9.2, 9.3, and 9.4). Posteriorly, from the

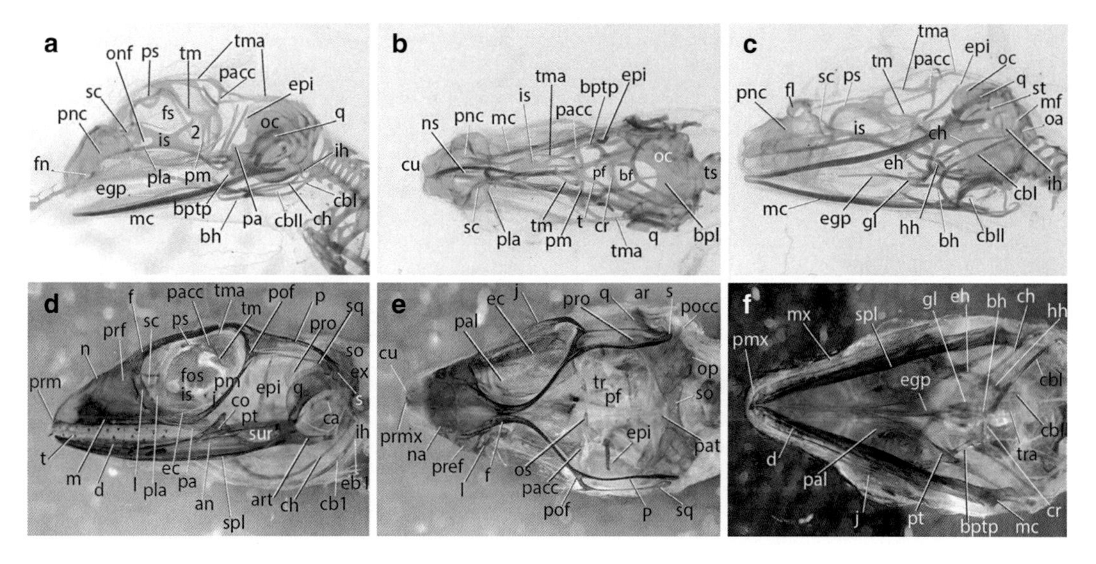

Fig. 9.2 The desert grassland whiptail lizard (*Aspidoscelis uniparens*; Teiidae) shown in lateral, dorsal, and ventrolateral view (**a–c**) of an alcian-stained chondrocranium (29–30 dpo embryo) and a lateral, dorsal, and ventral view (**d–f**) of an older specimen (51 dpo) undergoing ossification. See Fig. 9.1 of abbreviation explanations

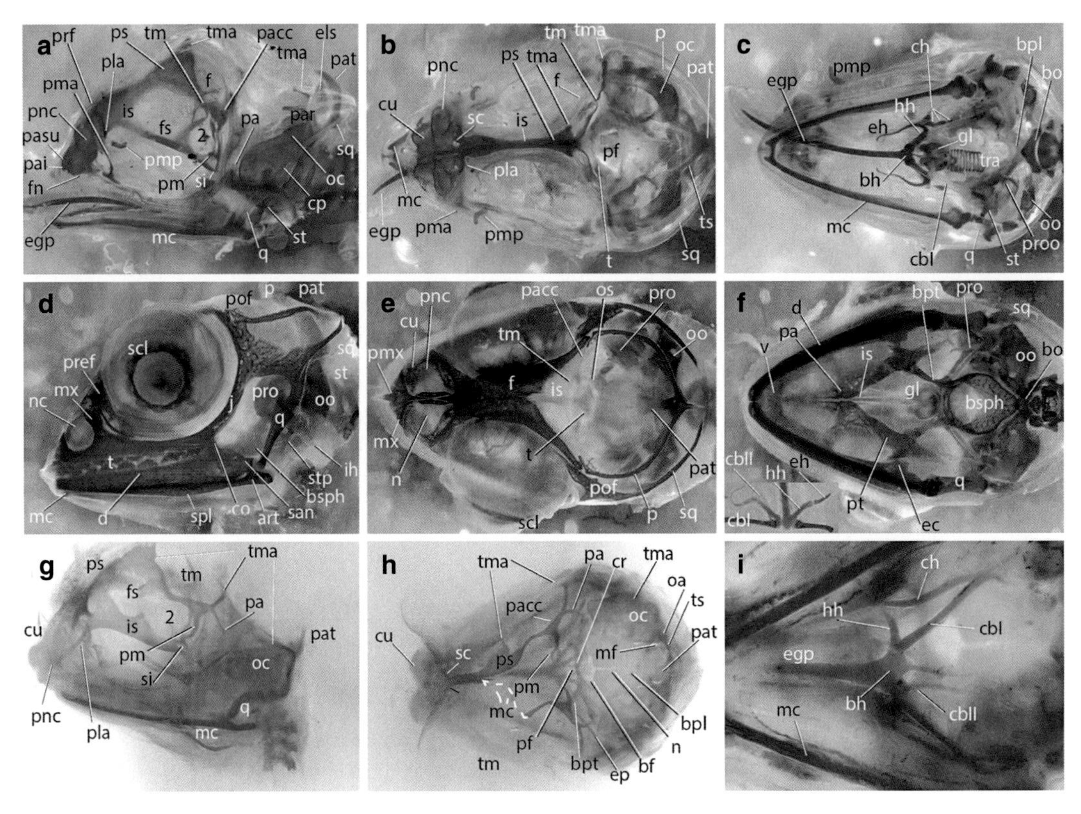

Fig. 9.3 Veiled chameleon (*Chamaeleo calyptratus*; Chamaeleonidae) alcian-stained chondrocrania in lateral, dorsal, and ventral views (**a**–**c**; 140 dpo) and an older individual in lateral, dorsal, and ventral view showing ossification (**d**–**f**; 167 dpo). A bearded dragon (*Pogona vitticeps*; Agamidae) is presented in **g**–**i** (45 dpo) as a representative of the sister family of Chamaeleonidae. See Fig. 9.1 for abbreviation explanations

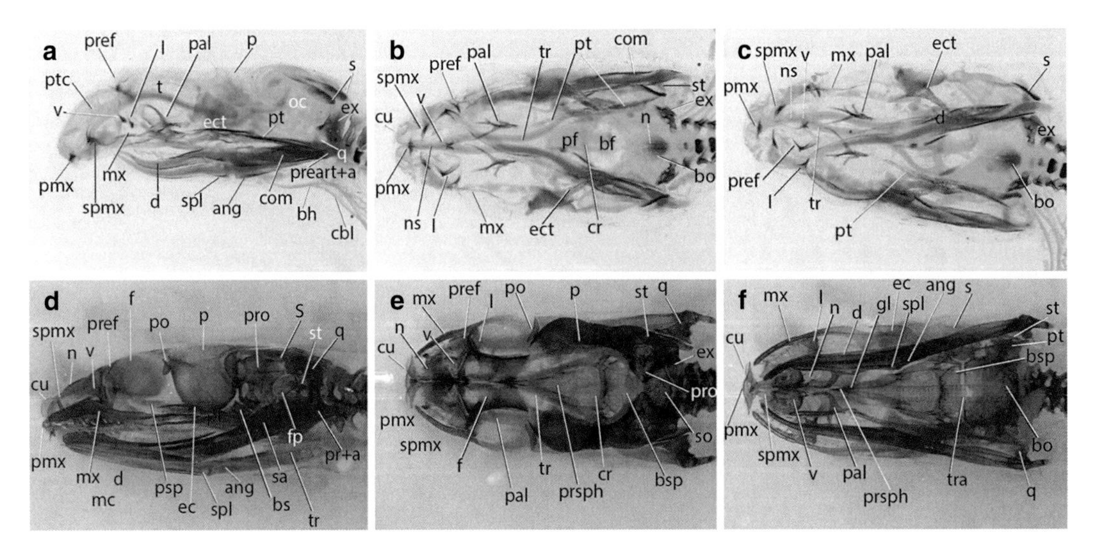

Fig. 9.4 African house snake [*Boaedon* (*Lamprophis*) *fuliginosus*; Colubroidea: Lamprophiidae] chondrocrania are shown in lateral, dorsal, and ventral views (respectively, **a**–**c**; 22–23 days post-oviposition – dpo) and an older specimen undergoing later stage ossification (**d**–**f**; 44–45 dpo). See Fig. 9.1 for abbreviation explanations

dorsal surface of the posterior margin of the otic capsules, a continuous bridge crosses the midline as the tectum synoticum (ts), which, from along its dorsomedial width, projects the processus anterior tecti (pat), whose anterior and distal tip appears to correspond with the posterior midline margin of the parietals and ultimately ossifies into a supraoccipital crest of the occipital. A columella/extrastapes project laterally posterior to the quadrate and paroccipital processes but does not terminate at a tympanum. Rather, the auditory system of the tuatara appears to use the overlying skin, the muscle depressor mandibulae, and a tendinous **aponeurosis**. Gans and Wever (1976) identified the auditory sensitivity of the tuatara as being optimized for reception of the tuatara's species-specific croak.

9.4 Squamate Chondrocranial and Osteocranial Diversity

As discussed previously, the **squamate** skull is a "modified" diapsid condition ranging from a single fenestra due to a loss of the lower temporal bar as in most lizards (complete in *Sphenodon*) to a loss of both bars along with the bones associated with them (snakes and some lizards). In addition, the species examined interact with their environment differently. Those observations lead to the question of whether chondrocranial differences correlate with differences in calvarial intramembranous bone shapes, patterns, or loss.

The genus *Aspidoscelis* (and Teiidae in general) have relatively narrow and anteriorly pointed crania. In addition, they are active foragers always on the move searching for food through olfaction, and visually through lifting items with their snout or limbs (pers. obs.; Pough et al. 2004; Pianka and Vitt 2006). With a strong dependence on olfaction, the nasal capsules in *A. uniparens* are as wide as the posterior cranium (Fig. 9.2a–c). While most aspects of the chondrocranium are exhibits as were seen in *Sphenodon*, the width of the basal plate and otic capsules are smaller relative to the length of the cranial base of the head. Additionally, *A. uniparens* presents a complete taenia marginalis, a shorter tectum synoticum, and a shorter pat (processus anterior tecti). The effect of the shorter pat is a more posterior placement of the posterior margin of the parietal, giving the back of the cranium a square shape rather than an anteriorly directed "^" as in the tuatara's parietal. As in the chondrocranium of *Sphenodon*, the frontal and parietal intramembranous slivers of bone correspond to the arc created by the taenia marginalis, but no chondrocranial elements are present that correspond to the areas of lower temporal bar formation (the jugal-quadratojugal in *Sphenodon*) or the upper temporal bar (postorbital-squamosal) of *A. uniparens*. The vertebrate chondrocranium lacks a cartilaginous framework in the temporal and signals underlying the formation of intramembranous bones in this region remain to be identified.

Of note, the ecologically equivalent and morphologically convergent genus *Acanthodactylus* (*Ac. boskianus*) has been well studied with respect to its chondrocranial and osteological morphology. Both lineages present with a head that is triangular and elongate in overall shape with, convergent scales in shape and pattern across the body. Both lineages exhibit an active foraging lifestyle where tongue flicking is combined with exploring their environment in search for food. As redrawn by Bellairs and Kamal (1981: their Figs. 17–20) from the works of Kamal and Abdeen (1972) and El-Toubi and Soliman (1967), both the chondrocranium and the osteocranium are almost identical to *A. uniparens*. Subtle differences in *Aspidoscelis* such as absence of a complete pila antotica (pa) and a cartilaginous posterior maxillary process (pmp) from the paranasal cartilages (pnc) exist, despite *Acanthodactylus* belonging to the Lacertidae which inhabit diverse open areas in the old world (Africa, Asia, Europe), whereas Teiidae are strictly new world (North America through South America) and have diverged for 139–159 MYA (Kumar et al. 2017).

Veiled chameleons (*Chamaeleo calyptratus*) exhibits a relatively more compact chondrocranium with a short anterior nasal capsule domain, which also corresponds to a short snout in this lineage and a postorbital domain that begins as a very wide space (between the two paccs). Unlike other reptiles, the basal plates narrow posteriorly, and this leads to a rather long and posteriorly projected tectum synoticum (atypical, generally directly connects as a bridge between otic capsules) and the processus anterior tecti of the tectum projects anterodorsally a long distance. Thus, the posterior margin of the chameleon chondrocranium is teardrop in shape (Fig. 9.3a–c) with incomplete taenia marginalis (which does not alter the conserved path of the frontal and parietal bones as in *Sphenodon*). Unlike *Aspidoscelis*, the chameleon skull is relatively akinetic (Haas 1973) and also lacks the epipterygoid cartilage (missing the ascending process of the pterygoquadrate, pas).

The chameleon hyoid skeleton (part of the viscerocranium) has been a structure of interest among comparative morphologists for over a century, especially given the recent discovery of a "proto" chameleon trapped in Burmese amber that appears to have a rodlike entoglossal (lingual) process of the hyoid (Daza et al. 2016). In addition, the chameleon hyoid skeleton, as described in the literature, lacks elements common to other lizard hyoids such as the ceratobranchial II and an epihyal. In the embryo shown in Fig. 9.3c a series of three degenerating cartilaginous nodules are visible as the remnants of a ceratohyal cartilage, which allows one to identify the "W"-shaped cartilages as the posterolaterally projecting hypohyal and an anterolaterally projecting epihyal, two elements present in most lizards. Additionally, Fig. 9.3d shows the remnant interhyal present in association with the middle ear cartilage, as is also present in *Sphenodon* (Fig. 9.1d) and in the whiptail (Fig. 9.2a, d). In addition (inset in Fig. 9.3f) chameleons are also shown to have a transient pair of ceratobranchial II cartilages attached medially to the basihyoid as they are in other lizards. Thus, what has been identified as the ceratohyal in the veiled chameleon literature is the hypohyal + epihyal with a (most likely) apoptotic loss of both the ceratohyal and ceratobranchial II, of which the latter may be sufficiently reduced to be missed during specimen preparation (see Fig. 2.2 in Anderson and Higham 2014). Variation in the degree of retainment of these skeletal elements should be looked at further, especially with reference to homology in adult structures both within and between lineages (Langebartel 1968; Tanner and Avery 1982).

To better understand the evolution of the unique cranium of chameleons relative to their "normal" sister family the Agamidae, Fig. 9.3g–i presents a 45 days post-oviposition (dpo) stage chondrocranium from *Pogona vitticeps* (the bearded dragon). While *P. vitticeps* lacks a complete tma, as does *C. calyptratus*, the occipital portion of the chondrocranium presents with a much wider distance between the tma insertions onto the otic capsule surface, which is due to a wider pair of basal plates lateral to the notochord (visible in Fig. 9.3i). This modification also leads to a lengthened tectum synoticum and larger relative foramen magnum. Thus, it appears that modularity in the size of the posterior (and mesoderm derived) cranial base can have a significant effect on the architecture of cranial vault intramembranous bones (adult *P. vitticeps* skull, not shown), a relationship observed in human congenital malformations and suggested in the development of the mouse cranial base (see McBratney-Owen et al. 2008).

Fewer studies have been conducted on snake chondrocranial development (summarized in Bellairs and Kamal 1981), with the most detailed recent studies by Haluska and Alberch (1983) for *Elaphe* and Khannoon and Evans (2015) for *Naja*. The African house snake [*Boaedon* (*Lamprophis*) *fuliginosus*] exhibits the typical serpentine chondrocranium, where the dorsal half of the cartilaginous architecture is reduced/missing. While Boback et al. (2012) also looked at *B. fuliginosus*, the skeletal anatomy was not the focus of their manuscript. The first detailed chondrocranial study of the genus *Lamprophis* *(Boaedon)* was conducted by Pringle (1952) utilizing *L. inornatus*, which is very similar cranioskeletally to our *B. fuliginosus* but is clearly a congeneric species given the differences in incubation and natural history described in that manuscript.

With respect to the chondrocranium, the primary sensory capsules and the associated elements of the cranial base are retained (trabecular and parachordals) and an internasal septum is formed; however, the interorbital septum and a midline fusion of the trabeculae cranii are lacking in all snakes studied (see review by Bellairs and Kamal 1981). Work from the Kuratani Lab (Wada et al. 2011, in the chick) highlights that there are two premandibular cranial neural crest streams associated with the eye in the early embryo, with the dorsal stream migrating over the eye and then ventromedially to form the interorbital septal population of cells, whereas the postorbital stream migrates below the eye and laterally to form the trabeculae. Thus, previous studies describing the trabeculae as growing medially to form a plate may not be accurate. Given the simplified anatomy of the eye due to a fossorial origin of snakes (Yi and Norell 2015), it is plausible that the anterodorsal stream of cranial neural crest cells may simply have been lost. Despite the significantly reduced chondrocranium above the cranial base and dorsal to the brain, the nasals, frontals, and parietals still form along the dorsal midline of the cranial vault but grow ventrally to form lateral walls over the neurocranium. The dorsal and lateral elements of the chondrocranium (is, tm, tma, pa, pacc, epi) may not only be serving as future sites guiding the appearance of intramembranous cranial vault bones but may also play a role in spacing cranioskeletal elements between the neurocranium and the calvarial surface. Their loss allows for a more cylindrical-shaped cranium but are clearly not necessary for midline intramembranous bone formation. Most lateral dermatocranial elements are missing in snakes (j, qj, sq, epi), though they retain the supratemporal bone in association with the dorsal articulation of the quadrate instead of the squamosal as well as maintain loosely joined cranial bones for maximum cranial kinesis with the maxilla and ectopterygoid forming the primary bones of the upper jaw.

While our images of embryos at two developmental time points illustrate a mature chondrocranium and ossifying osteocranium, several features of snake cranial anatomy can be addressed. We found no evidence for the formation of a squamosal, with the bone articulating with the quadrate supported as being the supratemporal due to its formation along the posteromedial and dorsal aspect of the quadrate cartilage. Though it has also been called the tabular, (see discussion by Pringle 1952 and Haluska and Alberch 1983) it elongates in an anterodorsal direction through ontogeny to topographically and functionally replace the squamosal bone as is seen in lizards. While snake crania seem to be simplified with respect to the presence of many cranial bones (see Cundall and Irish 2008), we found an independent intramembranous ossification along the anteromedial wall of the developing orbit. After subsequent analysis of embryos at later stages of ossification, this ossification expanded dorsally where it contacted a ventral extension of the prefrontal bone. Upon contact, both ossifications fused to form what has been typically regarded as the prefrontal bone of snakes, which in *B. fuliginosus* is clearly a cryptic fusion of a lacrimal (what we consider the ventral element to be homologous with, based on its intramembranous nature and position relative to the orbit) and the prefrontal (dorsal element). Our lacrimal is not to be confused with the lamina transversalis anterior described in Haluska and Alberch (1983), which forms the Jacobson's organ cartilage. Our study shows that this bone is further posterior along the anterior wall of the orbit and does not undergo a cartilaginous intermediate step (endochondral ossification) and thus is not an ossified Jacobson's cartilage. In addition, Fig. 7 of Haluska and Alberch (1983) shows the embryonic prefrontal bone of *Elaphe obsoleta* (Colubridae) as an element with dorsal and ventral ossification centers and two parallel slivers of mineralized matrix connecting them (thus leaving a gap between the main bones). Such an intermediate stage is visible in *B. fuliginosus* (Lamprophiidae) as the ventral bone (what we identify as the lacrimal) fuses with the dorsal bone (what we consider the prefrontal). Thus, given our skeletal developmental series for *B. fuliginosus* and the illustration presented for *Elaphe* by Haluska and Alberch (1983), it appears that derived snakes possesses a lacrimal that cryptically fuses with the dorsal prefrontal. This is significant given that a lacrimal has never been identified in a snake, but appears to be present in a derived snake lineage (Lamprophiidae) and *Elaphe*, thus supporting the idea that skeletal elements may not truly be lost but fuse with adjacent elements or are reduced in their differentiation (see Diaz and Trainor 2015). A broader examination of this bone in serpentes is warranted.

Another outstanding issue in snake evolution is whether the snake orbit contains a postfrontal, postorbital, or postorbitofrontal (see Haluska and Alberch 1983). Recent work by Palci and Caldwell (2013) phylogenetically tested the identity of the postorbital element in snakes and found that it should be considered to be the postfrontal rather than the postorbital as has been classically considered in snakes. Our embryonic work as well as their own data supports a postfrontal placement adjacent to the posterolateral margin in the frontal bone with the postorbital present posterior to the frontal bone and the postfrontal and not posterolaterally adjacent. Thus, in our *B. fuliginosus* embryo, the postorbital elements appear as distant and independent intramembranous ossifications posterior and lateral to the frontal bones with a significant gap of separation, unlike what we see in *A. uniparens* and *C. calyptratus* embryos or in adult crania as presented in Palci and Caldwell (2013). Thus, we believe that the postorbital element of *B. fuliginosus*, and potentially of snakes, represents a true postorbital bone.

9.5 Cranial Muscles and the Homology of the M. Levator Anguli Oris

Understanding cranioskeletal development across the diverse lineages within Lepidosauria provides a foundation to interpret the homology of muscles with respect to their skeletal attachments. Recent advances in our knowledge of the development and differentiation of tendons and their muscle or skeletal attachments (**myotendinous junction** and **enthesis**, respectively) can assist in interpreting homology statements through a mechanistic framework. Herein we use the antibody MF20 (Developmental Studies Hybridoma Bank) to identify myosin heavy chain (differentiated muscle cells) in embryos of *A. uniparens*, *C. calyptratus*, and *B. fuliginosus* while comparing to the extensively studied muscles of *S. punctatus* as reviewed by Haas (1973) and recently by Johnston (2014), Curtis et al. (2009), Jones et al. (2009), Daza et al. (2011), and Wu (2003). We focus on the jaw adductor musculature which has received much attention with regard to their homology to the jaw adductor musculature of *Sphenodon* and "lizards" as well as in comparison to serpentes and even within serpentes. This contention primarily surrounds the presence of the M. levator anguli oris in cranioskeletally divergent taxa.

There are two primary divisions of the **cranial muscles** of lepidosaurs, diagnosed by their innervation by cranial nerve V (**trigeminal,** CNV) or VII (**facial,** CNVII) (Luther 1914; Lakjer 1926; see Haas 1973 and Johnston 2014 for a review). Muscles arise as a "mass" of myoblast cells that, upon reaching a target region, segment into separate muscles or into distinct heads and insert onto target surfaces. Thus, muscles are described as belonging to particular groups but have very dynamic patterns of migration and differentiation. Following the divisions described by Haas (1973), muscles innervated by CNV include (1) constrictor internus dorsalis group; (2) M. adductor mandibulae group: (a) M. adductor mandibulae externus, (b) M. adductor mandibulae internus group, and (c) M. adductor mandibulae posterior; and (3) constrictor ventralis trigemini (the M. intermandibularis). Those innervated by CNVII include (1) those inserting onto the articular process of the lower jaw (M. depressor mandibulae, M. cervicomandibularis, etc.), (2) superficial constrictors (M. sphincter colli), and (3) the more posterior transverse portion (M. intermandibularis facialis). In this study only the most superficial muscles could be visualized in whole mount (constrictor internus dorsalis group was not visible externally), with our attention given primarily to the contentious adductor mandibulae group. The origin of these groups and their subsequent innervation by CNV or CNVII reflects the pharyngeal arch from which they originated, as those derived from pharyngeal arch I are innervated by CNV and from pharyngeal arch II by CNVII (Sambasivan et al. 2011; Diogo et al. 2015). Due to divergence in feeding mode (no longer dependent on mastication or utilizing the snout for subduing of prey due to shifting to live ingestion, envenomation, or constriction), the muscles associated with the jaw have undergone a reduction in number and size. Yet, they are required to remain flexible enough to allow for a highly kinetic skull to move during feeding.

Jaw adductor musculature relies heavily, in lizards and *Sphenodon*, on insertion onto the tendinous rictal plate and aponeuroses associated with the jawbones. Phylogenetically, basal alethinophidian snakes have more soft tissue components involved in jaw adduction in common with lizards and *Sphenodon*, with more derived snakes lacking the above mentioned insertion tissues. This has led to confusion of muscle terminology and homology, with the M. levator anguli oris inserting onto the rictal plate in some snakes as well as the venom gland and compound bones and the coronoid, but arising from the postorbital and/or the parietal (Johnston 2014). Discussions in the literature attempt to place identified jaw musculature of lizards and *Sphenodon* onto the highly variable cranium of serpentes, with the differential loss of these muscles not considered equally in these papers. Presenting terminology of a lizard with different insertion points (and origins) onto the serpent cranium is misleading.

We look at the of development of jaw adductor muscles in the context of a single species of snake (*B. fuliginosus*, a derived species) that lacks tendinous insertion tissues and attempt to bridge our knowledge of muscle development with adult muscle anatomy. We also aim to introduce current thinking in developmental biology about how muscles become segmented during development as a potentially new model of cranial muscle evolution. Briefly, rather than looking at homology across differentiated muscles (which are quite variable), we look at the different muscles arising from highly conserved embryonic muscle cell precursors which are homologous between the lizards and snake in the hope it serves as a foundation for future projects. Previous work on development of snake embryos also share some of these conclusions (Kochva 1963; Rieppel 1988). Our figures also present a lot more information (throughout the chapter) than we can possibly cover in the text presented here.

Cranial muscle cells are of mesodermal origin, but in the pharyngeal region of the cranium, neural crest cells differentiate into the connective tissue of the muscle and are responsible for forming the muscle fascia while also playing a role in their bundle formation (McClearn and Noden 1988; reviewed by Noden and Trainor 2005). Tongue muscle precursor cells migrate in the head from anterior trunk somites and extraocular muscle precursor cells migrate in the head from paraxial mesoderm (Noden and Trainor 2005; Noden and Francis-West 2006; Sambasivan et al. 2011; Michailovici et al. 2015). Recent work in mice highlights that in the cranium, tendon precursors are already established on the surface of skeletal tissue and only mature into differentiated tendons upon forming the myotendinous junction (reviewed in Schweitzer et al. 2010). Thus, the attachment of a muscle onto a particular skeletal element or region is predetermined by the tendon not intrinsically by the muscle cells. This is important as it shows that muscles cannot easily alter their origins, and/or insertions during evolution because of the developmental constraints imposed by interactions between neural crest cells (tendon, attachment sites) and mesodermal cells (muscle). Thus, future work should examine the patterns of differentiation of tendinous tissue in the skull through markers such as ***Scleraxis*** or ***Tenomodulin*** which are specific to tendons.

As introduced above, issues of homology for adductor musculature have been prevalent in the comparative literature for snakes (within Lepidosauria) since the work of Lakjer (1926) through to the most recent reevaluation by Daza et al. (2011) and Johnston (2014). The presence of an upper temporal bar in non-serpent squamates along with the tendinous rictal plate and basal aponeuroses (both lacking in non-scolecophidian snakes) provides a framework for the differentiation and insertion of the M. levator anguli oris (LAO), primarily originating from the upper temporal bar (postorbital and squamosal) and inserting onto the rictal plate ventrolaterally in the temporal region (Edgeworth 1935; Kochva 1963; Haas 1973; Cundall and Gans 1979; Rieppel 1980, 1981, 1988; Cundall 1986; Wu 2003; Daza et al. 2011; Johnston 2011). Rather than taking a topographic approach at mature differentiated cranial musculature, we attempted to follow the formation of individual myoblast (MF20+) masses as they grow, migrate, and ultimately separate into individual bundles. In *A. uniparens* (Fig. 9.5a–c), we see three MF20 positive "streams" (or masses, labeled 1–3) which are posteroventral to the orbit in the pharyngeal arches and ventral to the otic capsule. The most anterior, 1, if followed through the ascending embryonic stages, appears at the junction of the maxillary and mandibular prominences of pharyngeal arch I and retains MF20+ cells at this junction while expanding dorsally and behind the orbit in the temporal region. In Fig. 9.5d, e the LAO appears as a lateral and superficial mass that has an insertion onto the rictal plate along its anteroventral margin and anterodorsally along the postorbitofrontal bone (pof). Deep to the LAO is a large rectangular mass (the M. adductor mandibulae externus superficialis [MAMES]) and dorsal to this mass (across the upper temporal bar) is the M. adductor mandibulae externus medialis (MAMEM) which is visible as a separate mass in Fig. 9.5c. Along stream 2, from Fig. 9.5a–d, the most ventral MF20+ population

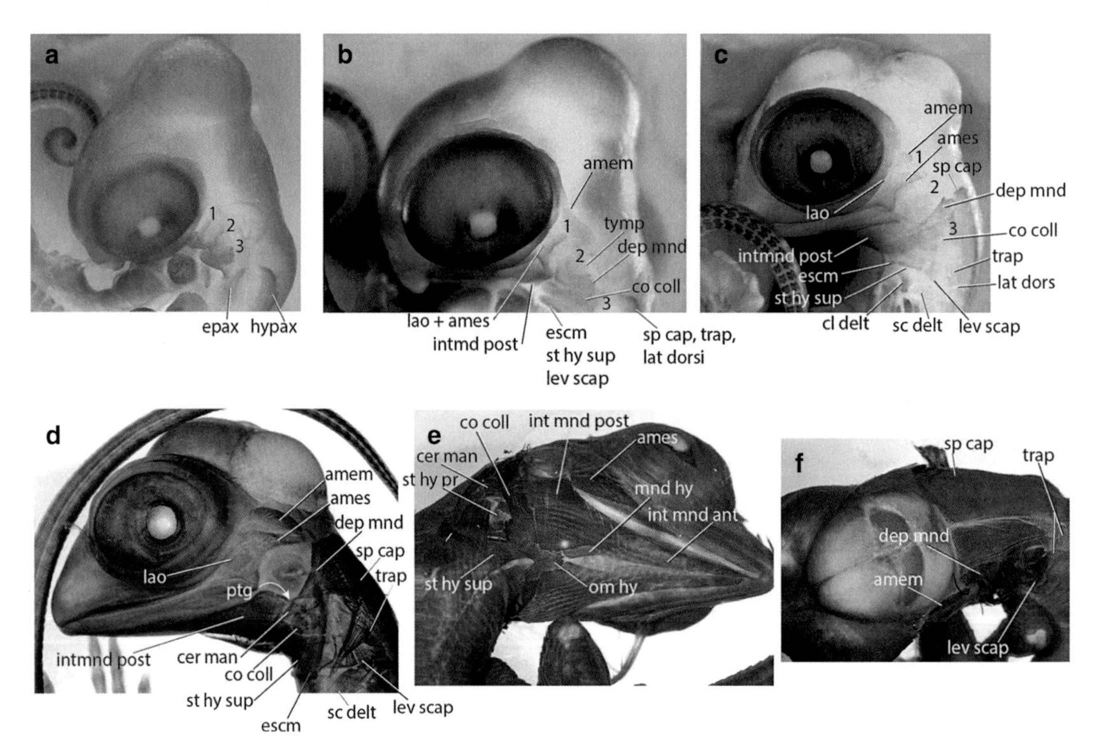

Fig. 9.5 *Aspidoscelis uniparens* MF20 (DSHB) antibody-stained embryonic series showing differentiation and bundle formation of cranial muscles. Lateral view of embryos at (**a**) 9–10 days post-oviposition (dpo), (**b**) 13–14 dpo, and (**c**) 16–17 dpo. A 25–26 dpo embryo is shown in (**d**) lateral, (**e**) ventrolateral, and (**f**) posterodorsal view. See Fig. 9.1 for abbreviation explanations

will give rise to the M. intermandibularis posterior (appearing along the ventral surface of pharyngeal arch I), while the dorsal portion seems to be non-myosin-specific staining of what begins to differentiate as the tympanic membrane proximal to pharyngeal cleft I. The stream labeled 3 arises from pharyngeal arch II and in following ontogenetic stages appears to give rise to the M. depressor mandibulae and the M. constrictor colli. The space between MF20+ streams 1 and 3 is taken by the developing and enlarging tympanic membrane in *A. uniparens*.

The veiled chameleon (Fig. 9.6) and African house snake (Fig. 9.7) both converged on losing their tympanic membrane, so the more proximal non-specific MF20+ patch of cells between stream 1 and 3 is not present in these two latter taxa, and thus stream 3 migrates anterodorsally toward stream 1 unimpeded by the tympanic membrane. Unlike serpentes, of which the majority lack aponeuroses, *C. calyptratus* MF20+ stream 1 differentiates into a more ventral LAO which lies superficial to the MAMES but is not as visible as that in *A. uniparens* due to the LAO seeming to fill the entire lower temporal fenestra (as illustrated in Rieppel 1980 for chameleons) and obscuring the MAMES below. Dorsally the MAMEM is already visible as a distinct element medial to the upper temporal bar (Fig. 9.6b, c). Proximal MF20+ cells give rise to the M. intermandibularis posterior while 3 differentiates into the M. depressor mandibulae and the M. constrictor colli (Fig. 9.6b, f). The African house snake (Fig. 9.7) also shows three MF20+ streams within the pharyngeal arches (Fig. 9.7b) while also showing, at an earlier stage (Fig. 9.7a), a population of scattered MF20+ cells anterior to the somites and dorsal to all pharyngeal arches that are not seen (or potentially missed) in lizards. Figure 9.7b illustrates that stream 1 has two distinct dorsal projections; the anterior gives rise to the MAMES, while the posterior extension becomes the MAMEM. Stream 2, as a separate MF20+ mass forms a separate adductor muscle, the M. adductor mandibulae posterior (MAMP). This is not the M. adductor mandibulae externus profundus of Lakjer (1926), Haas (1973), or Rieppel (1980), as the profundus is considered to belong to the M. adductor mandibulae externus group, which, as shown here, is a separate anterior

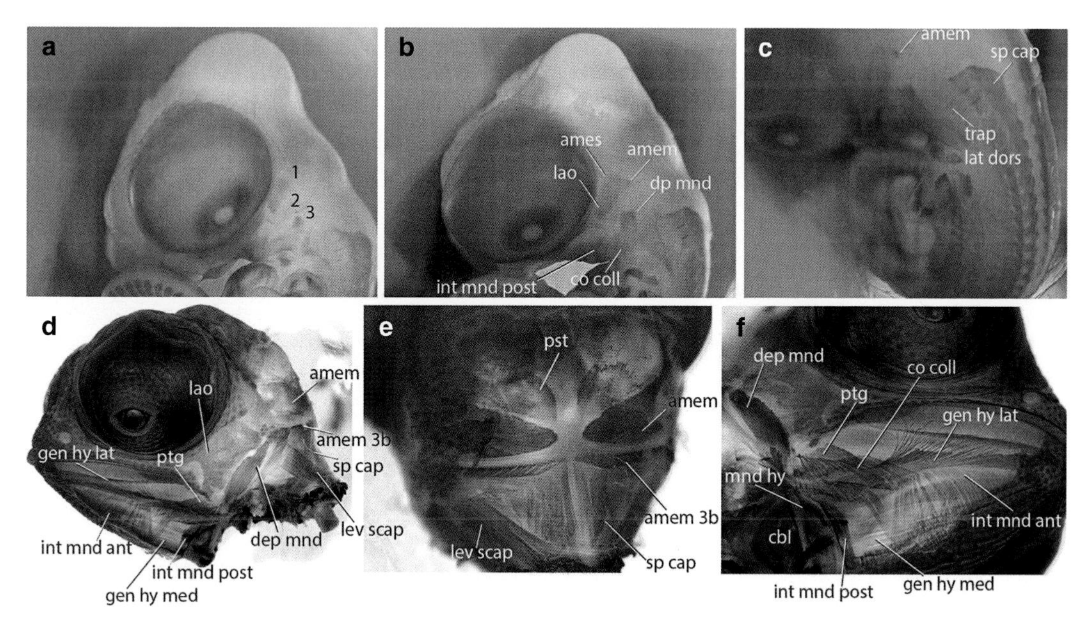

Fig. 9.6 *Chamaeleo calyptratus* MF20 (DSHB) antibody-stained embryonic series showing differentiation and bundle formation of cranial muscles. Lateral view of embryos at (**a**) 95 days post-oviposition (dpo), (**b**) 100 dpo, and posterolateral (**c**) 100 dpo. A 150 dpo embryo is shown in (**d**) lateral, (**e**) posterodorsal, and (**f**) ventrolateral view. See Fig. 9.1 for abbreviation explanations

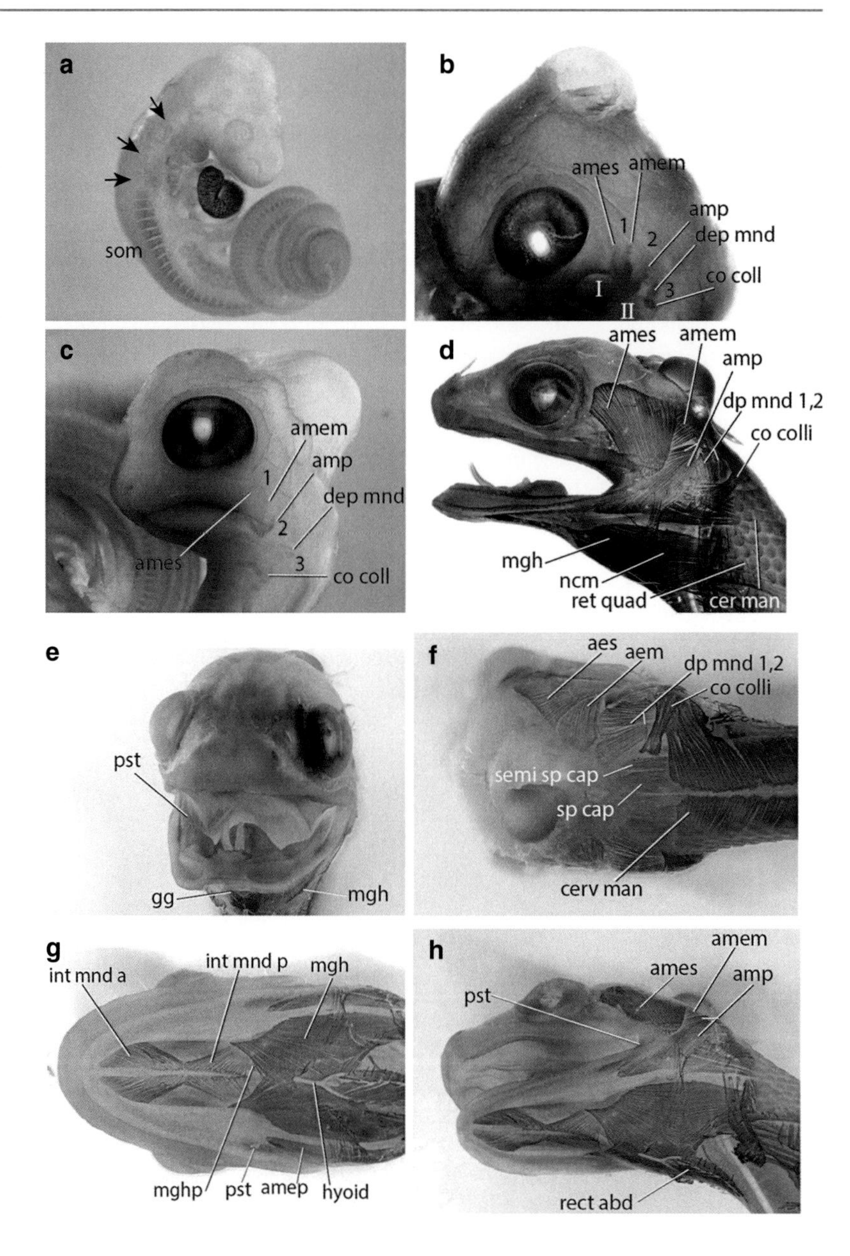

Fig. 9.7 *Boaedon* (*Lamprophis*) *fuliginosus* MF20 (DSHB) antibody-stained embryonic series showing differentiation and bundle formation of cranial muscles. Lateral view of embryos at (**a**) 0 days post-oviposition (dpo) whole embryo, (**b**) 8–9 dpo left lateral, and (**c**) 11–12 dpo left lateral. A 25–26 dpo embryo is shown in (**d**) left lateral, (**e**) frontal, (**f**) posterodorsal, (**g**) ventral, and (**h**) ventrolateral. See Fig. 9.1 for abbreviation explanations

mass at the junction of the maxillary and mandibular prominences ("1"). Stream 3, as in the two previous squamates, gives rise to the M. depressor mandibulae and the M. constrictor colli.

In summary, within a developmental framework (unlike previous topological frameworks), we do not see the presence of a LAO in serpentes and find support for the MAMES and MAMEM to be separate from the MAMP present as an adductor from the quadrate. Muscle precursor populations in the head mostly interact with signals from neural crest cells during their specification to form muscles with connective tissue and tendons that attach to specific skeletal sites that also derive from neural crest cells (Schweitzer et al. 2010). Thus, the retainment of an LAO is largely dependent on whether the rictal plate and elements of the upper temporal fenestra are present and produce signals for muscle attachment. Future work should focusing on examining the embryonic development of muscles in snakes that are phylogenetically, ecologically, and morphologically divergent.

9.6 Cranial Nerves

For the most part, there is high conservation across Lepidosauria with respect to the spatial origin of the cranial nerves and their targets (for the most recent and detailed review see ten Donkelaar 1998), with a few exceptions. Detailed past works on the cranial nerves have been more extensive on lizards (Fischer 1852; Watkinson 1906; Willard 1915; Oelrich 1956; Islam and Ashiq 1972; Szekely and Matesz 1988, 1993) than snakes (Auen and Langebartel 1977), and many of the cranial nerves in *Sphenodon* were illustrated during the course of describing other aspects of their biology (Fig. 9.8a, b of *Sphenodon* were presented during the illustration of cranial vasculature by Dendy 1909). The cranial nerve (CN) I (olfactory) is present in the olfactory epithelium and is more extensive in serpents, especially in regard to having a vomeronasal component (Fig. 9.8c–e). CNII (optic) axons arise from the retina and migrate through the optic tract, which is an extension of the base of the diencephalon. Cranial nerves III, IV, and VI innervate the extraocular muscles. CNIII (oculomotor) innervates the contralateral superior rectus muscles, the anterior and inferior recti, and the inferior oblique muscles. CNIV (trochlear) innervates the contralateral superior oblique muscles, while CNVI (abducens) innervates the retractor bulbi muscles and the posterior rectus. CNV (trigeminal) innervates the greatest number of tissues in the cranium and comprises an ophthalmic branch (CNV1) which extends anteromedially to the orbits, a maxillary branch (CNV2) for the upper jaw, and a mandibular branch (CNV3) for the lower jaw. In snakes a fourth branch to the palatopterygoid region innervates muscles of the internus dorsalis group. CNVII exits ventral to CNVIII and splits into two branches, an anterior ramus palatinus and a posterior hyomandibularis, with the latter giving rise to the chorda tympani, and then continues as a branchiomotor nerve, which serves to control the M. depressor mandibulae (Haas 1973; ten Donkelaar 1998). CNVIII (statoacoustic or vestibulocochlear) has two roots, the posterior is involved in transmitting auditory receptor information, while the anterior division transmits sensory information from the utricle and saccule. CNIX (glossopharyngeal) exits alongside CNXII (hypoglossal) and also joins CNX (and CNXI?) along the neck. CNX (vagus) arises from several rootlets that remain separate until entry through the metotic fissure posterior to the otic capsule (Fig. 9.8a–d). In snakes and anolis, CNIX, X, (XI?), and XII form a common cervical trunk (Willard 1915; Auen and Langebartel 1977). Contention seems to only surround CNXI (spinal accessory), which is considered to be missing in

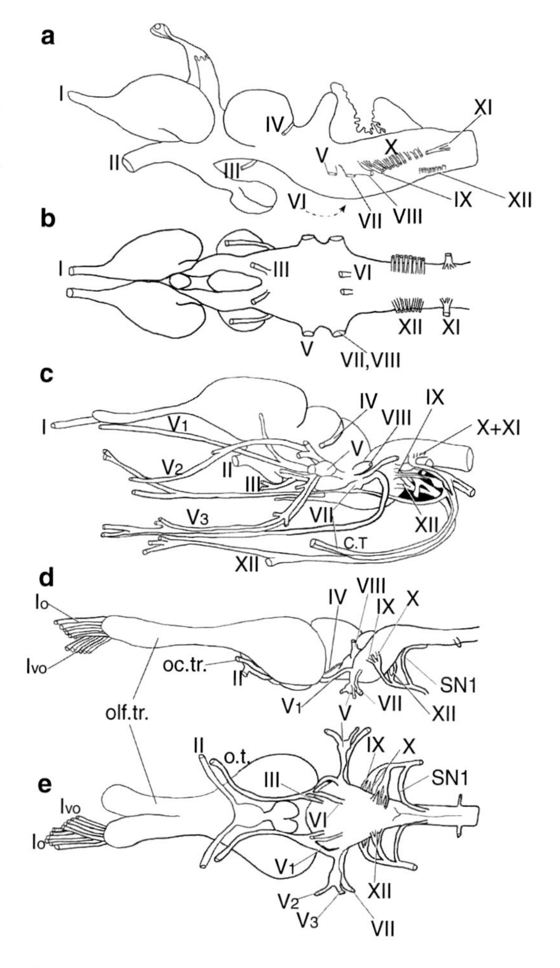

Fig. 9.8 Illustrations of the brains and associated cranial nerve organization. (**a**) and (**b**) are of *Sphenodon punctatus*, modified from Dendy (1909), with (**b**) not illustrating CNII in ventral view. (**c**) Lizard nervous system redrawn Romer and Parsons (1977) based on Willard (1915). Lateral (**d**) and ventral (**e**) view of a snake brain and cranial nerves from Auen and Langebartel (1977). See Fig. 9.1 for abbreviation explanations

all snakes (Auen and Langebartel 1977) and lizards (Willard 1915; Islam and Ashiq 1972), with its description by Watkinson (1906) in a monitor lizard (*Varanus bivittatus*) actually being a mistaken posterior rootlet of CNX. Dendy's illustration of the CNXI for *Sphenodon* (Dendy 1909) shows a distinct CNXI with two rootlets posterior to CNX (Fig. 9.8a, b). While most, if not all, descriptions are done in very late-stage embryos or adults, our work here (Fig. 9.9b, using a neuronal antibody Tuj1) shows a distinct CNXI in the African house snake that appears to fuse quite early with CNX (this is also seen in the veiled chameleon, manuscipt in prep.), and it is topologically where it is expected to be based on non-squamates reptiles (such as in *Sphenodon*). Further fine-scale examination of these rootlets and their trajectory into the brain are needed to support their identity as vestigial CNXI. A vestigial CNXI is not surprising given its innervation of the M. cucullaris derivatives (the M. trapezius and the M. episternocleidomastoideus) which are apparently missing in snakes which have completely lost their shoulder girdles and associated musculature.

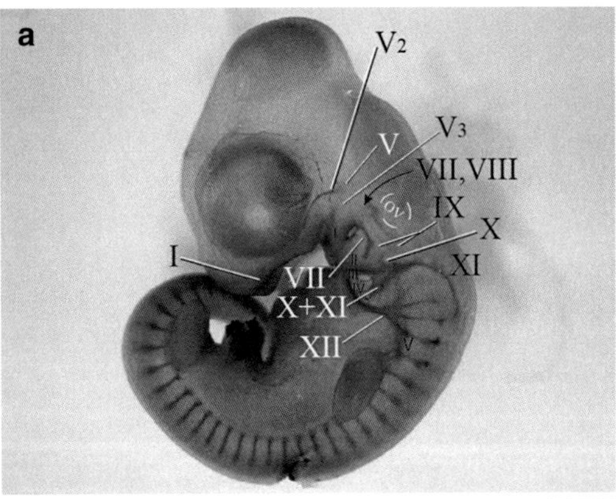

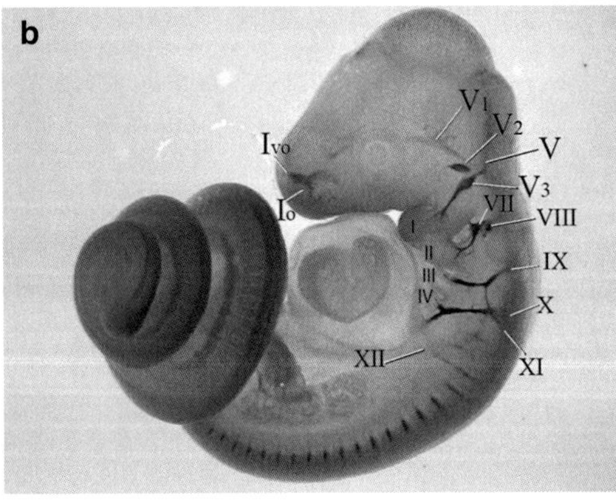

Fig. 9.9 Neuronal antibody staining with Tuj1 (Covance) on a (**a**) 90 days post-oviposition (dpo) veiled chameleon embryo and (**b**) a 0 dpo African house snake embryo. See text for details and Fig. 9.1 for abbreviation explanations

9.7 Future Directions

Given the large number of extant lepidosaurs (>10 K), confusion is expected to arise as to the homology of morphological structures, especially in the case of muscles that display an incredible plasticity in attachments (even within the same species and the same individual as has been recorded for *Sphenodon*; Wu 2003). One powerful method of studying the homology of features, when feasible, is to study embryonic development with both classic methods (histology, alizarin/alcian staining) as well as with newer methods of mRNA and protein expression visualization. Utilizing extant taxa as models to understand the molecular underpinnings of how unique features arose in extinct forms is in line with the current growth of the field of **evolutionary developmental biology** which allows us to better understand how organisms have evolved to adapt to their environment and thus understand biodiversity.

References

Anderson CV, Higham TE (2014) Chameleon anatomy. In: The biology of chameleons. University of California Press, London, pp 7–56

Apesteguía S, Jones MEH (2012) A late cretaceous "tuatara" (Lepidosauria: Sphenodontinae) from South America. Cretac Res 34:154–160 http://www.sciencedirect.com/science/article/pii/S0195667111001649

Apesteguia S, Novas FE (2003) Large cretaceous sphenodontian from Patagonia provides insight into lepidosaur evolution in Gondwana. Nature 425:609–612

Auen EL, Langebartel DA (1977) The cranial nerves of the colubrid snakes Elaphe and Thamnophis. J Morphol 154:205–222

Bellairs A d'A, Kamal AM (1981) The chondrocranium and the development of the skull in recent reptiles. In:

Gans C, Parsons TS (eds) Biology of the reptilia, vol 11. Academic Press, London, pp 1–263
Bhatt S, Diaz R, Trainor PA (2013) Signals and switches in mammalian neural crest cell differentiation. Cold Spring Harb Perspect Biol 5:a008326
Boback SM, Dichter EK, Mistry HL (2012) A developmental staging series for the African house snake, Boaedon (Lamprophis) fuliginosus. Zoology 115:38–46 http://www.sciencedirect.com/science/article/pii/S0944200611000894
Brandley MC, Huelsenbeck JP, Wiens JJ (2008) Rates and patterns in the evolution of snake-like body form in squamate reptiles: evidence for repeated re-evolution of lost digits and long-term persistence of intermediate body forms. Evolution 62:2042–2064
Chai Y, Jiang X, Ito Y, Bringas P, Han J, Rowitch DH, Soriano P, McMahon AP, Sucov HM (2000) Fate of the mammalian cranial neural crest during tooth and mandibular morphogenesis. Development 127:1671–1679
Couly GF, Coltey PM, Le Douarin NM (1993) The triple origin of skull in higher vertebrates: a study in quail-chick chimeras. Development 117:409–429 http://www.ncbi.nlm.nih.gov/pubmed/8330517
Cree A (2014) Tuatara: biology and conservation of a venerable survivor. Canterbury University Press, Christchurch
Cundall D (1986) Variations of the cephalic muscles in the colubrid snake genera Entechinus, Opheodrys, and Symphimus. J Morphol 187:1–21
Cundall D, Gans C (1979) Feeding in water snakes: an electromyographic study. J Exp Zool 209:189–207
Cundall D, Irish F (2008) The snake skull. In: Biology of the reptilia, Volume 20. The skull of Lepidosauria. Society for the Study of Amphibians and Reptiles, Ithaca, pp 349–392 http://scholar.google.com/scholar?hl=en&btnG=Search&q=intitle:The+Snake+Skull#1
Curtis N, Jones ME, Evans SE, O'Higgins P, Fagan M (2009) Visualizing muscle anatomy using three-dimensional computer models—an example using the head and neck muscles of Sphenodon. Palaeontol Electron 12:7T
Daza JD, Diogo R, Johnston P, Abdala V (2011) Jaw adductor muscles across Lepidosaurs: a reappraisal. Anat Rec 294:1765–1782
Daza JD, Stanley EL, Wagner P, Bauer AM, Grimaldi DA (2016) Mid-cretaceous amber fossils illuminate the past diversity of tropical lizards. Sci Adv 2:e1501080
De Beer G (1937) Development of the vertebrate skull. Oxford University Press, New York, NY
Dendy A (1909) The intracranial vascular system of sphenodon. Proc R Soc B 81:403–427
Diaz RE, Bertocchini F, Trainor PA (2017) Lifting the veil on reptile embryology: the veiled chameleon (Chamaeleo calyptratus) as a model system to study reptilian development. Methods Mol Biol 1650:269–284
Diaz RE, Trainor PA (2015) Hand/foot splitting and the "re-evolution" of mesopodial skeletal elements during the evolution and radiation of chameleons. BMC Evol Biol 15:184 http://www.ncbi.nlm.nih.gov/pubmed/26382964
Diogo R, Abdala V (2010) Muscles of vertebrates: comparative anatomy, evolution, homologies and development. Science Publishers, Enfield
Diogo R, Kelly RG, Christiaen L, Levine M, Ziermann JM, Molnar JL, Noden DM, Tzahor E (2015) A new heart for a new head in vertebrate cardiopharyngeal evolution. Nature 520:466–473 http://www.ncbi.nlm.nih.gov/pmc/articles/PMC4851342/
Edgeworth FH (1935) The cranial muscles of vertebrates. Cambridge University Press, London
El-Toubi MR, Soliman MA (1967) Studies on the osteology of the family Lacertidae in Egypt. I. The skull. Proc Zool Soc United Arab Repub 2:219–257
Estes R, De Queiroz K, Gauthier JA (1988) Phylogenetic relationships within Squamata. In: Phylogenetic relationships of the lizard families essays commemorating Charles L. Camp. Stanford University Press, Stanford, pp 119–281
Evans SE (2003) At the feet of the dinosaurs: the early history and radiation of lizards. Biol Rev Camb Philos Soc 78:513–551
Evans S, Jones M (2010) The origins, early history and diversification of lepidosauromorph reptiles. New Asp Mesozoic Biodivers 132:105–126 http://link.springer.com/10.1007/978-3-642-10311-7%5Cn, http://discovery.ucl.ac.uk/1308892/
Fischer JG (1852) Die Gehirnnerven der saurier anatomisch untersucht. Abh Naturw Ver Hambg 2:115–212
Fraser NC, Benton MJ (1989) The Triassic reptiles Brachyrhinodon and Polysphenodon and the relationships of the sphenodontids. Zool J Linnean Soc 96:413–445
Gans C, Gaunt AS, Adler K (2008) Biology of the reptilia. Volume 20. Morphology H. The skull of Lepidosauria. Society for the Study of Amphibians and Reptiles, New York
Gans C, Northcutt RG, Ulinski P (1979) Biology of the reptilia, Volume 10. Neurology B. Academic Press, London, New York and San Francisco
Gans C, Wever EG (1976) Ear and hearing in sphenodon puncatus. Proc Natl Acad Sci U S A 73:4244–4246
Gauthier JA, Kearney M, Maisano JA, Rieppel O, Behlke ADB (2012) Assembling the squamate tree of life: perspectives from the phenotype and the fossil record. Bull Peabody Museum Nat Hist 53:3–308
Gignac PM, Kley NJ (2014) Iodine-enhanced micro-CT imaging: methodological refinements for the study of the soft-tissue anatomy of post-embryonic vertebrates. J Exp Zool Part B Mol Dev Evol 322:166–176
Gignac PM, Kley NJ, Clarke JA, Colbert MW, Morhardt AC, Cerio D, Cost IN, Cox PG, Daza JD, Early CM, Echols MS, Henkelman RM, Herdina AN, Holliday CM, Li Z, Mahlow K, Merchant S, Müller J, Orsbon CP, Paluh DJ, Thies ML, Tsai HP, Witmer LM (2016) Diffusible iodine-based contrast-enhanced computed tomography (diceCT): an emerging tool for rapid, high-resolution, 3-D imaging of metazoan soft

tissues. J Anat 228:889–909. https://doi.org/10.1111/joa.12449
Gisi J (1907) Das gehirn von Hatteria punctata. Zool Jahrb (Anat) 25:1–166
Greer AE (1991) Limb reduction in squamates: identification of the lineages and discussion of the trends. J Herpetol 25:166 http://www.jstor.org/stable/1564644?origin=crossref
Gunther A (1867) Contribution to the anatomy of Hatteria. Philos Trans R Soc London 157:1–34
Haas G (1973) Muscles of the jaws and associated structures in the Rhynchocephalia and Squamata. In: Gans C, Parsons TS (eds) Biology of the reptilia, vol 4. Academica Press, New York and London, pp 285–490
Hall BK (2008) The neural crest and neural crest cells: discovery and significance for theories of embryonic organization. J Biosci 33:781–793
Haluska F, Alberch P (1983) The cranial development of Elaphe obsoleta (Ophidia, colubridae). J Morphol 178:37–55. https://doi.org/10.1002/jmor.1051780104
Hay JM, Sarre SD, Lambert DM, Allendorf FW, Daugherty CH (2010) Genetic diversity and taxonomy: a reassessment of species designation in tuatara (Sphenodon: Reptilia). Conserv Genet 11:1063–1081. https://doi.org/10.1007/s10592-009-9952-7
Heckert AB, Lucas SG, Rinehart LF, Hunt AP (2008) A new genus and species of sphenodontian from the ghost ranch coelophysis quarry (upper triassic: apachean), rock point formation, New Mexico, USA. Palaeontology 51:827–845. https://doi.org/10.1111/j.1475-4983.2008.00786.x
Herrera-Flores JA, Stubbs TL, Benton MJ (2017) Macroevolutionary patterns in Rhynchocephalia: is the tuatara (Sphenodon punctatus) a living fossil? Palaeontology 60:319–328
Hutchinson MN, Skinner A, Lee MSY (2012) Tikiguania and the antiquity of squamate reptiles (lizards and snakes). Biol Lett 8:665–669 http://rsbl.royalsocietypublishing.org/cgi/doi/10.1098/rsbl.2011.1216
Islam A, Ashiq S (1972) The cranial nerves of Uromastix hardwicki gray. Biologia (Bratisl) 18:51–73
Jiang X, Iseki S, Maxson RE, Sucov HM, Morriss-Kay GM (2002) Tissue origins and interactions in the mammalian skull vault. Dev Biol 241:106–116 http://www.sciencedirect.com/science/article/pii/S0012160601904877
Johnston P (2011) Cranial muscles of the anurals Leiopelma hochstetteri and Ascaphus truei and the homologies of the mandibular adductors in Lissamphibia and other gnathostomes. J Morphol 272:1492–1512
Johnston P (2014) Homology of the jaw muscles in lizards and snakes-a solution from a comparative gnathostome approach. Anat Rec 297:574–585
Jones MEH, Anderson CL, Hipsley CA, Müller J, Evans SE, Schoch RR (2013) Integration of molecules and new fossils supports a Triassic origin for Lepidosauria (lizards, snakes, and tuatara). BMC Evol Biol 13:208. https://doi.org/10.1186/1471-2148-13-208
Jones MEH, Curtis N, O'Higgins P, Fagan M, Evans SE (2009) Head and neck muscles associated with feeding in Sphenodon (Reptilia: Lepidsauria: Rhynchocephalia). Palaeontol Electron 12:1–56
Kamal AM, Abdeen AM (1972) The development of the chondrocranium of the lacertid lizard, Acanthodactylus boskiana. J Morphol 137:289–333
Kardong KV (2012) Vertebrates: comparative, function, evolution. http://www.ncbi.nlm.nih.gov/pubmed/15003161%5Cn, http://cid.oxfordjournals.org/lookup/doi/10.1093/cid/cir991%5Cn, http://www.scielo.cl/pdf/udecada/v15n26/art06.pdf%5Cn, http://www.scopus.com/inward/record.url?eid=2-s2.0-84861150233&partnerID=tZOtx3y1
Khannoon ER, Evans SE (2015) The development of the skull of the Egyptian Cobra Naja h. haje (Squamata: Serpentes: Elapidae). PLoS One 10:e0122185. https://doi.org/10.1371/journal.pone.0122185
Kochva E (1963) Development of the enom gland and trigeminal muscles in Vipera palestinae. Acta Anat 52:49–89
Kumar S, Stecher G, Suleski M, Hedges SB (2017) TimeTree: a resource for timelines, timetrees, and divergence times. Mol Biol Evol 34:1812–1819. https://doi.org/10.1093/molbev/msx116
Lakjer T (1926) Studien uber die trirgeminus-versorgte Kaumuskulatur der Sauropsiden
Langebartel DA (1968) The hyoid and its associated muscles in snakes. Illinois Biological Monographs 38. University of Illinois Press, Champaign, IL
Le Douarin NM, Dupin E (2014) The neural crest, a fourth germ layer of the vertebrate embryo: significance in chordate evolution. In: Neural crest cells: evolution, development and disease. Elsevier Inc., New York, pp 3–26
Luther A (1914) Uber die vom N. trigeminus versorgte Muskulatur der Amphibien, mit einen vergleichenden Ausblick uber den Adductor mandibulae der Gnathostomen, und einem Beitrag zu Verstandnis der organisation der Anuranlarven. Acta Soc Sci Fenn 44:1–151
McBratney-Owen B, Iseki S, Bamforth SD, Olsen BR, Morriss-Kay GM (2008) Development and tissue origins of the mammalian cranial base. Dev Biol 322:121–132 http://www.sciencedirect.com/science/article/pii/S0012160608010701
McClearn D, Noden DM (1988) Ontogeny of architectural complexity in embryonic quail visceral arch muscles. Am J Anat 183:277–293. https://doi.org/10.1002/aja.1001830402
Metscher BD (2009) MicroCT for developmental biology: a versatile tool for high-contrast 3D imaging at histological resolutions. Dev Dyn 238:632–640
Michailovici I, Eigler T, Tzahor E (2015) Chapter One—Craniofacial muscle development. In: Chai YBT (ed) Current topics in developmental biology, Craniofacial development, vol 115. Academic Press, Cambridge, MA, pp 3–30 http://www.sciencedirect.com/science/article/pii/S0070215315000587

Molnar JL, Diaz RE, Skorka T, Dagliyan G, Diogo R (2017) Comparative musculoskeletal anatomy of chameleon limbs, with implications for the evolution of arboreal locomotion in lizards and for teratology. J Morphol 278:1241–1261
Noden DM, Francis-West P (2006) The differentiation and morphogenesis of craniofacial muscles. Dev Dyn 235:1194–1218
Noden DM, Trainor PA (2005) Relations and interactions between cranial mesoderm and neural crest populations. J Anat 207:575–601. https://doi.org/10.1111/j.1469-7580.2005.00473.x
Oelrich TM (1956) The anatomy of the head of Ctenosaura pectinata (Iguanidae). Misc Publ Mus Zool Univ Mich 94:1–122
Palci A, Caldwell MW (2013) Primary homologies of the circumorbital bones of snakes. J Morphol 274:973–986. https://doi.org/10.1002/jmor.20153
Pianka ER, Vitt L (2006) Lizards: windows to the evolution of diversity. University of California Press, Berkeley, CA
Pough FH, Andrews RM, Cadle JE, Crump ML, Savitsky AH, Kentwood DW (2004) Herpetology as a field of study. In: Herpetology, 3rd edn. Prentice Hall, New York, pp 3–20
Pringle JA (1952) The cranial development of certain South African snakes and the relationship of these groups. Proc Zool Soc London 123:813–866. https://doi.org/10.1111/j.1096-3642.1954.tb00206.x
Pyron RA, Burbrink FT, Wiens JJ (2013) A phylogeny and revised classification of Squamata, including 4161 species of lizards and snakes. BMC Evol Biol 13:93
Reeder TW, Townsend TM, Mulcahy DG, Noonan BP, Wood PL Jr, Sites JW Jr, Wiens JJ (2015) Integrated analyses resolve conflicts over Squamate reptile phylogeny and reveal unexpected placements for fossil taxa. PLoS One 10:e0118199. https://doi.org/10.1371/journal.pone.0118199
Rieppel O (1980) The trigeminal jaw adductors of primitive snakes and their homologies with the lacertilian jaw adductors. J Zool 190:447–471
Rieppel O (1981) The skull and jaw adductor musculature in chameleons. Rev Suisse Zool 88:433–445
Rieppel O (1988) The development of the trigeminal jaw adductor musculature in teh grass snake Natrix natrix. J Zool 216:743–770
Romer A (1956) Osteology of the reptiles. University of Chicago Press, Chicago
Romer AS, Parsons TS (1977) The vertebrate body, 5th edn. W. B. Saunders, Philadelphia
Sambasivan R, Kuratani S, Tajbakhsh S (2011) An eye on the head: the development and evolution of craniofacial muscles. Development 138:2401 LP–2402415 http://dev.biologists.org/content/138/12/2401.abstract
Schweitzer R, Zelzer E, Volk T (2010) Connecting muscles to tendons: tendons and musculoskeletal development in flies and vertebrates. Development 137:3347–3347 http://dev.biologists.org/cgi/doi/10.1242/dev.057885
Sefton EM, Bhullar B-AS, Mohaddes Z, Hanken J (2016) Evolution of the head-trunk interface in tetrapod vertebrates. elife 5:e09972 http://www.ncbi.nlm.nih.gov/pmc/articles/PMC4841772/
Shapiro MD (2002) Developmental morphology of limb reduction in Hemiergis (Squamata: Scincidae): Chondrogenesis, osteogenesis, and heterochrony. J Morphol 254:211–231
Starck D (1979) Cranio-cerebral relations in recent reptiles. In: Biology of the reptilia, Neurology, vol 9. Academic Press, London, New York and San Francisco, pp 1–38
Szekely G, Matesz C (1988) Topography and organization of cranial nerve nuclei in the sand lizard, Lacerta agilis. J Comp Neurol 267:525–544
Szekely G, Matesz C (1993) The efferent system of cranial nerve nuclei: a comparative neuromorphological study. Springer-Verlag, Berlin, Heidelberg, New York
Tanner WW, Avery DF (1982) Buccal floor of reptiles, a summary. Gt Basin Nat 42:273–349 http://www.jstor.org/stable/41711932
ten Donkelaar HJ. 1998. Reptiles. Cent Nerv Syst Vertebr:1315–1524. http://link.springer.com/10.1007/978-3-642-18262-4_20
Trainor PA (2014) Neural crest cells: evolution, development and disease. Academic Press, Amsterdam
Uetz P, Freed P, Hosek J (2018) The reptile database. http://www.reptile-database.org
Wada N, Nohno T, Kuratani S (2011) Dual origins of the prechordal cranium in the chicken embryo. Dev Biol 356:529–540
Watkinson GB (1906) The cranial nerves of Varanus bivittatus. Morphol Jahrb 35:450–472
Werner G (1962) Das cranium der Bruckenechse, Sphenodon punctatus Gray, von 58 mm Gesamtlange. Z Anat EntwGesch 123:323–368
Wever EG (1968) The ear of the chameleon: Chamaeleo senegalensis and Chamaeleo quilensis. J Exp Zool 168:423–436
Whiteside DI (1986) The head skeleton of the Rhaetian Sphenodontid Diphydontosaurus Avonis Gen. Et Sp. Nov. and the modernizing of a living fossil. Philos Trans R Soc B Biol Sci 312:379–430
Willard WA (1915) The cranial nerves of Anolis carolinensis. Bull Mus Comp Zool Harv 49:15–116
Woltering JM (2012) From lizard to snake; behind the evolution of an extreme body plan. Curr Genomics 13:289–299 http://www.eurekaselect.com/openurl/content.php?genre=article&issn=1389-2029&volume=13&issue=4&spage=289
Wu X-C (2003) Functional morphology of the temporal region in the rhynchocephalia. Can J Earth Sci 40:589–607 http://www.nrcresearchpress.com/doi/abs/10.1139/e02-049
Yi H, Norell MA (2015) The burrowing origin of modern snakes. Sci Adv 1:e1500743 http://advances.sciencemag.org/content/1/10/e1500743.abstract

10 The Skull and Head Muscles of Archosauria

Daniel Smith-Paredes and Bhart-Anjan S. Bhullar

10.1 Introduction

Although seemingly different in many aspects of their external appearance, **birds** and **crocodylians** are each other's closest living relatives. This clade, Archosauria (Fig. 10.1) (Gauthier 1986; Gauthier et al. 1988), comprises all the descendants of the most recent common ancestor of a Nile crocodile and a sparrow and dates back to the Triassic, some 250 million years ago, when this ancestor lived (Nesbitt et al. 2013). From there, the archosaur tree split into two branches, one leading to modern-day crocodylians and the other to modern birds. Archosaur ancestors, Archosauriformes, were large-bodied predators living in the Late Permian (Ezcurra et al. 2013), and their descendants, archosaurs, were at the top of the food chain in Triassic ecosystems worldwide. Crocodylians, the crown group composed by and including all the descendants of the most recent common ancestor of the three living groups, Alligatoridea, Crocodyloidea, and *Gavialis gangeticus* (Benton and Clark 1988), now occupy this place in freshwater environments throughout the tropics. Crocodylians are one offshoot of a successful radiation of archosaurs that were both very distinct from the modern crocodile image based on the few living species, showing much more disparity in their morphology and way of life, including sail-backed herbivores, small cursorial (Irmis et al. 2013) or large armored armadillo-like animals (Desojo et al. 2013), bipedal dinosaur-like carnivores (Gauthier et al. 2011), and even crocodile-like semiaquatic carnivorous or piscivorous predators that evolved these ways of living independently (Stocker and Butler 2013). Crocodylians are latecomers in evolutionary terms; they evolved near the end of the Mesozoic, which ended by the time an asteroid hit the Earth and non-avian dinosaurs went extinct 66 million years ago. Even though crocodylians are successful and ancient, they are not quite the living fossil, relict species from the "Age of Reptiles," as often portrayed.

Avialae, the clade of flying dinosaurs containing living **birds** (**Aves**), and other extinct forms, such as the iconic *Archaeopteryx*, evolved from within the branches of the **dinosaur** tree around 150 million years ago (Ostrom 1976). Dinosauria consists of two main clades: **Ornithischia**, the horned, the "duck-billed," and the armored herbivores (Fig. 10.1), and **Saurischia**, including the long-necked herbivorous **sauropods** and the carnivorous **theropods** (Fig. 10.1) (Gauthier 1986). Living alongside extinct non-avian dinosaurs and stemming from the very base of the avian branch were also the pterosaurs, the first flying reptiles of the Mesozoic (Fig. 10.1) (Dalla Vecchia 2013). Modern living birds, Aves, arose probably by the

D. Smith-Paredes (✉) · B.-A. S. Bhullar
Department of Geology and Geophysics, Yale University, New Haven, CT, USA
e-mail: Dsmithparedes@yale.edu; bhart-anjan.bhullar@yale.edu

J. M. Ziermann et al. (eds.), *Heads, Jaws, and Muscles*, Fascinating Life Sciences,
https://doi.org/10.1007/978-3-319-93560-7_10

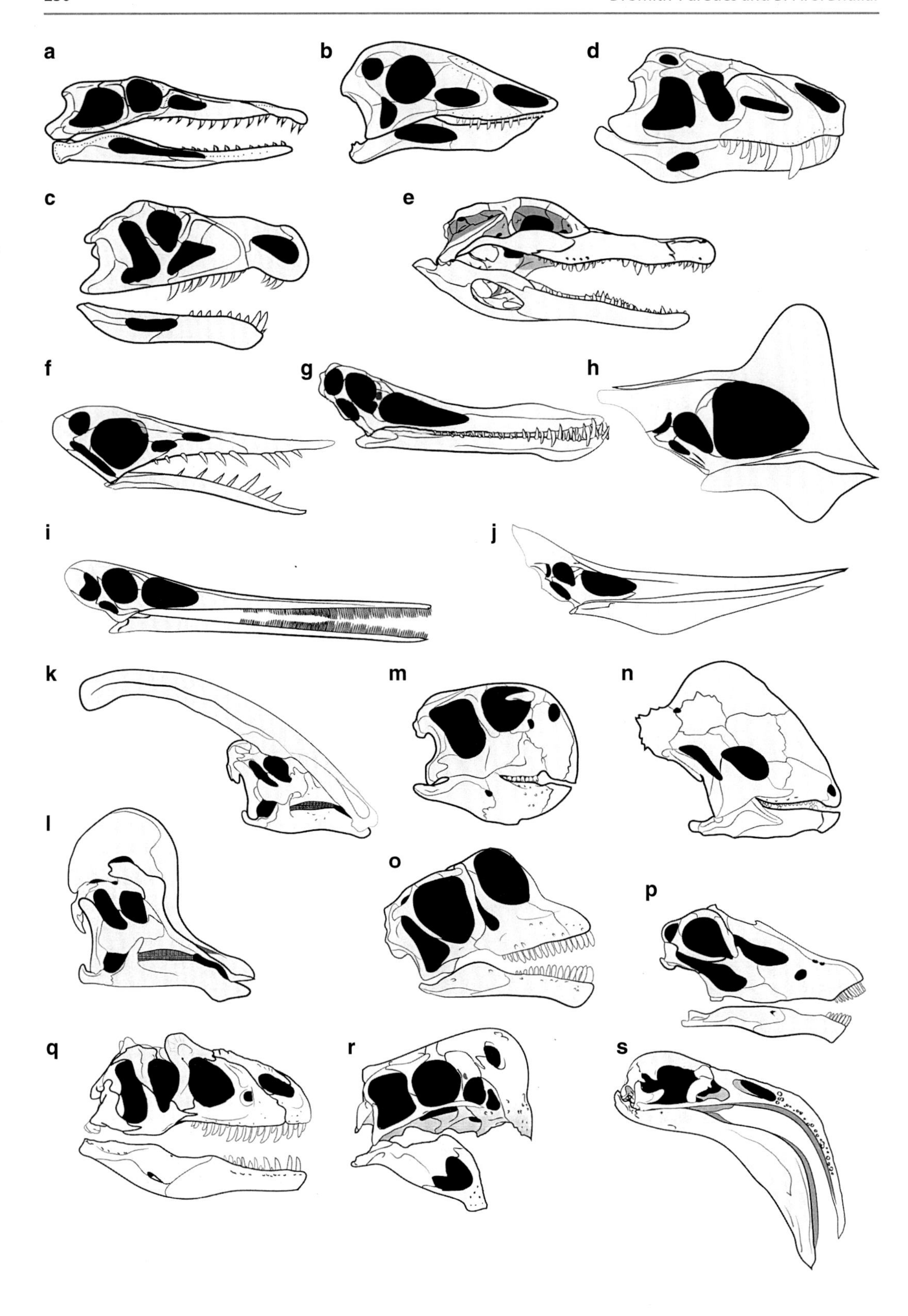
a
b
d
c
e
f
g
h
i
j
k
m
n
l
o
p
q
r
s

very end of the Cretaceous (Berv and Field 2017), diverging into two main branches: **Palaeognathae** and **Neognathae** (Pycraft 1900). Palaeognaths are not a very specious group of birds, consisting of 6 modern genera and about 60 living species, including the flightless ostrich, kiwi, and cassowary, and the flying tinamous (Jarvis et al. 2014; Prum et al. 2015). Neognaths on the other hand comprise most of the more than 10,000 species of Aves living today, including almost every kind of bird we can imagine, from a chicken to a hammerkop, a sheathbill, or a cock-of-the-rock or from a penguin to a hummingbird (Jarvis et al. 2014; Prum et al. 2015). All these archosaur groups, even without considering the incredibly diverse crown avians, display a huge variety of ecologies, diets, sizes, and probably behaviors, which are in some part responsible for the equally astonishing disparity of morphologies we see in their skulls.

Compared to early reptiles and archosauriforms, the skull of early archosaurs had already lost several bones (**septomaxillae**, **postfrontal**, **supratemporal**, **postparietal**, and **tabular** bones). The **parietal foramen** and the **palatal dentition** were also lost, and the posttemporal fenestrae were reduced that lead to a nearly complete enclosing of the brain anteriorly by the ossification of the **laterosphenoid**. They evolved their iconic serrated "thecodont" dentition and antorbital fenestrae bordered anteriorly by the maxilla and posteriorly, separating it from the orbit by the lacrimal (Romer 1956). The lower jaws were distinctive in having a fang near the tip, a mandibular fenestra posteriorly, and an external, oval opening between the **dentary**, **angular**, and **surangular** bones (Romer 1956). The skulls of extinct archosaurs exhibited an incredible disparity of forms and shapes (Fig. 10.1). The skulls of **pterosaurs** were lightly built with bones that in some cases fused together in the adults, as in modern birds, and often sported a variety of crests and enormous antorbital fenestrae, sometimes so big they could be confused with orbits housing huge eyes. In different groups, and again as in **Aves**, teeth were lost (Romer 1956). Within **dinosaurs**, ornithischians showed the most striking kinds of skull ornamentation, with the evolution of horns and frills, crests, dermal armor, and head domes (Romer 1956). Some **ornithischians** reduced or lost their **antorbital fenestrae**, evolved tall (i.e., high in the dorsoventral dimension) jaws articulating below the level of tooth occlusion, batteries of small teeth that were continuously replaced, and toothless premaxillae probably forming a beak matching the one supported by the neomorphic (evolutionary new) predentary bone in the anterior lower jaw (Weishampel and Witmer 1990; Horner et al. 2004). **Sauropod** skulls have a characteristic dorsal displacement of the narial openings, sometimes even moved to the top of the skull roof. The dentition of sauropods is extremely variable in shape, ranging from stout high spoon-shaped crowns to pencil-like teeth, probably adapted to specific plant material-based diets

Fig. 10.1 Archosauria. A comparison between the skulls of different stem archosaurs and members of Archosauria shows how diverse they have been in terms of morphology, diet, ecology, and behavior. (**a–e**) Archosauriformes and crocodylian-line archosaurs. (**a**) *Chanaresuchus bonapartei* (redrawn from Nesbitt 2011). (**b**) *Aetosaurus ferratus* (redrawn from Brusatte et al. 2010). (**c**) *Riojasuchus tenuisceps* (redrawn from Nesbitt 2011). (**d**) *Prestosuchus chiniquensis* (redrawn from Brusatte et al. 2010). (**e**) *Alligator mississippiensis* (redrawn from Kardong 2006). (**f–j**) Pterosauria. (**f**) *Rhamphorhynchus muensteri* and (**g**) *Anhanguera santanae* (redrawn from Witmer et al. 2003). (**h**) *Tapejara wellnhoferi*, (**i**) *Ctenochasma gracile*, and (**j**) *Pteranodon longiceps* (redrawn from Unwin 2003). (**k–n**) Ornithischian dinosaurs. (**k**) *Parasaurolophus walkeri* (redrawn from Evans et al. 2007). (**l**) *Corythosaurus casuarius* (redrawn from Horner et al. 2004). (**m**) *Psittacosaurus mongoliensis* (redrawn from Sereno et al. 1988). (**n**) *Stegoceras validum* (redrawn from Sues and Galton 1987). (**o–p**) Sauropoda. (**o**) *Europasaurus holgeri* (redrawn from Marpmann et al. 2015). (**p**) *Diplodocus carnegii* (redrawn from Whitlock 2011). (**q–s**) Theropoda. (**q**) *Allosaurus fragilis* (redrawn from Madsen 1976). (**r**) *Nemegtomaia barsboldi* (redrawn from Fanti et al. 2012). (**s**) *Phoenicopterus chilensis*

(McIntosh 1990). Early **theropods** retained the large, elongated skulls with big antorbital fenestrae, tall orbits and lower temporal fenestrae, and the large, recurved, sharply pointed, and serrated teeth of their ancestors but evolved a hinge in the lower jaw to increase the gape. Some clades showed trends toward reduction or loss of teeth during ontogeny (Wang et al. 2017a, b), and others, like oviraptorosaurs or avians, lost teeth altogether (Fig. 10.1) (Barsbold and Osmolska 1990; Barsbold et al. 1990).

The skull of both **crocodylians** and **birds** is highly modified from that of their ancestors (Fig. 10.2). Crocodylians (Fig. 10.2a, b) posscss flat heavy skulls and elongated jaws with numerous conical teeth lacking serrations (Iordansky 1973). The **external nares** are close-spaced and displaced dorsally, and their internal opening is displaced backward to the posterior border of the **pterygoids**, as the palatal processes of the **premaxilla** and **maxilla** fuse in the midline, forming, together with the palatines and pterygoids, a secondary palate (Iordansky 1973). The pterygoid and **quadrate bone**s attach firmly to the lateral wall of the braincase, rendering the skull highly rigid or akinetic (Romer 1956). No **antorbital fenestra** is present externally and the prearticulars and epipterygoids are lacking [although an embryonic structure homologue to the **epipterygoid** has been reported to form part

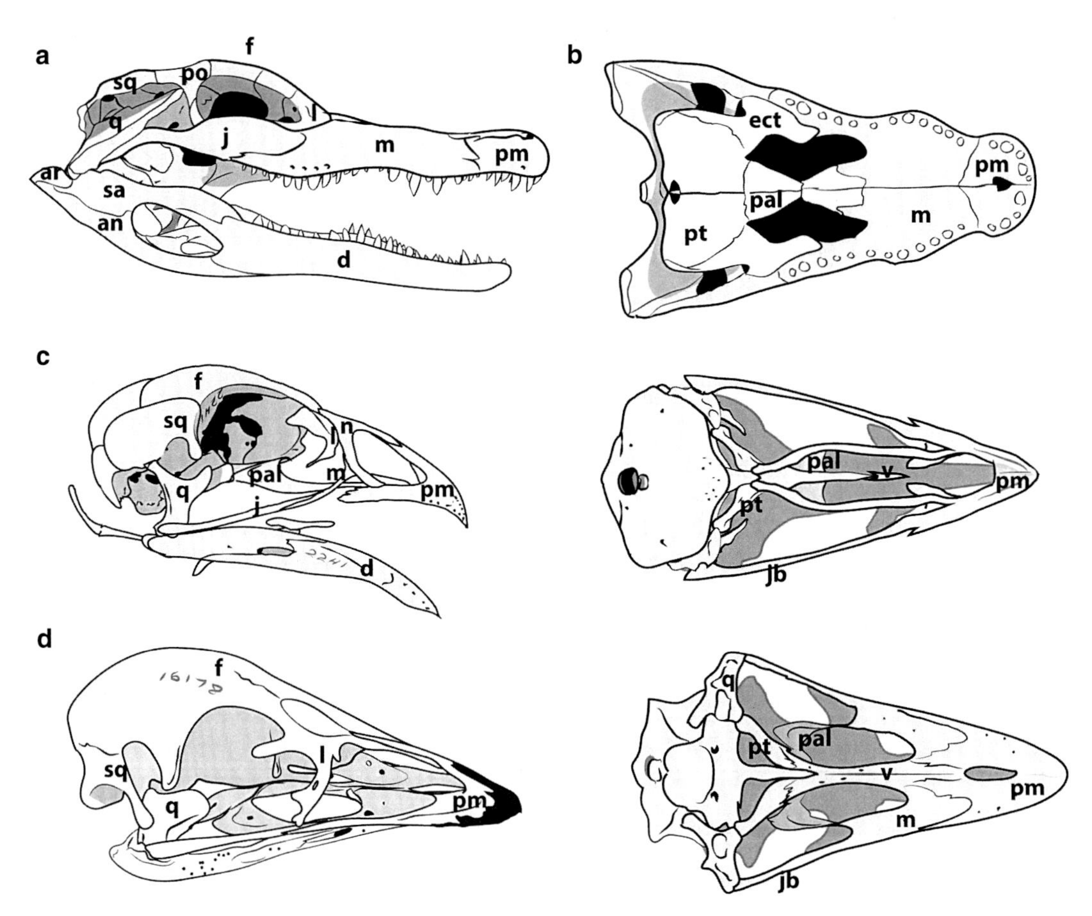

Fig. 10.2 Comparison between the skulls of archosaurs in lateral and ventral view. (**a**) *Alligator mississippiensis*, redrawn from Kardong (2006). (**b**) *Osteolaemus tetraspis*, redrawn from Iordansky (1973). (**c**) Juvenile *Gallus gallus*. (**d**) *Dromaius novaehollandiae*. Note the contact of the maxillae closing the secondary palate of crocodylians, also the absence of teeth and fusion of almost all skull bones in adult birds. *An* angular, *ar* articular, *d* dentary, *ect* ectopterygoid, *f* frontal, *j* jugal, *jb* jugal bar, *m* maxilla, *n* nasal, *pal* palatine, *po* postorbital, *pt* pterygoid, *q* quadrate, *sa* surangular, *sq* squamosal, *v* vomer

of the quadrate in *Alligator* (Klembara 2004) and vestiges of the element have been found in fossil close relatives of crocodylians, both outside and inside of the crown group (Holliday and Witmer 2009)]. Many **dermal bones** are sculpted, and the skull is pneumatized by the development of air spaces (Romer 1956). The skull of crocodylians tends to grow at a higher rate than their brains do (Fabbri et al. 2017), and within the skull the snout region grows faster than the postorbital portion (Grigg 2015), resulting in marked changes in morphology between the skulls of young and adult animals (Bhullar et al. 2012).

Avians (Fig. 10.2c, d), on the other hand, have paedomorphic skulls (Bhullar et al. 2012), i.e., their skulls retain in the adult proportions characteristic of the juvenile stages of their ancestors. The eyes and brain are huge, and the surrounding bones have been modified accordingly, following these expanded organs, as the frontal bone expands posteriorly accompanying the enlargement of the forebrain, pushing the parietal into a retracted position. All archosaurs have relatively large **premaxillae**, but within avialans the premaxillary bones grow to be the predominant element of the rostrum, fusing into a single element with long nasal processes extending posteriorly to separate the **nasal bones** in the midline and contacting the small maxilla that links it to the jugal bar ventrally. Avian snouts and lower jaws have turned into toothless beaks covered by a keratinous sheath (the **ramphotheca**), and most of their skull bones fuse together as the animal reaches adulthood. With the enlargement of the eyes and brain, accompanied by the loss of the upper temporal and postorbital bars, the two temporal fenestrae (characteristic of diapsid amniotes and housing important cranial muscle attachments) became reduced and confluent with the orbit space. The thin **jugal bar** (jugal and quadratojugal bones) connects the anterior tip of the snout with a mobile quadrate bone free of its connections to the palate, and as the jugal bar is no longer linked to the skull roof by the postorbital bone, the beak can move more independently relative to the rest of the skull. This form of **cranial kinesis**, characteristic of all birds, is further differentiated in the neognathous birds (Zusi 1993). In **Palaeognathae**, the palate remains more static, and the bones of the jugal bar and base of the beak are bent slightly as the quadrate moves. In **Neognathae** however, the **pterygoid** and **palatine bones** articulate in a mobile joint, enabling movement of the palate and beak as the quadrate is pushed forward or backward.

The visceral skeleton, or **hyoid apparatus**, of crocodylians is simple in contrast with that of other amniotes. It consists of a mostly rectangular cartilaginous **basihyal**, the **hyoid** body with different kinds of small fenestrations or notches and with angulated anterior and posterior corners varying among different species, and a pair of **ceratobranchial** (*cornu branchiale I*) articulating to the anterolateral portion of the basihyal, connecting in their distal portions with short epibranchials (Schumacher 1973). In birds the hyoid apparatus is more elaborated consisting of up to seven individual elements. In **neognaths**, there are three medial elements: the basihyal at the base of the tongue, which articulates via a joint to an anterior arrow-shaped **paraglossal** lying within the tongue itself, and a caudal **urohyal**, anteroventral to the larynx (Tomlinson 2000). In **palaeognaths**, there is no separation between the basihyal and a urohyal, which in all species, except *Rhea,* form a continuous element called the **basiurohyal** (Tomlinson 2000). The paraglossal of palaeognaths is broad, except in the ostrich that has two separate narrow cartilaginous elements named paraglossalia, that do not meet in the midline (Tomlinson 2000). The hyoid horns of birds consist of two elongated elements, the ceratobranchials and the epibranchials; the former articulate with the basihyal and the latter with the posterior end of the ceratobranchial. In neognaths the hyoid horns elongate caudally and curve around the base of the skull, arch upward at the base of the jaw, and attach to the occipital region by connective tissue (Tomlinson 2000). In contrast, the hyoid horns of palaeognaths are usually short, except for those of the ostrich that are elongated and bend downward along the muscles of the neck (Tomlinson 2000). In tinamous they can also be long as in *Tinamus* (Parker 1866;

Tomlinson 2000) or short and attaching to the ear region as in *Nothoprocta* via connective tissue as in the neognaths (Tomlinson 2000). The elements of the hyoid apparatus show different degrees of ossification and many remain cartilaginous throughout life. In **crocodylians** the **basihyal** itself is cartilaginous and the **ceratobranchial** is ossified (Schumacher 1973). In palaeognaths the **urohyal**, **basihyal**, and **paraglossal** elements remain cartilaginous, whereas in some neognaths they ossify. The ceratobranchials are the only elements to ossify in palaeognaths, and in general the epibranchials remain cartilaginous in neognaths too (Tomlinson 2000). As most of the hyoid apparatus is cartilaginous in crocodylians and palaeognaths, it is understandable that their fossil record is so poor; ceratobranchials have been the only elements found in fossil Archosauriformes and crocodylian-line archosaurs (Li and Clarke 2015), **pterosaurs** (Wang et al. 2007; Vullo et al. 2012; Wang et al. 2012), and **dinosaurs** of various clades including ornithischians (Ősi 2005; Vickaryous 2006; Liyong et al. 2010) and **non-avian theropods** (Pérez-Moreno et al. 1994; Chiappe et al. 1998; Dal Sasso and Signore 1998). An ossified basihyal is described in fossil **Avialae** such as *Confuciusornis* and *Hongshanornis*; the latter also have ossified epibranchials (Li 2015).

Vertebrae comprising the backbone were originally composed of four discrete cartilaginous elements, a **pleurocentrum** and **intercentrum** forming the vertebral body and paired neural arches enclosing the neural cord dorsally (Romer 1956). In reptiles, the vertebral body or centrum is formed by the pleurocentrum, with the intercentrum being absent, except in the **atlas-axis** complex and tail in archosaurs. Holding the head in position and connecting it to the rest of the body, cervical vertebrae facilitate motion of the skull with respect to the trunk. The first two vertebrae of the neck (the atlas-axis complex) are highly modified to provide stability while permitting movement of the head at their articulation with the single occipital condyle on the back of the skull. As in other amniotes, in the archosaurian first vertebrae, or atlas, the centrum is reduced and attached to the second vertebrae as the odontoid process, and its broad intercentrum is in contact with the intercentrum of the axis, the second vertebra. Together with the proatlas, the atlas forms a bony ring surrounding the **occipital condyle** (Romer 1956). The ancestral condition for archosaurs seems to be an atlas-axis complex in which all vertebral elements are discrete and separate, except for the axis centrum and its neural arch, though variations of this pattern occur and all the elements become fused together as in some sauropods, pterosaurs, or ornithischians (Romer 1956). In **crocodylians** the atlas centrum and the axis intercentrum are fused, and they can fuse to the axis centrum to form an anteriorly pointing fingerlike odontoid process (Romer 1956). **Dinosaurs**, including **Aves**, also have an odontoid process, called the "dens" (George and Berger 1966). The following vertebrae of the cervical region are typically distinguished from the trunk vertebrae by their plowshare-shaped rib heads and in having their rib shafts running parallel to the cervical centra. Cervical vertebrae counts vary among archosaurs; crocodilians stick to the ancestral numbers around 8 or 9, while **pterosaurs** also have 8 or 9 vertebrae, but these are elongated and show very little development of neural spines or rib articulations (Romer 1956). Dinosaurs have a variable number of cervical vertebrae with **ornithischians** numbers ranging from 9 to 15, **sauropods** between 12 and 17, and **theropods** from 9 or 10 in non-avian forms (Romer 1956) to up to as many as 25 in Aves (George and Berger 1966).

10.2 Embryology of the Head

In the head, many different tissues interact during the **embryogenesis**, contributing to the development of one of the most complex anatomical regions of the vertebrate body. A huge amount of our understanding on how cranial structures are formed and patterned comes from the use of chicken, in combination with other birds as developmental biology models, which allow for grafting tissues, labeling and mapping the fate of cell populations, and experimentally intervening with the normal development to understand the consequences of cell behaviors, tissue interactions, gene products, etc.

At its earliest stages (Fig. 10.3a, b), the head consists of no more than a slightly inflated neural tube, the brain, and surrounding layers of pharyngeal endoderm, ectodermal epidermic cells and mesodermal cells. As the **embryo** grows, the head region becomes more and more complex; eyes develop, **pharyngeal pouches** grow from the sides of the oral region, and **neural crest cells** (NCC) invade the head. The neural crest is a region of ectoderm that differentiates dorsal to the neural tube in vertebrates. The two most important characteristics of neural crest cells are

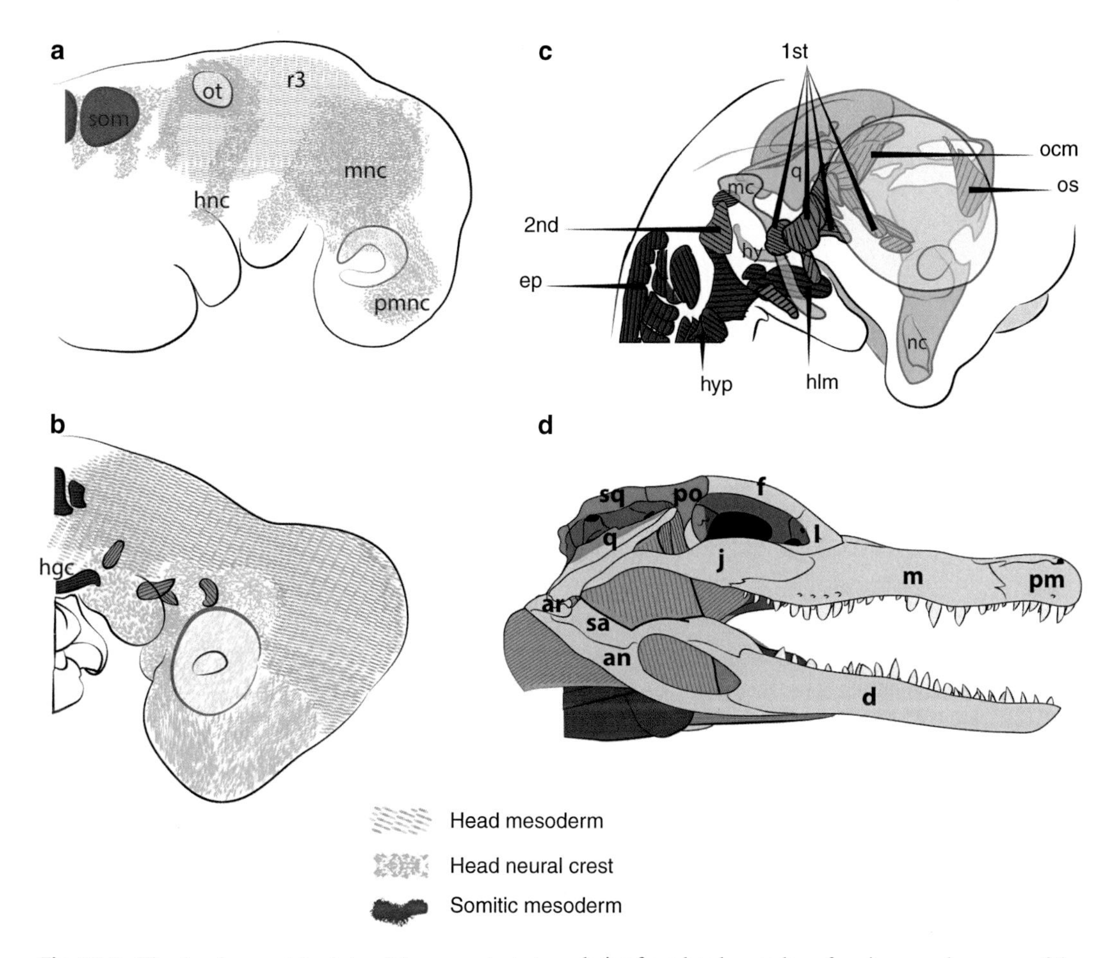

Fig. 10.3 The developmental origin of the musculoskeletal tissues in archosaurs, exemplified by schematic crocodylian development. (**a** and **b**) Early patterning of the head involves the migration of neural crest cells in separate hyoid, maxillar, and premaxillar streams (hnc, mnc and pmnc, respectively), extension of the hypoglossal cord (hgc) from the somites (som), and the differentiation of groups of myogenic populations from the head mesoderm (modified from Kundrát 2008, 2009). (**c**) The differentiation of muscle masses from the head mesoderm and somitic mesoderm occurs later in development, as does the early development of the skull components. The chondrocranium develops from a double embryonic origin, as the anterior portion, including the quadrate (*q*), Meckel's cartilage (*mc*), and hyoid apparatus, (*hy*), develops from cranial neural crest cells, while the posterior portion originates from head mesoderm. Muscles of the head derive from head mesoderm forming muscle masses of the first (1st), second (2nd), and third arches, the ocular muscle mass (*ocm*), and from somatic mesoderm giving rise to epaxial (*ep*) and hypaxial (*hyp*) primaxial muscles, and the abaxial muscles of the hypoglossal cord, like the hyolingual muscle mass (*hlm*). (**d**) As in the case of the chondrocranium, the anterior dermatocranium (including the bones forming the jaws, palate, and the anterior skull roof) also derives from cranial neural crest, while its posterior portion derives from mesodermal cells (modified from Kardong 2006; Holliday and Witmer 2007; chondrocranium portions modified from Bellairs and Kamal 1981). *An* angular, *ar* articular, *d* dentary, *ect* ectopterygoid, *f* frontal, *j* jugal, *jb* jugal bar, *m* maxilla, *n* nasal, *nc* nasal capsule, *pal* palatine, *po* postorbital, *pt* pterygoid, *q* quadrate, *sa* surangular, *sq* squamosal, *v* vomer

their migratory capacity and their pluripotency. Migrating first between the neural tube and the dorsal ectoderm, the cells later abandon this region, detaching from one another and move into the rest of the body, where they give rise to many different tissues, such as bone, teeth, glial cells, connective tissue, etc. In the head, the cranial neural crest cells migrates in spatially segregated streams (George and Berger 1966) to invade the anterior rostrum and fill the branchial arches, forming skeletal elements of the skull, jaws, middle ear, and neck (Santagati and Rijli 2003).

The elements of the skull have mixed origins (Fig. 10.3c, d); the more anterior portion of the **dermatocranium**, as well as the **hyoid apparatus**, and Meckel's cartilage of the jaw are derived from **neural crest cells**, while the posterior portion of the dermatocranium and the chondrocranium originate from mesoderm mesenchymal cells (Le Lièvre and Le Douarin 1975; Le Lièvre 1978; see also Chap. 7). The muscles of the head also derive from distinct groups of mesodermal cells, including trunk **somitic mesoderm** and cranial **prechordal and paraxial mesoderm** (Noden and Francis-West 2006; Shih et al. 2008). Head mesoderm is uncompartmentalized and unsegmented and squeezed between the epidermis, brain, and pharyngeal endoderm; it can be further subdivided into prechordal and cranial paraxial mesoderm. Even if there is no evident compartmentalization of the mesoderm, populations of cells end up making distinct myogenic masses, eventually giving rise to individualized muscles (Noden and Francis-West 2006). From the cranial mesoderm, four masses originate (Fig. 10.3c): three within the paraxial mesoderm which will give rise to the first, second, and third arch muscle groups, while the last, the eye muscles group, derives at least in part from the prechordal portion of mesoderm (Noden et al. 1999; Evans and Noden 2006; Noden and Francis-West 2006).

Posterior to the head, the paraxial mesoderm flanking the neural tube periodically segments into epithelialized compartments, the **somites**, which are important structures as they function as signaling centers involved in the induction and development of limbs but also give rise to many different skeletal and muscular elements. Somite derivatives include the vertebrae and ribs, the dermis, tendons, and axial and appendicular muscles. Muscle can develop in two contrasting ways from the somites; cells can first escape the somite and migrate out of its limits, after which they commit and differentiate into myoblasts (**abaxial development**) or they can commit into a muscle cell fate within the boundaries of the somite before extending to specific attachment points (**primaxial development**) (Burke and Nowicki 2003). **Laryngeal, tracheal, and tongue muscles** develop in the former way, moving out of the somite and forming a hypoglossal cord that extends from the trunk into the head before actually forming any muscle, while most of the muscles of the neck develop in the later fashion, differentiating within the somites and then extending into their specific attachment sites (Noden et al. 1999; Evans and Noden 2006).

10.3 Muscles

It can be confusing to realize many different names one muscle was given by different scientists through time; this situation was in part standardized for birds by the establishment of a common nomenclature for avian anatomy (Baumel et al. 1993; Van den Berge and Zweers 1993). The relatively recent understanding of the close relationship between **birds** and **crocodylians** probably explains the lack of a consistent muscle terminology across Archosauria, and their sometimes too-derived morphologies surely make the study and comparison of previous anatomical descriptions between these different groups challenging. While in some cases no easy comparison can link two muscles as the same, in others the suggestion of homology can be tempting, but – as always – caution and testing are required. Here, we will review the comparative muscular anatomy of archosaurs by trying to link the work of those who studied them with or without considering the close relationship between crocodylians and avians.

The head can be subdivided into several regions, depending on the criteria used. Sometimes the muscles found in one particular region are not homogenous in their origin, nerve innervation, or function, reflecting the dynamic patterning processes that take place during the building of the head.

- Nostril musculature

The external nostril openings are situated on a nasal disk in the tip of the **crocodylian** snout (Grigg 2015). Usually they are open, as when the animal rests on land or the surface of the water, but can be tightly closed or widened. Two unstriated muscles perform the opposed actions; a ring of muscular fibers which do not attach to any skeletal element, the (Musculus) **M. constrictor naris**, through which passes a longitudinal group of fibers originating from the premaxilla, the **M. dilator naris** (Bellairs and Shute 1953). The dilator naris contracts and retracts the posterior border of the nostril back, opening it, while the constrictor naris appears to close the opening by squeezing the dilator naris, pushing the posterior wall of the nostrils rostrally (Bellairs and Shute 1953). Both muscles are unique to crocodylians and are innervated by sympathetic neurons whose bodies are located in the superior cervical ganglion and reach the tip of the snout via the **cranial sympathetic nerve** (Bellairs and Shute 1953).

- Orbit and orbitotemporal musculature (Fig. 10.4)

Six **extraocular muscles** are responsible of the movements of the eyeball (Fig. 10.4). In crocodylians the **M. obliquus superior** originates from the descending frontal bone, and the **M. obliquus inferior** arises immediately ventral to it, extending dorsally and ventrally to attach onto the eyeball, respectively (Underwood 1970). The **M. rectus anterior** arises from the **interorbital septum**, and the (Musculi) **Mm. rectus superior, rectus inferior**, and **rectus posterior** origi-

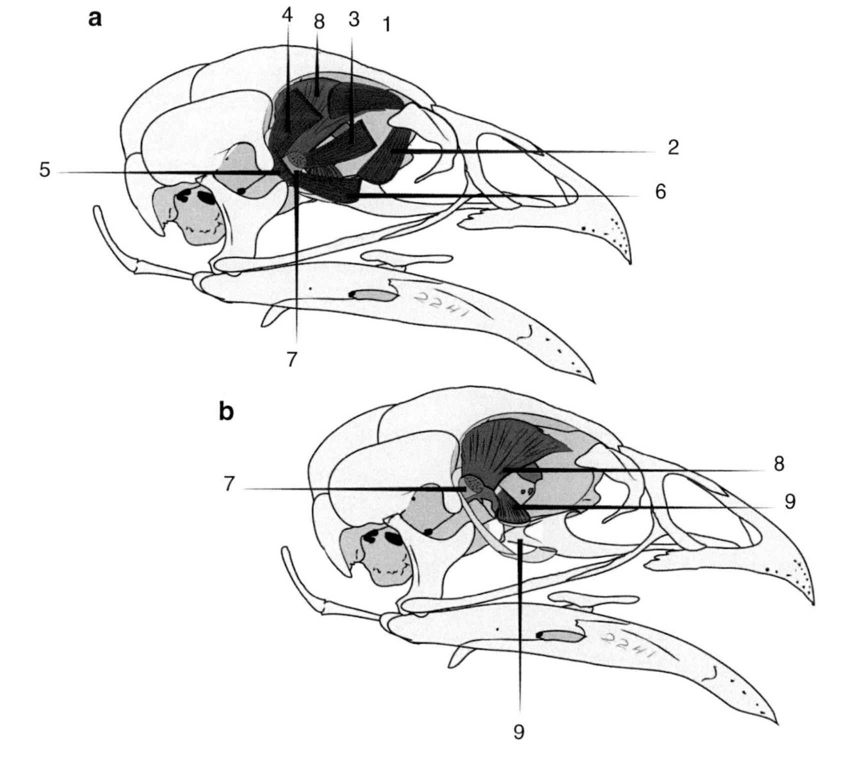

Fig. 10.4 Extraocular muscles of the orbit in a chicken, *Gallus gallus* in superficial lateral view (**a**) and deep lateral view, after removal of recti and obliqui muscles (**b**). Based on Chamberlain (1943). 1. M. obliquus superior; 2. M. obliquus inferior; 3. M. rectus medialis; 4. M. rectus dorsalis; 5. M. rectus lateralis; 6. M. rectus ventralis; 7. Optic nerve; 8. M. quadratus membranae nictitans; 9. M. pyramidalis

nate from the **basisphenoid bone** (Underwood 1970). All the extraocular muscles of birds (Fig. 10.4), that is, the Mm. obliquus superior and obliquus inferior, Mm. **rectus dorsalis** (superior), **rectus ventralis** (inferior), **rectus lateralis** [(posterior of crocodylians (Underwood 1970)], and **rectus medialis** [anterior of crocodylians (Underwood 1970)], originate from the interorbital septum and insert onto the sclera (Van den Berge and Zweers 1993). The slight change in terminology between some muscles might be related to the changes of the position of the eyes relative to the skull.

The **oculomotor nerve** (cranial nerve III, CN III) innervates the rectus dorsalis, rectus ventralis, rectus medialis, and obliquus ventralis muscles, while the obliquus dorsalis muscle is innervated by the **trochlear nerve** (CN IV) and the rectus lateralis by the **abducens nerve** (CN VI).

In birds, two muscles are responsible for the movement of the **nictitating membrane** (Fig. 10.4, bottom), a semitransparent membrane acting as the third eyelid, the dorsal **M. quadratus membranae nictitans** and the ventral **M. pyramidalis**, both innervated by the **abducens nerve** (CN VI) (George and Berger 1966; Van den Berge and Zweers 1993). The two muscles converge close to the equator of the eyeball and near the exit of the optic nerve, and the border of M. quadratus folds forming a sheath through which the tendon of the M. pyramidalis passes (Fig. 10.4; George and Berger 1966; Van den Berge and Zweers 1993). Contraction of these muscles sweeps the nictitating membrane horizontally posteriorly, over the eye, while their relaxation allows elastic tissue within the membrane to return it to its anterior position (Van den Berge and Zweers 1993).

The **M. retractor bulbi** was lost in the lineage leading to birds (Rieppel 2000), but it is present in crocodylians, pulling the eye into the horizontal plane of the skull and protecting it from damage when the animal is feeding or being attacked (Schwab and Brooks 2002). The muscle is innervated by the **abducens nerve** (CN VI). All muscles innervated by the CN VI in birds (**Mm. rectus lateralis, pyramidalis, and quadratus membranae nictitans**) develop from the same muscular mass in the embryo, including M. retractor bulbi in crocodylians (Wedin 1953). The M. pyramidalis of birds develops from the same portion of this muscle mass that gives rise to the M. retractor membranae nictitans of crocodylians; M. quadratus membranae nictitans of birds develops from the other portion of this muscle mass, which in *Alligator* corresponds to the anterior portion of the division of the muscle mass which originates M. retractor bulbi of crocodylians (Wedin 1953).

Birds and crocodylians have a levator of the upper eyelid and a depressor of the lower eyelid (George and Berger 1966; Underwood 1970). In crocodylians the upper eyelid is armored by a bony plate, the tarsus, and its movement is responsible for most of the eye closing (Underwood 1970; Grigg 2015), while the upper eyelid seems to have little movement in most birds and is immobile in others (Underwood 1970; Grigg 2015). The **M. levator palpebrae dorsalis** [**M. levator palpebralis** in crocodylians (Underwood 1970)] is responsible for moving the upper eyelid and is innervated by the **occulomotor nerve** (CN III), while the **M. depressor palpebrae ventralis** [**M. depressor palpebrae** in crocodylians (Underwood 1970)] moves the lower eyelid and is innervated by the **mandibular nerve** (CN V_3) (Van den Berge and Zweers 1993).

In birds, the **M. orbicularis palpebrarum**, composed of nonstriated muscular fibers, is an intrinsic muscle of the eyelids, is arranged parallel to the eyelid border, and functions as a sphincter, closing the lids (George and Berger 1966; Van den Berge and Zweers 1993). The **M. tensor periorbitale**, also innervated by CN V_3, forms a muscular separation between the orbit and jaw muscles (Van den Berge and Zweers 1993). Both birds and crocodylians have a **M. levator bulbi** or M. tensor periorbitae (Holliday and Witmer 2007), originating from the lateral surface of the **laterosphenoid** and attaching to the orbital septum or the preotic pillars in birds and crocodylians, respectively. The **M. protractor pterygoideus** [or M. protractor pterygoideus et quadrati (Van den Berge and Zweers 1993)] has been described as two muscle parts originating from the base of the orbital septum and attaching to the pterygoid and quadrate, protracting the palatoquadrate bridge, and therefore pushing the tip of the upper jaw upward (Elzanowski 1987; Van den Berge and Zweers 1993). In crocodylians the rigid structure of the palate is sutured to the braincase, and the M. protractor pterygoideus has been lost (Holliday 2009). These muscles derive from the **first mandibular arch muscle mass** (Noden et al. 1999).

- Auricular musculature

The external opening of the ear in Crocodylia is visible as a longitudinal slit, extending from behind the orbit toward the back of the head; it is covered entirely by a movable and scale covered dorsal skin flap (Shute and Bellairs 1955; Grigg 2015). A lower flap bounds the ventral margin of the external opening and its movement is entirely responsible for the opening or closing of the ear, as the dorsal flap undergoes very little movement (Shute and Bellairs 1955). The superior flap is moved by a superficial **M. levator auriculae superior** and a deep **M. depressor auriculae superior**, which are derived from the M. depressor mandibulae and innervated by the **facial nerve** (CN VII) (Shute and Bellairs 1955). A Y-shaped dense fibrous structure (the **Ypsilon**) attaches to the lower flap, the postorbital bone and the **M. depressor auriculae inferior**. This muscle, which derives from the same primordium that gives rise to the M. depressor palpebrae and **M. levator bulbi**, passes from the floor of the orbit, is responsible for movements of the lower earflap, and is innervated by a branch of the **mandibular nerve** (CN V_3) (Shute and Bellairs 1955). In general, the posterior portions of the external openings are closed, while the anterior ones stay opened and are closed when the animal submerges (Shute and Bellairs 1955; Grigg 2015). In birds, three muscles were described to arise from the surroundings of the external ear opening and inserting onto the skin covering or bordering it (George and Berger 1966).

- The temporal and palatal musculature (Figs. 10.5 and 10.6)

The majority of the muscles in this region primarily close the jaw and are innervated by the **trigeminal nerve** (CN V). They originate from the temporal or palatal region of the skull, attach

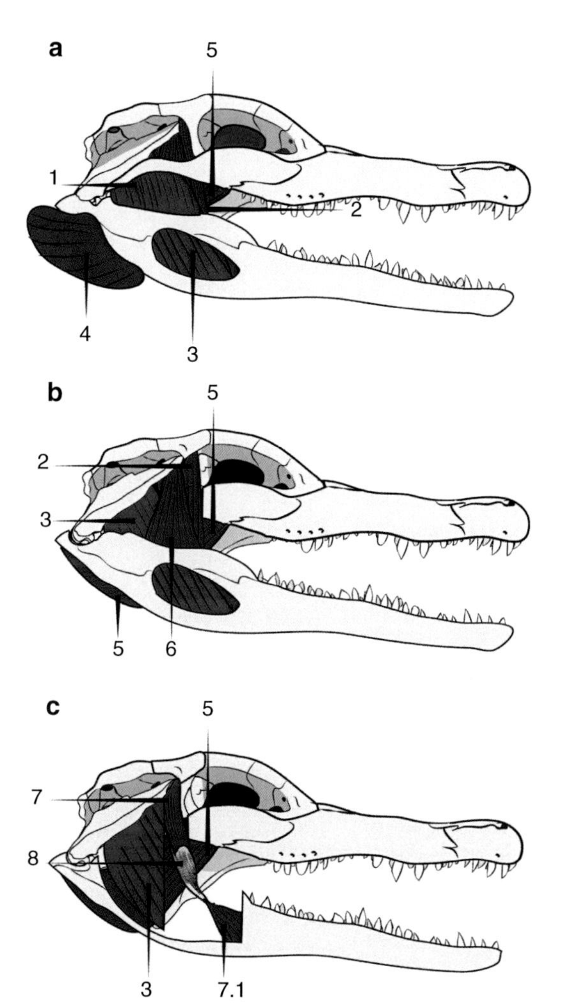

Fig. 10.5 The jaw muscles of *Alligator mississippiensis* (based on Schumacher 1973; Kardong 2006; Holliday and Witmer 2007). (**a**) Superficial, (**b**) intermediate, and (**c**) deep muscles. 1. M. adductor mandibulae externus superficialis; 2. M. adductor mandibulae externus profundus; 3. M. adductor mandibulae posterior; 4. M. pterygoideus ventralis; 5. M. pterygoideus dorsalis; 6. M. adductor mandibulae externus medialis; 7. M. pseudotemporalis superficialis; 7.1 M. intramandibularis; 8. Cartilago transiliens

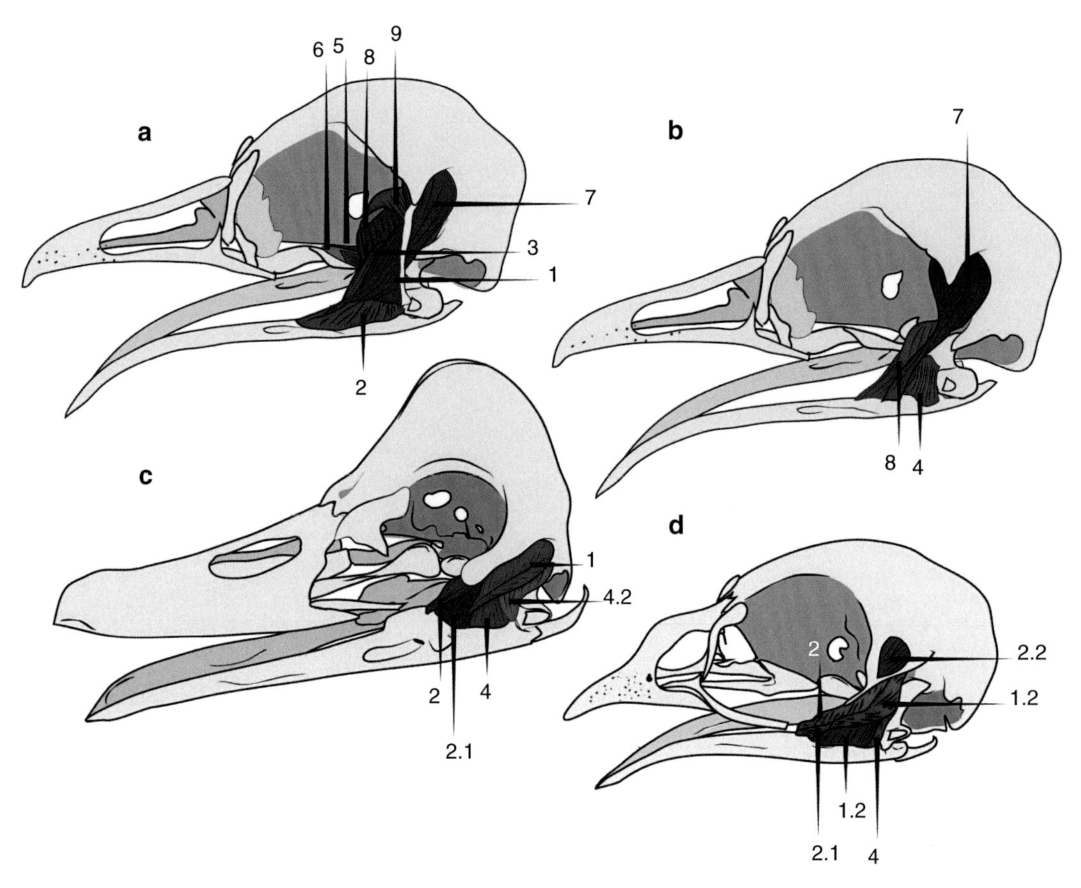

Fig. 10.6 The jaw muscles of modern birds. (**a**) Superficial jaw muscles and (**b**) deeper jaw muscles of the tinamou *Nothoprocta cinerascens,* modified from Elzanowski (1987). 1: M. adductor mandibulae externus superficialis; 2. M. adductor mandibulae externus profundus; 3. M. adductor mandibulae externus medialis; 4. M. adductor mandibulae posterior; 5. M. pterygoideus ventralis; 7. M. pseudotemporalis superficialis; 8. M. pseudotemporalis profundus; 9. M. protractor pterygoidei et quadrati. (**c** and **d**) Comparison of the M. adductor mandibulae externus complex in *Anseranas semipalmata* (Anseriformes) and *Alectornis gracea* (Galliformes), modified from Zusi and Livezey (2000). Note the differences in organization of the different subdivisions of the adductor muscles. 1.2 M. adductor mandibularis externus superficialis (m. AME articularis externus); 2.1 M. adductor mandibulae externus profundus (m. AME zygomaticus); 2.2 M. adductor mandibulae externus profundus (m. AME coronoideus); 4.2 M. adductor mandibulae posterior lateralis (m. AME articularis internus)

to the lower jaw, and elevate the jaw as they contract. As said before, a recurrent problem in comparative anatomy, terminology, and interpretation of identity of archosaur muscles has generally not considered the shared ancestry of birds and crocodylians. The most comprehensive studies on the **adductor chamber** and **jaw muscles** in archosaurs have been the ones by Holliday and Witmer (Holliday and Witmer 2007, 2009; Holliday 2009), not only providing needed common terminology but also both an extensive review and new anatomical insights critical for understanding archosaur jaw muscles. For this, their terminology and works are vastly used and cited here.

The **adductor chamber** is split into compartments separated by the passing of the three branches of the **trigeminal nerve** (CN V): the **ophthalmic** (CN V_1), the **maxillary** (CN V_2), and the **mandibular branches** (CN V_3) (Luther 1914; Holliday and Witmer 2007). According to their position relative to these branches, the

adductor muscles are divided into the **M. adductor mandibulae posterior**, **M. adductor mandibulae externus** (which is further subdivided into the superficialis, medialis, and profundus portions), and **M. adductor mandibulae internus** (which is subdivided into the **pseudotemporalis** and the **pterygoideus muscles**) (Fig. 10.5; Schumacher 1973; Holliday and Witmer 2007). The M. adductor mandibulae posterior lies caudal to the passing of CN V_3, and muscles of the M. adductor mandibulae externus group are located rostral to CN V_3 and lateral to CN V_2, while muscles of the M. adductor mandibulae internus group lie medial to CN V_2 and lateral to CN V_1. In crocodylians (Fig. 10.5), these spatial relationships, more or less conserved across reptiles, changed following the modifications of the adductor chamber muscles, bones, and nerves imposed by the suturing of the palate to the braincase (Holliday and Witmer 2007; Holliday 2009). In crocodylians, the **M. pseudotemporalis superficialis**, part of the M. adductor mandibulae internus group, separates early during ontogeny from this muscle mass and lies lateral to the CN V_2, instead of medial to it (Holliday 2009).

The **M. adductor mandibulae externus** (Fig. 10.5a) is divided into **M. adductor mandibulae externus superficialis, medialis, and profundus** (Fig. 10.5a, b) portions in crocodylians (Holliday and Witmer 2007). Modern birds display varying arrays of this muscle complex (Fig. 10.6); the ostrich, *Struthio*, has a seemingly undivided adductor mandibulae externus (Webb 1957), while in tinamous the superficialis and profundus portions can be fused or not. In general, other **palaeognaths** seem to have a simple adductor mandibulae externus structure (Elzanowski 1987). Within **neognaths**, many groups present further subdivision of the complex, with several identifiable bellies (Holliday and Witmer 2007).

In crocodylians, the **adductor musculature** is highly modified in relation to the ancestral condition; origins of the crocodylian **M. adductor mandibulae externus profundus and medialis** shifted from the surface of the **parietal** and **squamosal** to the surface of the parietal alone and from the lateral surface of the squamosal to the rostromedial surface of the **quadrate**, respectively (Holliday and Witmer 2007; Holliday 2009). Also, the **M. adductor mandibulae externus superficialis** retained an origin from the squamosal in birds, in addition to the quadrate; but it shifted entirely to the surface of the quadrate and **quadratojugal** in Crocodylia (Holliday and Witmer 2007). The **M. adductor mandibulae posterior** originates from the rostral surface of the quadrate bone and attaches to the medial portion of the jaw in **crocodylians** and in birds (Holliday and Witmer 2007). In crocodylians its insertion occupies most of the mandibular fossa, and in **palaeognaths** it attaches to the dorsomedial surface of the lower jaw, while in many **neognaths** it attaches onto its dorsolateral surface.

The **adductor mandibulae internus** is divided into the **pterygoideus muscles** (dorsalis and ventralis) and the **pseudotemporalis muscles** (profundus and superficialis). The **M. pterygoideus ventralis** originates from the caudal surfaces of the crocodylian palate, extends lateral to the neck musculature and folds around the retroarticular process of the mandible, and attaches to the lateral side of the jaw. In most neognaths, the attachment of the M. pterygoideus ventralis is simply onto the ventromedial border of the mandible, though in some orders the muscle wraps around the jaw, attaching onto the lateral surface, as in **crocodylians** (Holliday and Witmer 2007; Holliday 2009). The **M. pterygoideus dorsalis** of crocodylians originates from the dorsal surface around the suborbital region of the palate and passes caudal and ventral to attach onto the medial surface of the mandible, ventral to the jaw joint. In **neognaths** it splits into separate bellies that originate from the palatine and pterygoid, pass through the dorsal side of the palate, and attach to the medial side of the mandible (Holliday and Witmer 2007). The **M. pseudotemporalis superficialis** originates from the dorsotemporal fossa in ratites as in **lepidosaurs**, while in crocodylians it shifted to originate from the caudal portion of the **laterosphenoid** (Holliday 2009). The muscle is small and originates from the laterosphenoid in the temporal region (Holliday and Witmer 2007; Holliday 2009). The muscle then attaches to the **transiliens cartilage** (Fig. 10.5c), a fibrocartilaginous sesamoid that slides between the pterygoid buttress and the coronoid (Tsai and Holliday 2011). In both crocodylians and palaeognaths, a similar cartilaginous structure (more developed in crocodylians) connects the M. pseudotemporalis with the **M. intramandibularis** (Holliday and Witmer 2007; Holliday 2009), which have been shown to correspond to a single muscular unit in crocodylians (Tsai and Holliday 2011). In birds, the **M. pseudotemporalis profundus** originates from the orbital process of the quadrate and attaches to the dorsomedial surface of the mandible; however, in crocodylians this muscle is reduced, with a distinct origin from the laterosphenoid but then merging with fibers of other temporal muscles (Holliday 2009).

- The lower jaw musculature and the tongue (Figs. 10.7 and 10.8)

M. depressor mandibulae opens the mouth; it originates from the **parietal, supraoccipital**, and **quadrate** in **crocodylians** (Schumacher 1973) and from the subtemporal and temporal fossae in birds (Van den Berge and Zweers 1993); it extends inferiorly to the caudal portion of the lower jaw. This muscle develops from the **hyoid arch muscle mass** and is innervated by the **hyomandibular branch of the facial nerve** (CN

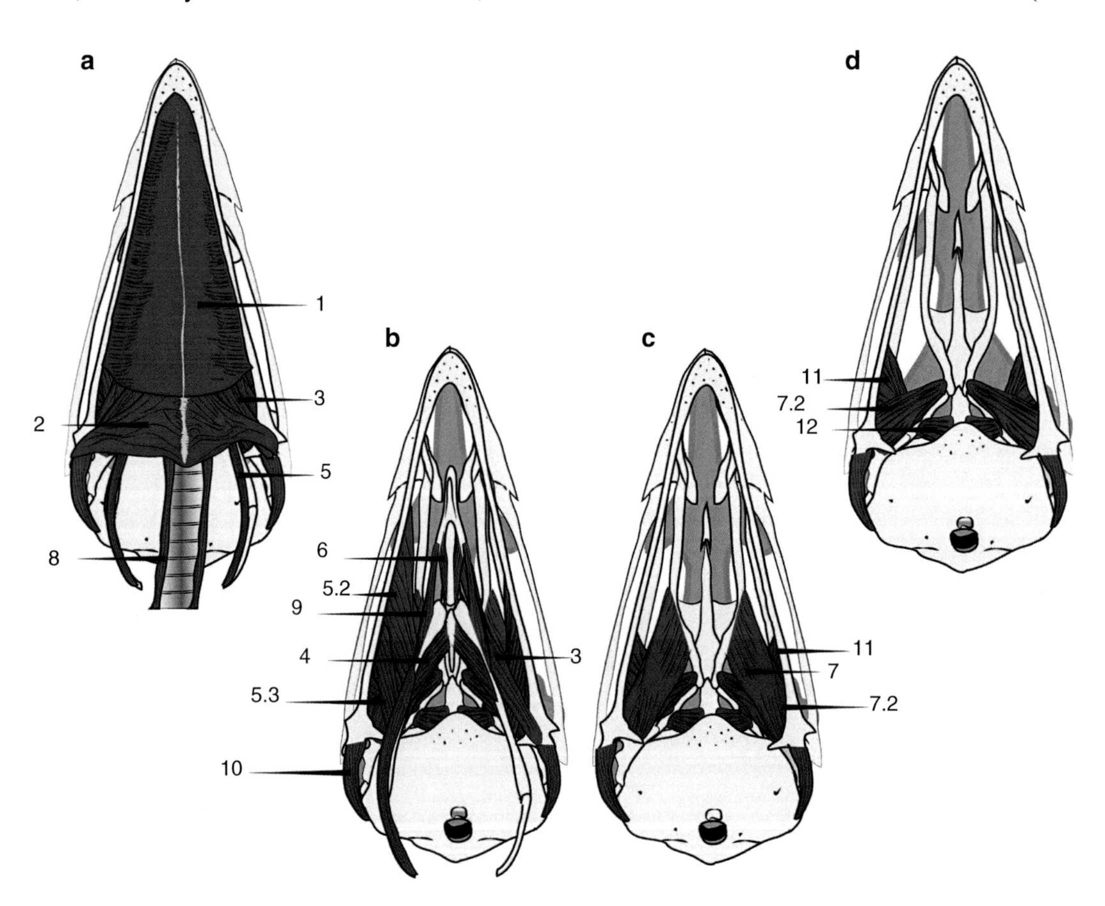

Fig. 10.7 The lower jaw, hypobranchial and tongue muscles of a Galliform bird from superficial (**a**), intermediate (**b**), deep (**c**), and deepest layers (**d**) (based on Chamberlain 1943; Fitzgerald 1969; and Homberger and Meyers 1989). (**a**) and (**b**) show muscles of the lower jaw, while (**c**) and (**d**) show muscles of the palate in ventral view. 1. M. intermandibularis; 2. M. constrictor colli intermandibularis (other portions of M. constrictor colli were not drawn); 3. M. stylohyoideus; 4. M. serpihyoideus; 5. M. branchiomandibularis; 5.2 M. branchiomandibularis anterior; 5.3 M. branchiomandibularis posterior; 6. M. hyoglossus; 7. M. pterygoideus ventralis; 7.2 M. pterygoideus dorsalis; 8. M. tracheohyoideus; 9. M. ceratoglossus; 10. M. depressor mandibulae; 11. M. pseudotemporalis profundus; 12. M. protractor pterygoid et quadratus

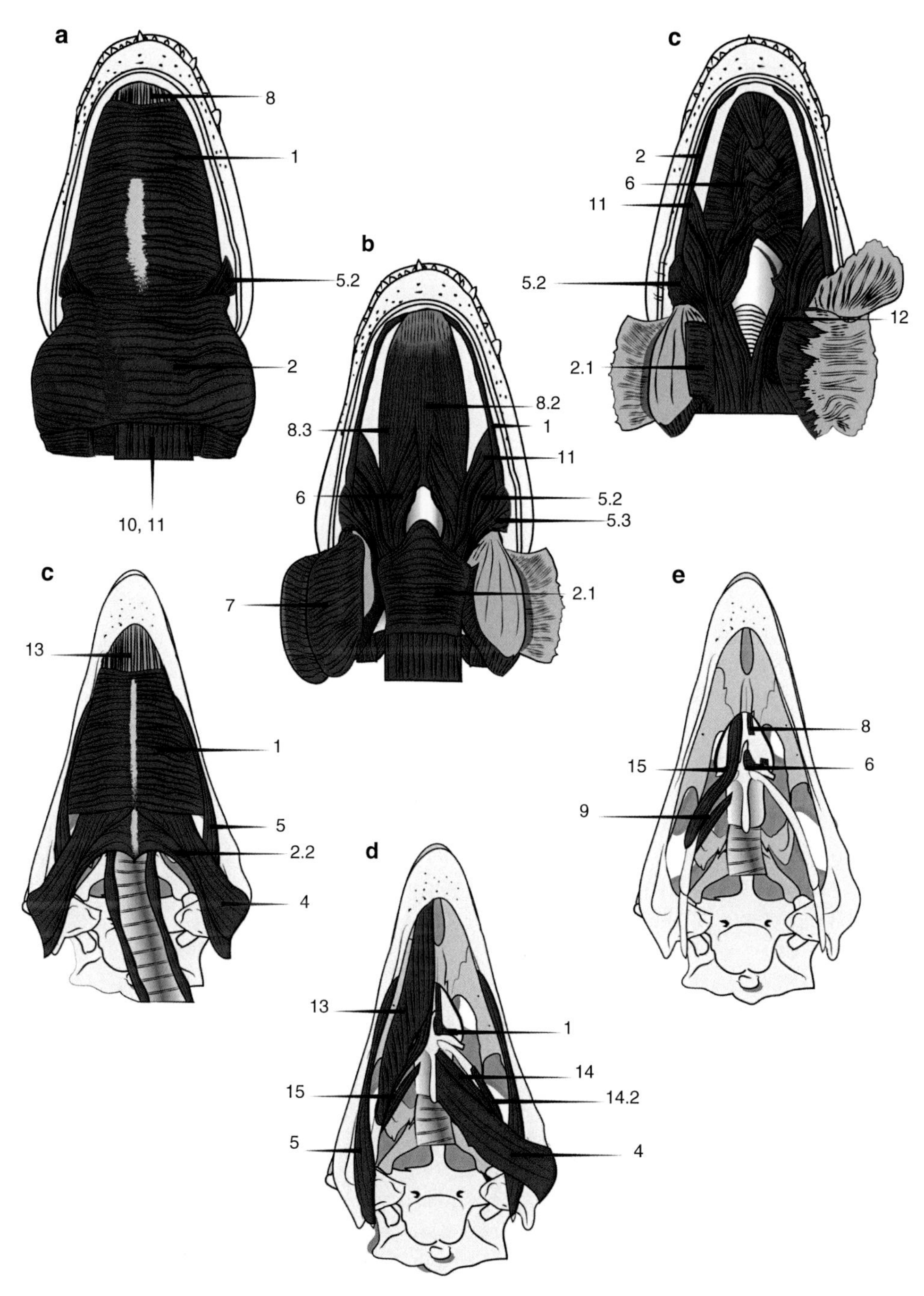

Fig. 10.8 The lower jaw, hypobranchial and tongue muscles of *Alligator mississipiensis* (Crocodylia) (redrawn from Schumacher 1973) and *Dromaius novaehollandiae* (Palaeognathae) (modified from Tomlinson 2000) showing a superficial ventral view of the muscles of the lower jaw (**a** and **d**), a superficial view of the muscles associated with the tongue and base of the jaw, in ventral view (**b** and **e**), and a deeper layer viewing the intrinsic muscles of the tongue (**c** and **e**). 1. M. intermandibularis; 2 M. constrictor colli; 2.1 M. constrictor colli profundus; 2.2 M. constrictor colli intermandibularis; 4. M. serpihyoideus 5. M. brachiomandibularis; 5.2 M. branchiomandibularis spinalis; 5.3 M. branchiomandibularis visceralis; 6. M. hyoglossus; 7 M. pterygoideus ventralis; 7.2 M. pterygoideus; 8. M. genioglossus; 8.2 M. genioglossus medialis; 8.3 M. genioglossus lateralis; 9. M. ceratoglossus; 10. M. coracohyoideus; 11. M. episternobrachiotendineus; 12. M. episternobrachialis; 13. M. genioceratohyoideus; 14. M. hyomandibularis medialis; 14.2 M. hyomandibularis lateralis; 15. M. ceratocricohyoideus

VII). The **M. intermandibularis** (Figs. 10.7a and 10.8a, c) is innervated by the mandibular branch of the trigeminal nerve (CN V_3) (Van den Berge and Zweers 1993) and is the most superficial muscle of the floor of the mouth, spanning the space between the lower jaw and inserting medially in a raphe (Schumacher 1973; Homberger and Meyers 1989; Grigg 2015). This muscle, occasionally called **M. mylohyoideus** in birds, is sometimes divided into two independent portions, anterior and posterior (Van den Berge and Zweers 1993). Posteriorly, M. intermandibularis overlaps the anterior portion of and inserts in a raphe continuous with that of **M. constrictor colli intermandibularis** (Homberger and Meyers 1989) (Figs. 10.7 and 10.8; more below).

Living archosaurs show derived feeding modes when compared to other amniotes; whereas many reptiles use their tongues to transport the food items either to or within the mouth (**lingual transport**), **crocodylians** and **palaeognaths** rely on inertial feeding mechanism (Cleuren and De Vree 2000; Tomlinson 2000). The food item is moved backward along with the head and then released as the head moves forward, using the food inertia to move it back through the oropharyngeal cavity. In **neognaths**, there is no obligate inertial feeding except for some species, but it is generally used for manipulating large food items. The kind of lingual transport neoganths display seems to be very different from that of other amniotes, as the tongue itself is a very different structure (Tomlinson 2000) (Fig. 10.7). In crocodylians and palaeognaths, the **tongue** is a small, stunt structure. The main muscular portion of the tongue in crocodylians (Fig. 10.8c) corresponds to the **M. hyoglossus**, which originates from the **hyoid apparatus** and extends anteriorly as medial and lateral bellies attach to the base of the tongue in a connective tissue raphe (Schumacher 1973). Pars medialis, the bigger portion of the muscle, originates from the ventral side of the hyoid and its fibers intercross in the midline from one side to the other (Schumacher 1973). In **Aves**, the tongue is not a fleshy structure. In palaeognaths (Fig. 10.8) it is limited to the epithelium and connective tissue attaching to the broad **paraglossal**, which forms the main skeletal body of the tongue. In the ostrich, the space left between the two paraglossalia forms a lingual pocket, an invagination of the distal tip of the tongue, which attaches to the **basiurohyal** (Tomlinson 2000). In neognaths the tongue is also mostly formed by the elongated paraglossal and the associated connective tissue, epithelial and glandular tissue.

Both crocodylians and palaeognaths possess a large muscle extending between the **mandibular symphysis**, where left and right dentary bones meet, and the body of the **hyoid apparatus**. This is the **M. genioglossus** in crocodylians and the **M. genioceratohyoideus in palaeognaths** (Fig. 10.8), which in addition have a small **M. genioglossus**. Together with the **Mm. branchiomandibularis** (Mm. **branchiomandibularis spinalis** and **visceralis** in Crocodylians), these muscles act as **protractors of the hyoid**. Neognathous protraction is mainly effected by the Mm. branchiomandibularis, as M. genioglossus is either reduced or absent, and M. genioceratoglossus is absent (Fig. 10.8) (Tomlinson 2000). Retractors in crocodylians, **Mm. coracohyoideus, episternobrachialis, and episternobrachiotendinus**, as in other reptiles originate from the **pectoral girdle** and extend rostrally to attach to the **hyoid apparatus** and **lower jaw** (Schumacher 1973). In birds, on the other hand, the **hyoid retractors**, **M. serpihyoideus**, and **M. stylohyoideus** originate from the medial surfaces of the lower jaw (Tomlinson 2000). The serpihyoideus, which arises from the basitemporal region of the skull, medial to **M. depressor mandibulae**, and inserts ventral onto the **urohyal** (George and Berger 1966), is exclusive of Aves (Tomlinson 2000). The M. stylohyoideus, named after presumed homology with the mammalian muscle, arises from the posterior portion of the lower jaw, anterior to the insertion of M. depressor mandibulae, inserts onto the **basihyal** (George and Berger 1966), and is only present in neognaths (Tomlinson 2000).

Birds are unique among vertebrates in the possession of a **syrinx**, a specialization in the junction between the posterior **trachea** and the bronchi involved in the production of sound (Prum 1992). The time of origin of the syrinx as an evolutionary novelty in the lineage leading to **Aves** is not known. It is also unknown if there is a relationship between its evolution and the independence of the **hyoid apparatus** from the retractor action of muscles with origin from the pectoral girdle, though a relationship between the two probably exists. In **non-avian reptiles**, the **larynx** is usually found anterior to the hyoid apparatus, while in birds it is found posterior to the **basihyal**, and the anterior shifting of the **ceratobranchials** into a birdlike position can be seen in **non-avian theropods** (Li 2015). The syrinx is operated by a series of muscles categorized as intrinsic and extrinsic. Intrinsic muscles of the syrinx correspond to those with both their origin and insertions within the syrinx itself, while extrinsic muscles of the syrinx originate from various places, like the larynx or the pectoral girdle (Prum 1992; Prum et al. 2015). **Extrinsic muscles of the syrinx** include the **M. sternotrachealis** (usually present in all birds), **M. tracheolateralis** (claimed to be lost or reduced in ratites, storks, New World vultures, and some galliformes), and **M. cleidotrachealis** (described in many waterfowl, penguins, and species of tinamous, curassows, and hornbills) (King 1993). The M. tracheolateralis extends anteriorly from the syrinx to attach onto the larynx, while the sternotrachealis and cleidotrachealis originate from the **sternum** and **clavicle,** respectively (King 1993). If these muscles correspond to homologues of reptilian **retractors of the hyoid apparatus** has not been addressed embryologically, and if they are, their release from the hyoid could have been a necessary step for their recruitment to the **syrinx**.

In birds, the **Mm. stylohyoid** and **serpihyoid** and **M. depressor mandibulae** derive from the **second** (hyoid) **arch muscle group**, while the **intrinsic muscles of the tongue**, **larynx**, or **trachea** develop from posterior arches and the **hypoglossal cord**. The hypoglossal cord is an **abaxial** group of **somitic** cells that migrate into the ventral region of the neck and into the floor of the mouth, and gives rise to tongue and laryngeal muscles (Noden et al. 1999). A detailed study on the origin of **hyoid retractors** in crocodylians and the differentiation of the hypoglossal cord and its derivatives in archosaurs in general could help solving the homology and evolutionary story of the muscles of the avian tongue or syrinx.

- The neck musculature

The **M. constrictor colli cervicalis** is the most superficial muscle of the neck; a superficial muscular layer immediately deep to the skin of the neck extends along the sides of the neck and meets in both its dorsal and ventral midlines (Homberger and Meyers 1989; Van den Berge and Zweers 1993). A deeper and more anterior layer is composed of longitudinal fibers originating from the retroarticular process of the mandible and inserting onto a midline raphe, continuous with that of the **M. intermandibularis** (Homberger and Meyers 1989; Van den Berge and Zweers 1993) (see Figs. 10.7 and 10.8). Both muscles are innervated by the **facial nerve**, CN VII (Van den Berge and Zweers 1993). In crocodylians, a superficial **M. constrictor colli** extends around the neck meeting its contralateral muscle in the ventral midline, while the **M. constrictor colli profundus** lies deeper and functions as a pharyngeal constrictor (Schumacher 1973).

The **primaxial muscles** (derived from the somites and differentiated within the somite boundaries) are further subdivided into two categories, the dorsal **epaxial** and the ventral **hypaxial muscles**, according to their position relative to the horizontal septum and their innervation by dorsal or ventral rami of the spinal nerves (Burke and Nowicki 2003). Hypaxial muscles extend from their site of origin, surrounding the lateral walls of the body, lining up the internal cavities, or making up the muscles in between individual ribs.

Along the vertebral column, the **epaxial muscles** are divided into three muscular groups, the **Mm. transversospinalis**, the **Mm. longissimus**, and the **Mm. iliocostalis group**, which are covered and separated from each other by complex fascia (Cleuren and De Vree 2000). These muscles correspond to serial homologues or the anterior portions of muscles also running parallel to the trunk vertebral column, and they are differentiated by the use of the appendix dorsi, cervicis, and capitis for muscles of the trunk, intrinsic of the neck, or attaching to the skull, respectively (Cleuren and De Vree 2000). In crocodylians, these muscles are relatively conservative and similar in anatomy to other diapsids. The deepest muscular layer of the Mm. transversospinalis group corresponds to **Mm. interspinales** and **interarticulares**, small muscles that span the distance between one vertebrae and the next one, connecting successive vertebral elements (Tsuihiji 2005). Lateral to these muscles, a group of muscles with four longitudinal series of tendons and no definite boundaries form the **M. transversospinalis cervicis** in the neck (Tsuihiji 2005). The **M. transversospinalis capitis** is the dorsalmost superficial muscle of the neck and arises from the tips of the neural spines of various vertebrae from the neck and trunk; the **M. spinocapitis posticus** arises from vertebrae of the neck and inserts onto the back of the skull (Tsuihiji 2005). From vertebrae 1 and 2, another muscle, the **M. atlanto-capitis**, extends to attach onto the back of skull (Tsuihiji 2005). The **Mm. transversospinalis** are separated from the muscles of the longissimus group by a fascia called the **septum intermusculare dorsi** (Tsuihiji 2005). The **Mm. longissimus dorsi and cervicis** are continuous muscles in the trunk and posterior neck region, respectively; they connected to a series of tendons that attach to the transverse processes of consecutive vertebrae and the septum intermuscularis dorsi, while the **Mm. longissimus capitis superficialis and profundus** arise from the anterior portion of the neck and attach to the back of the skull (Tsuihiji 2007). The **iliocostalis cervicis** is also a muscle attached to consecutive vertebrae via tendons; in its most anterior portion, it gives rise to the **M. iliocostalis capitis** (Tsuihiji 2007).

In birds the neck myology is more complex and establishing its homologies has been a problem for comparative anatomists, which have also used a nomenclature completely different from the one used for other reptiles (Tsuihiji 2007). For example, Tsuihiji (2007), after studying and comparing the anatomy of **lepidosaurs** and both clades of archosaurs, homologized the **M. longus colli dorsalis** pars cranialis and **M. splenius anticus** of birds (Van den Berge and Zweers 1993) to the **M. spinocapitis posticus** of crocodylians, the M. longus colli dorsalis pars profunda and **M. splenius accesorius** of birds (Van den Berge and Zweers 1993) to the medial portion of **M. transversospinalis cervicis**, and the **M. ascendens cervicalis** of birds (Van den Berge and Zweers 1993) to the lateral portion of the same muscle (Tsuihiji 2005). Many muscles described as separate structures in the nomina anatomica avium (Van den Berge and Zweers 1993) are included as part of the M. transversospinalis group, including **Mm. biventer cervicis**, **complexus**, or **splenius capitis** (Tsuihiji 2005) Also, portions of the **M. iliocostalis** of birds (Van den Berge and Zweers 1993) apparently correspond to Mm. longissimus dorsi and iliocostalis dorsi of crocodylians, while portions of **Mm. intertransversarii** and **flexor colli lateralis** of birds (Van den Berge and Zweers 1993) correspond to Mm. longissimus cervicis and iliocostalis cervicis (Tsuihiji 2007). **Hypaxial muscles of the neck** include separate muscles: the **M. longus colli** and **M. rectus capitis anticus** in crocodylians and **M. longus colli ventralis** and **M. rectus capitis ventralis** and lateralis in birds (Tsuihiji 2007) (Fig. 10.9).

- Extinct archosaur muscle anatomy

The capability of correctly inferring the anatomy of an extinct animal depends on both the **fossil** record and our knowledge of present-day species in an appropriate phylogenetic context. This applies to hard tissue anatomy, like bones, or soft tissue anatomy, like muscles; the latter is much harder, as they are not as easily preserved as fossils. For many years, the antorbital cavity, the space opened by the antorbital fenestrae in front of the eye, was thought to house the **M. pterygoideus dorsalis**, allowing for huge muscles closing the jaw to excerpt enormous bite forces in big **theropods**; however, the cavity contains paranasal air sinuses and the muscle only occupies the floor of the space (Witmer 1995, 1997; Dilkes et al. 2012). Based on osteological correlates and the principle of phylogenetic bracketing (scoring a level of certainty to an anatomical inference based

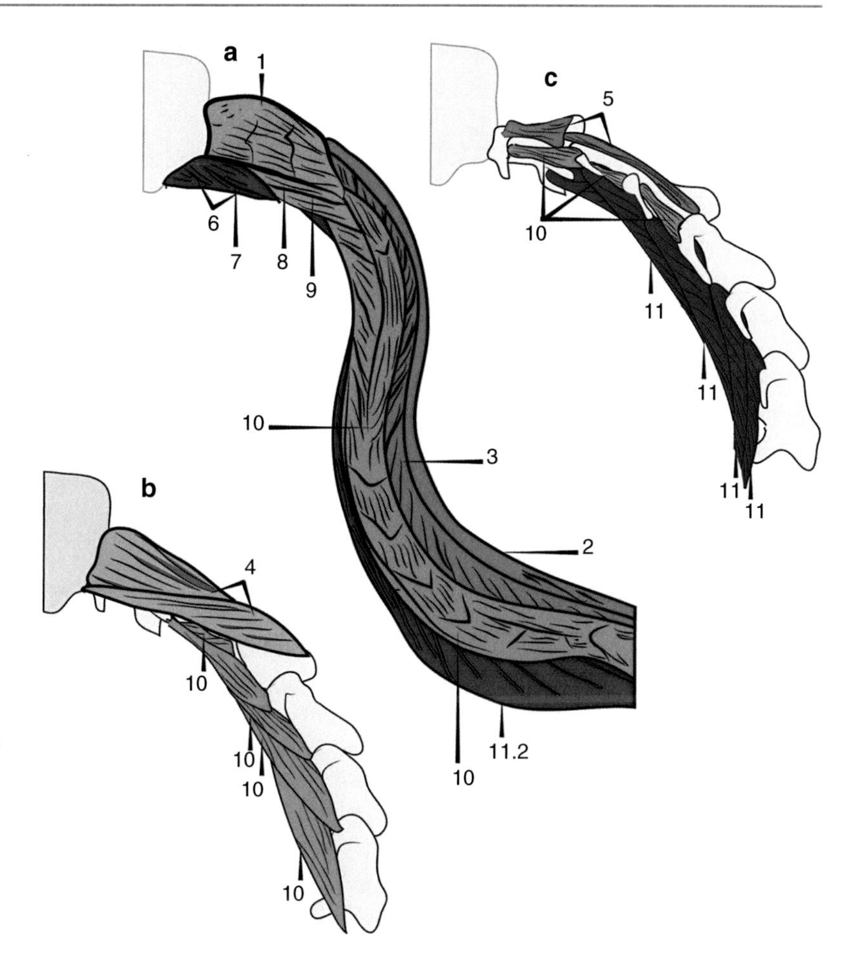

Fig. 10.9 Lateral view of the neck of a grebe, *Podilymbus podiceps*, modified from Van den Berge and Zweers (1993). (**a**) Superficial view, (**b**) intermediate view, and (**c**) deep view of neck muscles. Hypaxial muscles are colored red and epaxial muscles are colored orange, after Tsuihiji (2005, 2007). 1. M. complexus; 2. M. longus colli dorsalis caudalis; 3. M. ascendis cervices; 4. M. splenius capitis; 5. Mm. intercristales; 6. M. rectus capitis ventralis; 7. M. rectus capitis lateralis; 8. M. rectus capitis dorsalis; 9. M. flexor colli lateralis; 10. Mm. intertransversarii; 11. M. longus colli ventralis; 11.2 M. longus colli ventralis caudalis

on whether a structural correlate is present or absent in living species related to the extinct ones and depending on the position of the species on a tree), the muscle anatomy of extinct forms can be studied, and hypotheses regarding the evolution of archosaur soft anatomy can be tested. For example, the **M. levator pterygoideus** is absent in both groups of living archosaurs, but likely its loss is due to independent events. The M. levator pterygoideus attaches to the epipterygoid bone, and birds and archosaurs lost the epipterygoids independently, as the bone is present in non-avian dinosaurs. Although it shows a weak inference level, as it is absent in both living clades, varying correlates, like small fossae or tuberosities in different non-avian dinosaurs, justify its presence in the extinct taxa (Holliday 2009). Recent work dealing with reconstructing the muscles of dinosaurs in this way includes the aforementioned work on the jaw muscles (Holliday 2009) or the neck muscles (Tsuihiji 2010). There is much to learn, both from the past and the present, from reconstructing the musculature of extinct species, but as these studies do, caution is required for not incurring in misinterpretation, error, and misunderstanding.

10.4 Future Directions

It might be clear by now that the lineages of archosaurs differ drastically in their morphologies, and even though they conserve many shared features, they have also evolved particular anatomical innovations according to their different modes of life. Understanding how these innovative morphologies evolved might be a puzzle, that to solve requires the integration of different lines of evidence, drawing information from **fossils, embryos**, molecules, and behavior. Without clear understanding of the phyletic relationships between different vertebrates or without taking into consideration the anatomy of other groups, one might fail to solve this kind of evolutionary questions. For example, consider the constrictor and **dilator naris muscles** of crocodylians. These smooth muscles open and close the narial openings and might be regarded as evolutionary innovations arising without any seeming homologue, when considering that birds do not have them. However, some turtles and amphibians possess smooth musculature that closes the nares (Kingsley 1912; Winokur 1982) and so could be thought to simply have been lost in birds. These muscles seem to serve a function, as in crocodylians, of preventing the passage of water when submerged, and their presence in the long line of terrestrial ancestor of modern-day aquatic crocodylians can neither be observed in their fossil record nor justified given their lifestyle. Here we find a problematic end to the speculations, as we truly do not know if these muscles were continuously present in the lineage leading to modern-day crocs, coming from an ancestor shared with **turtles** or **amphibians**, nor do we know how these muscles develop or where are they derived from in these lineages, truncating any well-informed guess on their homology. For such a case, the presence or absence of other muscles cannot be easily asserted in fossil taxa when they don't leave any specific osteological correlate. Moreover, their similarity cannot be fully evaluated when embryonic origin and development remain unknown.

The most intriguing archosaur evolutionary story, however, could be the origin of the avian **syrinx** and its muscles, probably tightly related to the evolution of the avian tongue. As members of the avian lineage acquired a longer, S-shaped neck, the function of the retractor muscles connecting the hyoid apparatus to the pectoral girdle could have been impaired, and their importance reduced as new muscles became more relevant for the task, being released from their function and allowed to take on a new duty, the control of the syrinx. Thus, the evolution of a new vocal organ in birds would have not been directly dependent on the evolution of longer sinuous necks or cranially suspended hyoid apparatus but enabled by the series of transformations that eventually released specific muscles from their function. The "free" **hyoid apparatus**, now controlled by different groups of muscles, evolved on its own, allowing the origin of structures as astonishing as the responsible for the tongue feeding mechanisms of hummingbirds or woodpeckers. The ossification of **epibranchials** and **basihyal** in some early **Avialae** might indicate a degree of hyoid suspension (Li 2015), but their ossification does not seem to relate to the independence from the **pectoral girdle**, as these hyoidal elements are also cartilaginous in **palaeognaths**. The fossil record might show hints related to when and how many times did these transformations occurred, but embryological information helping to identify the nature and origin of the muscular elements is still key to decipher the complete series of steps leading to such different patterns. Future studies must integrate these different lines of evidence in order to provide a clear picture and understanding of the transformations that occurred during Archosaur evolution, including anatomical embryological and paleontological studies of diverse archosaurian clades.

References

Barsbold R, Osmolska H (1990) Ornithomimosauria. In: The Dinosauria. University of California Press, Berkeley, pp 225–244

Barsbold R, Maryanska T, Osmolska H (1990) Oviraptorosauria. In: The Dinosauria, vol 1. University of California Press, Berkeley, pp 249–258

Baumel JJ, King AS, Breazile JE, Evans HE, Vanden Berge JC (1993) Handbook of avian anatomy: nomina anatomica avium. Publications of the Nuttall Ornithological Club, Cambridge

Bellairs AdA, Kamal A (1981) The chondrocranium and the development of the skull in recent reptiles. Biol Reptilia 11:1–264

Bellairs AD, Shute CC (1953) Observations on the narial musculature of Crocodilia and its innervation from the sympathetic system. J Anat 87:367

Benton MJ, Clark JM (1988) Archosaur phylogeny and the relationships of the Crocodylia. In: The phylogeny

and classification of the tetrapods, vol 1. Clarendon Press, Oxford, pp 295–338
Berv JS, Field DJ (2017) Genomic signature of an avian Lilliput effect across the K-Pg extinction. Syst Biol 67(1):1–13
Bhullar BA, Marugan-Lobon J, Racimo F et al (2012) Birds have paedomorphic dinosaur skulls. Nature 487:223–226
Brusatte SL, Benton MJ, Desojo JB, Langer MC (2010) The higher-level phylogeny of Archosauria (Tetrapoda: Diapsida). J Syst Palaeontol 8(1):3–47
Burke AC, Nowicki JL (2003) A new view of patterning domains in the vertebrate mesoderm. Dev Cell 4:159–165
Chamberlain FW (1943) Atlas of avian anatomy; osteology, arthrology, myology. Michigan State College, Agricultural Experiment Station, East Lansing
Chiappe LM, Norell MA, Clark JM (1998) The skull of a relative of the stem-group bird Mononykus. Nature 392:275–278
Cleuren J, De Vree F (2000) Feeding in crocodilians. In: Feeding: form, function, and evolution in tetrapod vertebrates. Academic Press, New York, pp 337–358
Dal Sasso C, Signore M (1998) Exceptional soft-tissue preservation in a theropod dinosaur from Italy. Nature 392:383–387
Dalla Vecchia FM (2013) Triassic pterosaurs. Geol Soc Lond, Spec Publ 379:119–155
Desojo JB, Heckert AB, Martz JW et al (2013) Aetosauria: a clade of armoured pseudosuchians from the Upper Triassic continental beds. Geol Soc Lond, Spec Publ 379:203–239
Dilkes DW, Hutchinson JR, Holliday CM, Witmer LM (2012) Reconstructing the musculature of dinosaurs. In: The complete dinosaur. Indiana University Press, Bloomington, pp 151–190
Elzanowski A (1987) Cranial and eyelid muscles and ligaments of the tinamous. Zoologische Jahrbücher (Abteilung für Anatomie) 116:63–118
Evans DJ, Noden DM (2006) Spatial relations between avian craniofacial neural crest and paraxial mesoderm cells. Dev Dyn 235:1310–1325
Evans DC, Reisz RR, Dupuis K (2007) A juvenile Parasaurolophus (Ornithischia: Hadrosauridae) braincase from Dinosaur Provincial Park, Alberta, with comments on crest ontogeny in the genus. J Vertebr Paleontol 27:642–650
Ezcurra MD, Butler RJ, Gower DJ (2013) 'Proterosuchia': the origin and early history of Archosauriformes. Geol Soc Lond, Spec Publ 379:9–33
Fabbri M, Koch NM, Pritchard AC et al (2017) The skull roof tracks the brain during the evolution and development of reptiles including birds. Nat Ecol Evol 1(10):1543
Fanti F, Currie PJ, Badamgarav D (2012) New specimens of Nemegtomaia from the Baruungoyot and Nemegt formations (Late Cretaceous) of Mongolia. PLoS One 7:e31330
Fitzgerald, T.C. (1969). The coturnix quail: anatomy and histology. The coturnix quail: anatomy and histology.
Gauthier J (1986) Saurischian monophyly and the origin of birds. In: Memoirs of the California Academy of Sciences, vol 8. University of California, Berkeley, pp 1–55
Gauthier J, Kluge AG, Rowe T (1988) Amniote phylogeny and the importance of fossils. Cladistics 4:105–209
Gauthier JA, Nesbitt SJ, Schachner ER, Bever GS, Joyce WG (2011) The bipedal stem crocodilian Poposaurus gracilis: inferring function in fossils and innovation in archosaur locomotion. BullPeabody Mus Nat Hist 52:107–126
George JC, Berger AJ (1966) Avian myology. Academic Press, New York
Grigg GC (2015) Biology and evolution of crocodylians. CSIRO, Collingwood
Holliday CM (2009) New insights into dinosaur jaw muscle anatomy. Anat Rec 292:1246–1265
Holliday CM, Witmer LM (2007) Archosaur adductor chamber evolution: integration of musculoskeletal and topological criteria in jaw muscle homology. J Morphol 268:457–484
Holliday CM, Witmer LM (2009) The epipterygoid of crocodyliforms and its significance for the evolution of the orbitotemporal region of eusuchians. J Vertebr Paleontol 29:715–733
Homberger DG, Meyers RA (1989) Morphology of the lingual apparatus of the domestic chicken, Gallus gallus, with special attention to the structure of the fasciae. Dev Dyn 186:217–257
Horner JR, Weishampel DB, Forster C (2004) Hadrosauridae. In: The dinosauria, vol 2. University of California Press, Berkeley, pp 438–463
Iordansky NN (1973) The skull of the crocodilia. In: Biology of the reptilia, vol 4. Academic Press, London, pp 201–262
Irmis RB, Nesbitt SJ, Sues H (2013) Early crocodylomorpha. Geol Soc Lond, Spec Publ 379:275–302
Jarvis ED, Mirarab S, Aberer AJ et al (2014) Whole-genome analyses resolve early branches in the tree of life of modern birds. Science 346:1320–1331
Kardong KV (2006) Vertebrates: comparative anatomy, function, evolution. McGraw-Hill, New York
King AS (1993) Apparatus respiratorius (systema respiratorium). In: Handbook of avian anatomy: nomina anatomica avium. Nuttall Ornithological Club, Cambridge, pp 257–299
Kingsley JS (1912) Comparative anatomy of vertebrates, 1912. P. Blakiston's, Philadelphia [591.4]
Klembara J (2004) Ontogeny of the palatoquadrate and adjacent lateral cranial wall of the endocranium in prehatching Alligator mississippiensis (Archosauria: Crocodylia). J Morphol 262:644–658
Kulesa P, Ellies DL, Trainor PA (2004) Comparative analysis of neural crest cell death, migration, and function during vertebrate embryogenesis. Dev Dyn 229:14–29
Kundrát M (2008) HNK-1 immunoreactivity during early morphogenesis of the head region in a nonmodel vertebrate, crocodile embryo. Naturwissenschaften 95:1063

Kundrát M (2009) Heterochronic shift between early organogenesis and migration of cephalic neural crest cells in two divergent evolutionary phenotypes of archosaurs: crocodile and ostrich. Evol Dev 11:535–546

Le Lièvre CS (1978) Participation of neural crest-derived cells in the genesis of the skull in birds. Development 47:17–37

Le Lièvre CS, Le Douarin NM (1975) Mesenchymal derivatives of the neural crest: analysis of chimaeric quail and chick embryos. Development 34:125–154

Li Z (2015) Evolution of the hyoid apparatus in Archosauria: implications for the origin of avian tongue function. The University of Texas, Austin

Li Z, Clarke JA (2015) New insight into the anatomy of the hyolingual apparatus of Alligator mississippiensis and implications for reconstructing feeding in extinct archosaurs. J Anat 227:45–61

Liyong J, Jun C, Shuqin Z, Butler RJ, Godefroit P (2010) Cranial anatomy of the small ornithischian dinosaur Changchunsaurus parvus from the Quantou formation (Cretaceous: Aptian-Cenomanian) of Jilin Province, northeastern China. J Vertebr Paleontol 30:196–214

Luther A (1914) Über die vom n. Trigeminus versorgte Muskulatur der Amphhibien. Druckerei der Finnischen Literaturgesellschaft, Helsingfors

Madsen JH Jr (1976) Allosaurus fragilis: a revised osteology. Utah Geol Min Survey Bull 109:1–163

Marpmann JS, Carballido JL, Sander PM, Knötschke N (2015) Cranial anatomy of the Late Jurassic dwarf sauropod Europasaurus holgeri (Dinosauria, Camarasauromorpha): ontogenetic changes and size dimorphism. J Syst Palaeontol 13:221–263

McIntosh JS (1990) Sauropoda. In: The dinosauria, vol 1. University of California Press, Berkeley, pp 345–401

Nesbitt SJ (2011) The early evolution of archosaurs: relationships and the origin of major clades. Bull Am Mus Nat Hist 352:1–292

Nesbitt SJ, Desojo JB, Irmis RB (2013) Anatomy, phylogeny and palaeobiology of early archosaurs and their kin. Geol Soc Lond, Spec Publ 379:1–7

Noden DM, Francis-West P (2006) The differentiation and morphogenesis of craniofacial muscles. Dev Dyn 235:1194–1218

Noden DM, Marcucio R, Borycki AG, Emerson CP (1999) Differentiation of avian craniofacial muscles: I. Patterns of early regulatory gene expression and myosin heavy chain synthesis. Dev Dyn 216:96–112

Ősi A (2005) Hungarosaurus tormai, a new ankylosaur (Dinosauria) from the Upper Cretaceous of Hungary. J Vertebr Paleontol 25:370–383

Ostrom JH (1976) Archaeopteryx and the origin of birds. Biol J Linn Soc 8:91–182

Parker WK (1866) On the structure and development of the skull in the ostrich tribe. Philos Trans R Soc Lond 156:113–183

Pérez-Moreno P, Sanz JL, Buscalioni AD et al (1994) A unique multitoothed ornithomimosaur dinosaur from the lower cretaceous of Spain. Nature 370:363–367

Prum RO (1992) Syringeal morphology, phylogeny, and evolution of the neotropical manakins (Aves, Pipridae). American Museum novitates; no. 3043

Prum RO, Berv JS, Dornburg A et al (2015) A comprehensive phylogeny of birds (Aves) using targeted next-generation DNA sequencing. Nature 526:569–573

Pycraft WP (1900) On the morphology and phylogeny of the Palæognathæ (Ratitæand Crypturi) and Neognathæ (Carinatæ). J Zool 15:149–290

Rieppel O (2000) Turtles as diapsid reptiles. Zool Scr 29:199–212

Romer AS (1956) Osteology of the reptiles. University of Chicago Press, Chicago

Santagati F, Rijli FM (2003) Cranial neural crest and the building of the vertebrate head. Nat Rev Neurosci 4:806–818

Schumacher GH (1973) The head muscles and hyolaryngeal skeleton of turtles and crocodilians. In: Biology of the reptilia, vol 4. Academic Press, London, pp 101–200

Schwab IR, Brooks DE (2002) He cries crocodile tears…. Br J Ophthalmol 86:23–23

Sereno PC, Shichin C, Zhengwu C, Chenggang R (1988) Psittacosaurus meileyingensis (Ornithischia: Ceratopsia), a new psittacosaur from the Lower Cretaceous of northeastern China. J Vertebr Paleontol 8:366–377

Shih HP, Gross MK, Kioussi C (2008) Muscle development: forming the head and trunk muscles. Acta Histochem 110:97–108

Shute CCD, Bellairs A (1955) The external ear in crocodilia. J Zool 124:741–749

Stocker MR, Butler RJ (2013) Phytosauria. Geol Soc Lond, Spec Publ 379:91–117

Sues H-D, Galton PM (1987) Anatomy and classification of the North American Pachycephalosauria (Dinosauria: Ornithischia). E. Schweizerbart'sche, Stuttgart

Tomlinson B (2000) Feeding in paleognathous birds. In: Feeding: form, function, and evolution in tetrapod vertebrates. Academic Press, San Diego, pp 359–394

Tsai HP, Holliday CM (2011) Ontogeny of the alligator cartilago transiliens and its significance for sauropsid jaw muscle evolution. PLoS One 6:e24935

Tsuihiji T (2005) Homologies of the transversospinalis muscles in the anterior presacral region of Sauria (crown Diapsida). J Morphol 263:151–178

Tsuihiji T (2007) Homologies of the longissimus, iliocostalis, and hypaxial muscles in the anterior presacral region of extant Diapsida. J Morphol 268:986–1020

Tsuihiji T (2010) Reconstructions of the axial muscle insertions in the occipital region of dinosaurs: evaluations of past hypotheses on Marginocephalia and Tyrannosauridae using the extant phylogenetic bracket approach. Anat Rec 293:1360–1386

Underwood G (1970) The eye. In: Biology of the reptilia, vol 2. Academic Press, London, pp 1–97

Unwin DM (2003) On the phylogeny and evolutionary history of pterosaurs. Geol Soc Lond Spec Publ 217:139–190

Van den Berge C, Zweers GA (1993) Myologia. In: Handbook of avian anatomy: nomina anatomica avium. Nuttall Ornithological Club, Cambridge, pp 189–247

Vickaryous MK (2006) New information on the cranial anatomy of Edmontonia rugosidens Gilmore, a late cretaceous nodosaurid dinosaur from Dinosaur Provincial Park, Alberta. J Vertebr Paleontol 26:1011–1013

Vullo R, Marugán-Lobón J, Kellner A et al (2012) A new crested pterosaur from the early cretaceous of Spain: the first European tapejarid (Pterodactyloidea: Azhdarchoidea). PLoS One 7:e38900

Wang X, Kellner A, Zhou Z, de Almeida Campos D (2007) A new pterosaur (Ctenochasmatidae, Archaeopterodactyloidea) from the lower Cretaceous Yixian Formation of China. Cretac Res 28:245–260

Wang X, Kellner AW, Jiang S, Cheng X (2012) New toothed flying reptile from Asia: close similarities between early cretaceous pterosaur faunas from China and Brazil. Naturwissenschaften 99:249–257

Wang S, Stiegler J, Amiot R et al (2017a) Extreme ontogenetic changes in a ceratosaurian theropod. Curr Biol 27:144–148

Wang S, Stiegler J, Wu P et al (2017b) Heterochronic truncation of odontogenesis in theropod dinosaurs provides insight into the macroevolution of avian beaks. Proc Natl Acad Sci U S A 114(41):10930–10935

Webb M (1957) The ontogeny of the cranial bones, cranial peripheral and cranial parasympathetic nerves, together with a study of the visceral muscles of Struthio. Acta Zool 38:81–203

Wedin B (1953) The development of the eye muscles in Ardea cinerea L. Cells Tissues Organs 18:30–48

Weishampel DB, Witmer LM (1990) Heterodontosauridae. In: The dinosauria, vol 1. University of California Press, Berkeley, pp 486–497

Whitlock JA (2011) A phylogenetic analysis of Diplodocoidea (Saurischia: Sauropoda). Zool J Linn Soc 161:872–915

Winokur RM (1982) Erectile tissue and smooth muscle in snouts of Carettochelys insculpta, trionchids and other chelonia. Zoomorphology 101:83–93

Witmer LM (1995) Homology of facial structures in extant archosaurs (birds and crocodilians), with special reference to paranasal pneumaticity and nasal conchae. J Morphol 225:269–327

Witmer LM (1997) The evolution of the antorbital cavity of archosaurs: a study in soft-tissue reconstruction in the fossil record with an analysis of the function of pneumaticity. J Vertebr Paleontol 17:1–76

Witmer LM, Chatterjee S, Franzosa J, Rowe T (2003) Neuroanatomy of flying reptiles and implications for flight, posture and behaviour. Nature 425:950

Zusi RL (1993) Patterns of diversity in the avian skull. The Skull 2:391–437

Zusi R, Livezey B (2000) Homology and phylogenetic implications of some enigmatic cranial features in galliform and anseriform birds. Ann Carnegie Mus 69:157–193

11 The Origin and Evolution of Mammalian Head Muscles with Special Emphasis on the Facial Myology of Primates and Modern Humans

Rui Diogo and Vance Powell

11.1 Introduction

The **Mammalia** is a diverse clade (Fig. 11.1) that includes **monotremes** (**Prototheria**, e.g., echidnas and platypus; see, e.g., Figs. 11.2, 11.3, and 11.4), **marsupials** (**Metatheria**, e.g., opossums; see, e.g., Fig. 11.5), and **placentals** (**Eutheria**, e.g., humans; see, e.g., Figs. 11.6, 11.7, 11.8, 11.9, 11.10, 11.11, 11.12, 11.13, 11.14, 11.15, 11.16, 11.17, 11.18, 11.19, and 11.20) and that is frequently the focus of evolutionary and developmental studies. This is because, apart from the interest in this group per se due to its fascinating diversity and evolutionary history, the Mammalia includes primates and therefore our own species, as well as model organisms, in particular mice, that are often used in genetic and medical studies. Figure 11.1 shows just a few among many examples of the striking diversity of the mammalian head and jaws. Among these and other examples, one can refer, for instance, to the heads of elephants, with more than 200 bundles of muscles of facial expression, the heads of whales that can grow to over 200 tons, the minuscule heads of 2 g bumblebee bats (*Craseonycteris thonglongyai*), and the extremely derived and edentulous jaws of anteaters (suborder Vermilingua) (see, e.g., Surlykke et al. 1993).

Among the numerous studies that have been done concerning the adult cephalic muscles of mammals, a major weakness of most of them is that they do not provide comparisons between monotremes, marsupials, and placentals or between these groups and other tetrapods, therefore making it difficult to understand the origin and evolutionary history of mammalian head muscles. Numerous of these studies are cited in Diogo and Abdala's (2010) book *Muscles of Vertebrates*, and in Diogo et al.'s (2018) book *Muscles of Chordates*, which provides the most recent review on the subject. Traditionally, omissions or even factual errors have been common in the literature concerning mammalian head muscles, often due to a strong historical bias regarding mammals as an example of a *scala naturae* leading from monotremes to marsupials and then to placentals, culminating in humans. This notion of a *scala naturae*, which dates back to thinkers such as Aristotle, represents an evolutionary trend in complexity from "lower" to "higher" taxa, with *Homo sapiens* as the end stage (discussed in Diogo et al. 2015a, 2016). Many

R. Diogo (✉)
Department of Anatomy, Howard University, Washington, DC, USA
e-mail: rui.diogo@howard.edu

V. Powell
CASHP, Department of Anthropology, George Washington University, Washington, DC, USA

J. M. Ziermann et al. (eds.), *Heads, Jaws, and Muscles*, Fascinating Life Sciences,
https://doi.org/10.1007/978-3-319-93560-7_11

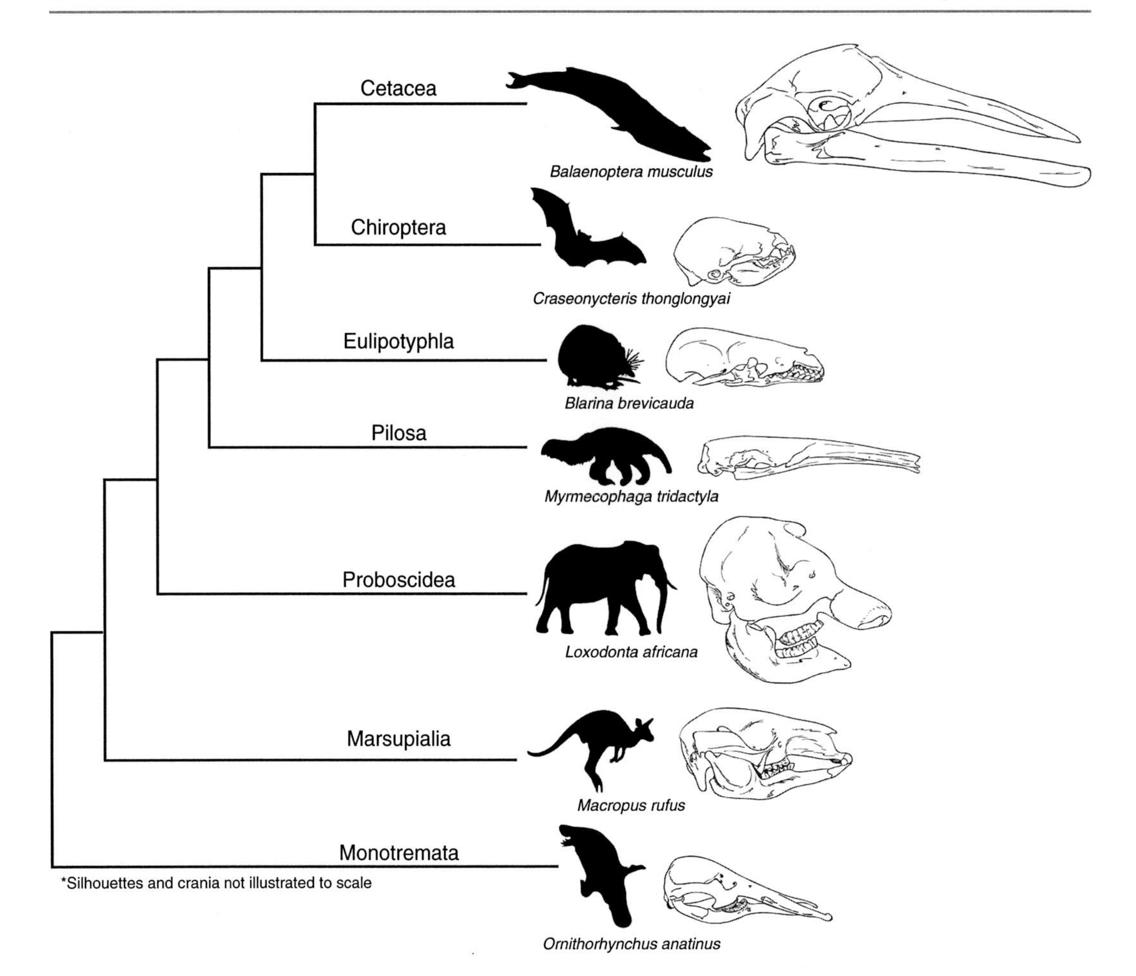

Fig. 11.1 Examples of mammalian cranio-mandibular diversity using a simplified consensus phylogeny: blue whale (*Balaenoptera musculus*), bumblebee bat (*Craseonycteris thonglongyai*), Northern short-tailed shrew (*Blarina brevicauda*), giant anteater (*Myrmecophaga tridactyla*), African elephant (*Loxodonta africana*), red kangaroo (*Macropus rufus*), and platypus (*Ornithorhynchus anatinus*). Mammals occupy disparate subterranean, arboreal, and marine habitats, which are related to remarkable size differences among the smallest (i.e., the bumblebee bat and short-tailed shrews) and the largest taxa (i.e., the blue whale). Unique dietary specializations are also reflected in dentognathic adaptations like the edentulous maxillae and mandibles of blue whales and anteaters that consume exceedingly small prey (e.g., krill and termites, respectively) and the duck-like electroreceptive rostrum of the platypus used in foraging

authors have, for instance, reported only a few undifferentiated muscles of facial expression—which are special muscles derived from the 2nd (hyoid) branchial arch that attach to, and move, the skin—in marsupials, more muscles in placentals such as rats, and a "most complex" facial musculature in humans (e.g., Huber 1930a, b; Huber 1931; Lightoller 1940a, 1942). Such notions of *scala naturae* are found not only in works from the nineteenth and early twentieth centuries but even in publications from the late twentieth century. For instance, Minkoff et al.'s (1979) study of the facial muscles of *D. virginiana* describes 21 muscles of facial expression in this species, including extrinsic ear muscles, i.e., only about 2/3 of the 31 facial muscles found in humans (25 + 6 extrinsic ear muscles: see Table 11.2). In contrast, in our recent dissections, we found exactly the same number of facial muscles in *D. virginiana* as in placentals such as rats, a number that is moreover very similar to that found in humans, as we will explain below (Table 11.2; Fig. 11.5).

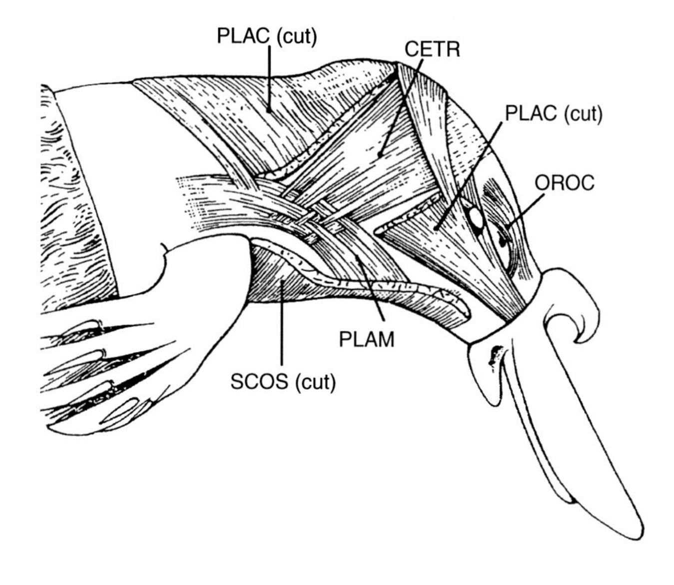

Fig. 11.2 *Ornithorhynchus anatinus* (Mammalia, Monotremata): lateral view of the deep facial musculature; muscles such as the interhyoideus profundus, buccinatorius, orbicularis oris, and mentalis are not shown (modified from Saban 1971; the nomenclature of the structures illustrated basically follows that used in the present work; anterior is to the right). *CETR* cervicalis transversus; *OROC* orbicularis oculi; *PLAC*, *PLAM* platysma cervicale and platysma myoides; *SCOS* sphincter colli superficialis

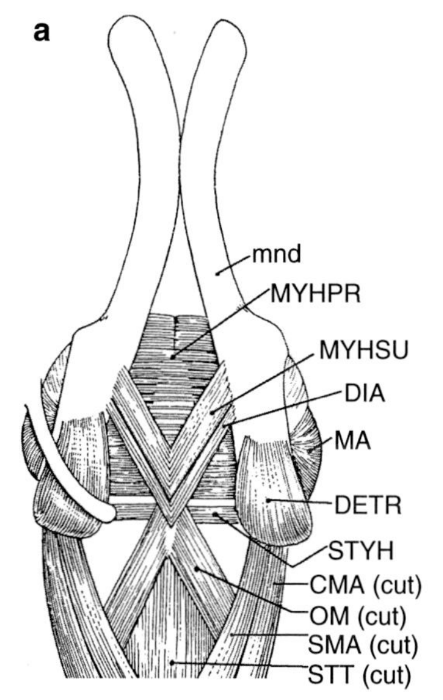

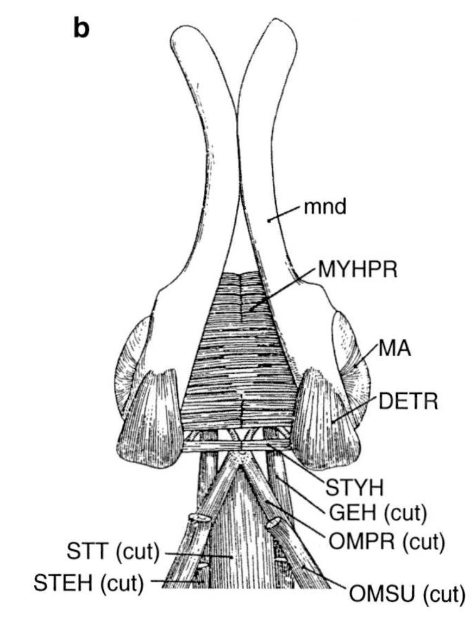

Fig. 11.3 *Ornithorhynchus anatinus* (Mammalia, Monotremata): (**a**) ventral view of the head and neck musculature, muscles such as the geniohyoideus and sternohyoideus are not shown; (**b**) same view, but the digastricus anterior, superficial part of the mylohyoideus, sternomastoideus, and cleidomastoideus were removed, and the anterior portion of the sternohyoideus and of the superficial part of the omohyoideus were partially cut (modified from Saban 1971; the nomenclature of the structures illustrated basically follows that used in the present work; anterior is to the top). *CMA* cleidomastoideus; *DETR* detrahens mandibulae; *DIA* digastricus anterior; *GEH* geniohyoideus; *MA* masseter; *mnd* mandible; *MYHPR*, *MYHSU* pars profunda and pars superficialis of mylohyoideus; *OM* omohyoideus; *OMPR*, *OMSU* pars profunda and pars superficialis of omohyoideus; *SMA* sternomastoideus; *STEH* sternohyoideus; *STT* sternothyroideus; *STYH* styloideus

In this chapter we briefly summarize what is known about the deep evolutionary origin—all the way back to fishes—about the evolution of the head muscles of marsupials, placentals, and monotremes by providing comparisons with other vertebrates (Tables 11.1, 11.2, 11.3, and 11.4). We also provide notes on the development of these muscles, particularly from descriptive and experimental works in mice and other taxa. Specifically, we provide a list and brief description of all head muscles of representatives of the three major extant mammalian clades, as well as of other tetrapods and fish using an updated, unifying vertebrate myological nomenclature to allow more straightforward comparison between all these taxa. This summary is based on our own dissections of thousands of specimens from these vertebrate taxa (e.g., Diogo 2007, 2008, 2009; Diogo and Abdala 2007, 2010; Diogo et al. 2008a, 2009a, b, 2013a, b, 2014, 2015a, b, c, d, 2016, 2017; Diogo and Wood 2012; Diogo and Molnar 2014; Diogo and Tanaka 2014; Diogo and Ziermann 2014) as well as a detailed review of the literature (see, e.g., reviews by Diogo and Wood 2012 and Diogo and Abdala 2010). It should be noted that we do not include here details about the extraocular muscles. This is because all six oculorotatory muscles (inferior and superior oblique muscles and inferior, lateral, medial, and superior rectus muscles) and the levator palpebrae superioris muscle are fairly constant across extant mammals and have been moreover well described by numerous previous authors (see, e.g., review by Haugen 2002).

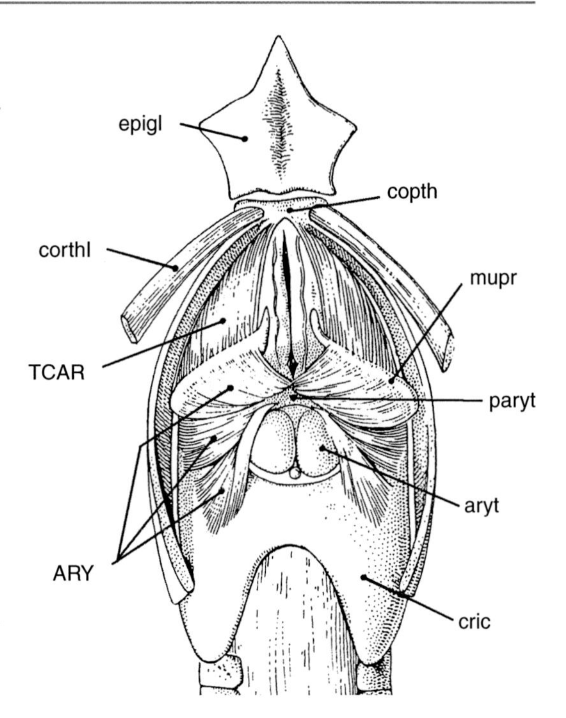

Fig. 11.4 *Ornithorhynchus anatinus* (Mammalia, Monotremata): dorsal view of the laryngeal musculature; the cricoarytenoideus posterior is not shown (modified from Saban 1971; the nomenclature of the structures illustrated basically follows that used in the present work; anterior is to the top). *ARY* arytenoideus, *aryt* arytenoid cartilage, *copth* copula thyroidea, *corthI* cornua thyroidea I, *cric* cricoid cartilage, *epigl* epiglottis, *mupr* muscular process, *paryt* proarytenoid cartilage, *TCAR* thyrocricoarytenoideus

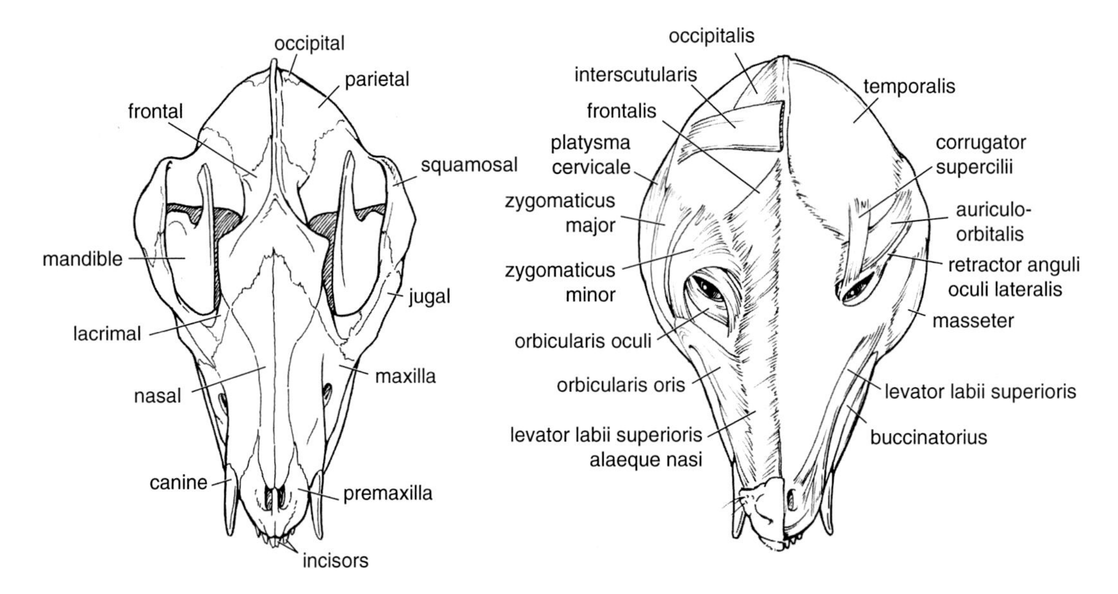

Fig. 11.5 *Didelphis virginiana* (Mammalia, Marsupialia): dorsofrontal view of head skeleton and muscles (schematic drawing by J. Molnar; modified from Diogo et al. 2016)

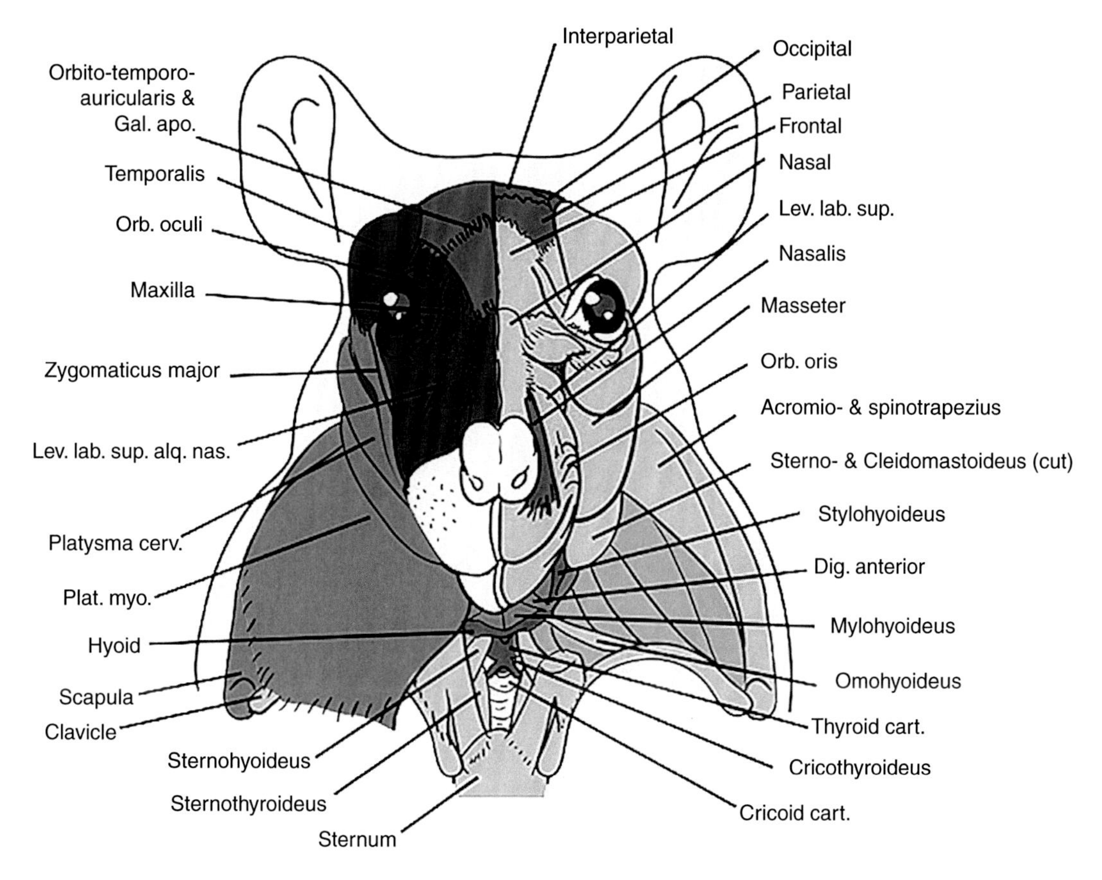

Fig. 11.6 *Mus musculus* (Mammalia, Rodential): frontal view of head muscles (schematic drawing by V. Powell). *cart.* cartilage, *cerv.* cervicalis, *dig.* digastricus, *gal. apo* galea aponeurotica, *lev. lab. sup.* levator labii superioris, *lev. lab. sup. alq. nas.* levator labii superioris alaeque nasi, *orb.* orbicularis, *plat. myo.* platysma myoides

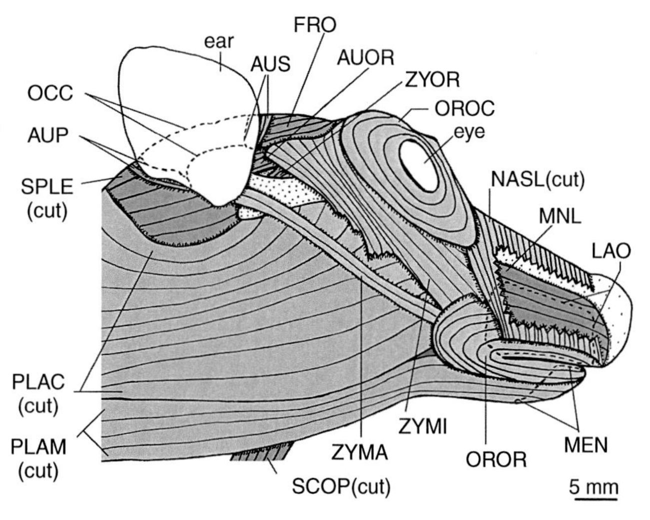

Fig. 11.7 *Cynocephalus volans* (Mammalia, Dermoptera): lateral view of the facial muscles, the splenius capitis is also shown; anteriorly, the nasolabialis was partially cut in order to show the maxillonasolabialis [anterior is to the right; muscles shown in darker gray are deeper than (medial to) those shown in lighter gray]. *AUOR* auriculo-orbitalis, *AUP* auricularis posterior, *AUS* auricularis superior, *FRO* frontalis, *LAO* levator anguli oris facialis, *MEN* mentalis, *MNL* maxillonasolabialis, *NASL* nasolabialis, *OCC* occipitalis, *OROC* orbicularis occuli, *OROR* orbicularis oris, *PLAC* platysma cervicale, *PLAM* platysma myoides, *SCOP* sphincter colli profundus, *SPLE* splenius capitis, *ZYMA* zygomaticus major, *ZYMI* zygomaticus minor, *ZYOR* zygomatico-orbitalis

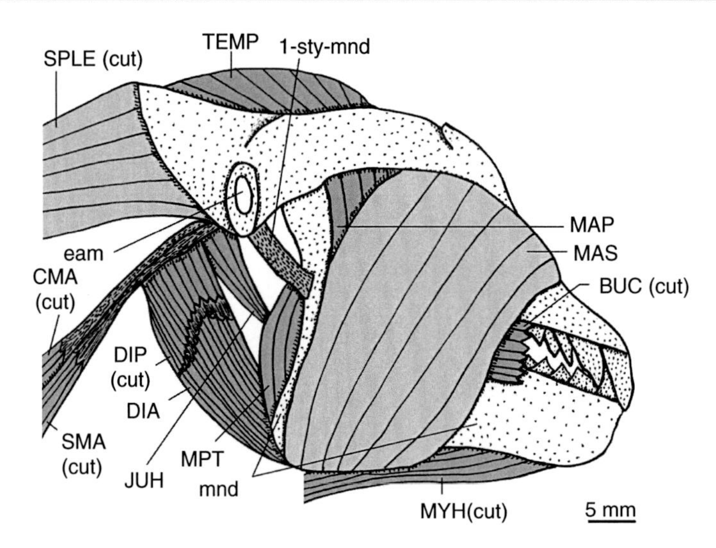

Fig. 11.8 *Cynocephalus volans* (Mammalia, Dermoptera): postero-ventro-lateral view [anterior is to the right; muscles shown in darker gray are deeper than (medial to) those shown in lighter gray]. *BUC* buccinatorius; *CMA* cleidomastoideus; *DIA*, *DIP* digastricus anterior and digastricus posterior; *eam* external auditory meatus; *JUH* jugulohyoideus; *l-sty-mnd* stylomandibular ligament; *MAP*, *MAS* pars profunda and pars superficialis of masseter; *mnd* mandible; *MPT* pterygoideus medialis; *MYH* mylohyoideus; *SMA* sternomastoideus; *SPLE* splenius capitis; *TEMP* temporalis

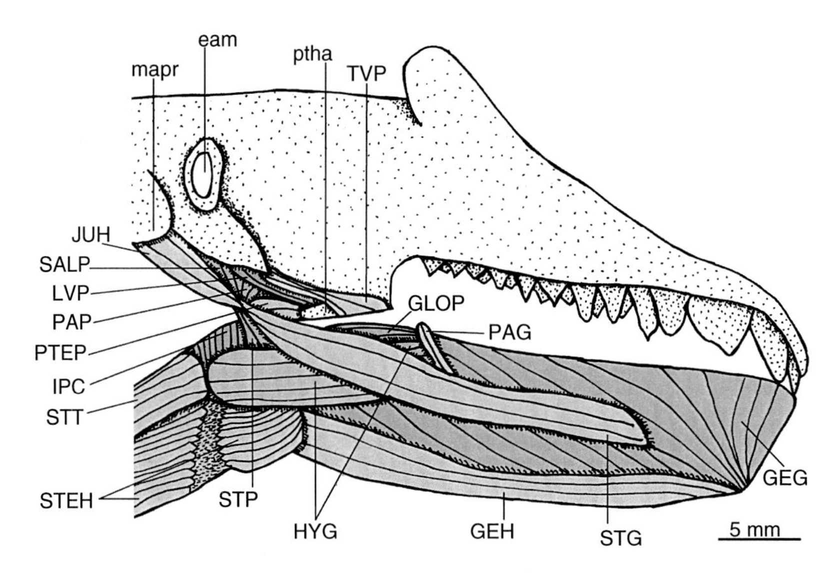

Fig. 11.9 *Cynocephalus volans* (Mammalia, Dermoptera): ventro-lateral view the mandible, zygomaticus arch, and part of the orbit were removed [anterior is to the right; muscles shown in darker gray are deeper than (medial to) those shown in lighter gray]. *eam* external auditory meatus, *GEG* genioglossus, *GEH* geniohyoideus, *GLOP* glossopharyngeus, *HYG* hyoglossus, *IPC* constrictor pharyngis inferior, *JUH* jugulohyoideus, *LVP* levator veli palatini, *mapr* mastoid process, *PAG* palatoglossus, *PAP* palatopharyngeus, *PTEP* pterygopharyngeus, *ptha* pterygoid hamulus, *SALP* salpingopharyngeus, *STEH* sternohyoideus, *STG* styloglossus, *STP* stylopharyngeus, *STT* sternothyroideus, *TVP* tensor veli palatini

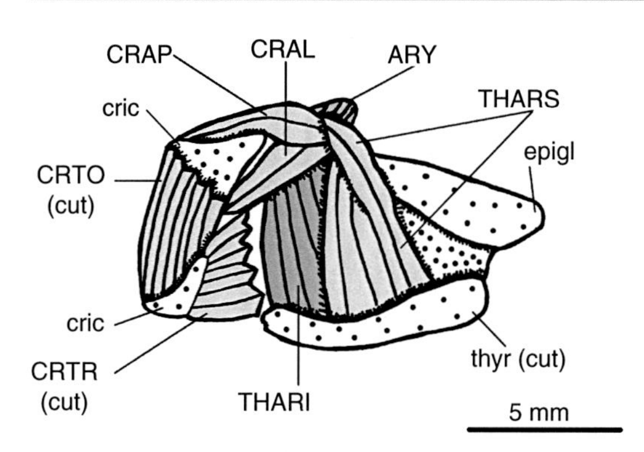

Fig. 11.10 *Cynocephalus volans* (Mammalia, Dermoptera): lateral view of the laryngeal muscles and of the pharyngeal muscle cricothyroideus; this latter muscle and the lateral surface of the thyroid cartilage were partially cut in order to show the deeper (more medial) muscles [anterior is to the left, dorsal to the top; muscles shown in darker gray are deeper than (medial to) those shown in lighter gray]. *ARY* arytenoideus; *CRAL* cricoarytenoideus lateralis; *CRAP* cricoarytenoideus posterior; *cric* cricoid cartilage; *CRTO*, *CRTR* pars obliqua and pars recta of cricothyroideus; *epigl* epiglottis; *THARI*, *THARS* pars intermedia and pars superioris of thyroarytenoideus; *thyr* thyroid cartilage

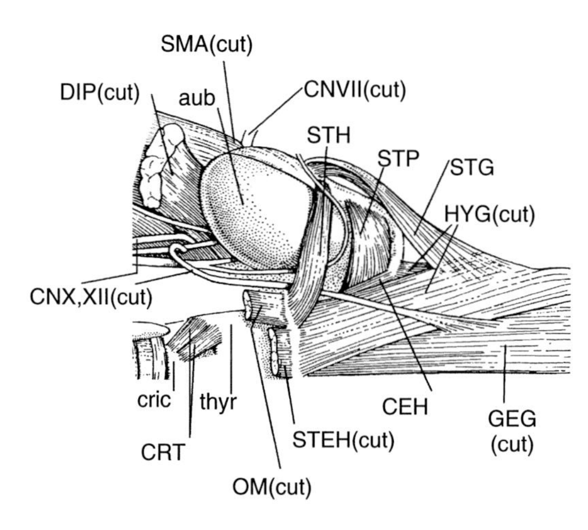

Fig. 11.11 *Ptilocercus lowii* (Mammalia, Scandentia): ventral view of the musculature of the hyoid region of the right side of the body; muscles such as the geniohyoideus, sternothyroideus, and thyrohyoideus are not shown (modified from Saban 1968; the nomenclature of the structures illustrated basically follows that used in the present work; anterior is to the right). *aub* auditory bulla; *CEH* ceratohyoideus; *CNVII, X, XII* cranial nerves VII, X, and XII; *cric* cricoid cartilage; *CRT* cricothyroideus; *DIP* digastricus posterior; *GEG* genioglossus; *HYG* hyoglossus; *OM* omohyoideus; *SMA* sternomastoideus; *STEH* sternohyoideus; *STG* styloglossus; *STH* stylohyoideus; *STP* stylopharyngeus; *thyr* thyroid cartilage

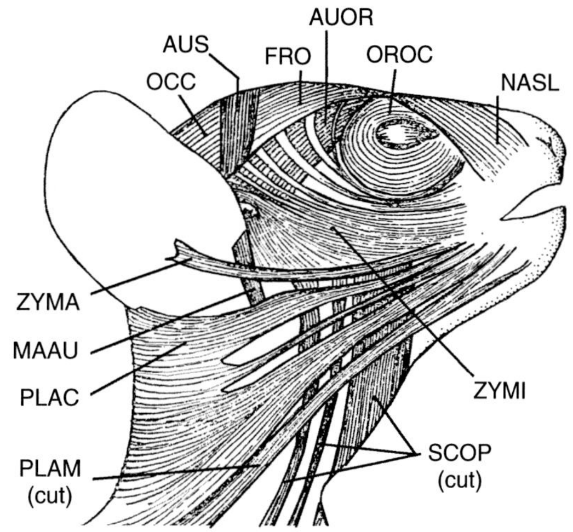

Fig. 11.12 *Lepilemur* sp. (Mammalia, Primates): lateral view of the facial musculature; muscles such as the buccinatorius, orbicularis oris, and mentalis are not shown (modified from Jouffroy and Saban 1971; the nomenclature of the structures illustrated basically follows that used in the present work; anterior is to the right). *AUOR* auriculo-orbitalis, *AUS* auricularis superior, *FRO* frontalis, *MAAU* mandibulo-auricularis, *NASL* nasolabialis, *OCC* occipitalis, *OROC* orbicularis oculi, *PLAC, PLAM* platysma cervicale and platysma myoides, *SCOP* sphincter colli profundus, *ZYMA*, *ZYMI* zygomaticus major and zygomaticus minor

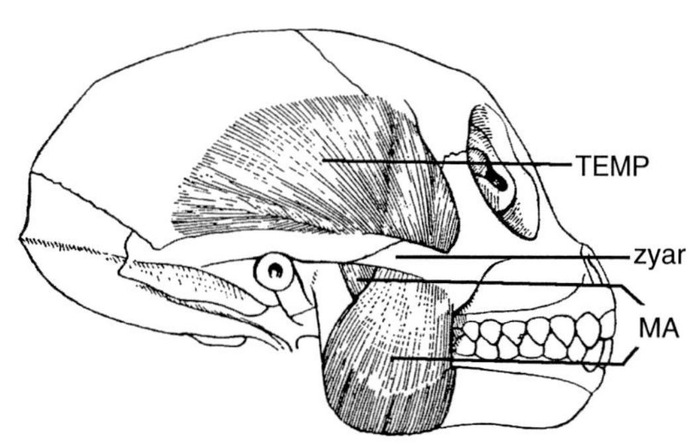

Fig. 11.13 *Macaca mulatta* (Mammalia, Primates): lateral view of the masseter and temporalis (modified from Saban 1968; the nomenclature of the structures illustrated basically follows that used in the present work; anterior is to the right). *MA* masseter, *TEMP* temporalis, *zyar* zygomatic arch

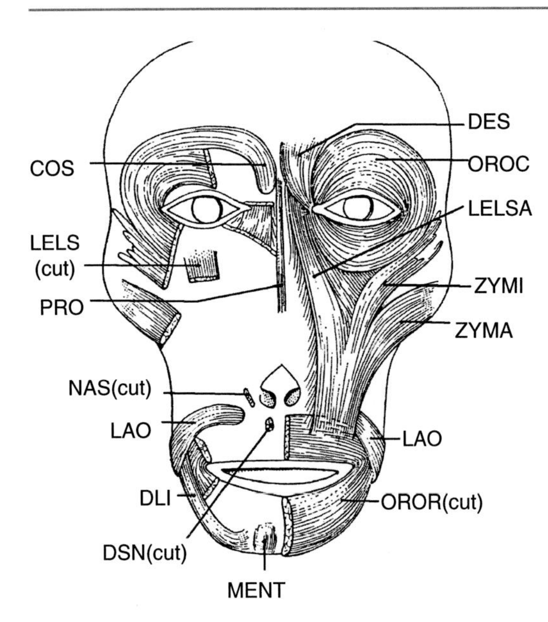

Fig. 11.14 *Macaca cyclopis* (Mammalia, Primates): anterior view of the facial musculature; muscles such as the buccinatorius, platysma, frontalis, and occipitalis are not shown; on the left side, the depressor supercilii was removed, and the orbicularis oculi, zygomaticus minor, zygomaticus major, levator labii superioris, and levator labii superioris alaeque nasi were partially cut (modified from Jouffroy and Saban 1971; the nomenclature of the structures illustrated basically follows that used in the present work). *COS* corrugator supercilii, *DES* depressor supercilii, *DLI* depressor labii inferioris, *DSN* depressor septi nasi, *LAO* levator anguli oris facialis, *LELS* levator labii superioris, *LELSA* levator labii superioris alaeque nasi, *MENT* mentalis, *NAS* nasalis, *OROC* orbicularis oculi, *OROR* orbicularis oris, *PRO* procerus, *ZYMA, ZYMI* zygomaticus major and zygomaticus minor

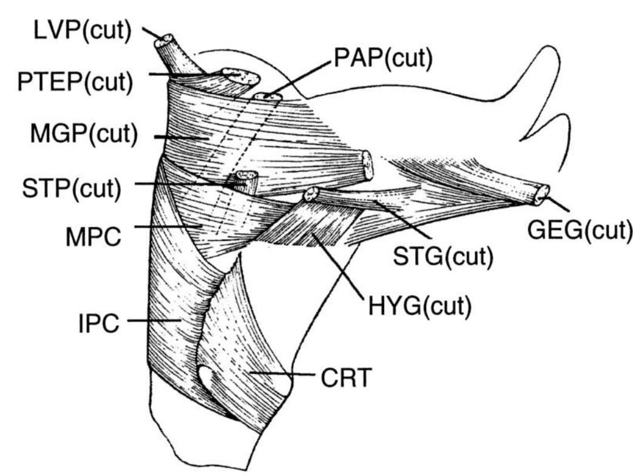

Fig. 11.15 *Hylobates hoolock* (Mammalia, Primates): lateral view of the pharyngeal musculature (modified from Saban 1968; the nomenclature of the structures illustrated basically follows that used in the present work; anterior is to the top; dorsal is to the left: see text). *CRT* cricothyroideus, *GEG* genioglossus, *HYG* hyoglossus, *IPC* constrictor pharyngis inferior, *LVP* levator veli palatini, *MGP* mylo-glossopharyngeus, *MPC* constrictor pharyngis medius, *PAP* palatopharyngeus, *PTEP* pterygopharyngeus, *STG* styloglossus, *STP* stylopharyngeus

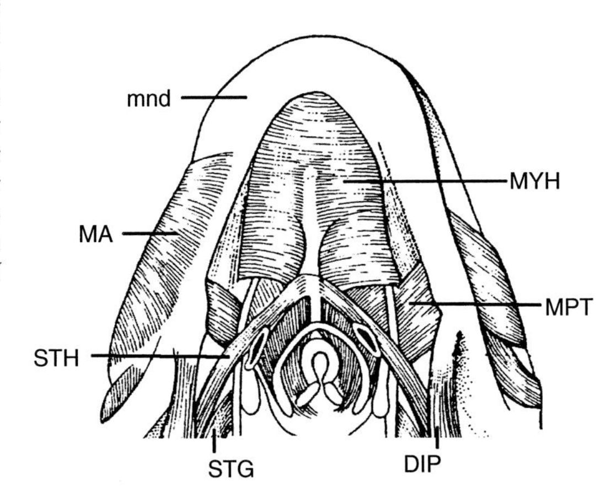

Fig. 11.16 *Pongo pygmaeus* (Mammalia, Primates): ventral view of the head musculature; on the right side the superficial portion of the masseter was removed (modified from Saban 1968; the nomenclature of the structures illustrated basically follows that used in the present work; anterior is to the top). *DIP* digastricus posterior, *mnd* mandible, *MA* masseter, *MPT* pterygoideus medialis, *MYH* mylohyoideus, *STG* styloglossus, *STH* stylohyoideus

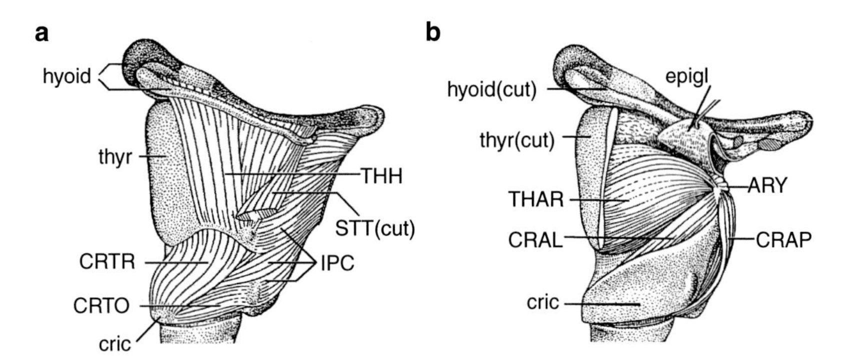

Fig. 11.17 *Pan troglodytes* (Mammalia, Primates): (**a**) lateral view of the laryngeal musculature; (**b**) same view, but the thyrohyoideus, sternothyroideus, constrictor pharyngis inferior, and cricothyroideus were removed, and the lateral portions of the thyroid cartilage and hyoid bone were partially cut (modified from Saban 1968; the nomenclature of the structures illustrated basically follows that used in the present work; anterior is to the top; dorsal is to the right: see text). *ARY* arytenoideus; *CRAL* cricoarytenoideus lateralis; *CRAP* cricoarytenoideus posterior; *cric* cricoid cartilage; *CRTO*, *CRTR* pars obliqua and pars recta of cricothyroideus; *epigl* epiglottis; *IPC* constrictor pharyngis inferior; *STT* sternothyroideus; *THAR* thyroarytenoideus; *THH* thyrohyoideus; *thyr* thyroid cartilage

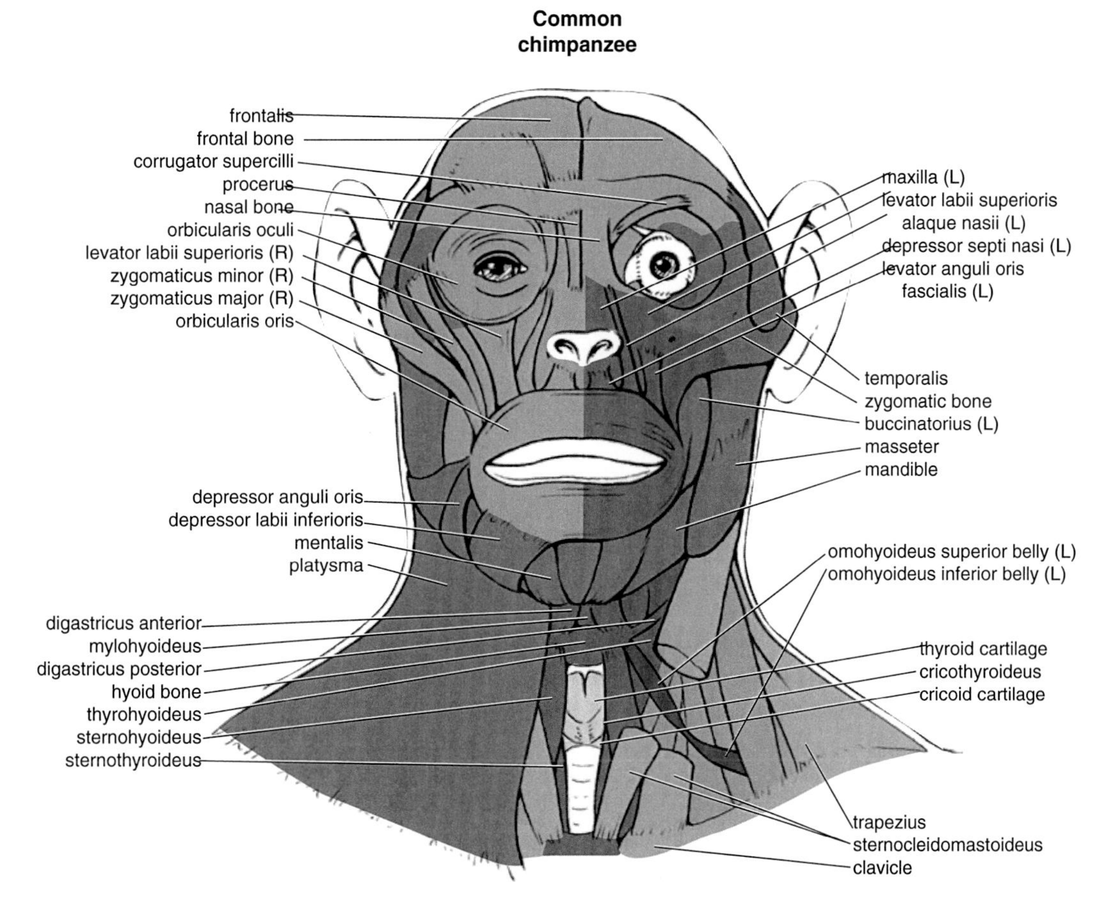

Fig. 11.18 *Pan troglodytes* (Mammalia, Primates): frontal view of the head muscles, with the more superficial muscles removed on the left side of the head to show the deeper muscles (schematic drawing by J. Molnar; modified from Diogo et al. 2017, with permission)

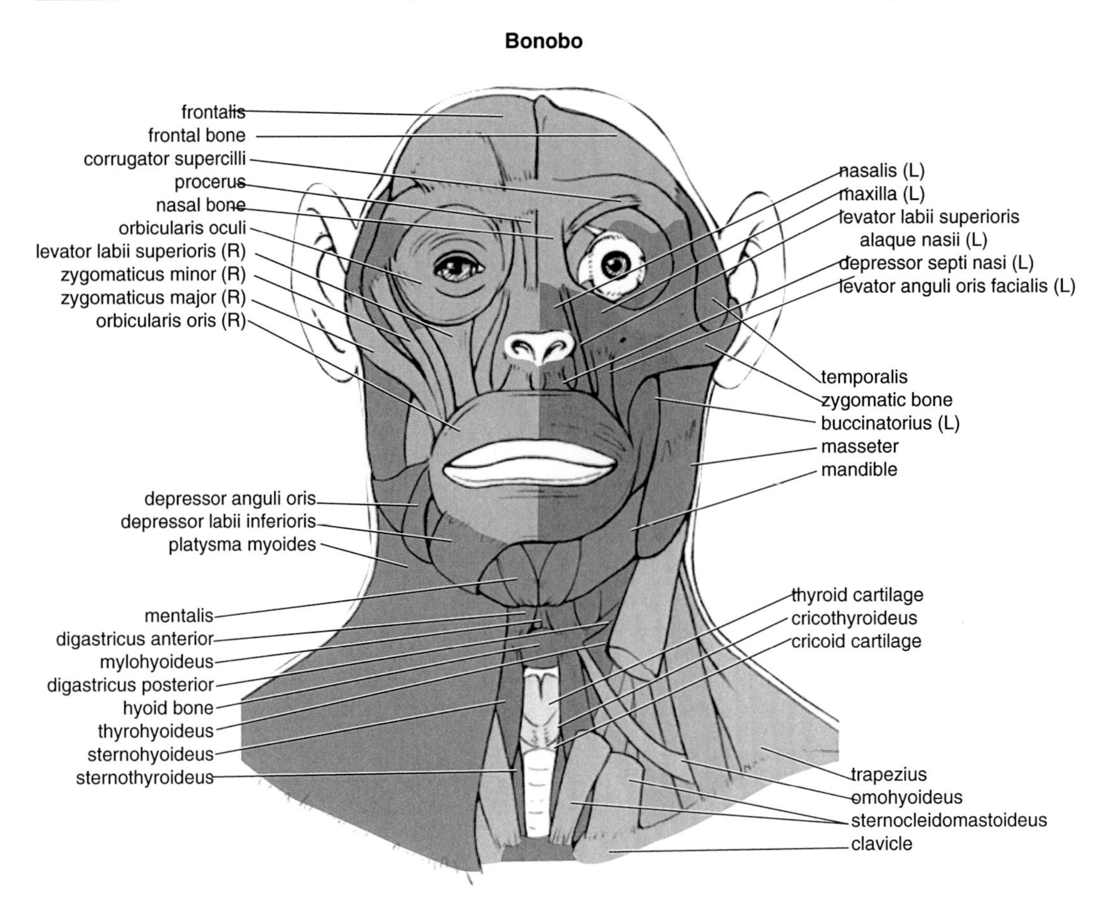

Fig. 11.19 *Pan paniscus* (Mammalia, Primates): frontal view of the head muscles, with the more superficial muscles removed on the left side of the head to show the deeper muscles (schematic drawing by J. Molnar; modified from Diogo et al. 2017, with permission)

11.2 Origin and Evolution of the Mammalian Mandibular Muscles (Table 11.1)

The plesiomorphic condition for **sarcopterygians**, i.e., the clade that includes sarcopterygian fishes and all tetrapods, is that two ventral **mandibular muscles—intermandibularis anterior** and **intermandibularis posterior**—connect the hemimandibles. The **mylohyoideus** and **digastricus anterior** of mammals (Figs. 11.6, 11.8, 11.16, 11.18, 11.19, and 11.20) correspond to the intermandibularis posterior of other sarcopterygians (e.g., Bryant 1945; Jarvik 1963; Saban 1971; Diogo and Abdala 2010). Contrary to the condition in monotremes (Fig. 11.3), in most marsupials and placentals, including modern humans, the digastricus anterior and **digastricus posterior** (a dorsomedial hyoid muscle; see below) form a compound structure ("digastricus") that is often related to the depression of the mandible. According to Edgeworth (1935), various tetrapod groups have independently acquired different mechanisms for depressing the mandible (i.e., to open the mouth) that use muscles other than the hypobranchial ones: amphibians and reptiles usually have a **depressor mandibulae** (which is a modified dorsomedial hyoid muscle; see Table 11.2), monotremes have a **detrahens mandibulae** (which is a new division of the adductor mandibulae: Table 11.1; Fig. 11.3), and marsu-

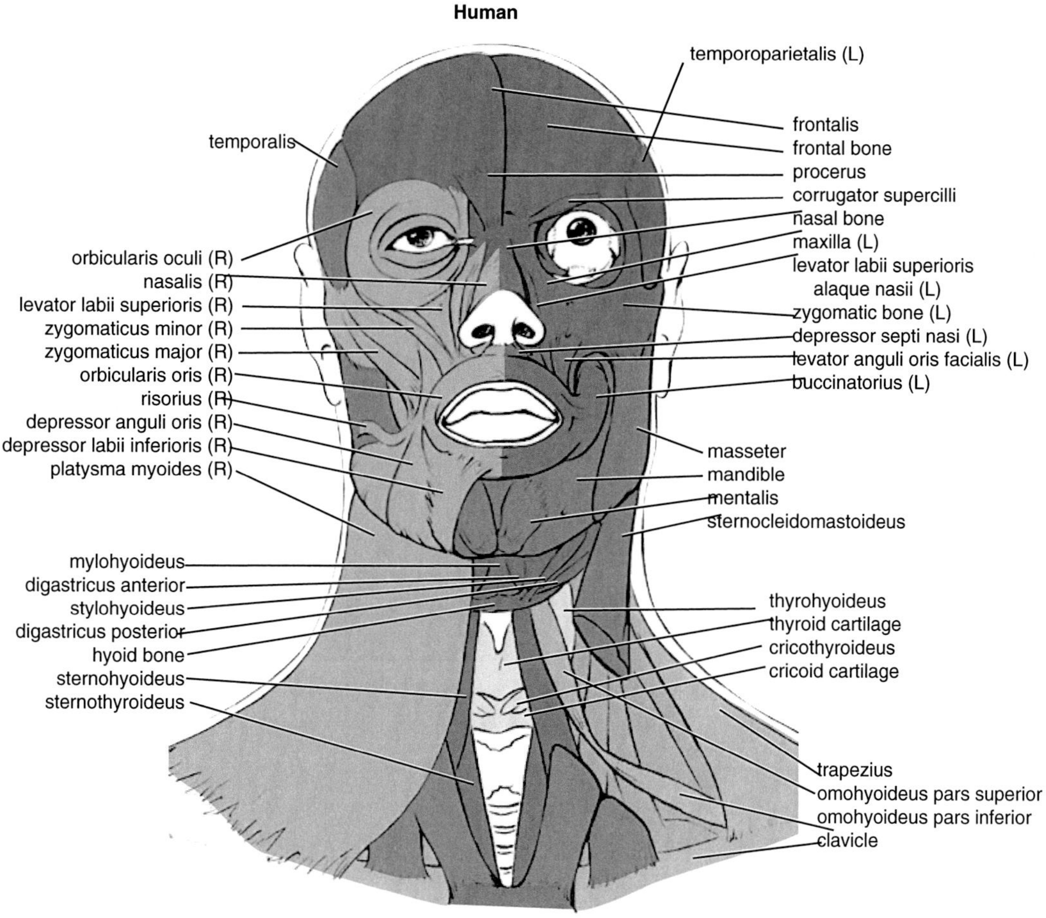

Fig. 11.20 *Homo sapiens* (Mammalia, Primates): frontal view of the head muscles, with the more superficial muscles removed on the left side of the head to show the deeper muscles (schematic drawing by J. Molnar; modified from Diogo et al. 2017, with permission)

pial and placental mammals usually have the "digastricus" (see just above).

The plesiomorphic condition for the sarcopterygian adductor mandibular muscles is seemingly that in which there is an **adductor mandibulae** A2, an adductor mandibulae Aω, an adductor mandibulae A3′, and possibly an adductor mandibulae A3″ (Diogo and Abdala 2010). The adductor mandibulae Aω was not present as an independent muscle in any of the mammals we dissected, and to our knowledge it has also not been found in any extant mammals described in the literature. The **masseter**, **temporalis**, **pterygoideus lateralis**, and **detrahens mandibulae** of monotremes (Fig. 11.3) and the masseter, temporalis, and pterygoideus lateralis of other extant mammals (Figs. 11.5, 11.6, 11.8, 11.13, 11.18, 11.19, and 11.20) apparently correspond to the A2 of reptiles such as the lizard *Timon* (Table 11.1). However, it should be noted that although the mammalian temporalis seemingly corresponds to a part of the A2 of other tetrapods, it might also include part of other adductor mandibulae structures such as the **pseudotemporalis** (see, e.g., Barghusen 1968). In two previous papers (Diogo et al. 2008a, b), we stated that the **tensor tympani** and **tensor veli palatini** of mammals were probably derived from the adductor mandibulae A2-PVM, as proposed by authors such as Edgeworth (1935) and Saban (1971) but that they

Table 11.1 Scheme illustrating the authors' hypotheses regarding the homologies of the mandibular muscles of adults of representative sarcopterygian taxa and the probable condition for the last common ancestor (LCA) of marsupials + placentals

	Latimeria (7 m.)	*Lepidosiren* (5 m.)	*Ambystoma* (6 m.)	*Timon* (11 m.)	*Ornithorhynchus* (9 m.)	*LCA marsupials + placentals* (9 m.)	*Didelphis* (8 m.)	*Rattus* (9 m.)	*Tupaia* (9 m.)	*Cynocephalus* (8 m.)	*Homo* (8 m.)
VENTRAL	Interm. posterior	Interm.	Interm. posterior	Interm. posterior	Mylohyoideus	Mylohyoideus	Mylohyoideus	Mylohyoideus	Mylohyoideus	Mylohyoideus	Mylohyoideus
	---	---	---	---	Dig. anterior	Dig. anterior	Dig. anterior	Dig. anterior	Dig. anterior	Dig. anterior	Dig. anterior
	Interm. anterior	---	Interm. anterior	Interm. anterior	---	Interm. anterior	---	Interm. anterior	Interm. anterior	---	---
ADDUCTOR MANDIBULAE	Ad. man. A2	Ad. man. A2	Ad. man. A2	Ad. man. A2	Masseter	Masseter	Masseter	Masseter	Masseter	Masseter	Masseter
	---	---	---	---	Detrahens man.	---	---	---	---	---	---
	---	---	---	---	Temporalis	Temporalis	Temporalis	Temporalis	Temporalis	Temporalis	Temporalis
	---	---	---	---	Pterygoideus lat.	Pterygoideus lat.	Pterygoideus lat.	Pterygoideus lat.	Pterygoideus lat.	Pterygoideus lat.	Pterygoideus lat.
	---	Ad. man. A2-PVM	Ad. man. A2-PVM	Ad. man. A2 -PVM	---	---	---	---	---	---	---
	---	Retractor ang. oris	---	---	---	---	---	---	---	---	---
	---	---	---	Le. anguli oris mandib.	---	---	---	---	---	---	---
	Ad. mand. A3'	Ad. man. A3'	Pseudotemporalis	Pseudotemporalis	---	---	---	---	---	---	---
	Ad. mand. A3''	---	---	---	---	---	---	---	---	---	---
	---		**--- [psoa.]**	Pterygomandibularis	---	---	---	---	---	---	---
	---		---	---	Pterygoideus med.	Pterygoideus med.	Pterygoideus med.	Pterygoideus med.	Pterygoideus med.	Pterygoideus med.	Pterygoideus med.
	---	---	---	---	Tensor tympani	Tensor tympani	Tensor tympani	Tensor tympani	Tensor tympani	Tensor tympani	Tensor tympani
	---		---	---	Tensor veli palatini	Tensor veli palatini	Tensor veli palatini	Tensor veli palatini	Tensor veli palatini	Tensor veli palatini	Tensor veli palatini
	Ad. mand. Aω	---	---	Ad. man. Aω	---	---	---	---	---	---	---
DORSAL	Le. arcus palatini	---	---	Le. pterygoidei	---	---	---	---	---	---	---
	---	---	---	Protractor pterygoidei	---	---	---	---	---	---	---
	---	---	Le. bulbi	Le. bulbi	---	---	---	---	---	---	---

Data from evidence provided by our own dissections and comparisons and by a review of the literature; muscles and other terms in bold and red highlight cases in which the homology and/or evolutionary hypotheses of Diogo and Abdala (2010) were updated, in the present chapter, in face of the new data obtained in our dissections, and in the review of the literature. The black arrows indicate the hypotheses that are most strongly supported by the evidence available; the gray arrows indicate alternative hypotheses that are supported by some of the data, but overall they are not as strongly supported by the evidence available as are the hypotheses indicated by black arrows. VENTRAL, DORSAL means ventral musculature and dorsal constrictor musculature of the mandibular arch *sensu* Edgeworth (1935). *ad.* adductor, *dig.* digastricus, *interm.* intermandibularis, *lat.* lateralis, *le.* levator, *m.* muscles, *man.* mandibulae, *mandib.* mandibularis, *psoa.* present in some amphibians, *psor.* present in some other reptiles

Table 11.2 Scheme illustrating the authors' hypotheses regarding the homologies of the hyoid muscles (see caption of Table 11.1)

	Latimeria (5 m.)	*Lepidosiren* (3 m.)	*Ambystoma* (4 m.)	*Timon* (3 m.)	*Ornithorhynchus* (12 m. - not ex. ear/int. snout*)	*LCA marsupials + placentals* (27 m. - not ex. ear/int. snout*)	*Didelphis* (25 m. - not ex. ear/int. snout*)	*Rattus* (25 m. -not ex. ear/int. snout*)	*Tupaia* (26 m. -not ex. ear/int. snout*)	*Cynocephalus* (23 m. -not ex. ear/int. snout*)	*Homo* (27 m.-not ex. ear*)
DORSO-MEDIAL HYOID MUS.	'Ad.hyo. Y'	---	---	---	---	---	---	---	---	---	---
	Ad. arcus palatini	---	---	---	---	---	---	---	---	---	---
	---	De. man.	De. man.	De. man.	Styloideus	Stylohyoideus	Stylohyoideus	Stylohyoideus	Stylohyoideus	---	Stylohyoideus
	---	---	---	---	---	---	---	---	Jugulohyoideus	Jugulohyoideus	---
	---	---	---	---	---	Digastricus posterior	Digastricus posterior	Digastricus posterior	Digastricus posterior	Digastricus posterior	Digastricus posterior
	Latimeria 's 'le.oper'	---	---	---	---	---	---	---	---	---	---
	Ad. operculi	---	---	---	---	---	---	---	---	---	---
	---	---	Bra.+Cerat.	---	---	---	---	---	---	---	---
	---	Le. hyoideus	Le. hyoideus	---(Le.hyoideus poar.)	Stapedius	Stapedius	Stapedius	Stapedius	Stapedius	Stapedius	Stapedius
	---	---	---	Cervicomandibularis	Platysma cervicale	Platysma cervicale	Platysma cervicale	Platysma cervicale	Platysma cervicale	Platysma cervicale	---
	---	---	---	---	Platysma myoides	Platysma myoides	Platysma myoides	Platysma myoides	Platysma myoides	Platysma myoides	Platysma myoides
	---	---	---	---	---	Occipitalis	Occipitalis	Occipitalis	Occipitalis	Occipitalis	Occipitalis
	---	---	---	---	---	Auricularis posterior	Auricularis posterior	Auricularis posterior	Aur. posterior	Auricularis posterior	Auricularis posterior
	---	---	---	---	Ex. ear mus. *	Ex. ear mus. *	Ex. ear mus. *	Ex. ear mus. *	Ex. ear mus. *	Ex. ear mus. *	Ex. ear mus. *
	---	---	---	---	---	Mandibulo -auricularis	Mandibulo -auricularis	Mandibulo -auricularis	Mandibulo -auricularis	---	---
	---	---	---	---	---	---	---	---	---	---	Risorius
VENTRAL HYOID MUS.	Interhyoideus	Interhyoideus	Interhyoideus	Interhyoideus	Interhyoideus prof.	---	---	---	---	---	---
	---	---	---	---	Sphincter colli supe.	Sphincter colli supe.	---	Sphincter colli supe.	Sphincter colli supe.	---	---
	---	---	---	---	---(colli prof. in Echidna)	Sphincter colli prof.	Sphincter colli prof.	Sphincter colli prof.	Sphincter colli prof.	Sphincter colli prof.	---
	---	---	---	---	---	---	---	Sternofacialis	---	---	---
	---	---	---	---	Cervicalis tra.	---	---	---	---	---	---
	---	---	---	---	---	Interscutularis	Interscutularis	Interscutularis	---	---	---
	---	---	---	---	---	Zygomaticus major	Zygomaticus major	Zygomaticus major	Zygomaticus major	Zygomaticus major	Zygomaticus major
	---	---	---	---	---	Zygomaticus minor	Zygomaticus minor	Zygomaticus minor	Zygomaticus minor	Zygomaticus minor	Zygomaticus minor
	---	---	---	---	---	Frontalis	Frontalis	Orbito -temporo -aur.	Frontalis	Frontalis	Frontalis
	---	---	---	---	---	Auriculo-orbitalis	Auriculo-orbitalis	---	Auriculo-orbitalis	Auriculo-orbitalis	Temporoparietalis
	---	---	---	---	---	---	---	---	---	---	Auricularis anterior
	---	---	---	---	---	Auricularis superior	Auricularis superior	---	Auricularis superior	Auricularis superior	Auricularis superior
	---	---	---	---	Orbicularis oculi	Orbicularis oculi	Orbicularis oculi	Orbicularis oculi	Orbicularis oculi	Orbicularis oculi	Orbicularis oculi
	---	---	---	---	---	---	---	---	Zygomatico-orbicularis	Zygomatico-orbicularis	---
	---	---	---	---	---	---	---	---	---	---	De. supercilii
	---	---	---	---	---	Corrugator supercilii	Corrugator supercilii	---	Corrugator supercilii	Corrugator supercilii	Corrugator supercilii
	---	---	---	---	---	Re. anguli oculi lateralis	Re. anguli oculi lateralis	--- (poap.)	---	---	---
	---	---	---	---	Naso -labialis	**Le. labii sup. al. nasi**	Le. labii sup. al. nasi	Le. labii sup. al. nasi	Le. labii sup. al. nasi	Le. labii sup. al. nasi	Le. labii sup. al. nasi
	---	---	---	---	---	---	---	**Procerus?**	---	---	Procerus
	---	---	---	---	Buccinatorius	Buccinatorius	Buccinatorius	Buccinatorius	Buccinatorius	Buccinatorius	Buccinatorius
	---	---	---	---	---	Dilatator nasi	Dilatator nasi	Dilatator nasi	---	---	---
	---	---	---	---	---	**Le. labii sup.**	Le. labii sup.	Le. labii sup.	Le. labii sup.	Le. labii sup.	Le. labii sup.
	---	---	---	---	---	**Nasalis**	Nasalis	Nasalis	Nasalis	Nasalis	Nasalis
	---	---	---	---	---	---	---	**Depressor septi nasi**	--- (really absent?)	--- (really absent?)	De. septi nasi
	---	---	---	---	---	**In. snout mus.***	In. snout mus.*	In. snout mus.*	In. snout mus.*	In. snout mus.*	---
	---	---	---	---	---	Le. anguli oris facialis	Le. anguli oris facialis	Le. anguli oris facialis	Le. anguli oris facialis	Le. anguli oris facialis	Le. anguli oris facialis
	---	---	---	---	Orbicularis oris	Orbicularis oris	Orbicularis oris	Orbicularis oris	Orbicularis oris	Orbicularis oris	Orbicularis oris
	---	---	---	---	---	---	---	---	---	---	Depressor labii inf.
	---	---	---	---	---	---	---	---	---	---	De. anguli oris
	---	---	----	---	Mentalis	Mentalis	--- (poam.)	---	Mentalis	Mentalis	Mentalis

ad. adductor, *al.* alaeque, *aur.* auricularis, *bra. + cerat.* branchiohyoideus + ceratomandibularis, *de.* depressor, *ex.* extrinsic, *hyo.* hyomandibulae, *inf.* inferioris, *int.* intrinsic, *le.* levator, *m.* muscles, *man.* depressor mandibulae, *operc.* operculi, *poam.* present in other adult marsupials, *poap.* present in some other adult placentals, *poar.* present in other adult reptiles, *prof.* profundus, *sup.* superioris, *supe.* superficialis, *tra.* transversus

Table 11.3 Scheme illustrating the authors' hypotheses regarding the evolution and homologies of the branchial, pharyngeal, and laryngeal muscles of adults of representative sarcopterygian taxa (see caption of Table 11.1)

		Latimeria (4 m.-not st.*)	*Lepidosiren* (3 mus.-not st.*)	*Ambystoma* (4 m.-not str.*)	*Timon* (6 m. total)	*Ornithorhynchus* (14 m. total)	*LCA marsupials + placentals* (19 m.total)	*Didelphis* (17 m. total)	*Rattus* (21 m. total)	*Tupaia* (17 m. total)	*Cynocephalus* (19 m. total)	*Homo* (17 m. total)
'TRUE' BRANCHIAL MUS.	STRICTO*	Fu.m.br.ap.*	Fu.m.br.ap.*	Fu.m.br.ap..	--- (ab. as a group)	--- (ab. as group)	--- (ab. as a group)	--- (ab. as a group)	--- (ab. as a group)	--- (ab. as a group)	--- (abs. as a group)	--- (abs. as a group)
		---	---	---	Hyobranchialis	Stylopharyngeus	Stylopharyngeus	Stylopharyngeus	Stylopharyngeus	Stylopharyngeus	Stylopharyngeus	Stylopharyngeus
		---	---	---	'Ceratohyoideus'	Ceratohyoideus	Ceratohyoideus	Ceratohyoideus	Ceratohyoideus	Ceratohyoideus	Ceratohyoideus	---
		---	---	---	---	Subarcualis rectus III	---	---	---	---	---	---
	OTHER	---	Pro. pectoralis	Pro. pectoralis	Trapezius	Acromiotrapezius	Acromiotrapezius	Acromiotrapezius	Acromiotrapezius	Trapezius	Acromiotrapezius	Trapezius
		---	---	---	---	Spinotrapezius	Spinotrapezius	Spinotrapezius	Spinotrapezius	---	Spinotrapezius	---
		---	---	---	---	Dorsocutaneous	Dorsocutaneous	--- (psom.)	--- (psop.)	---	---	---
		---	---	---	---	---	Cleido -occipitalis	Cleido -occipitalis	Cleido -occipitalis	Cleido -occipitalis	---	---
		---	---	---	Sternocleidomastoideus	Cleidomastoideus	Cleidomastoideus	Cleidomastoideus	Cleidomastoideus	Cleidomastoideus	Cleidomastoideus	Sternocleidomastoideus
		---	---	---	---	Sternomastoideus	Sternomastoideus	Sternomastoideus	Sternomastoideus	Sternomastoideus	Sternomastoideus	---
PHARYNGEAL MUS.		---	---	---	---	Co. pharyngis	Co.pharyngis medius	Co.pharyngis medius	Co.pharyngis medius	Co. pharyngis medius	Co. pharyngis medius	Co. pharyngis medius
		----	---	---	---	---	Co.pharyngis inferior	Co.pharyngis inferior	Co.pharyngis inferior	Co. pharyngis inferior	Co. pharyngis inferior	Co. pharyngis inferior
		----	---	---	---	Cricothyroideus	Cricothyroideus	---(lost in marsupials)	Cricothyroideus	Cricothyroideus	Cricothyroideus	Cricothyroideus
		---	---	---	---	---	---	---	Co. pharyngis superior	Co. pharyngis superior	Co. pharyngis sup.	Co. pharyngis superior
		---	---	---	---	---	**Palatoglossus**	Palatoglossus	---	---	Palatoglossus	Palatoglossus
		----	---	---	---	---	**Pterygopharyngeus**	Pterygopharyngeus	Pterygopharyngeus	---	Pterygopharyngeus	---
		----	---	---	---	Palatopharyngeus	Palatopharyngeus	Palatopharyngeus	Palatopharyngeus	Palatopharyngeus	Palatopharyngeus	Palatopharyngeus
		---	---	---	---	---	**Musculus uvulae**	Musculus uvulae	**Musculus uvulae**	**--- ? (really absent?)**	**--- ? (really absent?)**	Musculus uvulae
		----	---	---	---	---	---	---	Le. veli palatini	Le. veli palatini	Le. veli palatini	Le. veli palatini
		----	---	---	---	---	---	---	Salpingopharyngeus	Salpingopharyngeus	Salpingopharyngeus	Salpingopharyngeus
LARYNGEAL MUS.		---? (see DA)	Co. laryngis	Co. laryngis	Co. laryngis	---	---	---	---	---	---	---
		---	---	Laryngeus	---	Thyrocricoarytenoideus	Thyroarytenoideus	Thyroarytenoideus	Thyroarytenoideus	Thyroarytenoideus	Thyroarytenoideus	Thyroarytenoideus (+vocalis)
		---	---	---	---	---	Cricoarytenoideus lat.	Cricoarytenoideus lat.	Cricoarytenoideus lat.	Cricoarytenoideus lat.	Cricoarytenoideus lat.	Cricoarytenoideus lat.
		---	---	---	---	Arytenoideus	Arytenoideus	Arytenoideus	Arytenoideus	Arytenoideus	Arytenoideus	Arytenoideus transversus
		---	---	---	---	---	---	---	---	---	---	Arytenoideus obliquus
		---	---	---	---	---	---	---	**Cricoarytenoideus alaris**	---	---	---
		---? (see DA)	Dilator laryngis	Dilator laryngis	Dilator laryngis	Cricoarytenoideus post.	Cricoarytenoideus post.	Cricoarytenoideus post.	Cricoarytenoideus post.	Cricoarytenoideus post.	Cricoarytenoideus post.	Cricoarytenoideus post.

ab. absent; *ap.* apparatus; *br.* branchial; *co.* constrictor; *fu.* functional; *le.* levator; *m.* muscles; *post.* posterior; *pro.* protractor; *psom.* present some other marsupials; *psop.* present some other placentals; see Diogo and Abdala (2010; *st. sensu stricto*; *sup.* superior

Table 11.4 Scheme illustrating the authors' hypotheses regarding the evolution and homologies of the hypobranchial muscles of adults of representative sarcopterygian taxa (see caption of Table 11.1)

	Latimeria (2 m.)	*Lepidosiren* (2 m.)	*Ambystoma* (6 m. - not in.to.*)	*Timon* (5 m. - not in. to.*)	*Ornithorhynchus* (6 m. - not in. to.*)	*LCA marsupials + placentals* (8 m. - not in. to.*)	*Didelphis* (8 m. - notin. to.*)	*Rattus* (8 m. - not in. to.*)	*Tupaia* (8 m. - not in. to.*)	*Cynocephalus* (6 m. - not in. to.*)	*Homo* (8 m. - not in. to.*)
'GENIOHYOIDEUS'	Coracomandibularis	Coracomandibularis	Geniohyoideus	Geniohyoideus	Geniohyoideus	Geniohyoideus	Geniohyoideus	Geniohyoideus	Geniohyoideus	Geniohyoideus	Geniohyoideus
	---	---	Genioglossus	Genioglossus	Genioglossus	Genioglossus	Genioglossus	Genioglossus	Genioglossus	Genioglossus	Genioglossus
	---	---	---	In. mus. tongue*	In. mus. tongue*	In. mus. tongue*	In. mus. tongue*	In. mus. tongue*	In. mus. tongue*	In. mus. tongue*	In. mus. tongue*
	---	---	Hyoglossus	Hyoglossus	Hyoglossus	Hyoglossus	Hyoglossus	Hyoglossus	Hyoglossus	Hyoglossus	Hyoglossus
	---	---	---	---	---	Styloglossus	Styloglossus	Styloglossus	Styloglossus	Styloglossus	Styloglossus
	---	---	Interradialis	---	---	---	---	---	---	---	---
'RECTUS CERVICIS'	Sternohyoideus	Sternohyoideus	Sternohyoideus	Sternohyoideus	Sternohyoideus	Sternohyoideus	Sternohyoideus	Sternohyoideus	Sternohyoideus	Sternohyoideus	Sternohyoideus
	---	---	Omohyoideus	Omohyoideus	Omohyoideus	Omohyoideus	Omohyoideus	Omohyoideus	Omohyoideus	---	Omohyoideus
	---	---	---	---	Sternothryroideus	Sternothryroideus	Sternothryroideus	Sternothryroideus	Sternothryroideus	Sternothryroideus	Sternothryroideus
	---	---	---	---	---	Thyrohyoideus	Thyrohyoideus	Thyrohyoideus	Thyrohyoideus	---	Thyrohyoideus

"GENIOHYOIDEUS", "RECTUS CERVICIS" mean "geniohyoideus" and "rectus cervicis" groups *sensu* Edgeworth (1935). *in. to.* intrinsic muscles of the tongue, *m.* muscles

could also have been derived from the **pterygomandibularis** instead (see, e.g., Diogo et al. 2008b). However, the dissections, comparisons, and review of the literature available on this subject that we have been doing after writing those papers indicate that the most likely hypothesis is that the tensor tympani and tensor veli palatini correspond to a part of, or are derived from, the pterygomandibularis, as is in fact accepted by most anatomists (e.g., Adams 1919; Brock 1938; Goodrich 1958; Barghusen 1986; Smith 1992, 1994; Witmer 1995) (Table 11.1).

This subject is related to the fascinating origin of parts of the **mammalian inner ear** from the jaws of their nonmammalian ancestors. Well-supported theories like those of Reichert (1937) and Gaupp (1912) demonstrated on comparative anatomical and developmental evidence that the middle ear bones (i.e., **malleus**, **incus**, **gonial**, and **tympanic ring**) of modern mammals evolved from repurposed accessory jawbones (i.e., **articular**, **quadrate**, **prearticular**, and **angular**, respectively) of their nonmammalian tetrapod ancestors, which are maintained in extant reptilian and avian relatives. Fossils show that the mammalian exaptations of ancestral bony anatomy detectable in *Morganucodon oehleri* were preceded by topological and mechanical alterations to preexisting masticatory and pharyngeal musculature that began within the Eucynodontia (Crompton 1963; Crompton and Parker 1978; Lautenschlager et al. 2017). The evolution of the auditory system thus relied not only on osteological reorganization but also on myological differentiation of the pterygomandibularis portion of the adductor mandibulae. In particular, the main arguments supporting the differentiation of the tensor tympani and tensor veli palatini from the pterygomandibularis have been clearly summarized in works such as Barghusen (1986), among others. Apart from the arguments summarized in such works, there is also developmental data supporting this hypothesis. For instance, in Smith's (1994) detailed work on the development of the craniofacial musculature of marsupials, she found that the pterygoideus medialis, the tensor tympani, and the tensor veli palatini of these mammals develop ontogenetically from the same medial anlage, which seems to correspond to the anlage that forms the pterygomandibularis + pseudotemporalis in reptiles (while the pterygoideus lateralis derives instead from the anlage that gives rise to the temporalis). Interestingly, in Fig. 3 of Smith's (1994) paper, there appears to be a thin, small muscle connecting the malleus and the incus, which could possibly be a "remnant of the PVM" (Peter Johnson, pers. comm.), but this latter hypothesis clearly needs to be investigated in much more detail. Actually, one of the main arguments that authors such as Saban (1971) provided in favor of a derivation of the mammalian tensor veli palatini and tensor tympani from the A2-PVM was that Edgeworth (1935) stated that his developmental work clearly showed that the two former muscles were derived ontogenetically from the "levator mandibulae posterior," which is the name that is often used in the literature to designate the A2-PVM *sensu* the present work (see Table 11.1). However, there is actually much confusion in the literature about the identity and homologies of the components of the "adductor mandibulae complex" of tetrapods and particularly of the structures that are often named "adductor mandibulae posterior" in different nonmammalian tetrapod clades. It is possible that in this specific case, Edgeworth (1935) used the name "levator mandibulae posterior" to designate the pterygomandibularis (and not the A2-PVM) *sensu* the present work, as suggested by authors such as Goodrich (1958). Goodrich (1958) stated that a correct interpretation of Edgeworth's data actually supports the idea that the tensor tympani and tensor veli palatini are derived from the pterygomandibularis because those data show that the mammalian tensor tympani, tensor veli palatini, and pterygoideus medialis derive from the same anlage, as found in more recent works (e.g., Smith 1994) (see above and Table 11.1). Rodríguez-Vázquez et al. (2016) recently showed that in early human development, the tensor tympani and tensor veli palatini are (1) tendinously connected to each other, forming a single "digastric" muscle complex and (2) are effectively also closely related to the pterygoideus medialis, as described in developmental works about other mammalian taxa.

The mammalian **pterygoideus medialis** (Fig. 11.8) may derive from the pseudotemporalis but may also derive from a part of the **pterygomandibularis** of nonmammalian tetrapods (Table 11.1). Furthermore, extant mammals lack any **dorsal mandibular muscles**, *sensu* Edgeworth (1935), which is uncommon among vertebrates (e.g., Saban 1968, 1971; Kardong 2002; this chapter). Interestingly, tetrapod mandibular muscles (derived from the "adductor mandibulae" plate *sensu* Edgeworth, such as the masseter, temporalis, pterygoideus medialis, and pterygoideus lateralis) of mice display engrailed immunoreactivity (Knight et al. 2008). Teleost fish (e.g., the zebrafish) also exhibit engrailed immunoreactivity that is only been detected in dorsal mandibular muscles (i.e., in the levator arcus palatini and dilatator operculi) *sensu* Edgeworth (Knight et al. 2008). This means that the muscles that arise from cells expressing the same gene in two different vertebrate taxa are not necessarily homologous among those taxa, thus supporting the idea that no single criterion (including the expression of genes such as engrailed) is enough to establish myological homologies (for more details on this subject, see Diogo and Abdala 2010).

11.3 Hyoid Muscles (Table 11.2)

Edgeworth (1935) and Huber (1930a, b, 1931) divided the **hyoid muscles** into two main groups: dorsomedial and ventral (Table 5.4). The plesiomorphic configuration for sarcopterygians as a whole is a single ventral hyoid muscle, the **interhyoideus**, and two dorsomedial hyoid muscles, the **adductor arcus palatini** and the **adductor operculi** (note that the "adductor hyomandibulae Y" and "**levator operculi**" of *Latimeria* are not homologues of the **adductor hyomandibulae** and levator operculi of actinopterygians such as teleosts: see Diogo and Abdala 2010). The **depressor mandibulae**, **levator hyoidei**, **branchiohyoideus**, and **cervicomandibularis** of extant dipnoans, amphibians, and reptiles seem to develop from the anlage that give rise to the adductor arcus palatini in other osteichthyans (Table 11.2). The **adductor operculi** is not present as an independent muscle in extant dipnoans, amphibians, and reptiles; at least in dipnoans, it is most likely fused with the ventral hyoid muscle interhyoideus (Table 11.2). This comparison with nonmammalian taxa is helpful to understand the origin of the mammalian muscles as well as to emphasize that the number of hyoid muscles found in extant mammals, and particularly in therians (placentals + marsupials), is much greater than that found in extant nonmammalian tetrapods (Table 11.2). Also, in nonmammalian vertebrates the hyoid muscles are mainly restricted to the region of the second branchial arch and occasionally to the mandibular and/or neck regions, whereas in extant mammals these muscles extend more anteriorly, covering much of the anterior region of the head (Figs. 11.2, 11.5, 11.6, 11.7, 11.12, 11.14, 11.18, 11.19, and 11.20).

Importantly, with the exception of the **styloideus**, **stylohyoideus**, **digastricus posterior**, **jugulohyoideus**, and **stapedius**, all the mammalian hyoid muscles listed in Table 11.2 are usually designated as **facial muscles** because they attach to freely movable skin and are associated with the display of **facial expressions** (Figs. 11.2, 11.5, 11.6, 11.7, 11.12, 11.14, 11.18, 11.19, 11.20, and 11.21) (e.g., Ruge 1885, 1897, 1910; Boas and Paulli 1908; Lightoller 1928a, b, 1939, 1940a, b, 1942; Huber 1930a, b, 1931; Edgeworth 1935; Andrew 1963; Gasser 1967; Jouffroy and Saban 1971; Saban 1971; Seiler 1971a, b, c, d, e, 1974a, b, 1975, 1979, 1980; Minkoff et al. 1979; Preuschoft 2000; Schmidt and Cohn 2001; Burrows and Smith 2003; Burrows et al. 2006; Burrows 2008; Diogo et al. 2009a; Diogo and Wood 2012; Santana et al. 2014; Diogo and Santana 2017). Some researchers have suggested that the mammalian facial muscles are derived exclusively from the interhyoideus of nonmammalian tetrapods (e.g., Huber 1930a, b, 1931), but our dissections and comparisons support the ideas of authors such as Lightoller (1942) and Jouffroy and Saban (1971), who claimed that at least some of these muscles (e.g., the extrinsic ear muscles and/or the **platysma cervicale**, **platysma myoides**, and **mandibulo-auricularis**) correspond to a part of the dorsomedial hyoid musculature (e.g., cervicomandibularis) of other tetrapods (Table 11.2). The evolution and homologies of the mammalian facial muscles have been, and continue to be, controversial. In light of the overall analysis of the data obtained by our dissections and comparisons and by a review of the literature, it can be said that some of the hypotheses proposed in Table 11.2 (black arrows) are in fact well supported by the data that are now available. For instance, data available on topology, functional morphology, development, and innervation strongly suggest that the platysma cervicale, platysma myoides, **occipitalis, auricularis posterior**, and some of the extrinsic muscles of the ear (e.g., **antitragicus, helicis**, and/or **transversus and obliquus auriculae**) of mammals (Figs. 11.2, 11.5, 11.6, 11.7, 11.12, 11.14, 11.18, 11.19, 11.20, and 11.21) have a common phylogenetic and ontogenetic origin (e.g., Boas and Paulli 1908; Huber 1930a, b, 1931; Gasser 1967; Jouffroy and Saban 1971; Saban 1971; Diogo et al. 2009a; this chapter). These same lines of evidence also suggest that the **interhyoideus profundus**, **sphincter colli superficialis, sphincter colli profundus, nasolabialis, levator labii superioris, levator labii superioris alaeque nasi, buccinatorius, dilatator nasi, maxillonasolabialis, nasalis, depressor septi nasi, levator anguli oris facialis, orbicularis oris, depressor labii inferioris, depressor anguli oris**, and **mentalis** of mammals (Figs. 11.2, 11.5, 11.6, 11.7, 11.12, 11.14, 11.18, 11.19, 11.20, and 11.21) derive from the interhyoideus of nonmammalian taxa (see also, e.g., Gasser 1967; Jouffroy and Saban 1971; Saban 1971; Seiler 1971a, b, c, d, e, 1974a, b, 1975, 1979, 1980). Moreover, these hypotheses have been supported by further developmental studies of mice. For instance, Carvajal et al. (2001) have shown that hyoid arch muscle progenitors that migrate out of the hyoid arch from 10.5 dpc (days post coitum) split into dorsal and ventral branches by 11.5 dpc. The dorsal domain divides further at 12.5 dpc and gives rise to the extrinsic facial muscles of the ear and to some auricular muscles (e.g., auricularis posterior; plus the occipitalis according to our interpretation of their figures). The ventral domain elongates rostrally and separates into dorsal and ventral branches that do not divide into different muscle masses until 13.5 dpc and that expand toward the snout and the eye regions where they then form most other facial muscles. However, Carvajal et al.'s (2001) figures seem to indicate that at least part of the platysma myoides derives from the ventral domain, and not from the dorsal domain as hypothesized in our Table 11.2, although it is not clear from those figures if at least a part of the platysma myoides, and/or a part or the totality of the platysma cervicale, derive from the dorsal domain.

It is also still not completely clear if, for instance, the therian mandibulo-auricularis (a muscle that is usually deep to all the other mammalian facial muscles) is phylogenetically more closely related to the other facial muscles than to deeper dorsomedian hyoid muscles such as the stylohyoideus, digastricus posterior, jugulohyoideus, and stapedius (e.g., Lightoller 1934; Jouffroy and Saban 1971; Seiler 1971a, b, c, d, e, 1974a, b, 1975, 1979, 1980; this chapter) (Table 11.2). Also, it is commonly accepted that muscles such as the **zygomaticus major, zygomaticus minor, orbito-temporo-auricularis, frontalis, auriculo-orbitalis, temporoparietalis, auricularis anterior**, and **auricularis superior** (Figs. 11.2, 11.5, 11.6, 11.7, 11.12, 11.14, 11.18, 11.19, 11.20, and 11.21) derive from the sphincter colli profundus and/or superficialis, but Seiler (1971a, b, c, d, e, 1974a, b, 1975, 1979, 1980), based on his comparative and developmental studies, argues that at least some of these muscles may derive from the platysma cervicale and/or myoides (Table 11.2).

Seiler did an impressive series of works on the facial muscles of mammals, which are, unfortunately, often neglected by non-German-speaking authors. However, some of Seiler's methods and interpretations are questionable. For example, in his 1980 developmental study of primates and treeshrews, he argues that the facial muscles that are more superficial in early developmental stages are necessarily a part of a "platysma anlage" and thus derived phylogenetically from an "ancestral platysma," whereas the majority of the other facial muscles are a part of a "sphincter colli profundus" anlage and thus are derived phylogenetically from a "primitive sphincter colli profundus."

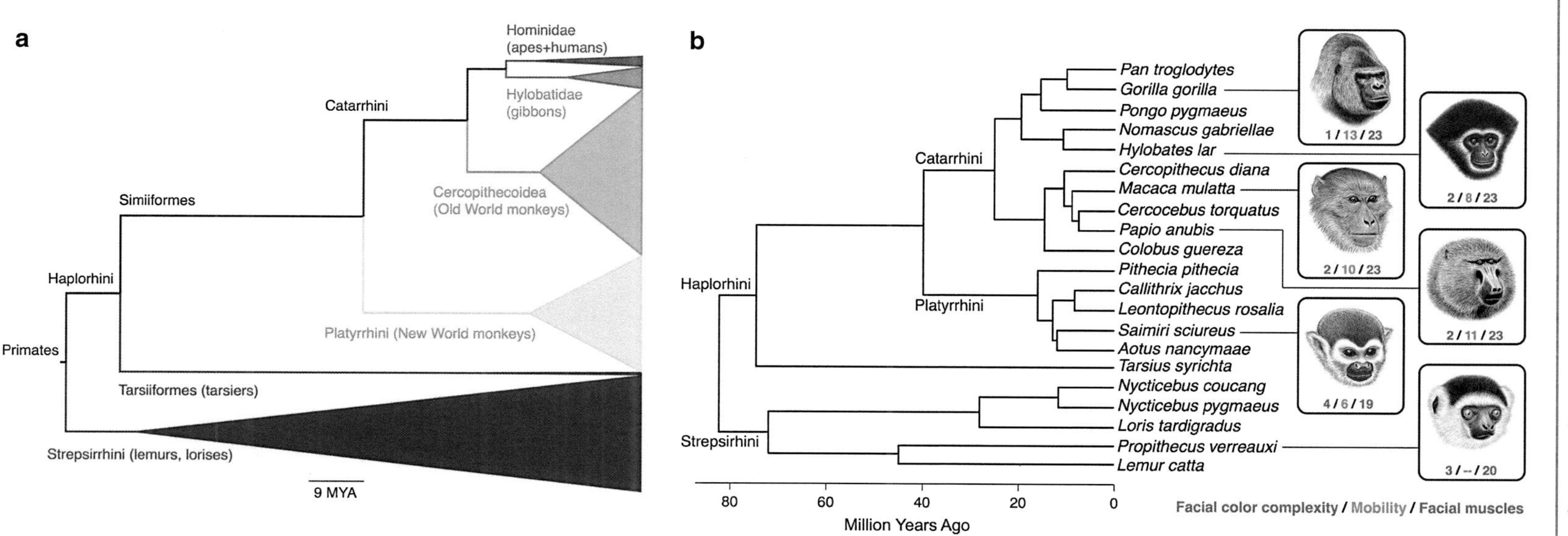

Fig. 11.21 Phylogenies of (**a**) the order Primates, showing the major lineages in proportion to their numbers of species, and (**b**) the primate species included in Santana et al.'s (2014) and Diogo and Santana's (2017) works, with examples illustrating major trends in facial color pattern complexity, mobility, and facial muscles; species that are larger and have more plainly colored faces tend to have a larger repertoire of facial expressions (©2012 Stephen D. Nash/IUCN/SSC Primate Specialist Group; modified from Diogo and Santana 2017)

This contrasts with Gasser's (1967) study of the ontogeny of the facial muscles of modern humans, in which various other anlagen are recognized in early developmental stages. Also, it should be stressed that in adult mammals, including monotremes, at least some portions of the platysma (cervicale and/or myoides) lie deep to facial muscles such as the sphincter colli superficialis and even to facial muscles that Seiler categorizes as "sphincter colli profundus derivatives" (e.g., part of the orbicularis oris and/or levator labii superioris) (see, e.g., Fig. 11.2). The majority of researchers consider that the sphincter colli of mammals derives from the interhyoideus of other tetrapods, so it is likely that the mammalian sphincter colli was plesiomorphically mainly superficial, and not deep, to the other hyoid muscles (the interhyoideus of other tetrapods is usually superficial not only to the other hyoid muscles but to all the other muscles of the head). Monotremes are phylogenetically basal, and also anatomically plesiomorphic, mammals, and both the platypus and the echidna have a well-developed, broad sphincter colli superficialis that is superficial to most of the other facial muscles. Note that the platypus actually lacks a sphincter colli profundus, although it has an interhyoideus profundus that seems to be derived from the deeper part of the interhyoideus (Fig. 11.2); in the echidna most of the sphincter colli is superficial to the other facial muscles, but part of it passes deep to these muscles, forming a sphincter colli profundus (e.g., Huber 1930a; Lightoller 1942; Jouffroy and Saban 1971; this chapter). A more detailed comparative analysis of the development and innervation of the hyoid group of muscles in vertebrates, including various key mammalian groups such as monotremes, is needed to clarify these and other controversial issues regarding the origin, homologies, and evolution of the mammalian facial muscles and to test the hypotheses proposed in Table 11.2. Below we will further discuss the evolution of the mammalian facial muscles and focus in particular on the fascinating case study provided by primate and human evolution.

11.4 Branchial, Pharyngeal, and Laryngeal Muscles (Table 11.3)

The muscles listed in Table 11.3 correspond to the **branchial muscles** *sensu lato* of Edgeworth (1935), which can be divided into three groups: (1) the "true" branchial muscles, which are subdivided into the branchial muscles *sensu stricto* and the **protractor pectoralis** and its derivatives, e.g., the trapezius and sternocleidomastoideus; (2) the **pharyngeal muscles**, which are only present as independent structures in extant mammals; and (3) the **laryngeal muscles** (e.g., Figs. 11.4, 11.6, 11.9, 11.10, 11.11, 11.15, 11.17, 11.18, 11.19, and 11.20).

Sarcopterygians such as coelacanths, dipnoans, and many amphibians retain various branchial muscles *sensu stricto* (Table 11.3) (e.g., Bischoff 1840; Owen 1841; Cuvier and Laurillard 1849; Pollard 1892; Gaupp 1896; Danforth 1913; Lubosch 1914; Sewertzoff 1928; Edgeworth 1935; Brock 1938; Piatt 1938; Millot and Anthony 1958; Osse 1969; Larsen and Guthrie 1975; Greenwood 1977; Wiley 1979a, b; Jollie 1982; Lauder and Shaffer 1985; Bemis 1984, 1986; Reilly and Lauder 1989, 1990, 1991; Miyake et al. 1992; Wilga et al. 2000; Kardong 2002, 2011; Carroll and Wainwright 2003; Johanson 2003; Kleinteich and Haas 2007; Diogo and Abdala 2010). Most authors agree that the branchial muscles *sensu stricto* are not present as a group in extant reptiles and extant mammals (Table 11.3). For instance, many adult reptiles have only one branchial muscle *sensu stricto*, the **hyobranchialis** (which is often named "branchiohyoideus" or "branchiomandibularis" in the literature; Diogo and Abdala 2010). The two branchial muscles *sensu stricto* seen in adult reptiles such as the "lizard" *Euspondylus*, the **hyobranchialis** and "ceratohyoideus," seem to be the result of a subdivision of the subarcualis rectus I *sensu* Edgeworth (1935). That is, the "ceratohyoideus" found in these reptiles seems to correspond to/derive from a part of the hyobranchialis of other reptiles. This comparison with other vertebrates shows that adult extant mammals lack all the branchial muscles *sensu stricto* except the **subarcualis rectus I** *sensu* Edgeworth (1935) (present in most adult mammals, being often divided into a **ceratohyoideus** and a **stylopharyngeus**; see below), the **subarcualis rectus II** (usually present only in adult marsupials), and the **subarcualis rectus III** (usually present only in adult monotremes) (e.g., Edgeworth 1935; Smith 1992).

Edgeworth (1935) claimed that the pharyngeal muscles of mammals (Figs. 11.9 and 11.15) are not derived from branchial muscle plates but from a separate *de novo* condensation of myoblasts surrounding the pharyngeal epithelium. He did not consider the mammalian pharyngeal muscles to be homologous with the "pharyngeal muscles" of some amphibians (which probably correspond to branchial muscles *sensu stricto*, such as the **levatores arcuum branchialum** and/or the **transversus ventralis**) and some reptiles (which are seemingly derived from the hyoid musculature) (e.g., Piatt 1938; Schumacher 1973; Smith 1992; Diogo and Abdala 2010). Authors such as Jouffroy and Lessertisseur (1971) and Smith (1992) contradicted Edgeworth (1935) by suggesting that the mammalian pharyngeal musculature may derive from the amphibian "pharyngeal musculature" and therefore that this pharyngeal musculature innervated by the vagus nerve (CNX) was lost in reptiles. They state that "pharyngeal muscles" innervated by this nerve are present in larval amphibians and adult salamanders such as *Ambystoma punctatum* (e.g., "cephalo-dorso-subpharyngeus" *sensu* Piatt 1938) and *Thorius dubius* ("dorso-pharyngeus"). They argue that these muscles are similar to those of mammals because they lie between the hyoid apparatus and pharyngeal wall, are innervated by the vagus nerve, and have two layers—a more or less longitudinal (oblique) layer and a circular muscle layer (Smith 1992). Moreover, they state that monotremes have five branchial arches (mandibular, hyoid, and three more), while no extant reptile has more than four arches (mandibular, hyoid, and two more), and that the laryngeal muscles of mammals and amphibians are innervated by two homologous branches of the vagus nerve, the superior and inferior (or recurrent) laryngeal nerves. In contrast, in reptiles the innervation of the larynx is via a single laryngeal nerve that is a branch of CN IX (glossopharyngeal nerve) instead. Further studies of amphibians, reptiles, and mammals, using state-of-the-art developmental techniques, are needed to test these interesting hypotheses and thus to shed light on the origin of the mammalian pharyngeal musculature.

Our dissections, comparisons, and literature review do support Smith's (1992) and Noden and Francis-West's (2006) claims that at least one of the mammalian muscles included in Edgeworth's pharyngeal group, the **stylopharyngeus**, is derived from the true branchial musculature of basal tetrapods. That is, this mammalian muscle is not a *de novo* structure, being instead a derivative of the branchial musculature *sensu stricto* and, namely, of the subarcualis rectus I (Table 11.3). The mammalian stylopharyngeus and the reptilian "subarcualis rectus I" are among the few muscles in either taxon innervated by the glossopharyngeal nerve (CN IX): most of the mammalian pharyngeal muscles are innervated, instead, by the vagus nerve (CN X). In fact, in many mammals, including primates such as *Macaca*, the **ceratohyoideus** and stylopharyngeus are closely related and are innervated by the same ramus of the glossopharyngeal nerve (buccal ramus *sensu* Sprague 1944a, b; see also Saban 1968). Developmental data from monotremes and marsupials show that early in development the stylopharyngeus is similar to the nonmammalian "subarcualis rectus I" in position, function, and connections. As stressed by Smith (1992), although Edgeworth (1935) did not accept that the stylopharyngeus was derived from the branchial musculature *sensu stricto*, he did state that it develops from a muscle primordium that differs from the one that gives rise to the other pharyngeal muscles.

The homology between the mammalian stylopharyngeus and part of the "subarcualis rectus I" of other tetrapods is further supported by the results of a comparison between adult mammals and adult nonmammalian tetrapods. The stylopharyngeus of mammals usually originates from the styloid process, which is derived from a portion of the second (hyoid) arch; the "subarcualis rectus I" of nonmammalian taxa is usually associated with this arch (Saban 1971; Smith 1992). Also, as explained above, some reptiles (e.g., "lizards") have two branchial muscles *sensu stricto*, which apparently are the result of a subdivision of the "subarcualis rectus I" *sensu* Edgeworth 1935 (Table 11.3). The more anterior of these muscles, the **hyobranchialis** (often called "**branchiohyoideus**" or "branchiomandibularis"), usually originates from the hyoid arch (as does the mammalian stylopharyngeus; see above) and connects the hyoid cornu to the epihyal, although in various reptilian groups it extends anterolaterally to attach on the lower jaw (that is why it is usually named "**branchiomandibularis**" in these groups). We refer to this muscle as the hyobranchialis, because it is not homologous with the hyoid muscle branchiohyoideus of amphibians nor with the hypobranchial muscle branchiomandibularis of actinopterygian fish such as cladistians, chondrosteans, and *Amia* (Diogo and Abdala 2010). The most posteriorly situated muscle in "lizards," often named the "ceratohyoideus," usually connects the hyoid arch to other (more posterior) branchial arches, as does the mammalian ceratohyoideus (Table 11.3). It should be noted that in various mammals, such as colugos and treeshrews, the stylopharyngeus does not reach the styloid process, i.e., it may originate from more distal hyoid structures such as the epihyal (as does the reptilian hyobranchialis; e.g., Sprague 1942, 1943, 1944a, b; Saban 1968). This observation, together with the other data available (see above), suggests that the combination of stylopharyngeus and the ceratohyoideus in mammals and the combination of hyobranchialis and the "ceratohyoideus" in reptiles are both the result of the subdivision of the "subarcualis rectus I" *sensu* Edgeworth 1935 (Table 11.3). However, this does not mean that the stylopharyngeus of mammals is necessarily the homologue of the reptilian hyobranchialis, for one cannot refute the hypothesis that the subdivision of the "subarcualis rectus I" into two muscles occurred more than once within the amniotes resulting in the hyobranchialis and "ceratohyoideus" of "lizards" and in the stylopharyngeus and ceratohyoideus of mammals. But it is important to note that the mammalian stylopharyngeus is innervated by the glossopharyngeal nerve (CN IX) and not by the vagus nerve (CN X) as noted above, so it is not truly a vagus nerve pharyngeal muscle, such as the other longitudinal pharyngeal constrictors (e.g., palatopharyngeus) and the circular constrictors (middle and inferior and, in placentals, also superior) of mammals.

The mammalian **acromiotrapezius**, **spinotrapezius**, **dorsocutaneous**, **cleido-occipitalis**, **sternocleidomastoideus**, **cleidomastoideus**, and **sternomastoideus** correspond to the reptilian **trapezius** and **sternocleidomastoideus** and thus to the **protractor pectoralis** of amphibians, of dipnoans, and of bony fishes (see recent reviews by Ziermann et al. 2014; Diogo and Ziermann 2015). The protractor pectoralis of non-amniote taxa is not a branchial muscle *sensu stricto* because it is mainly involved in the movements of the pectoral girdle and not of the branchial arches. Interestingly, the results of recent developmental and molecular studies indicate that the protractor pectoralis of *Ambystoma* and the trapezius of chickens and mice are at least partially derived from somites (e.g., Köntges and Lumsden 1996; Matsuoka et al. 2005; Noden and Francis-West 2006; Piekarski and Olsson 2007; Shearman and Burke 2009). These studies have also shown that during the ontogeny of mice, some of the cells of the trapezius that are originated from the somites pass the lateral somitic frontier in order to develop within lateral plate-derived connective tissue of the forelimb (e.g., Shearman and Burke 2009). That is, the mammalian trapezius is a rather peculiar muscle that is directly associated with three different types of connective tissue:

connective tissue derived from branchial arch crest cells, somite-derived connective tissue, and lateral plate-derived (forelimb) connective tissue. Therefore, it has been controversial whether the protractor pectoralis was primarily derived from the paraxial mesoderm, as suggested by Edgeworth (1935), and only later became ontogenetically associated with the cranialmost somites and even with lateral plate-derived connective tissue of the forelimb, or if it was instead primarily derived from somites.

Interestingly, recent works have shown that apart from branchial muscles (*sensu* Edgeworth 1935, Diogo et al. 2008b, and Diogo and Abdala 2010) such as the protractor pectoralis and the laryngeal muscles **constrictor laryngis** and **dilatator laryngis**, even branchial muscles *sensu stricto* such as the **levatores arcuum branchialium** and hyoid muscles such as the **interhyoideus** are also partially derived from somites in tetrapods such as amphibians (e.g., Piekarski and Olsson 2007). Thus, the fact that muscles such as the protractor pectoralis have a partial somitic origin does not necessarily mean that they cannot be considered to be part of the branchial musculature. Matsuoka et al. (2005) recognize that the amniote trapezius is partially derived from somites but also argue that the sum of the data available (i.e., innervation, topology, development, and phylogeny) provides more support for grouping this muscle with the "true" branchial musculature than for including it in the hypobranchial musculature or in the postcranial axial musculature *sensu* Jouffroy (1971). In fact, it is important to stress that lineage tracing analyses in transgenic mice provide some support for the idea that the trapezius is a branchial muscle: these analyses reveal that neural crest cells from a caudal pharyngeal arch travel with the trapezius myoblasts and form tendinous and skeletal cells within the spine of the scapula (see, e.g., Noden and Schneider 2006). According to Noden and Schneider (2006: 14), "this excursion seemingly recapitulates movements established ancestrally, when parts of the pectoral girdle abutted caudal portions of the skull."

However, it should be noted that Epperlein et al.'s (2012) developmental study of salamanders and review of the literature suggested that neural crest actually does not serve a general function in vertebrate shoulder muscle attachment sites as predicted by the "muscle scaffold theory" and that it is not necessary to maintain connectivity of the endochondral shoulder girdle to the skull. According to these authors, the contribution of the neural crest to the endochondral shoulder girdle observed in the mouse probably arose *de novo* in mammals as a developmental basis for their skeletal synapomorphies. On the other hand, the innervation of the trapezius by the spinal accessory nerve (CN XI) and, in many cases, by C3 and C4 spinal cord segments adds weight to the argument that the muscle is derived from both the paraxial mesoderm and somites. According to a recent study, the spinal accessory nerve might be a novel structure specific to living gnathostomes that arose through the repatterning of preexisting spinal motoneurons in the hypothetical ancestor; by *de novo* upregulation of cranial nerve-specific regulatory genes, the ancestral spinal accessory nerve would have acquired intermediate branchiomeric motoneuron properties (Tada and Kuratani 2015). According to this view, it would not be possible to characterize the spinal accessory nerve based on a simple head/trunk dualism but rather in a third category of peripheral nerve; it would thus be conceivable that the gnathostome cucullaris muscle would represent a similar intermediate or mixed nature (Tada and Kuratani 2015). However, the innervation of the cucullaris might alternatively provide support for a branchial component because of the position of the accessory nucleus in the ventral horn of the spinal cord, which is in line with the more cranial branchiomotor nuclei (see, e.g., Butler and Hodos 2005).

But, as stressed by Ericsson et al.'s (2013) review on the origin and evolution of the neck muscles, the strongest evidence provided so far to support the idea that the trapezius and other derivatives of the cucullaris are following a head muscle developmental program was provided by Tajbakhsh et al. (1997) and Theis et al. (2010). As noted by Ericsson et al., myogenic differentiation in the branchial muscles is regulated by *Pitx2* and *Tbx1*; these two transcription factors act to regulate expression of the myogenic regulatory factors *Myf5* and *Myod*, while *Myf5* and *Myod* expression in the somitic mesoderm are regulated by *Pax3*. Mice lacking Pax3 function thus show a loss of several trunk and limb muscles, but head muscles are unaffected; *Pax3:Myf5* double mutant mice lack all somitic-derived muscles, including some (but, interestingly, not all: see below) tongue and infrahyoid muscles, but branchial muscles and the trapezius and sternocleidomastoideus are still present (Tajbakhsh et al. 1997). This suggests that the cucullaris derivatives, trapezius and sternocleidomastoideus, are not developing under the myogenic program functioning in the somites but instead are following a head muscle program. Similar evidence for a head muscle program operating in some neck muscles is observed in *Tbx1* mutant mice, because posterior branchial muscles are absent, as are the trapezius and sternocleidomastoideus (Theis et al. 2010). In contrast, as noted by Ericsson et al. (2013), these mice have been reported to lack muscle defects in somite-derived muscles of the limb. Moreover, as also noted by these authors, human patients with DiGeorge syndrome have point mutations in the *Tbx1* gene and show many similar features to those observed in the mouse *Tbx1* mutants, and, intriguingly, these patients also display sloping shoulders due to small shoulder and pectoral muscles, suggesting that *Tbx1* is also important for the development of these somitic-derived muscles in humans. Furthermore, the trapezius and sternocleidomastoideus receive cells from an *Isl1*-expressing lineage, providing evidence that these muscles are following a head muscle program (Theis et al. 2010) because fate mapping of this lateral cells at the posterior extent of the cranial mesoderm expressing *Isl1* in the mouse has revealed that they also contribute to branchial muscles (Theis et al. 2010).

But the strongest direct, empirical support for the cucullaris and its derivatives being true head (branchial muscles) came from the recent retrospective clonal studies done in mice by Lescroart et al. (2015), showing that the cardiopharyngeal field does includes the trapezius and sternocleidomastoideus muscles (see Chap. 1 on this same book about the non-chordate taxa and Fig. 1.6 therein). Moreover, these clonal studies also supported a close relationship between the trapezius and sternocleidomastoideus and the laryngeal muscles, confirming the idea proposed by Edgeworth (1935) and in our previous works that the trapezius and sternocleidomastoideus are branchial muscles and thus true head (i.e., branchiomeric) muscles. The labeled cells in the laryngeal muscles that they analyzed (only scored in sectioned embryos) did seem to be ipsilateral to labeling in the trapezius; it is likely that a similar situation for unilateral-labeled first arch muscles might be also present but that their analyses just did not allow to discern which parts of the right ventricle are derived from the left and right sides of the embryo (Robert Kelly, pers. comm.). Interestingly, these clonal studies indicate that the extrinsic ocular muscles (no data was given about their right/left modularity in that study) are closely related to the group formed by the first arch muscles and by the right ventricle, supporting the idea, also proposed by Edgeworth (1935) and followed in our previous works, that the extrinsic eye muscles are linked developmentally to the first arch muscles. Another interesting result of more recent developmental works is that they have confirmed Edgeworth's (1935) hypothesis—based on his embryological observations—that at least part of the esophageal muscles, i.e., the "esophagus striated muscles," are branchiomeric, i.e., have an origin from the cranial mesoderm (Gopalakrishnan et al. 2015). Also worthy of mention is the fact that the mammalian splenius muscle, which is often seen as a somitic, epaxial muscle innervated by dorsal rami of spinal nerves, appears in these clonal studies as partially somitic and partially cranial, i.e., related to the trapezius/sternocleidomastoideus. Of the 22 embryos in which the splenius was labeled in the clonal study (Lescroart et al. 2015), 9 also had labeling in somitic neck muscles (40.9%), and 5 showed labeling in the trapezius muscle group (22.7%). Genetic tracing with a Pax3Cre allele in the same study also confirmed that some progenitor cells for the splenius had expressed Pax3 and are therefore somite-derived. That is, the splenius seems to be mainly of somitic origin, although it does seem to include a portion derived from the cardiopharyngeal field.

Importantly, our recent dissections of chondrichthyans and comparisons with other vertebrates (Diogo et al. 2015c, d; Ziermann et al. 2014) and updated review of the literature clearly support the idea that the cucullaris and its derivatives are mainly branchial muscles. In adult sharks such as *Squalus*, there is a single, continuous muscle protractor pectoralis inserting onto both the dorsal surface of the branchial arches and the posterior surface of the pectoral girdle. Edgeworth has shown that this condition is seen from the first embryonic stages until the adult stages of sharks, the protractor pectoralis developing from the dorsal portion of the branchial muscle plates. In adult holocephalans such as *Hydrolagus*, the protractor pectoralis has two bundles, one (superficial) inserting onto the pectoral girdle and the other (deep) inserting onto the dorsal portion of the branchial arches. Edgeworth has shown that in holocephalans the protractor pectoralis develops from the dorsal portion of the branchial muscle plates, forming first a single, continuous muscle and then separating during ontogeny into the deep and superficial bundles. Concerning batoids, the protractor pectoralis develops only from the dorsal portion of the last branchial muscle plate, forming a single muscle that then divides during ontogeny into three bundles, for example, in taxa such as *Leucoraja*, an inner bundle going to the suprascapula, a middle bundle going to the scapula, and an external bundle going to the branchial arches (according to, e.g., Marion 1905, such bundles were not found by authors such as Kesteven Kesteven 1942–1945; NB, as noted by Miyake et al. 1992, some authors argue that the inner and middle bundles of batoids are not really part of the protractor pectoralis because they do not attach onto the branchial arches and they receive innervation from spinal nerves, but as shown by, e.g., Edgeworth 1935 in other gnathostomes, the derivatives of the protractor pectoralis might also receive innervation from spinal nerves and/or not attach to branchial arches). According to Edgeworth in chondrichthyans, the protractor pectoralis is innervated by cranial nerve X; in osteichthyans there is usually a protractor pectoralis (or its derivatives, e.g., trapezius and sternocleidomastoideus) that derives ontogenetically exclusively from the last branchial muscle plate and that inserts exclusively to the pectoral girdle (in both bony fish and most tetrapods) and levatores arcuum branchialum developing from the dorsal portion of the branchial muscle plates and going exclusively to the dorsal surface of the branchial arches (in bony fish and amphibians, the levatores arcuum branchialum being absent in amniotes). Most authors do not describe levatores arcuum branchialum in chondrichthyans, but Kesteven (1942–1945) did describe these muscles as very thin structures in sharks such as *Mustelus* and *Orectolobus* and batoids such as *Dasyatis* and holocephalans such as *Callorhynchus* and *Hydrolagus* and as even thinner and seemingly vestigial structures in sharks such as *Squalus* and batoids such as *Leucoraja*, going to the dorsal surface of five branchial arches (he does not refer to holocephalans). Most authors do not seem to agree with Kesteven's descriptions of levatores arcuum branchialum in chondrichthyans. For instance, Didier (1987, 1995) does not describe these muscles in holocephalans and states that some of the levatores arcuum branchialum *sensu* Kesteven (1942–1945) correspond to part of the protractor pectoralis profundus, while others correspond to part of the epibranchial muscle subspinalis. However, it should be noted that Kesteven (1942–1945) described the subspinalis, the protractor pectoralis, and the levatores arcuum branchialum in chondrichthyans, so the synonymies proposed by Didier are questionable. However, the fact that to our knowledge no other author (apart from Kesteven 1942–1945) has described distinct, fleshy muscles levatores arcuum branchialum in both the elasmobranchs and the holocephalans, together with the fact that even Kesteven recognized that the structures that he designated under the name levatores arcuum branchialum in chondrichthyans are mainly innervated by "spinal" nerves, while those in osteichthyans are mainly innervated by nerves CNX and/or CNIX, also put in question the homology of the muscles of these two major gnathostomes taxa proposed by

Kesteven. That is, the levatores arcuum branchialum described by Kesteven in chondrichthyans might well be instead part of the epibranchial musculature, as suggested by Didier (1995), or even part of the axial (body) musculature, although one should not discard the hypothesis that they are homologous to the levatores arcuum branchialum of osteichthyans.

This comparison with nonmammalian taxa, including fishes, is crucial to understanding the origin of the mammalian head muscles. For instance, because Edgeworth regarded the dipnoans as the most plesiomorphic group of gnathostomes, he considered the condition found in dipnoans and thus other bony fish as the plesiomorphic gnathostome condition. However, the comparison with other fishes and other vertebrates in general clearly indicates that the plesiomorphic gnathostome condition is to have a single branchial muscle protractor pectoralis innervated by cranial nerves and inserting onto both the branchial arches and the pectoral girdle, as found in adult sharks and in the early development of sharks, batoids, and holocephalans. The fact that the protractor pectoralis only divides into bundles later during the ontogeny of batoids and holocephalans, together with the fact that in adult batoids the muscle usually has three bundles while in adult holocephalans the muscle usually has two bundles, clearly indicates that the division of the cucullaris into bundles or muscles is a derived condition independently acquired in batoids and holocephalans. The derived and homoplasic division of the protractor pectoralis in batoids and holocephalans into bundles going exclusively to the pectoral girdle and exclusively to the branchial arches seems to be the result of an evolutionary trend seen in gnathostomes. That is, in early gnathostomes the pectoral girdle basically is, at least functionally and often also anatomically, part of the head, while during gnathostome evolution the pectoral girdle often tended to become functionally and/or anatomically more separated from the head; as such it makes sense not to have a continuous muscle inserting onto both the branchial arches and the pectoral girdle. During the evolutionary transitions leading to the origin of osteichthyans, the cucullaris was differentiated into a protractor pectoralis attached onto the pectoral girdle only and levatores arcuum branchialium attached onto the branchial arches only (see also Ziermann et al. 2014). The protractor pectoralis is a neck muscle in amphibians and became later subdivided into the trapezius and sternocleidomastoideus in amniotes, therefore being deeply related to the origin and evolution of the tetrapod, and thus the mammalian, neck. A recent developmental study including computed tomography datasets, fate mapping, and various vertebrate model organisms provided further evidence confirming that the cucullaris is effectively a branchiomeric muscle (Sefton et al. 2016). Interestingly, their data suggest that, at least in the axolotl, somite 3 is the posterior limit of mesodermal contribution to cranial structures in both paraxial and lateral mesoderm. This could help explain why some authors have reported a partial contribution of the anterior somites to the cucullaris/protractor pectoralis.

The detailed analysis of the data obtained from our dissections, combined with the information provided in the literature, has also allowed us to develop robust hypotheses of homology for most of the pharyngeal and laryngeal muscles (Table 11.3). The monotreme pharyngeal muscle **constrictor pharyngis** corresponds to the **constrictor pharyngis medius + constrictor pharyngis inferior** and possibly the **constrictor pharyngis superior** and **pterygopharyngeus** of therian mammals, although it is more likely that the superior constrictor and pterygopharyngeus are derived instead from the **palatopharyngeus** (Table 11.3; see Figs. 11.9 and 11.15). The **pharyngeal muscles salpingopharyngeus + levator veli palatini + musculus uvulae + palatoglossus** + palatopharyngeus of therian mammals clearly seem to derive from the primordia that gives rise to the palatopharyngeus in monotremes (Table 11.3; see Figs. 11.9 and 11.15). In Diogo and Abdala (2010), we stated that the **palatoglossus** was most likely derived from the hypobranchial muscle hyoglossus and specifically from the styloglossus, as proposed by Edgeworth (1935). Edgeworth (1935) based on his own developmental studies, as well as on the data provided by other authors, stated that the palatoglossus is usually innervated by the hypoglossal nerve (CN XII), including in humans. Most current atlases of human anatomy however refer to an innervation by the vagus nerve (CN X) and therefore group this muscle with the true pharyngeal muscles and not with the tongue hypobranchial muscles; some atlases refer to an innervation by the cranial part of the spinal accessory nerve (CN XI) (see review of Diogo and Wood 2012). But it should be noted that, in support of Edgeworth's hypothesis of a tongue developmental origin, various detailed studies have suggested that in at least some mammals, including nonhuman primates, the palatoglossus is innervated by the hypoglossal nerve (reviewed by Sokoloff and Deacon 1992). However, in their careful study of *Macaca fascicularis*, Sokoloff and Deacon (1992) did not find a pattern of innervation truly similar to that found in other tongue muscles such as the styloglossus, which according to Edgeworth (1935) is the main muscle from which the palatoglossus is derived; in fact, Sokoloff and Deacon (1992) pointed out that based on their data and also on developmental data on mice, either a palatal or a tongue (or both) origin of the palatoglossus can be considered plausible hypotheses. In fact, apart from the well-studied innervation of the palatoglossus in humans indicating that the muscle is innervated by the vagus nerve, there are also developmental studies supporting the idea that the muscle does derive from the palatopharyngeus/superior constrictor musculature (Schaeffer 1929; Cohen et al. 1993, 1994). Authors such as House (1953), who studied and discussed in detail the pharyngeal region in mammals, suggested that the palatoglossus is derived specifically from the glossopharyngeal part of the superior constrictor of the pharynx, i.e., the part that inserts onto the tongue, through an anterior migration of the origin of the muscle from the pharyngeal wall/medial raphe to the soft palate/lateral wall of the oropharynx. More recent developmental studies, including detailed studies of human development (e.g., Cohen et al. 1993) and also molecular developmental studies of mice (Grimaldi et al. 2015), strongly support the idea that the palatoglossus is derived from the pharyngeal—and not the tongue—musculature and provided a stronger support for a closer developmental relationship between the palatoglossus and the palatopharyngeus, levator veli palatini and uvulae, than between the palatoglossus and the superior pharyngeal constrictor. This idea is furthermore supported by the fact that our dissections, and our review of the literature on marsupial myology, indicate that the palatoglossus is a well-developed muscle in marsupials and therefore that this muscle was already differentiated in the LCA of placentals + marsupials, while the superior constrictor only became differentiated in placentals. Therefore it is difficult to argue that the more recent superior constrictor muscle gave rise to the older muscle palatoglossus. Although we consider that more data is needed to settle once and for all the origin of the palatoglossus, the combination of the available literature and our own observations strongly supports a pharyngeal origin of the palatoglossus, specifically from the primordia that also give rise to the levator veli palatini, palatopharyngeus, and musculus uvulae (Table 11.3).

However, it should also be noted that some studies, particularly recent ones and/or focused on humans, have suggested that the levator veli palatini is at least partially innervated by the facial nerve (CN VII) and could even be primarily a hyoid muscle (i.e., of second branchial arch); most studies of nonhuman mammals such as *Macaca*, dogs, and cats do not refer to an innervation from the facial nerve, but a few studies in mice have suggested an innervation by the vagus (CN X), glossopharyngeal (CN IX), and facial nerve (reviewed by Kishimoto et al. 2016). Kishimoto et al. (2016), based on their study of human embryos and fetuses, defend that the levator veli palatini primordium does appear in the region of the hyoid arch and argue that the reason why most authors do not mention an innervation by the facial nerve is because the pharyngeal plexus (which includes branches of the vagus nerve) innervates most of the superior part (near the origin) of the levator veli palatini, while the lesser palatine nerve (related to the facial nerve) innervates only a very small inferior part of the muscle (near its insertion).

Another interesting aspect of Grimaldi et al.'s (2015) developmental study of mice is that it has shown that all the mesenchyme and tendons of the soft palate muscles were derived from the cranial neural crest, with exception to the posterior attachment of the palatopharyngeus (which extends posteriorly to the region of the larynx), which was anchored in a mesoderm-derived pharyngeal wall, constituting the posterior border of the (cranial neural crest) CNC-derived mesenchyme domain of the pharynx. The circular pharyngeal constrictor muscles (superior, middle, and inferior) were partially embedded in neural crest-derived mesenchyme, and the pharyngeal wall constituted the posterior border of the neural crest contribution to the craniofacial mesenchymal tissues. That is, the pharyngeal wall may represent an interface between CNC-derived mesenchyme and mesoderm-derived mesenchyme. In birds, the larynx and the area with efferent innervation from the vagus nerve (CN X) also mark the transition between the region of the head where the connective tissues are derived from neural crest cells and the muscles are patterned by these cells; in the region posterior to it, the connective tissues arise from somites (reviewed by Smith 1992).

With respect to the laryngeal muscles (Figs. 11.4, 11.10, and 11.17), the **thyroarytenoideus**, **vocalis**, **cricoarytenoideus lateralis**, and **arytenoideus** of therian mammals correspond to the **thyrocricoarytenoideus** and arytenoideus of monotremes and to the laryngeus of nonmammalian tetrapods such as salamanders. The mammalian **cricoarytenoideus posterior** corresponds to the **dilator laryngis** of other tetrapods (Table 11.3). It should be noted that in terms of both its ontogeny and phylogeny, the mammalian **cricothyroideus** is clearly a pharyngeal, and not a laryngeal, muscle as is sometimes suggested in the literature (e.g., Terminologia Anatomica 1998) (Table 11.3). It should also be noted that according to authors such as Smith (1994), marsupials have no levator veli palatini. Their "functional superior constrictor" is formed by an expansion of the stylopharyngeus and not from the same mass of muscles that give rise to the middle and inferior constrictors, as is the case in placental mammals.

11.5 Hypobranchial Muscles (Table 11.4)

According to Edgeworth (1935), the hypobranchial muscles (Figs. 11.3, 11.9, 11.11, 11.16, 11.18, 11.19, and 11.20) are divided into a "geniohyoideus" group and a "rectus cervicis" group (Table 11.4). However, it is not clear if Edgeworth's groups represent separate premyogenic condensations or whether they only become apparent at the later stages of muscle development. The plesiomorphic condition for sarcopterygians is seemingly that found in extant actinistians and dipnoans: there are two hypobranchial muscles that are mainly related to the opening of the mouth, the **coracomandibularis** and the **sternohyoideus** (Edgeworth 1935; Kesteven 1942–1945; Wiley 1979a, b; Jollie 1982; Mallat 1997; Wilga et al. 2000; Johanson 2003; this work) (Table 11.4). Amphibians and reptiles have various hypobranchial muscles (e.g., the omohyoideus and the specialized glossal muscles related to tongue movements) that are not present in sarcopterygian fish (Table 11.4). The **geniohyoideus**, **genioglossus**, **hyoglossus**, and intrinsic muscles of the tongue of nonmammalian tetrapods very likely correspond to the coracomandibularis of sarcopterygian fish, although it is possible that the "hyoglossus" of, e.g., salamanders is at least partially derived from the sternohyoideus (e.g., Edgeworth 1935; Jarvik 1963; Diogo and Abdala 2010) (Table 11.4). The **styloglossus** of therian mammals seems to correspond to part of the hyoglossus of monotremes (Table 11.4). The mammalian **thyrohyoideus** and **sternothyroideus** correspond to part of the sternohyoideus of reptiles such as *Timon* (Table 11.4; it should be noted that some authors described a "sternothyroideus" in a few reptilian taxa but that this muscle is probably not homologous to the mammalian sternothyroideus). See also the comments on the mammalian palatoglossus muscle that were provided on the section just above.

Interestingly, the clonal studies that were discussed in the above section (Lescroart et al. 2010, 2015) confirmed that the tongue muscles (which are normally grouped together with the hypobranchial musculature: see section below) are exclusively of somitic origin. This has been almost consensually accepted but was put in question in a recent study by Czajkowski et al. (2014), which argued that the extrinsic tongue muscles are actually mainly of cranial origin (the intrinsic tongue muscles being mainly of somitic origin, as expected). For instance, they have stated that Mesp1Cre-dependent lineage tracing on the *Met* mutant background demonstrated that residual muscle and myogenic progenitors in the tongue of *Met* mutants indeed derive from cranial mesoderm. However, studies such as Harel et al. (2009) have shown that all tongue muscles are derived from the Pax3 lineage and not from the Isl1 lineage and moreover that Mesp1 is not good for such analyses because it marks cranial muscles but also the occipital somites. Moreover, Kelly's work (Robert Kelly, pers. comm.) has shown that tongue muscles are largely unaffected in Tbx1 nulls and mostly missing in Pax3 nulls, though they do express *Tbx1* at later stages, showing the dangers of drawing conclusions from gene expression at a single developmental stage. Also interestingly, *Lbx1* expression in the tongue muscle precursors is only delayed, and not abolished, in the Pax 3 mutant Splotch (Brand-Saberi 2002). This suggests that, for the occipital somites that have been incorporated into the head during evolution, the loss of *Pax3* function is compensated and thus that additional genes may be involved in the mediation of the extrinsic signals (Brand-Saberi 2002).

The clonal studies of Lescroart et al. (2015) thus seem to provide further evidence for an exclusive somitic origin for both the extrinsic and intrinsic tongue muscles, as the tongue muscles never appear together with the branchiomeric muscles in that study. However, it should be noted that unpublished data by Carmen Birchmeier (pers. comm.) with lineage tracing with Pax3 has shown that some tongue muscles (namely, part of the extrinsic tongue muscles) do not seem to derive from the Pax3 somitic lineage. Therefore, it cannot be completely excluded that a part of the tongue muscles are not derived from somites. Recent studies have stressed that the tongue muscles do have, in some aspects, some hybrid characteristics between branchiomeric and somitic muscles. For instance, contrary to somitic muscles, their patterning and attachments seem to be deeply related to neural crest cells (Parada et al. 2012). For example, these cells are not required for myogenic progenitor migration toward their presumptive destinations in the branchial arches and tongue primordium, but, as these myogenic progenitors first enter the craniofacial region, they immediately establish intimate contact with these cells; this close association between the two cell types continues throughout the entire course of tongue morphogenesis and suggests that tissue-tissue interactions may play an important role in regulating cell fate determination (Parada et al. 2012). Also, craniofacial myogenesis depends on *Dlx5/6* expression by CNCC (cranial neural crest cells), because inactivation of *Dlx5* and *Dlx6* results in loss of jaw muscles and compromised tongue development; since *Dlx5/6* are not expressed by the myogenic component, this result indicates an instructive role for *Dlx5/6*-positive CNCC in muscle formation (Parada et al. 2012). In *Dlx5/6*$^{-/-}$, the intrinsic muscles of the tongue and sublingual muscles are severely affected, e.g., the genioglossus and geniohyoideus are absent, and intrinsic muscles of the tongue are reduced and disorganized, but the remaining tongue muscles express determination and differentiation markers. Because limb and trunk muscles in *Dlx5/6* mutant mice are not affected, this indicates a specific function of Dlx genes in tissue-tissue interactions involving neural crest derivatives, i.e., *Dlx5/6* expression by CNCC is necessary for interactions between CNCC-derived mesenchyme and mesoderm to occur, which result in myogenic determination, differentiation, and patterning (Parada et al. 2012). According to Parada et al. (2012), the importance of Dlx genes in tongue development is twofold: (1) they establish the dorsoventral pattern of the first BA and, indirectly, that of the tongue; and (2) they regulate myogenic determination and differentiation processes, including those affecting the tongue myogenic core. However, this might simply mean that Dlx5/6 is related to neural crests cells, and because these cells affect the patterning of the tongue muscles, they are affected, so this might have nothing to do with showing that these muscles are not completely somitic. Further developmental, experimental, and comparative studies are thus needed to clarify the specific origins of the mammalian tongue muscles.

11.6 An Emblematic Example of the Remarkable Diversity and Evolvability of the Mammalian Head: The Evolution of Primate Facial Expression, with Notes on the Notion of a *Scala Naturae*

The face of humans and other mammals is a complex morphological structure in which both external and internal parts function in conveying information relevant for social interactions. Externally, facial features bear signals that allow recognition of conspecifics, individuals within the social group and potential mates. This information is encrypted in traits such as the shape of facial parts and the complexity and hues of its color patterns (Fig. 11.21). Internally, the facial musculature (Fig. 11.2) and neural centers control how the external morphology is showcased to other individuals through the production of facial expressions, which are important in communicating behavioral intentions within a social context (e.g., bared teeth communicate the intent to withdraw from an agonistic encounter; Preuschoft 2000). Therefore, internal and external anatomical features of the face are not only in close physical proximity but are also tightly connected in their function.

Facial coloration patterns evolved in tandem with sociality and sympatry (when two species or populations exist in the same geographic area) in primates (Santana et al. 2014). In most primate radiations, highly social and sympatric species evolved multicolored faces, while less social species tend to have less colorful faces. Complex facial patterns potentially enable higher interindividual variation within social groups and among species, facilitating recognition at either of these levels. Facial expressions are also linked to sociality; highly gregarious species produce a wider variety of facial movements, which may function in group cohesion by enhancing communication during conflict management and bonding facial expressions result from the action of facial muscles that are controlled by neural pathways (facial nucleus of the pons—cranial nerve VII—and the primary motor cortex), and primate species with relatively large facial nuclei tend to have highly dexterous faces (Sherwood et al. 2005). The primate facial musculature is among the most complex across mammals (although not as complex as that of, e.g., elephants; Boas and Paulli 1908), but it has been unclear if and how it has evolved in response to functional demands associated with ecology and sociality (Burrows 2008; Diogo et al. 2009b; Diogo and Wood 2012).

The 1st (mandibular), 2nd (hyoid), and more posterior branchial arches are formed from bilateral swellings on either side of the pharynx. The muscles of facial expression (e.g., Fig. 11.20)—usually designated simply as "facial muscles" as noted above—are a subgroup of the hyoid (second arch) muscles and are innervated by the facial nerve (cranial nerve VII). This means that all other hyoid muscles (e.g., stapedius, stylohyoideus) are not designated as facial muscles, despite being also innervated by the facial nerve. Except for the buccinatorius (and the mandibulo-auricularis present in many nonhuman mammals), the facial muscles are mainly attached to the dermis of the skin and the elastic cartilage of the pinna. They are involved in generating facial expressions during social interactions among conspecifics, as well as in feeding, chemosensation, whisker motility, hearing, vocalization, and human speech. This section, which is a summary of one of the sections of Diogo and Santana's (2017) recent review, provides a short summary on the evolution of primate facial muscles as a case study showing the remarkable diversity and evolvability of the mammalian head musculature and its links with external features such as the color of the skin and hair as well as with behavior and ecology.

As explained above, the muscles of facial expression are only present in mammals, probably deriving from the ventral hyoid muscle interhyoideus and likely also from at least some dorsomedial hyoid muscles (e.g., cervicomandibularis) of other tetrapods. Monotremes such as the platypus only have ten distinct facial muscles (not including the extrinsic muscles of the ear) (Fig. 11.2). Rodents, such as rats, have up to 24 facial muscles (Fig. 11.6). The occipitalis + auricularis posterior, the procerus, and the dilatator nasi + levator labii superioris + levator anguli oris facialis of therian mammals (marsupials + placentals) probably correspond to part of the platysma cervicale (muscle connecting the back of the neck—nuchal region—to the mouth, different from platysma myoides connecting front of the neck and pectoral region to the mouth), of the levator labii superioris alaeque nasi, and of the orbicularis oris of monotremes, respectively. The sternofacialis, interscutularis, zygomaticus major, zygomaticus minor, and orbito-temporo-auricularis of therian mammals probably derive from the sphincter colli profundus, but it is possible that at least some of the former muscles derive from the platysma cervicale and/or platysma myoides. Colugos (Dermoptera or "flying lemurs") and treeshrews (Scandentia), the closest living relatives of primates, have a similar facial musculature (Fig. 11.6), but the former lack two muscles that are usually present in the latter, the sphincter colli superficialis and the mandibulo-auricularis. As both these muscles are found in rodents, as well as in treeshrews and at least some primates, they were likely present in the last common ancestor (LCA) of primates + Dermoptera + Sca ndentia. The frontalis, auriculo-orbitalis, and auricularis superior of this LCA very likely derived from the orbito-temporo-auricularis of other mammals, while the zygomatico-orbicularis and corrugator supercilii most likely derived from the orbicularis oculi.

The facial musculature of the LCA of primates was probably very similar to that seen in the extant treeshrew *Tupaia*. Muscles that have been described in the literature as peculiar to primates, e.g., the zygomaticus major and zygomaticus minor, are now commonly accepted as homologues of muscles of other mammals (e.g., of the "auriculolabialis inferior" and "auriculolabialis superior"). The only muscle that is actually often present as a distinct structure in strepsirhines (see Fig. 11.12)—i.e., the primate group including extant members such as lemurs and lorises (see Fig. 11.21)—but not in treeshrews or colugos, is the depressor supercilii, which derives from the orbicularis oris matrix. As the depressor supercilii is present in strepsirhine and non-strepsirhine primates, it is likely that this muscle was present in the LCA of primates. In summary, the ancestral condition predicted for the LCA of primates is probably similar to that found in some extant strepsirhines (e.g., *Lepilemur*). Importantly, the number of facial muscles present in living strepsirhines is higher than that originally reported by authors in the nineteenth and first decades of the twentieth century. For instance, Murie and Mivart (1869) reported only seven facial muscles in a lemur, grouping all the muscles associated with the nasal region into a single "nasolabial muscle mass." The supposed lack of complexity seen in strepsirhines was consistent with the anthropocentric, scala naturae, finalistic evolutionary paradigm subscribed to by many anatomists at that time (see above). However, it is now accepted that strepsirhines often have more than 20 facial muscles and that although humans have more facial muscles than most primates, the difference is minimal in general. In fact, the total number of facial muscles found in humans is similar to that found in rats, as shown in Table 11.2, contradicting one of the major myths of human complexity and exceptionalism (Table 11.2; see Diogo and Wood 2012, 2013 and Diogo et al. 2015b, c, d for more details on this subject).

In order to give a functional context for these descriptions of the evolution and comparative anatomy of the primate facial muscles, here we provide a brief account of the general function of the facial muscles that are present in strepsirhines (Fig. 11.12). Then, when we refer in the next section to a certain muscle that is not differentiated in strepsirhines but that is present in anthropoids (monkeys and apes, including humans), we will also briefly describe the general function of that muscle. The platysma myoides most likely draws the oral commissure posteroinferiorly, an action that may be used in social interactions as well as feeding, while the platysma cervicale most likely elevates the skin of the neck. The occipitalis draws the scalp posteriorly toward the nuchal region, while the frontalis elevates the skin/brow over the superciliary region. The auriculo-orbitalis may be used to draw the lateral corner of the eyelid posteroinferiorly or the external ear anterosuperiorly. The corrugator supercilii and the depressor supercilii are used to draw the medial edge of the superciliary region inferomedially and inferiorly, respectively. The mandibulo-auricularis may be used to approximate the superior and inferior edges of the external ear, as well as the internal ear and the mandible. The muscles clustered around the upper lip, including the zygomaticus major and zygomaticus minor, may be used to draw the upper lip and the posterior region of the mouth posterosuperiorly, functions which may be used in both social interactions and in the use of the vomeronasal organ. As their name indicate, the extrinsic muscles of the ear, as well as the auricularis posterior and auricularis superior, are mostly related to movement of the external ear, while the orbicularis oculi and orbicularis oris are primarily associated with movement of the eyes and of the lips, respectively. The buccinatorius mainly pulls the corner of the mouth laterally and presses the cheek against the teeth. The levator labii superioris alaeque nasi, levator labii superioris, and levator anguli oris facialis are most likely used together in drawing the upper lip and the posterior region of the mouth superiorly and medially, which most likely is used in social interactions and in feeding. The mentalis mainly elevates the skin ventral to the lower lip, while the sphincter colli profundus most likely draws the skin of the neck posterosuperiorly.

There are some notable differences between the ancestral condition described above for non-anthropoid primates such as *Lepilemur* (Fig. 11.12) and the condition found in New World and Old World monkeys (see Fig. 11.14). For example, the mandibulo-auricularis is usually not present as an independent, fleshy muscle in most anthropoids, although some of these primates have fleshy vestiges of this muscle as a rare variant. It likely corresponds to the stylomandibular ligament seen in hominoids (apes, including humans) such as humans and in some monkeys. The sphincter colli profundus is also normally absent in anthropoids, but fleshy vestiges of this muscle have been described in a few macaques as well. Anthropoids often have a depressor anguli oris and a depressor labii inferioris. These muscles are probably derived from the orbicularis oris matrix; some authors suggested that the depressor anguli oris might be the result of a ventral extension of the levator anguli oris. Generally, the depressor anguli oris and depressor labii inferioris function in anthropoids to draw the corner of the mouth posteroinferiorly and to draw the lower lip inferiorly, respectively. These movements are seen in some displays of facial expression and in some feeding contexts.

Within hominoids the platysma cervicale is usually present in hylobatids (lesser apes: gibbons and siamangs) and gorillas but is often highly reduced or absent in adult orangutans, chimpanzees, and humans (see Figs. 11.18, 11.19, and 11.20). The transversus nuchae, found as a variant in the three latter taxa, is often considered to be a vestigial remain/bundle of the platysma cervicale. Interestingly, the platysma cervicale is present early in the development of humans, but it normally disappears as an independent structure in later stages of development. Contrary to the platysma cervicale, the platysma myoides is usually present as a separate structure in adult members of all the major five extant hominoid taxa. The occipitalis is also usually present in these five, but the auricularis posterior is normally not differentiated in orangutans, although it has been described in a few species.

In humans the risorius (Fig. 11.20) is usually, but not always, present, pulling the lip corners backward, stretching the lips—a function that is, interestingly, usually associated with the display of fear—being likely derived from the platysma myoides, although it cannot be discarded that it is partly, or even wholly, derived from the zygomaticus major. Among primates, a "risorius" is sometimes found in some other hominoids, e.g., chimps, but it does not seem to be present in the fixed phenotype (i.e., >50% of the cases) of any of the four major nonhuman hominoid taxa. Moreover, some structures that are often named "risorius" in these hominoids are probably not homologous to the human risorius and even to each other, because some apparently derive from the platysma myoides, others from the depressor anguli oris, and others from muscles such as the zygomaticus major. All the other facial muscles that are present in macaques are normally present in extant hominoids, but contrary to monkeys and to other hominoids, humans—and possibly also gorillas—usually also have an auricularis anterior and a temporoparietalis. Both of these muscles are derived from the auriculo-orbitalis, which, in other hominoids such as chimpanzees, has often been given the name "auricularis anterior," although it actually corresponds to the auricularis anterior plus the temporoparietalis of humans and gorillas. When present, the temporoparietalis stabilizes the epicranial aponeurosis (a tough layer of dense fibrous tissue covering the upper part of the cranium: see, e.g., Fig. 11.6), whereas the auricularis anterior draws the external ear superoanteriorly, closer to the orbit.

It is interesting to note that each of the three non-primate taxa listed in Table 11.2 has at least one derived, peculiar muscle that is not differentiated in any other taxa listed in this Table. So, for instance, *Ornithorhynchus* has a cervicalis transversus (Fig. 11.2), *Rattus* has a sternofacialis and an interscutularis, and *Tupaia* has a zygomatico-orbicularis. This is an excellent example illustrating that evolution is not directed "toward" a goal and surely not "toward" primates and humans; each taxon has its own particular mix of conserved and derived anatomical structures, which is the result of its unique evolutionary history (Diogo and Wood 2013). This is why we encourage the use of the term *correspond* to describe evolutionary relationships among facial muscles, because muscles such as the zygomatico-orbicularis are not *ancestral* to the muscles of primates. The zygomatico-orbicularis simply corresponds to a part of the orbicularis oculi that, in taxa such as *Tupaia*, became sufficiently differentiated to deserve being recognized as a separate muscle. Also, strepsirhines and monkeys have muscles that are usually not differentiated in some hominoid taxa, e.g., the platysma cervicale (usually not differentiated in orangutans, chimps, and humans) and the auricularis posterior (usually not differentiated in orangutans).

Humans, together with gorillas, have the greatest number of facial muscles within primates, and this is consistent with the important role played by facial expression in anthropoids in general, and in humans in particular, for communication. Nevertheless, the evidence presented in this chapter, as well as in recent works by Burrows and colleagues (e.g., Burrows 2008), shows that the difference between the number of facial muscles present in humans and in hominoids such as hylobatids, chimpanzees, and orangutans and between the number of muscles seen in these latter hominoids and in strepsirhines is not as marked as previously thought. In fact, as will be shown below, the display of complex facial expressions in a certain taxon is not only related with the number of facial muscles but also with their subdivisions, arrangements of fibers, topology, biochemistry and microanatomical mechanical properties, as well as with the peculiar osteological and external features (e.g., color) and specific social group and ecological features of the members of that taxon.

For instance, from bright red to yellow, black, brown, and even blue, the faces of primates exhibit almost every possible hue in the spectrum of mammalian coloration (Fig. 11.21). In many species, such as mandrills and guenons, facial skin and hair colors are combined to create remarkably complex patterns that are unique to the species. Is there a functional significance to these colors and their patterns? Recently, researchers have harnessed the tools of modern comparative methods and computer simulation to answer this question and investigate the factors underlying the evolution of facial color diversity across primate radiations. Several lines of evidence suggest that facial colors are crucial to the ecology and social communication of primates. Variation in coloration within a species, such as the differences in brightness of red facial patches among male mandrills, appears to be used for assessment of overall health condition and potential mate quality. At a broader scale, differences across species in facial color patterns are hypothesized to enable individuals of sympatric and closely related species to identify one another and avoid interbreeding. Phylogenetic comparative studies have demonstrated that social recognition explains trends in the evolution of primate facial color patterns. In the New World primate radiation (Platyrrhini), species that live in small social groups or are solitary (e.g., owl monkeys, *Aotus*) have evolved more complexly patterned faces (Santana et al. 2014). In sharp contrast, diversity trends in Old World groups (Catarrhini) are the opposite, with highly gregarious species having more complexly patterned faces (Santana et al. 2014). These divergent trends may be explained by habitat differences and a higher reliance on facial expressions and displays for intraspecific communication in catarrhines, in which facial colors may be further advertised through stereotyped head movements during courtship or appeasement behaviors (Kingdon 1992, 2007).

Across all primates studied to date, the evolution of complexly patterned faces is also tightly linked to high levels of sympatry with closely related species (Santana et al. 2014). A face that is colorful may present features that are unique and more easily recognizable in the context of multiple sympatric species. Allen et al. (2014) used computational face recognition algorithms to model primate face processing. Their results demonstrated that the evolution of facial color patterns in guenons fits models of selection to become more visually distinctive from other sympatric guenon species. This indicates that facial color patterns function as signals for species recognition in primates and may promote and maintain reproductive isolation among species.

The degree of facial skin and hair pigmentation is also highly variable across primates, and comparative studies suggest that this diversity may illustrate adaptations to habitat. Darker, melanin-based colors in the face and body are characteristic of primate species that inhabit tropical, more densely forested regions (reviewed by Santana et al. 2014). It is hypothesized that these darker colors may reduce predation pressure by making individuals more cryptic to visually oriented predators and increase resistance against pathogens (reviewed by Santana et al. 2014). Darker facial colors may also offer protection against high levels of UV radiation and solar glare and aid in thermoregulation. However, the role of facial pigmentation in these functions remains unclear because primates may use behaviors to regulate their physiology (e.g., arboreal species can move from the upper canopy, which has the highest UV levels, to the middle and lower canopy, which are highly shaded). In catarrhines, ecological trends in facial pigmentation are only significant in African species (Santana et al. 2014), presumably because the African continent presents more distinct habitat gradients than South East Asia. In platyrrhines, darker faces are found in species that live in warmer and more humid areas, such as the Amazon, and darker eye masks are predominant in species that live closer to the equator. Eye masks likely function in glare reduction in habitats with high ultraviolet incidence, and similar trends in this facial feature have also been observed in carnivorans and birds (reviewed by Santana et al. 2014). The presence and length of facial hair is highly variable across primate species, but its role in social communication, besides acting as a vehicle to display color, has not been broadly investigated. In platyrrhines, species that live in temperate regions have longer and denser facial hair (Santana et al. 2014), which could aid in thermoregulation. Similar trends would be expected in other primate radiations.

To date, the evolutionary connections between external (coloration, facial shape) and internal (musculature) facial traits are poorly known. In recent studies (Santana et al. 2014; Diogo and Santana 2017), we contrasted two major hypotheses that could explain the evolution of primate facial diversity when these traits are integrated. First, if the evolution of facial displays has been primarily driven by social factors, highly gregarious primates would possess both complexly colored and highly expressive faces as two concurrent means for social communication. Alternatively, if external facial features influence the ability of primates to perceive and identify facial expressions, there would be a trade-off in the evolution of facial mobility and facial color patterning, such that highly expressive faces would have simpler color patterns. We used phylogenetic comparative analyses integrating data on facial mobility, facial musculature, facial color pattern complexity, body size, and orofacial motor nuclei across 21 primate species to test these hypotheses.

The results from our study indicated a significant association between the evolution of facial color patterns and facial mobility in primates. Supporting the second hypothesis, primates evolved plainly colored faces in tandem with an enhanced ability for facial expressions. Thus, while complex facial color patterns may be beneficial for advertising identity (Allen et al. 2014; Santana et al. 2014), a highly "cluttered" face may mask the visibility of facial expressions used to convey behavioral intention. Why a species may rely more on facial color patterns versus facial expressions for communication is still unclear, but it is possible that these different modalities may be differentially selected across primate lineages based on the species' habitat, social systems, or body size. Larger primates (e.g., apes), which have a larger facial nucleus, have more expressive faces than smaller species (e.g., marmosets), which in turn seem to use colorful facial patterns and head movements for communication. The evolution of larger bodies, potentially coupled with increased reliance on vision for other ecological tasks (e.g., finding food and avoiding predators) may have allowed a higher reliance on facial expressions, which was not possible at smaller body sizes due to physical constraints on the perception of facial movements. Smaller species are expected to have more difficulty discerning facial expressions because smaller mammalian eyes have lower visual acuity (reviewed by Santana et al. 2014).

Although the evolution of facial mobility is linked to facial coloration and body mass, we found that it is not directly related to the number of muscles that produce facial movements. The number of facial muscles per se is a slowly evolving trait that has strong phylogenetic inertia (Table 11.2; see Sect. 11.2 and also Diogo and Wood 2012, 2013). Conversely, the size of the facial nucleus has evolved rapidly in the sample of primates studied. These results indicate that changes in facial mobility are likely to evolve first via changes in neurophysiology and body mass, instead of muscle morphology, that is, through motor control of muscles instead of the creation of new divisions of preexisting musculature. That is, it is interesting to note that while the number of facial muscles is rather conservative within primates, the evolution of the facial musculature as a whole and in particular and of facial expression in general actually provides an emblematic case study of the diversity and evolvability of primates and of mammals. Moreover, these patterns of evolution and potential trade-offs give important insight into the simple organismal features, such as body mass, that have a strong relevance for which and how different types of facial cues evolve for social communication.

11.7 General Remarks on the Evolution of the Mammalian Head Muscles with Notes on Body Plans

This review, summarized in Tables 11.1, 11.2, 11.3, and 11.4, allows us to provide a very detailed list of muscle synapomorphies of mammals (therians + monotremes), of therians (marsupials + placentals), and of placentals. Based on this comparison, extant mammals share 34 muscle synapomorphies for the head. These numbers illustrate the utility of studying muscles to characterize certain clades and pave the way for paleontological, developmental, and functional works that investigate the specific evolutionary time of origin/loss and developmental mechanisms that led to the characteristic muscle anatomy of each clade and their functional implications. Specifically, there are ten synapomorphies of the mandibular muscles of extant mammals: differentiation of mylohyoideus, digastricus anterior, masseter, temporalis, pterygoideus lateralis, pterygoideus medialis, tensor tympani, and tensor veli palatini and loss of adductor mandibulae A2-PVM and of dorsal mandibular muscles (Table 11.1). There are 12 for the hyoid muscles: differentiation of styloideus, stapedius, platysma cervicale, platysma myoides, extrinsic ear muscles, sphincter colli superficialis, sphincter colli profundus, orbicularis oculi, nasolabialis/levator labii superioris alaeque nasi, buccinatorius, orbicularis oris, and mentalis (Table 11.2). There are 11 for the branchial muscles: differentiation of acromiotrapezius, spinotrapezius, dorsocutaneous, cleidomastoideus, sternomastoideus, constrictor pharyngis, cricothyroideus, palatopharyngeus, thyrocricoarytenoideus, and arytenoideus and loss of constrictor laryngis (Table 11.3). There is only one for the hypobranchial muscles: differentiation of sternothyroideus (Table 11.4). Therefore, in the transitions that led to the LCA of extant mammals, all major groups of head muscles experienced drastic changes with the exception of the much more conserved hypobranchial muscles of somitic origin, which experienced a single synapomorphic change. In addition to describing the drastic changes that occurred in both the head and limbs during the transitions leading to extant mammals, our results also show mosaic evolution because some subregions of the head, for example, the hypobranchial muscles, changed less than other head muscles.

There were only two head muscle synapomorphic changes from the LCA of extant mammals to the LCA of extant monotremes, both of them concerning mandibular muscles: the loss of intermandibularis anterior and the differentiation of detrahens mandibulae (Table 11.1). These numbers go against the idea, commonly defended in the literature (e.g., Saban 1968, 1971), that monotremes are an example of a "phylogenetically basal taxon" displaying a mix of plesiomorphic and highly derived, peculiar musculoskeletal features within the mammalian clade. In fact, these numbers instead show that monotremes are a very good models to study and discuss the origin and early evolution of the mammalian head and neck musculature (for more details, see Diogo et al. 2015a).

In contrast, there were 28 head muscle synapomorphic changes from the LCA of extant mammals to the LCA of extant therians. There were no changes within the mandibular muscles (Table 11.1), 18 within the hyoid muscles (differentiation of stylohyoideus, digastricus posterior, occipitalis, auricularis posterior, mandibulo-auricularis, interscutularis, zygomaticus major, zygomaticus minor, frontalis, auriculo-orbitalis, auricularis superior, corrugator supercilii, retractor anguli oculi lateralis, dilatator nasi, levator labii superioris, nasalis, and levator anguli oris facialis and loss of remaining of original interhyoideus: Table 11.2), 8 within the branchial muscles (differentiation of cleido-occipitalis, constrictor pharyngis medius, constrictor pharyngis inferior, palatoglossus, pterygopharyngeus, musculus uvulae, thyroarytenoideus, and cricoarytenoideus lateralis: Table 11.3), and only 2 within the hypobranchial muscles (differentiation of styloglossus and thyrohyoideus: Table 11.4). These results indicate that the origin of therians was particularly marked by evolutionary changes in facial muscles, pharyngeal muscles, and laryngeal muscles. These changes were probably related to specializations in facial and vocal communication through movements of both the larynx and pharynx and new ways of feeding, including mastication and suckling.

With respect to the clade including extant placentals, it can be diagnosed by three of four head muscle synapomorphies. None concerns the mandibular and hypobranchial muscles (Tables 11.1 and 11.4), none or just one concern the hyoid muscles (possibly the differentiation of depressor septi nasi: Table 11.2), and three concern the branchial muscles (differentiation of constrictor pharyngis superior, levator veli palatini, and salpingopharyngeus: Table 11.3). Thus, except for the pharyngeal muscles, the head muscles changed very little from the LCA of extant therians to the LCA of placentals. These changes probably related to further specializations of the movements of the larynx (moved by the pharyngeal muscle salpingopharyngeus) and pharynx for vocal communication and/or feeding mechanisms. Therefore, the few synapomorphic changes from the LCA of extant therians to the LCA of extant placentals are distributed more or less equally among the three major anatomical regions (head three or four, forelimb three, hindlimb four). These numbers provide empirical support for a well-defined therian body plan, which can still be easily recognized in most extant placentals and marsupials.

Of course, even among therians that conform to this characteristic therian Bauplan, there are minor differences in adult phenotype, particularly between taxa from different higher clades such as placentals vs. marsupials. For instance, the larynx in marsupials is clearly derived: the cricoid and thyroid cartilages are fused, leading to absence of the cricothyroid muscles and an articulation between the two arytenoid cartilages (N.B. the articulation between these cartilages and an interarytenoid cartilage seems to be plesiomorphic for mammals). Many of these specific, “minor” differences among adults of different taxa seem to be related to the needs of the embryos and/or neonates. For instance, Symington (1898) explained that these differences in larynx morphology might be related to the fact that marsupials remain in the pouch for a long time attached to the teat and thus need to, for instance, have safer ways to drink and breathe simultaneously. This requirement might also explain the expansion of the palatopharyngeus muscle/connective tissue and perhaps the expansion of the pars pharyngea of the stylopharyngeus, which are also derived characters within marsupials. It is hoped that this review, and our long-term project in general, will contribute toward the multidisciplinary data needed for an integrative synthesis of the anatomical macroevolution of the mammalian head and for future functional and developmental comparative studies.

References

Adams LA (1919) A memoir on the phylogeny of the jaw muscles in recent and fossils vertebrates. Ann NY Acad Sci 28:51–166

Allen WL, Stevens M, Higham JP (2014) Character displacement of Cercopithecini primate visual signals. Nat Commun 5:4266

Andrew RJ (1963) Evolution of facial expression. Science 142:1034–1041

Barghusen HR (1968) The lower jaw of cynodonts (Reptilia, Therapsida) and the evolutionary origin of mammal-like adductor jaw musculature. Postilla 116:1–49

Barghusen HR (1986) On the evolutionary origin of the therian tensor veli palatini and tensor tympani muscles. In: Hotton N, MacLean PD, Roth JJ, Roth EC (eds) The ecology and biology of mammal-like reptiles. Smithsonian Institution Press, Washington, DC, pp 253–262

Bemis WE (1984) Paedomorphosis and the evolution of Dipnoi. Paleobiology 10:293–307

Bemis WE (1986) Feeding mechanisms of living Dipnoi: anatomy and function. J Morphol 190(Suppl 1):249–275

Bischoff TLW (1840) Description Anatomique du Lepidosiren paradoxa. Ann Sci Nat 14(Ser 2):116–159

Boas JEV, Paulli S (1908) The elephant head – studies in the comparative anatomy of the organs of the head of the Indian elephant and other mammals, 1: the facial muscles and the proboscis. Gustav Fischer, Jena

Brand-Saberi B (2002) Vertebrate myogenesis. Springer-Verlag, Heidelberg, Berlin

Brock GT (1938) The cranial muscles of the Gecko – a general account with a comparison of muscles in other gnathostomes. Proc Zool Soc Lond (Ser B) 108:735–761

Bryant MD (1945) Phylogeny of Nearctic Sciuridae. Am Midland Nat 33:257–390

Burrows AM (2008) The facial expression musculature in primates and its evolutionary significance. BioEssays 30:212–225

Burrows AM, Smith TD (2003) Muscles of facial expression in Otolemur with a comparison to Lemuroidea. Anat Rec 274:827–836

Burrows AM, Waller BM, Parr LA, Bonar CJ (2006) Muscles of facial expression in the chimpanzee (Pan troglodytes): descriptive, comparative and phylogenetic contexts. J Anat 208:153–167

Butler AB, Hodos W (2005) Comparative vertebrate neuroanatomy: evolution and adaptation. Wiley Interscience, Hoboken

Carroll AM, Wainwright PC (2003) Functional morphology of feeding in the sturgeon, Scaphirhyncus albus. J Morphol 256:270–284

Carvajal JJ, Cox D, Summerbell D, Rigby PW (2001) A BAC transgenic analysis of the Mrf4/Myf5 locus reveals interdigitated elements that control activation and maintenance of gene expression during muscle development. Development 128(10):1857–1868

Cohen SR, Chen L, Trotman CA, Burdi AR (1993) Soft-palate myogenesis: a developmental field paradigm. Cleft Palate Craniofac J 30:441–446
Cohen SR, Chen LL, Burdi AR, Trotman CA (1994) Patterns of abnormal myogenesis in human cleft palates. Cleft Palate Craniofac J 31(5):345–350
Crompton AW (1963) The evolution of the mammalian jaw. Evolution 17:431–439
Crompton AW, Parker P (1978) Evolution of the mammalian masticatory apparatus: the fossil record shows how mammals evolved both complex chewing mechanisms and an effective middle ear, two structures that distinguish them from reptiles. Am Sci 66(2):192–201
Cuvier G, Laurillard L (1849) Recueil de planches de myologie, vol 3. Anatomie Comparée, Paris
Czajkowski MT, Rassek C, Lenhard DC, Bröhl D, Birchmeier C (2014) Divergent and conserved roles of Dll1 signaling in development of craniofacial and trunk muscle. Dev Biol 395(2):307–316
Danforth CH (1913) The myology of Polyodon. J Morphol 24:107–146
Didier DA (1987) Myology of the pectoral, branchial, and jaw regions of the ratfish Hydrolagus colliei (Holocephali). Honors Projects, paper 46, Illionois Wesleyan University
Didier DA (1995) Phylogenetic systematics of extant chimaeroid fishes (Holocephali, Chimaeroidei). Am Mus Novit 3119:86
Diogo R (2007) On the origin and evolution of higher-clades: osteology, myology, phylogeny and macroevolution of bony fishes and the rise of tetrapods. Science, Enfield
Diogo R (2008) Comparative anatomy, homologies and evolution of the mandibular, hyoid and hypobranchial muscles of bony fish and tetrapods: a new insight. Anim Biol 58:123–172
Diogo R (2009) The head musculature of the Philippine colugo (Dermoptera: Cynocephalus volans), with a comparison to tree-shrews, primates and other mammals. J Morphol 270:14–51
Diogo R, Abdala V (2007) Comparative anatomy, homologies and evolution of the pectoral muscles of bony fish and tetrapods: a new insight. J Morphol 268:504–517
Diogo R, Abdala V (2010) Muscles of vertebrates: comparative anatomy, evolution, homologies and development. Taylor & Francis, Oxford
Diogo R, Molnar JL (2014) Comparative anatomy, evolution and homologies of the tetrapod hindlimb muscles, comparisons with forelimb muscles, and deconstruction of the forelimb-hindlimb serial homology hypothesis. Anat Rec 297:1047–1075
Diogo R, Santana SE (2017) Evolution of facial musculature and relationships with facial color patterns, mobility, social group size, development, birth defects and asymmetric use of facial expressions. In: Russel R, Dols JMF (eds) The science of facial expression. Oxford University Press, Oxford, pp 133–152
Diogo R, Tanaka EM (2014) Development of fore- and hindlimb muscles in GFP-Transgenic axolotls: morphogenesis, the tetrapod Bauplan, and new insights on the forelimb-hindlimb enigma. J Exp Zool B Mol Dev Evol 322:106–127
Diogo R, Wood B (2012) Comparative anatomy and phylogeny of primate muscles and human evolution. Taylor & Francis, Oxford
Diogo R, Wood BA (2013) The broader evolutionary lessons to be learned from a comparative and phylogenetic analysis of primate muscle morphology. Biol Rev 88:988–1001
Diogo R, Ziermann JM (2014) Development of fore- and hindlimb muscles in frogs: morphogenesis, homeotic transformations, digit reduction, and the forelimb-hindlimb enigma. J Exp Zool B Mol Dev Evol 322:86–105
Diogo R, Ziermann JM (2015) Muscles of chondrichthyan paired appendages: comparison with osteichthyans, deconstruction of the fore-hindlimb serial homology dogma, and new insights on the evolution of the vertebrate neck. Anat Rec 298:513–530
Diogo R, Hinits Y, Hughes SM (2008a) Development of mandibular, hyoid and hypobranchial muscles in the zebrafish: homologies and evolution of these muscles within bony fishes and tetrapods. BMC Dev Biol 8:24–46
Diogo R, Abdala V, Lonergan N, Wood BA (2008b) From fish to modern humans – comparative anatomy, homologies and evolution of the head and neck musculature. J Anat 213:391–424
Diogo R, Abdala V, Aziz MA, Lonergan N, Wood BA (2009a) From fish to modern humans – comparative anatomy, homologies and evolution of the pectoral and forelimb musculature. J Anat 214:694–716
Diogo R, Wood BA, Aziz MA, Burrows A (2009b) On the origin, homologies and evolution of primate facial muscles, with a particular focus on hominoids and a suggested unifying nomenclature for the facial muscles of the Mammalia. J Anat 215:300–319
Diogo R, Linde-Medina M, Abdala V, Ashley–Ross MA (2013a) New, puzzling insights from comparative myological studies on the old and unsolved forelimb/hindlimb enigma. Biol Rev 88:196–214
Diogo R, Murawala P, Tanaka EM (2013b) Is salamander hindlimb regeneration similar to that of the forelimb? Anatomical and morphogenetic analysis of hindlimb muscle regeneration in GFP transgenic axolotls as a basis for regenerative and developmental studies. J Anat 10:459–468
Diogo R, Nacu E, Tanaka EM (2014) Is salamander limb regeneration really perfect? Anatomical and morphogenetic analysis of forelimb muscle regeneration in GFP-transgenic axolotls as a basis for regenerative, developmental and evolutionary studies. Anat Rec 297:1076–1089
Diogo R, Esteve-Altava B, Smith C, Boughner JC, Rasskin-Gutman D (2015a) Anatomical network comparison of human upper and lower, newborn and adult, and normal and abnormal limbs, with notes on development, pathology and limb serial homology vs. homoplasy. PLoS One 10:e0140030
Diogo R, Kelly R, Christiaen L, Levine M, Ziermann JM, Molnar J, Noden D, Tzahor E (2015b) A new heart for

a new head in vertebrate cardiopharyngeal evolution. Nature 520:466–473

Diogo R, Ziermann JM, Linde-Medina M (2015c) Is evolutionary biology becoming too politically correct? A reflection on the scala naturae, phylogenetically basal clades, anatomically plesiomorphic taxa, and 'lower' animals. Biol Rev 90:502–521

Diogo R, Ziermann JM, Linde-Medina M (2015d) Specialize or risk disappearance – empirical evidence of anisomerism based on comparative and developmental studies of gnathostome head and limb musculature. Biol Rev 90:964–978

Diogo R, Bello-Hellegouarch G, Kohlsdorf T, Esteve-Altava B, Molnar J (2016) Comparative myology and evolution of marsupials and other vertebrates, with notes on complexity, Bauplan, and "Scala Naturae". Anat Rec 299:1224–1255

Diogo R, Molnar JL, Wood BA (2017) Bonobo anatomy reveals stasis and mosaicism in chimpanzee evolution, and supports bonobos as the most appropriate extant model for the common ancestor of chimpanzees and humans. Nat Sci Rep 7:608

Diogo R, Ziermann JM, Molnar J, Siomava N, Abdala V (2018) Muscles of chordates: development, homologies and evolution. Taylor & Francis, Oxford

Edgeworth FH (1935) The cranial muscles of vertebrates. Cambridge University Press, Cambridge

Epperlein H-H, Khattak S, Knapp D, Tanaka EM, Malashichev Y (2012) Neural crest does not contribute to the neck and shoulder in the Axolotl (Ambystoma mexicanum). PLoS One 7:e52244

Ericsson R, Knight R, Johanson Z (2013) Evolution and development of the vertebrate neck. J Anat 222: 67–78

Gasser RF (1967) The development of the facial muscles in man. Am J Anat 120:357–376

Gaupp E (1896) A. Ecker's und R. Wiedersheim's Anatomie des Frosches, part 1. Friedrich Vieweg und Sohn, Braunschweig

Gaupp E (1912) De Reichertsche Theorie. (Hammer-, amboss- und Kerferfrage). Arch Anat Physiol Suppl:1–416

Goodrich ES (1958) Studies on the structure and development of vertebrates. Dover Publications, New York

Gopalakrishnan S, Comai G, Sambasivan R, Francou A, Kelly RG, Tajbakhsh S (2015) A cranial mesoderm origin for esophagus striated muscles. Dev Cell 34:694–704

Greenwood PH (1977) Notes on the anatomy and classification of elopomorph fishes. Bull Br Mus Nat Hist (Zool) 32:65–103

Grimaldi A, Parada C, Chai Y (2015) A comprehensive study of soft palate development in mice. PLoS One 10(12):e0145018

Harel I, Nathan E, Tirosh-Finkel L, Zigdon H, Guimarães-Camboa N, Evans SM, Tzahor E (2009) Distinct origins and genetic programs of head muscle satellite cells. Dev Cell 16:822–832

Haugen I-BK (2002) Neuromuscular organization of mammalian extraocular muscles. Rap Hogskolen Buskerud 36:1–56

House EL (1953) A myology of the pharyngeal region of the albino rat. Anat Rec 116:363–381

Huber E (1930a) Evolution of facial musculature and cutaneous field of trigeminus – Part I. Q Rev Biol 5:133–188

Huber E (1930b) Evolution of facial musculature and cutaneous field of trigeminus – Part II. Q Rev Biol 5:389–437

Huber E (1931) Evolution of facial musculature and expression. The Johns Hopkins University Press, Baltimore

Jarvik E (1963) The composition of the intermandibular division of the head in fishes and tetrapods and the diphyletic origin of the tetrapod tongue. Kungl Sven Veten Handl 9:1–74

Johanson Z (2003) Placoderm branchial and hypobranchial muscles and origins in jawed vertebrates. J Vert Paleont 23:735–749

Jollie M (1982) Ventral branchial musculature and synapomorphies questioned. Zool J Linnean Soc 75:35–47

Jouffroy FK (1971) Musculature des membres. In: Grassé PP (ed) Traité de Zoologie, XVI: 3 (Mammifères). Masson et Cie, Paris, pp 1–475

Jouffroy FK, Lessertisseur J (1971) Particularités musculaires des Monotrémes – musculature post-craniénne. In: Grassé PP (ed) Traité de Zoologie, XVI: 3 (Mammifères). Masson et Cie, Paris, pp 679–836

Jouffroy FK, Saban R (1971) Musculature peaucière. In: Grassé PP (ed) Traité de Zoologie, XVI: 3 (Mammifères). Masson et Cie, Paris, pp 477–611

Kardong KV (2002) Vertebrates: comparative anatomy, function, evolution, 3rd edn. McGraw-Hill, New York

Kardong KV (2011) Vertebrates: comparative anatomy, function, evolution, 7th edn. Mcgraw-Hill, Boston

Kesteven HL (1942–1945) The evolution of the skull and the cephalic muscles. Mem Austr Mus 8:1–361

Kingdon J (1992) Facial patterns as signals and masks. In: Jones S (ed) The Cambridge encyclopedia of human evolution. Cambridge University Press, Cambridge, pp 161–165

Kingdon J (2007) Primate visual signals in noisy environments. Folia Primatol (Basel) 78:389–404

Kishimoto H, Yamada S, Kanahashi T, Yoneyama A, Imai H, Matsuda T, Takeda T, Kawai K, Suzuki S (2016) Three-dimensional imaging of palatal muscles in the human embryo and fetus: development of levator veli palatini and clinical importance of the lesser palatine nerve. Dev Dyn 245:123–131

Kleinteich T, Haas A (2007) Cranial musculature in the larva of the caecilian, Ichthyophis kohtaoensis (Lissamphibia: Gymnophiona). J Morphol 268:74–88

Knight RD, Mebus K, Roehl HH (2008) Mandibular arch muscle identity is regulated by a conserved molecular process during vertebrate development. J Exp Zool Mol Dev Evol 310B:355–369

Köntges G, Lumsden A (1996) Rhombencephalic neural crest segmentation is preserved throughout craniofacial ontogeny. Development 122:3229–3242

Larsen JH, Guthrie DJ (1975) The feeding system of terrestrial tiger salamanders (Ambystoma tigrinum melanostictum Baird). J Morphol 147:137–154

Lauder GV, Shaffer HB (1985) Functional morphology of the feeding mechanism in aquatic ambystomatid salamanders. J Morphol 185:297–326
Lautenschlager S, Gill P, Luo ZX, Fagan MJ, Rayfield EJ (2017) Morphological evolution of the mammalian jaw adductor complex. Biol Rev Camb Philos Soc 92(4):1910–1940
Lescroart F, Kelly RG, Le Garrec JF, Nicolas JF, Meilhac SM, Buckingham M (2010) Clonal analysis reveals common lineage relationships between head muscles and second heart field derivatives in the mouse embryo. Development 137:3269–3279
Lescroart F, Hamou W, Francou A, Théveniau-Ruissy M, Kelly RG, Buckingham M (2015) Clonal analysis reveals a common origin between nonsomite-derived neck muscles and heart myocardium. Proc Natl Acad Sci 112(5):1446–1451
Lightoller GS (1928a) The action of the m. mentalis in the expression of the emotion of distress. J Anat 62:319–332
Lightoller GS (1928b) The facial muscles of three orang-utans and two cercopithecidae. J Anat 63:19–81
Lightoller GS (1934) The facial musculature of some lesser primates and a Tupaia. Proc Zool Soc Lond 1934:259–309
Lightoller GS (1939) V. Probable homologues. A study of the comparative anatomy of the mandibular and hyoid arches and their musculature – Part I. Comparative myology. Trans Zool Soc Lond 24:349–382
Lightoller GS (1940a) The comparative morphology of the platysma: a comparative study of the sphincter colli profundus and the trachelo-platysma. J Anat 74:390–396
Lightoller GS (1940b) The comparative morphology of the M. caninus. J Anat 74:397–402
Lightoller GS (1942) Matrices of the facialis musculature: homologization of the musculature in monotremes with that of marsupials and placentals. J Anat 76:258–269
Lubosch W (1914) Vergleischende anatomie der kaumusculatur der Wirbeltiere, in fünf teilen: 1 – die kausmukulatur der Amphibien. Jen Z Naturwiss 53:51–188
Mallat J (1997) Shark pharyngeal muscles and early vertebrate evolution. Acta Zool 78:279–294
Marion E (1905) Mandibular and pharyngeal muscles of Acanthias and Raia. Am Nat 39:891–924
Matsuoka T, Ahlberg PE, Kessaris N, Iannarelli P, Dennehy U, Richardson WD, McMahon AP, Koentges G (2005) Neural crest origins of the neck and shoulder. Nature 436:347–355
Millot J, Anthony J (1958) Anatomie de Latimeria chalumnae – I, squelette, muscles, et formation de soutiens. CNRS, Paris
Minkoff EC, Mikkelsen P, Cunningham WA, Taylor KW (1979) The facial musculature of the opossum (Didelphis virginiana). J Mammal 60:46–57
Miyake T, McEachran JD, Hall BK (1992) Edgeworth's legacy of cranial muscle development with an analysis of muscles in the ventral gill arch region of batoid fishes (Chondrichthyes: Batoidea). J Morphol 212:213–256
Murie J, Mivart ST (1869) On the anatomy of the Lemuroidea. Trans Zool Soc Lond 7:1–113
Noden DM, Francis-West P (2006) The differentiation and morphogenesis of craniofacial muscles. Dev Dyn 235:1194–1218
Noden DM, Schneider RA (2006) Neural crest cells and the community of plan for craniofacial development: historical debates and current perspectives. In: Saint-Jeannet J (ed) Neural crest induction & differentiation – advances in experimental medicine and biology, vol 589. Landes Bioscience, Georgetown, pp 1–31
Osse JWM (1969) Functional morphology of the head of the perch (Perca fluviatilis L.): an electromyographic study. Neth J Zool 19:289–392
Owen R (1841) Description of the Lepidosiren annectens. Trans Linn Soc Lond 18:327–361
Parada C, Han D, Chai Y (2012) Molecular and cellular regulatory mechanisms of tongue myogenesis. J Dent Res 91:528–535
Piatt J (1938) Morphogenesis of the cranial muscles of Ambystoma punctatum. J Morphol 63:531–587
Piekarski N, Olsson L (2007) Muscular derivatives of the cranialmost somites revealed by long-term fate mapping in the Mexican axolotl (Ambystoma mexicanum). Evol Dev 9:566–578
Pollard HB (1892) On the anatomy and phylogenetic position of Polypterus. Zool Jahrb 5:387–428
Preuschoft S (2000) Primate faces and facial expressions. Soc Res 67:245–271
Reichert KB (1937) Über die Visceralbogen der Wirbelthiere im Allgemeinen und deren Metamorphosen bei den Vögeln und Säugethieren. Arch Anat Physiol Wissensch Med 1837:120–220
Reilly S, Lauder GV (1989) Kinetics of tongue projection in Ambystoma tigrinum: quantitative kinematics, muscle function and evolutionary hypotheses. J Morphol 199:223–243
Reilly S, Lauder GV (1990) The evolution of tetrapod feeding behavior: kinematic homologies in prey transport. Evolution 44:1542–1557
Reilly S, Lauder GV (1991) Prey transport in the tiger salamander: quantitative electromyography and muscle function in tetrapods. J Exp Zool 260:1–17
Rodríguez-Vázquez JF, Sakiyama K, Abe H, Amano O, Murakami G (2016) Fetal Tendinous connection between the tensor tympani and tensor veli palatini muscles: a single digastric muscle acting for morphogenesis of the cranial base. Anat Rec 299:474–483
Ruge G (1885) Über die Gesichtsmuskulatur der Halbaffen. Morph Jahrb 11:243–315
Ruge G (1897) Über das peripherische Gebiet des nervus facialis boi Wirbelthieren, vol 3. Festschr f Gegenbaur, Leipzig
Ruge G (1910) Verbindungen des Platysma mit der tiefen Muskulatur des Halses beim Menschen. Morph Jahrb 41:708–724

Saban R (1968) Musculature de la tête. In: Grassé PP (ed) Traité de Zoologie, XVI: 3 (Mammifères). Masson et Cie, Paris, pp 229–472
Saban R (1971) Particularités musculaires des Monotrémes—musculature de la tête. In: Grassé PP (ed) Traité de Zoologie, XVI: 3 (Mammifères). Masson et Cie, Paris, pp 681–732
Santana SE, Dobson SD, Diogo R (2014) Plain faces are more expressive: comparative study of facial color, mobility and musculature in primates. Biol Letters 10:20140275
Schaeffer JP (1929) Some problems in genesis and development with special reference to the human palate. Int J Orthod 15:291–310
Schmidt KL, Cohn JF (2001) Human facial expressions as adaptations: evolutionary questions in facial expression research. Yearb Phys Anthropol 44:3–24
Schumacher GH (1973) The head muscles and hyolaryngeal skeleton of turtles and crocodilians. In: Gans C, Parsons TS (eds) Biology of the reptilia, vol 14. Academic Press, New York, pp 101–199
Sefton EM, Bhullar BAS, Mohaddes Z, Hanken J (2016) Evolution of the head-trunk interface in tetrapod vertebrates. eLife 5:e09972
Seiler R (1971a) A comparison between the facial muscles of Catarrhini with long and short muzzles. In: Proc. 3rd Int. Congr. Primat., Zürich 1970, vol l. Karger, Basel, pp 157–162
Seiler R (1971b) Facial musculature and its influence on the facial bones of catarrhinous Primates. I. Morphol Jahrb 116:122–142
Seiler R (1971c) Facial musculature and its influence on the facial bones of catarrhine Primates. II. Morphol Jahrb 116:147–185
Seiler R (1971d) Facial musculature and its influence on the facial bones of catarrhine Primates. III. Morphol Jahrb 116:347–376
Seiler R (1971e) Facial musculature and its influence on the facial bones of catarrhine Primates. IV. Morphol Jahrb 116:456–481
Seiler R (1974a) Muscles of the external ear and their function in man, chimpanzees and Macaca. Morphol Jahrb 120:78–122
Seiler R (1974b) Particularities in facial muscles of Daubentonia madagascariensis (Gmelin 1788). Folia Primatol 22:81–96
Seiler R (1975) On the facial muscles in Perodicticus potto and Nycticebus coucang. Folia Primatol 23:275–289
Seiler R (1979) Criteria of the homology and phylogeny of facial muscles in primates including man. I. Prosimia and Platyrrhina. Morphol Jahrb 125:191–217
Seiler R (1980) Ontogenesis of facial muscles in primates. Morphol Jahrb 126:841–864
Sewertzoff AN (1928) The head skeleton and muscles of Acipenser ruthenus. Acta Zool-Stockholm 9:193–319
Shearman RM, Burke AC (2009) The lateral somitic frontier in ontogeny and phylogeny. J Exp Biol Mol Dev Evol 312B:602–613
Sherwood CC, Hof PR, Holloway RL, Semendeferi K, Gannon PJ, Frahm HD, Zilles K (2005) Evolution of the brainstem orofacial motor system in primates: a comparative study of trigeminal, facial, and hypoglossal nuclei. J Hum Evol 48:45–84
Smith KK (1992) The evolution of the mammalian pharynx. Zool J Linnean Soc 104:313–349
Smith KK (1994) Development of craniofacial musculature in Monodelphis domestica (Marsupialia: Didelphidae). J Morphol 222:149–173
Sokoloff AJ, Deacon TW (1992) Musculotopic organization of the hypoglossal nucleus in the cynomolgus monkey (Macaca fascicularis). J Comp Neurol 324:81–93
Sprague JM (1942) The hyoid apparatus of Neotoma. J Mammal 23:405–411
Sprague JA (1943) The hyoid region of placental mammals with special reference to the bats. Am J Anat 72:385–472
Sprague JA (1944a) The hyoid region in the insectivora. Am J Anat 74:175–216
Sprague JA (1944b) The innervation of the pharynx in the rhesus monkey, and the formation of the pharyngeal plexus in primates. Anat Rec 90:197–208
Surlykke A, Miller LA, Møhl B, Andersen BB, Christensen-Dalsgaard J, Buhl Jørgensen M (1993) Echolocation in two very small bats from Thailand Craseonycteris thonglongyai and Myotis siligorensis. Behav Ecol Sociobiol 33(1):1–12
Symington J (1898) The marsupial larynx. J Anat Physiol 33:31–49
Tada M, Kuratani S (2015) Evolutional and developmental understanding of the spinal accessory nerve. Zool Lett 1:4
Tajbakhsh S, Rocancourt D, Cossu G et al (1997) Redefining the genetic hierarchies controlling skeletal myogenesis: Pax-3 and Myf-5 act upstream of MyoD. Cell 89:127–138
Terminologia Anatomica (1998) Federative Committee on anatomical terminology. Georg Thieme, Stuttgard
Theis S, Patel K, Valasek P et al (2010) The occipital lateral plate mesoderm is a novel source for vertebrate neck musculature. Development 137:2961–2971
Wiley EO (1979a) Ventral gill arch muscles and the interrelationships of gnathostomes, with a new classification of the vertebrata. J Linn Soc Lond (Zool) 67:149–179
Wiley EO (1979b) Ventral gill arch muscles and the phylogenetic interrelationships of Latimeria. Occ Pap Calif Acad Sci 134:56–67
Wilga CD, Wainwright PC, Motta PJ (2000) Evolution of jaw depression mechanisms in aquatic vertebrates: insights from Chondrichthyes. Biol J Linn Soc 71:165–185
Witmer LM (1995) Homology of facial structures in extant archosaurs (birds and crocodilians), with special reference to paranasal pneumaticity and nasal conchae. J Morphol 225:269–327
Ziermann J, Miyashita T, Diogo R (2014) Cephalic muscles of cyclostomes (hagfishes and lampreys) and chondrichthyes (sharks, rays, and holocephalans): comparative anatomy and early evolution of the vertebrate head muscles. Zool J Linnean Soc 172: 771–802

Index

J. M. Ziermann et al. (eds.), *Heads, Jaws, and Muscles*, Fascinating Life Sciences,
https://doi.org/10.1007/978-3-319-93560-7

T

PGMO 02/21/2019